Methods in Molecular

Series Editor
John M. Walker
School of Life and Medical Sciences
University of Hertfordshire
Hatfield, Hertfordshire, UK

For further volumes:
http://www.springer.com/series/7651

For over 35 years, biological scientists have come to rely on the research protocols and methodologies in the critically acclaimed *Methods in Molecular Biology* series. The series was the first to introduce the step-by-step protocols approach that has become the standard in all biomedical protocol publishing. Each protocol is provided in readily-reproducible step-by-step fashion, opening with an introductory overview, a list of the materials and reagents needed to complete the experiment, and followed by a detailed procedure that is supported with a helpful notes section offering tips and tricks of the trade as well as troubleshooting advice. These hallmark features were introduced by series editor Dr. John Walker and constitute the key ingredient in each and every volume of the *Methods in Molecular Biology* series. Tested and trusted, comprehensive and reliable, all protocols from the series are indexed in PubMed.

Muscular Dystrophy Therapeutics

Methods and Protocols

Edited by

Rika Maruyama and Toshifumi Yokota

Department of Medical Genetics, University of Alberta, Edmonton, AB, Canada

Humana Press

Editors
Rika Maruyama
Department of Medical Genetics
University of Alberta
Edmonton, AB, Canada

Toshifumi Yokota
Department of Medical Genetics
University of Alberta
Edmonton, AB, Canada

ISSN 1064-3745 ISSN 1940-6029 (electronic)
Methods in Molecular Biology
ISBN 978-1-0716-2774-7 ISBN 978-1-0716-2772-3 (eBook)
https://doi.org/10.1007/978-1-0716-2772-3

This Humana imprint is published by the registered company Springer Science+Business Media, LLC, part of Springer Nature.
The registered company address is: 1 New York Plaza, New York, NY 10004, U.S.A.

Preface

> "The greatest enemy of knowledge is not ignorance, it is the illusion of knowledge" -Stephen Hawking

With the development of novel targeted therapies, including antisense-mediated exon skipping and gene replacement therapy, the field of muscular dystrophy therapeutics has been drastically changed in the last several years. Until 2016, there had been no FDA-approved drug to treat Duchenne muscular dystrophy (DMD), the most common muscular dystrophy. However, several new therapies, including antisense oligonucleotides, small molecules, and gene replacement therapy, are now approved or under late-stage clinical trials. We are witnessing the start of a therapeutic revolution in muscular dystrophy. However, the studies also made clear that these therapies are still far from a cure.

This book presents a comprehensive collection of state-of-the-art protocols on muscular dystrophy therapeutics, covering therapeutics using antisense oligonucleotides, gene replacement, genome editing, small molecules, stem cells, and antibodies, from the leaders in the field. The goal of this book is to inspire readers and provide hands-on tips, strategies, and advice to develop new therapeutic technologies for muscular dystrophy.

We are very grateful to all authors for sharing their detailed methods. These detailed methods and tips are not easily available elsewhere. We also deeply thank Dr. John Walker, the series editor, for his excellent guidance, careful paper reading, and encouragement throughout the development of this book. We are very grateful to the Friends of Garrett Cumming Research, Muscular Dystrophy Canada, Henri M. Toupin Chair in Neurological Sciences, and the University of Alberta for the funding to establish a muscular dystrophy research lab at the University of Alberta. We are also thankful to the Canadian Institutes of Health Research (CIHR), Canada Foundation for Innovation (CFI), Japan Society for the Promotion of Science (JSPS), Alberta Enterprise and Advanced Education, BC Children's Hospital Foundation, FSH Society, FSHD Canada Foundation, Rare Disease Foundation, and Defeat Duchenne Foundation (Jesse's Journey) for their support in expanding our research in developing antisense therapies for muscular dystrophy. Lastly but most importantly, we are deeply grateful to the patients and family members; their support, inspiration, and participation in research have been critical to the development of novel therapies.

Edmonton, AB, Canada

Rika Maruyama
Toshifumi Yokota

Contents

Part V Genome Editing/CRISPR

Part VI Small Molecules

Part VII Cell Models and Stem Cells

Part VIII Antibody

Contributors

SARA AGUTI • *The Dubowitz Neuromuscular Centre, Molecular Neurosciences Section, Developmental Neurosciences Research and Teaching Department, Great Ormond Street Institute of Child Health, University College London, London, UK*

SAEED ANWAR • *Department of Medical Genetics, Faculty of Medicine and Dentistry, University of Alberta, Edmonton, AB, Canada*

YOSHITSUGU AOKI • *Department of Molecular Therapy, National Institute of Neuroscience, National Center of Neurology and Psychiatry (NCNP), Tokyo, Japan*

MAY T. AUNG-HTUT • *Centre for Molecular Medicine and Innovative Therapeutics, Health Futures Institute, Murdoch University, Murdoch, WA, Australia; Perron Institute for Neurological and Translational Science, Nedlands, WA, Australia; Centre for Neuromuscular and Neurological Disorders, University of Western Australia, Crawley, WA, Australia*

RHONDA BASSEL-DUBY • *Department of Molecular Biology, University of Texas Southwestern Medical Center, Dallas, TX, USA; Hamon Center for Regenerative Science and Medicine, University of Texas Southwestern Medical Center, Dallas, TX, USA; Senator Paul D. Wellstone Muscular Dystrophy Specialized Research Center, University of Texas Southwestern Medical Center, Dallas, TX, USA*

DANIEL C. BITTEL • *Center for Genetic Medicine Research, Children's National Research Institute, Washington, DC, USA*

VÉRONIQUE BOLDUC • *Neuromuscular and Neurogenetic Disorders of Childhood Section, National Institute of Neurological Disorders and Stroke, National Institutes of Health, Bethesda, MD, USA*

CARSTEN BÖNNEMANN • *Neuromuscular and Neurogenetic Disorders of Childhood Section, National Institute of Neurological Disorders and Stroke, National Institutes of Health, Bethesda, MD, USA*

MATTHHEW J. BURKE • *Department of Molecular Microbiology and Immunology, School of Medicine, The University of Missouri, Columbia, MO, USA*

WILLIAM CHEN • *Sprott Center for Stem Cell Research, Regenerative Medicine Program, Ottawa Hospital Research Institute, Ottawa, ON, Canada; Department of Cellular and Molecular Medicine, Faculty of Medicine, University of Ottawa, Ottawa, ON, Canada*

RACHELLE H. CROSBIE • *Department of Integrative Biology and Physiology, University of California Los Angeles, Los Angeles, CA, USA; Department of Neurology, David Geffen School of Medicine, University of California Los Angeles, Los Angeles, CA, USA*

ALEXIS R. DEMONBREUN • *Center for Genetic Medicine, Feinberg School of Medicine, Northwestern University, Chicago, IL, USA*

JOHN R. DODAM • *Department of Veterinary Medicine and Surgery, College of Veterinary Medicine, The University of Missouri, Columbia, MO, USA*

DONGSHENG DUAN • *Department of Molecular Microbiology and Immunology, School of Medicine, The University of Missouri, Columbia, MO, USA; Department of Biomedical, Biological & Chemical Engineering, College of Engineering, The University of Missouri, Columbia, MO, USA; Department of Neurology, School of Medicine, The University of Missouri, Columbia, MO, USA; Department of Biomedical Sciences, College of Veterinary Medicine, The University of Missouri, Columbia, MO, USA*

HANNAH F. DUGDALE • *Centre for Human and Applied Physiological Sciences, School of Basic and Medical Biosciences, Faculty of Life Sciences and Medicine, Guy's Campus, King's College London, London, UK; School of Sport, Exercise and Health Sciences, Loughborough University, Loughborough, UK*

YUSUKE ECHIGOYA • *Laboratory of Biomedical Science, Department of Veterinary Medicine, College of Bioresource Sciences, Nihon University, Fujisawa, Kanagawa, Japan; Nihon University Veterinary Research Center, Fujisawa, Kanagawa, Japan*

MATHIEU FISCHER • *Department of Paediatrics, Institute of Developmental and Regenerative Medicine (IDRM), University of Oxford, Oxford, UK; MDUK Oxford Neuromuscular Centre, Oxford, UK*

PETER GEE • *MaxCyte Inc., Gaithersburg, MD, USA*

FADY GUIRGUIS • *The Dubowitz Neuromuscular Centre, Molecular Neurosciences Section, Developmental Neurosciences Research and Teaching Department, Great Ormond Street Institute of Child Health, University College London, London, UK; Neuromuscular and Neurogenetic Disorders of Childhood Section, National Institute of Neurological Disorders and Stroke, National Institutes of Health, Bethesda, MD, USA*

CHADY H. HAKIM • *Department of Molecular Microbiology and Immunology, School of Medicine, The University of Missouri, Columbia, MO, USA*

KRISTIN A. HAM • *Centre for Molecular Medicine and Innovative Therapeutics, Health Futures Institute, Murdoch University, Murdoch, WA, Australia; Perron Institute for Neurological and Translational Science, Nedlands, WA, Australia; Centre for Neuromuscular and Neurological Disorders, University of Western Australia, Crawley, WA, Australia*

MARK HAMRICK • *Medical College of Georgia, Augusta University, Augusta, GA, USA*

GANG HAN • *School of Medical Laboratory & Department of Cell Biology, Tianjin Medical University, Tianjin, China*

PRADEEP HARISH • *College of Health, Medicine and Life Science, Biosciences, Brunel University London, Uxbridge, UK*

HIROMI HAYASHITA-KINOH • *Division of Molecular and Medical Genetics, Center for Gene and Cell Therapy, The Institute of Medical Science, The University of Tokyo, Tokyo, Japan*

ROLAND W. HERZOG • *Department of Pediatrics, Indiana University, Indianapolis, IN, USA; Herman B Wells Center for Pediatric Research, Indiana University, Indianapolis, IN, USA*

JAMES R. HOLLAND • *Department of Veterinary Medicine and Surgery, College of Veterinary Medicine, The University of Missouri, Columbia, MO, USA*

AKITSU HOTTA • *Center for iPS Cell Research and Application, Kyoto University, Kyoto, Japan; Takeda-CiRA Joint Program (T-CiRA), Fujisawa, Kanagawa, Japan*

YAHYA JAD • *Department of Paediatrics, Institute of Developmental and Regenerative Medicine (IDRM), University of Oxford, Oxford, UK; MDUK Oxford Neuromuscular Centre, Oxford, UK*

JYOTI K. JAISWAL • *Center for Genetic Medicine Research, Children's National Research Institute, Washington, DC, USA; Department of Genomics and Precision Medicine, George Washington University School of Medicine and Health Sciences, Washington, DC, USA*

BERNARD J. JASMIN • *Department of Cellular and Molecular Medicine, and the Eric Poulin Centre for Neuromuscular Disease, Faculty of Medicine, University of Ottawa, Ottawa, ON, Canada*

YUE JIN • *Medical College of Georgia, Augusta University, Augusta, GA, USA*
RUSSELL D. JOHNSEN • *Centre for Molecular Medicine and Innovative Therapeutics, Health Futures Institute, Murdoch University, Murdoch, WA, Australia; Perron Institute for Neurological and Translational Science, Nedlands, WA, Australia; Centre for Neuromuscular and Neurological Disorders, University of Western Australia, Crawley, WA, Australia*
TAHNEE L. KENNEDY • *Department of Physiology, Anatomy and Genetics, University of Oxford, Oxford, UK*
IL-MAN KIM • *Anatomy, Cell Biology & Physiology, Indiana University School of Medicine, Indianapolis, IN, USA*
SANDEEP R. P. KUMAR • *Department of Pediatrics, Indiana University, Indianapolis, IN, USA; Herman B Wells Center for Pediatric Research, Indiana University, Indianapolis, IN, USA*
KATSUHIKO KUNITAKE • *Department of Molecular Therapy, National Institute of Neuroscience, National Center of Neurology and Psychiatry (NCNP), Tokyo, Japan*
JIMMY C. LATTIMER • *Department of Veterinary Medicine and Surgery, College of Veterinary Medicine, The University of Missouri, Columbia, MO, USA*
STACY B. LEACH • *Department of Veterinary Medicine and Surgery, College of Veterinary Medicine, The University of Missouri, Columbia, MO, USA*
KENJI ROWEL Q. LIM • *Department of Medical Genetics, Faculty of Medicine and Dentistry, University of Alberta, Edmonton, AB, Canada; Department of Medical Genetics, Faculty of Medicine and Dentistry, Edmonton, AB, Canada*
CAORUI LIN • *School of Medical Laboratory & Department of Cell Biology, Tianjin Medical University, Tianjin, China*
ALBERTO MALERBA • *Department of Biological Sciences, Royal Holloway University of London, Egham, Surrey, UK*
ELIZABETH M. MCNALLY • *Center for Genetic Medicine, Feinberg School of Medicine, Northwestern University, Chicago, IL, USA*
DAVID SEOANE MIRAZ • *Department of Paediatrics, Institute of Developmental and Regenerative Medicine (IDRM), University of Oxford, Oxford, UK; MDUK Oxford Neuromuscular Centre, Oxford, UK*
DAIGO MIYAZAKI • *Department of Medicine (Neurology and Rheumatology), Shinshu University School of Medicine, Matsumoto, Japan; Shinshu Medical Care Collaboration Center (Division of Support for Intractable Disease), Shinshu University Hospital, Matsumoto, Japan*
KAREN MIZ • *Division of Molecular Cardiovascular Biology, Heart Institute, Cincinnati Children's Hospital Medical Center, Cincinnati, OH, USA*
EKATERINA MOKHONOVA • *Department of Integrative Biology and Physiology, University of California Los Angeles, Los Angeles, CA, USA; Department of Neurology, David Geffen School of Medicine, University of California Los Angeles, Los Angeles, CA, USA*
JEFFERY D. MOLKENTIN • *Division of Molecular Cardiovascular Biology, Heart Institute, Cincinnati Children's Hospital Medical Center, Cincinnati, OH, USA; Department of Pediatrics, University of Cincinnati College of Medicine, Cincinnati, OH, USA*
NORIO MOTOHASHI • *Department of Molecular Therapy, National Institute of Neuroscience, National Center of Neurology and Psychiatry (NCNP), Tokyo, Japan*
FRANCESCO MUNTONI • *The Dubowitz Neuromuscular Centre, Molecular Neurosciences Section, Developmental Neurosciences Research and Teaching Department, Great Ormond Street Institute of Child Health, University College London, London, UK; NIHR Great Ormond Street Hospital Biomedical Research Centre, London, UK*

AKINORI NAKAMURA • *Department of Medicine (Neurology and Rheumatology), Shinshu University School of Medicine, Matsumoto, Japan; Department of Clinical Research, National Hospital Organization Matsumoto Medical Center, Matsumoto, Japan*
MINAS NALBANDIAN • *Department of Clinical Application, Center for iPS Cell Research and Application (CiRA), Kyoto University, Kyoto, Japan*
TAKASHI OKADA • *Division of Molecular and Medical Genetics, Center for Gene and Cell Therapy, The Institute of Medical Science, The University of Tokyo, Tokyo, Japan*
ERIC N. OLSON • *Department of Molecular Biology, University of Texas Southwestern Medical Center, Dallas, TX, USA; Hamon Center for Regenerative Science and Medicine, University of Texas Southwestern Medical Center, Dallas, TX, USA; Senator Paul D. Wellstone Muscular Dystrophy Specialized Research Center, University of Texas Southwestern Medical Center, Dallas, TX, USA*
CHRISTINE PÉLADEAU • *Department of Cellular and Molecular Medicine, and the Eric Poulin Centre for Neuromuscular Disease, Faculty of Medicine, University of Ottawa, Ottawa, ON, Canada*
DENNIS PÉREZ-LÓPEZ • *Department of Molecular Microbiology and Immunology, School of Medicine, The University of Missouri, Columbia, MO, USA*
THEODORE J. PERKINS • *Sprott Center for Stem Cell Research, Regenerative Medicine Program, Ottawa Hospital Research Institute, Ottawa, ON, Canada; Department of Cellular and Molecular Medicine, Faculty of Medicine, University of Ottawa, Ottawa, ON, Canada; Department of Biochemistry, Microbiology, and Immunology, Faculty of Medicine, University of Ottawa, Ottawa, ON, Canada*
LINDA POPPLEWELL • *Department of Biological Sciences, Royal Holloway University of London, Egham, Surrey, UK; National Horizons Centre, Teesside University, Darlington, UK*
MATTIA QUATTROCELLI • *Division of Molecular Cardiovascular Biology, Heart Institute, Cincinnati Children's Hospital Medical Center, Cincinnati, OH, USA; Department of Pediatrics, University of Cincinnati College of Medicine, Cincinnati, OH, USA*
ROHINI ROY ROSHMI • *Department of Medical Genetics, Faculty of Medicine and Dentistry, University of Alberta, Edmonton, Canada*
MICHAEL A. RUDNICKI • *Sprott Center for Stem Cell Research, Regenerative Medicine Program, Ottawa Hospital Research Institute, Ottawa, ON, Canada; Department of Cellular and Molecular Medicine, Faculty of Medicine, University of Ottawa, Ottawa, ON, Canada; Department of Medicine, Faculty of Medicine, University of Ottawa, Ottawa, ON, Canada*
HIDETOSHI SAKURAI • *Department of Clinical Application, Center for iPS Cell Research and Application (CiRA), Kyoto University, Kyoto, Japan*
CHAITRA SATHYAPRAKASH • *Department of Molecular Therapy, National Institute of Neuroscience, National Center of Neurology and Psychiatry (NCNP), Tokyo, Japan*
MITSUTO SATO • *Department of Medicine (Neurology and Rheumatology), Shinshu University School of Medicine, Matsumoto, Japan*
MD NUR AHAD SHAH • *Department of Medical Genetics, Faculty of Medicine and Dentistry, University of Alberta, Edmonton, AB, Canada*
YAN SHEN • *Medical College of Georgia, Augusta University, Augusta, GA, USA*
NAOKO SHIBA • *Department of Regenerative Science and Medicine, Shinshu University School of Medicine, Matsumoto, Japan*
YUJI SHIBA • *Department of Regenerative Science and Medicine, Shinshu University School of Medicine, Matsumoto, Japan*

CYNTHIA SHU • *Department of Integrative Biology and Physiology, University of California Los Angeles, Los Angeles, CA, USA; Department of Neurology, David Geffen School of Medicine, University of California Los Angeles, Los Angeles, CA, USA*

JESSICA STOODLEY • *Department of Paediatrics, Institute of Developmental and Regenerative Medicine (IDRM), University of Oxford, Oxford, UK; MDUK Oxford Neuromuscular Centre, Oxford, UK*

YAOLIANG TANG • *Medical College of Georgia, Augusta University, Augusta, GA, USA*

MICHEL TCHAN • *Genetic Medicine, Westmead Hospital, Sydney, NSW, Australia; Sydney Medical School, The University of Sydney, Sydney, NSW, Australia*

JAMES TEIXEIRA • *Department of Molecular Microbiology and Immunology, School of Medicine, The University of Missouri, Columbia, MO, USA*

MIGUEL A. VARELA • *Department of Paediatrics, Institute of Developmental and Regenerative Medicine (IDRM), University of Oxford, Oxford, UK; MDUK Oxford Neuromuscular Centre, Oxford, UK*

KEI WATANABE • *Center for iPS Cell Research and Application, Kyoto University, Kyoto, Japan*

NEAL L. WEINTRAUB • *Medical College of Georgia, Augusta University, Augusta, GA, USA*

STEVE D. WILTON • *Centre for Molecular Medicine and Innovative Therapeutics, Health Futures Institute, Murdoch University, Murdoch, WA, Australia; Perron Institute for Neurological and Translational Science, Nedlands, WA, Australia; Centre for Neuromuscular and Neurological Disorders, University of Western Australia, Crawley, WA, Australia*

MICHELLE WINTZINGER • *Division of Molecular Cardiovascular Biology, Heart Institute, Cincinnati Children's Hospital Medical Center, Cincinnati, OH, USA*

MATTHEW J. A. WOOD • *Department of Paediatrics, Institute of Developmental and Regenerative Medicine (IDRM), University of Oxford, Oxford, UK; MDUK Oxford Neuromuscular Centre, Oxford, UK*

HAIFANG YIN • *School of Medical Laboratory & Department of Cell Biology, Tianjin Medical University, Tianjin, China*

TOSHIFUMI YOKOTA • *Department of Medical Genetics, Faculty of Medicine and Dentistry, University of Alberta, Edmonton, AB, Canada; The Friends of Garrett Cumming Research and Muscular Dystrophy Canada, Henry M. Toupin Neurological Science Research Chair, Edmonton, AB, Canada*

ALLEN YORK • *Division of Molecular Cardiovascular Biology, Heart Institute, Cincinnati Children's Hospital Medical Center, Cincinnati, OH, USA*

AMY R. ZALCMAN • *Department of Veterinary Medicine and Surgery, College of Veterinary Medicine, The University of Missouri, Columbia, MO, USA; VetCT, Orlando, FL, USA*

YU ZHANG • *Department of Molecular Biology, University of Texas Southwestern Medical Center, Dallas, TX, USA; Hamon Center for Regenerative Science and Medicine, University of Texas Southwestern Medical Center, Dallas, TX, USA; Senator Paul D. Wellstone Muscular Dystrophy Specialized Research Center, University of Texas Southwestern Medical Center, Dallas, TX, USA*

MINGMING ZHAO • *Department of Clinical Application, Center for iPS Cell Research and Application (CiRA), Kyoto University, Kyoto, Japan*

HAIYAN ZHOU • *NIHR Great Ormond Street Hospital Biomedical Research Centre, London, UK; Genetics and Genomic Medicine Research and Teaching Department, Great Ormond Street Institute of Child Health, University College London, London, UK*

Part I

Basics and Introduction

Chapter 1

Current Strategies of Muscular Dystrophy Therapeutics: An Overview

Kenji Rowel Q. Lim and Toshifumi Yokota

Abstract

Muscular dystrophies are a group of genetic disorders characterized by varying degrees of progressive muscle weakness and degeneration. They are clinically and genetically heterogeneous but share the common histological features of dystrophic muscle. There is currently no cure for muscular dystrophies, which is of particular concern for the more disabling and/or lethal forms of the disease. Through the years, several therapies have encouragingly been developed for muscular dystrophies and include genetic, cellular, and pharmacological approaches. In this chapter, we undertake a comprehensive exploration of muscular dystrophy therapeutics under current development. Our review includes antisense therapy, CRISPR, gene replacement, cell therapy, nonsense suppression, and disease-modifying small molecule compounds.

Key words Muscular dystrophy, Antisense oligonucleotides, CRISPR, Gene replacement, Cell therapy, Nonsense suppression, Small molecule compounds

1 Introduction

The muscular dystrophies are a group of genetic disorders characterized by varying degrees of progressive muscle weakness and degeneration. Mutations in more than 40 genes have been associated with muscular dystrophies, including those involved with cellular structure (muscle membrane, extracellular matrix, nuclear envelope), muscle contraction, mRNA splicing, and early embryonic development, among others [1]. This genetic diversity is accompanied by vast clinical heterogeneity, with factors such as the age of onset, severity, disease pattern, and presence of extramuscular symptoms widely varying across and within muscular dystrophies [1, 2]. Despite this complexity, the muscular dystrophies are histologically similar, with affected muscle showing features of necrosis, fibrosis, fat replacement, immune cell infiltration, variable fiber size/shape, and the presence of central nuclei [1].

Rika Maruyama and Toshifumi Yokota (eds.), *Muscular Dystrophy Therapeutics: Methods and Protocols*, Methods in Molecular Biology, vol. 2587, https://doi.org/10.1007/978-1-0716-2772-3_1,

Several muscular dystrophies are disabling or fatal to patients. For instance, in the most common form of muscular dystrophy, Duchenne muscular dystrophy (DMD), patients lose ambulation by their teens and typically suffer from cardiorespiratory complications in the third decade of life, which leads to death [1, 3]. Patients with certain forms of congenital muscular dystrophy (CMD) may not even get the chance to walk at all and could experience concerns such as difficulty swallowing, respiratory insufficiency, cardiomyopathy, and cognitive impairment [4]. While management strategies are being used to combat symptoms associated with muscular dystrophy, these do not ultimately deter progression in the more severe forms of the disease. There is also currently no cure for any of the muscular dystrophies.

Overall, these highlight the importance of developing effective, curative treatments for muscular dystrophies. Fortunately, our knowledge on the underlying genetics of this group of disorders has not only considerably advanced through the years but has also enabled the development of various therapeutic approaches. Certain therapies have made significant strides in human clinical trials, with a number already approved for patient use. We review current strategies for the treatment of muscular dystrophies in this chapter, covering a wide range of genetic, cellular, and pharmacological therapies (Fig. 1). Given the breadth of material, this exploration is meant to be more comprehensive than exhaustive, providing enough information to understand each therapeutic approach and appreciate their individual, unique characteristics.

2 Antisense Therapy

Antisense oligonucleotides (AOs) are short, single-stranded, synthetic nucleic acid analogs that bind mRNA sequences via Watson–Crick base-pairing [5–7]. This simple mechanism of action lends incredible versatility to AOs as a genetic therapy. Immediately apparent would be the fact that AOs can target any mRNA sequence of choice. By also altering the chemistry of an AO, it can be designed to induce post-transcriptional changes in one of two ways. AOs with bases of a single chemistry or of alternating chemistries throughout their length (the "mixmer" configuration) act as steric blockers, physically preventing access to targeted sequences [5]. Such AOs modulate splicing by targeting splice sites and splicing enhancer/silencer sequences or reduce gene expression by targeting sequences critical for mRNA processing and maturation. On the other hand, AOs with the "gapmer" configuration that consist of a central DNA segment flanked by bases of modified chemistry act by inducing the degradation of their target mRNAs [6]. Hybridization of gapmers to their targets forms an RNA/DNA hybrid in the middle that is recognized and cleaved by ribonuclease H (RNase H), effectively reducing target

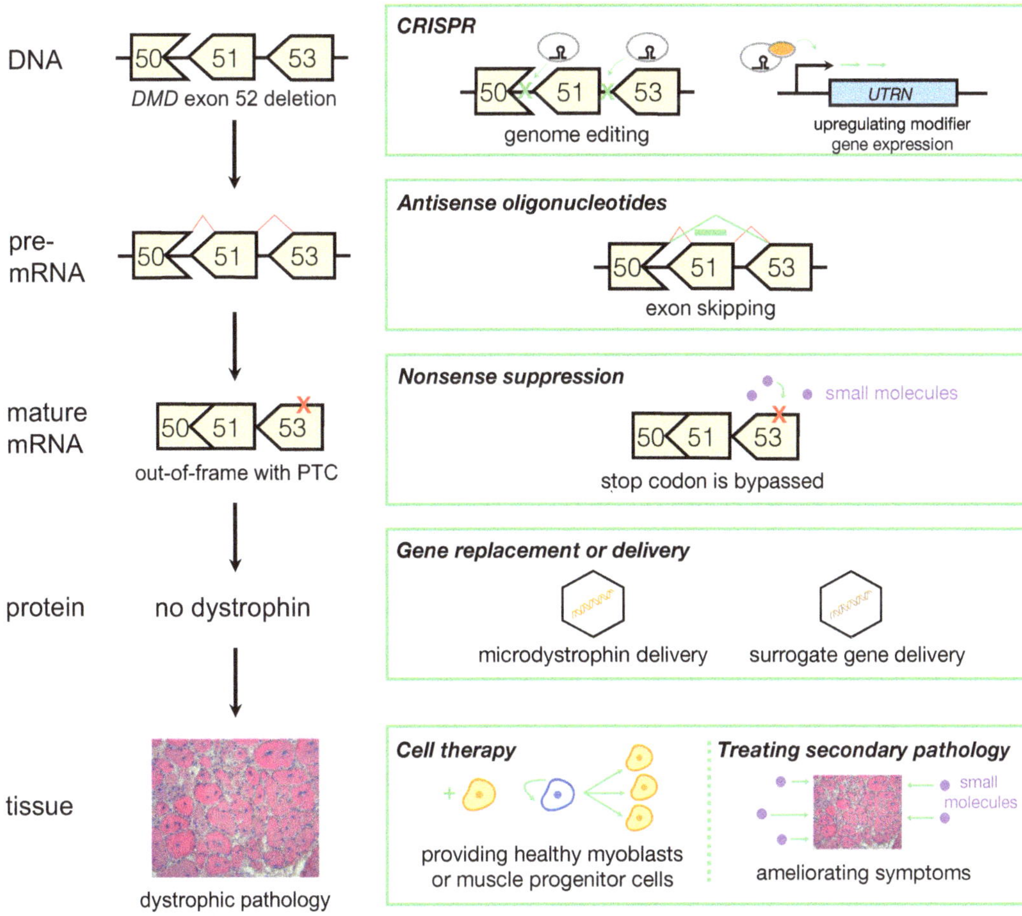

Fig. 1 Overview of muscular dystrophy therapeutics. A number of current therapeutic strategies covered in this chapter are depicted, using Duchenne muscular dystrophy (DMD) as an example. An out-of-frame deletion in *DMD* exon 52 leads to a lack of dystrophin in muscle, resulting in dystrophic histopathology as shown in a hematoxylin and eosin–stained image of the diaphragm from an exon 52–deleted DMD mouse model (hDMDdel52;*mdx*). Various approaches have been developed that either correct/address the genetic defect in DMD or ameliorate the observed dystrophic pathology and include the use of clustered regularly interspaced short palindromic repeats (CRISPR) for genome editing and gene modulation, antisense oligonucleotides for splice modulation (and gene knockdown for other muscular dystrophies), nonsense suppression for bypassing premature termination codons (PTCs) using pharmacological agents, gene replacement or delivery, cell therapy, and the treatment of secondary dystrophic pathology using pharmacological agents. These strategies are shown right next to the level (e.g., DNA, pre-mRNA, etc.) they target for therapy

mRNA expression [8, 9]. As we shall see, both kinds of AOs have been used in treating muscular dystrophies, with the choice depending on the genetic etiology of the disease under question.

2.1 Exon Skipping

Perhaps the most clinically advanced AOs in muscular dystrophy therapy are those that work by exon skipping. Interestingly, exon skipping is a strategy inspired by the genetics of dystrophinopathies, i.e., muscular dystrophies caused by mutations in the dystrophin gene (*DMD*) [10]. Dystrophin maintains sarcolemmal integrity and function by facilitating connections between the cytoskeleton of muscle cells with the surrounding extracellular matrix [11, 12]. This is done through the formation of a dystrophin-associated glycoprotein complex (DAGC) on the muscle cell membrane [13, 14]. Muscles with reduced levels of dystrophin are easily ruptured upon use, leading to weakness and degeneration. The two most common dystrophinopathies are the severe DMD and its considerably milder counterpart, Becker muscular dystrophy (BMD). Despite being caused by mutations in the same gene, it is in fact the type of mutation that determines whether a patient will develop DMD or BMD in ~90% of cases [15–17]. Generally speaking, BMD patients have in-frame *DMD* mutations, which allow for the production of truncated but partially functional dystrophin protein [15]. DMD patients on the other hand mostly have out-of-frame *DMD* mutations and are unable to produce dystrophin (with some exceptions, e.g., revertant fibers [18, 19], use of alternative translational start sites [20–22], and naturally occurring dystrophin isoforms [23]).

With these observations at hand, AOs were designed to target and exclude out-of-frame exons from DMD-associated dystrophin transcripts, converting out-of-frame mutations to in-frame mutations. From its first reported use for DMD in 1995 [24], exon skipping has since been extensively demonstrated to restore dystrophin production to clinically beneficial levels in patient-derived muscle cells and in the skeletal muscles of various DMD animal models [5]. Promising preclinical results eventually pushed exon skipping AOs into human clinical trials, which have led at present to the accelerated approval of four such therapeutics for DMD treatment by the US Food and Drug Administration (FDA). These are eteplirsen (Exondys 51, Sarepta; exon 51-skipping) [25], golodirsen (Vyondys 53, Sarepta; exon 53-skipping) [26], viltolarsen (Viltepso, NS Pharma; exon 53-skipping) [27], and casimersen (Amondys 45, Sarepta; exon 45-skipping) [28].

All the above AOs have the phosphorodiamidate morpholino oligomer (PMO) chemistry. PMOs are thought to be neutral molecules, but recent evidence indicates that they may possess a negative charge [29]. These four FDA-approved exon skipping AOs have each been shown to restore dystrophin production in patients, but at levels far below the target 10% of normal levels. Results from the latest clinical trials showed that eteplirsen, golodirsen, and casimersen could only restore up to 0.93–1.74% dystrophin of normal levels on average [25, 30, 31]; viltolarsen fared better, restoring up to 5.9% dystrophin of normal levels on average [32, 33]. There is limited evidence supporting functional benefit upon treatment with these AOs, but more information is expected once ongoing phase III trials are completed. Therapeutic outcomes in the heart have not been shown and should be a focus for future studies given the fatal nature of cardiac failure in DMD. Two other exon skipping

AOs, the ethylene-bridged nucleic acid DS-5141 (Daiichi Sankyo; exon 45-skipping) and the peptide-conjugated SRP-5051 (Sarepta; exon 51-skipping), are currently in phase I or II trials.

Exon skipping has also been applied to muscular dystrophies besides DMD, with the goal of bypassing exons that harbor small, deleterious mutations. Dysferlinopathies are a group of muscular dystrophies caused by dysferlin gene (*DYSF*) mutations, with its two most common types being limb girdle muscular dystrophy type 2B (LGMD2B) and Miyoshi myopathy [34, 35]. Dysferlin loss leads to defective plasma membrane repair in muscle fibers, making them susceptible to damage [36]. Skipping exons 28–29 or 32 of mutant *DYSF* transcripts rescued membrane resealing in patient-derived fibroblasts and muscle fibers, respectively, as observed via in vitro laser wounding assays [37, 38]. Other promising targets for *DYSF* exon skipping have been identified through patient genotype–phenotype correlation analyses and empirical observations [37, 39, 40].

Exon skipping has likewise been used to treat LGMD2C, a muscular dystrophy caused by mutations in the gene for γ-sarcoglycan (*SGCG*), which is an essential member of the DAGC in muscle cell membranes [41, 42]. Skipping *SGCG* exons 4–7 leads to the translation of an abbreviated, partially functional γ-sarcoglycan variant called Mini-Gamma [43]. Mini-Gamma transcripts were successfully generated after exons 4–7 skipping treatment in myogenic lineage-reprogrammed patient fibroblasts and urine-derived cells, as well as upon intramuscular injection of the exon skipping cocktail in a LGMD2C mouse model [43–45]. In vitro hypo-osmotic shock experiments suggest that exons 4–7 skipping has the potential to restore muscle cell membrane integrity [44]. On the other hand, exon 4 skipping of mutant *Lama2* transcripts improved lifespan in a mouse model of laminin-α2 chain-deficient muscular dystrophy (LAMA2 MD) [46]. Laminin-α2 is critical for the formation of laminin-211, which not only establishes key connections between muscle cell surface receptors and the basement membrane but also stabilizes the basement membrane itself [47–49]. Depending on the extent of laminin-α2 deficiency, LAMA2 MD presents as either a severe, fatal congenital form or a milder, late-onset form [49, 50].

Finally, exon skipping has also been applied to correct the aberrant splicing of *Clcn1* transcripts in a mouse model of myotonic dystrophy type 1 (DM1) [51, 52] as well as of *COL6A1* transcripts in Ullrich CMD patient fibroblasts [53, 54]. Interestingly, exon skipping has been used to disrupt the reading frame of dominant-negative mutant *COL6A2* transcripts in Ullrich CMD patient fibroblasts as well [55]. Both DM1 and Ullrich CMD will be discussed in more detail later. The potential clinical benefit of exon skipping in these other muscular dystrophies has not yet been sufficiently demonstrated however, and will require further in vivo testing before proceeding to human trials.

2.2 More Steric Blocking Approaches

Aside from exon skipping, steric-blocking AOs have been used in other applications to treat muscular dystrophies such as facioscapulohumeral muscular dystrophy (FSHD), DM1, and Fukuyama CMD. These applications are very much in the early stages of development – none have made it to clinical trials yet, but the preclinical work has yielded promising results. FSHD is caused by aberrant expression of the *DUX4* gene in skeletal muscle [56]. *DUX4* is normally only expressed during early embryonic development, after which its expression is silenced through methylation in most differentiated tissues [57–59]. In FSHD, this epigenetic silencing is disrupted, resulting in inappropriate *DUX4* expression and DUX4-mediated cytotoxicity [60, 61]. PMOs and 2′-*O*-methyl RNA AOs targeting the splice sites or the polyadenylation signal in *DUX4* transcripts effectively reduced *DUX4* expression in patient-derived muscle fibers and FSHD mouse models [62–67]. Systemic in vivo testing of such AOs has been limited to one study, which showed minimal but significant histological and functional improvements upon treatment [68].

In DM1, a CTG repeat expansion in the *DMPK* gene produces a toxic transcript that sequesters CUG-binding proteins in the nucleus (particularly those from the Muscleblind-like family) and somehow increases CUG-binding protein 1 (CUGBP1) levels in muscle [69–72]. Both events induce the aberrant splicing of various transcripts including those of *CLCN1*, *INSR*, and *cTNT*, leading to pathology [73, 74]. A number of studies using DM1 cell and animal models have consistently shown that AOs binding to expanded CUG repeats could prevent the sequestration of CUG-binding proteins, knock down expanded *DMPK* transcript levels, reverse abnormal splicing events, and reduce myotonia [75]. AOs against microRNAs (miRNAs) that inhibit the translation of Muscleblind-like proteins 1 and/or 2 have likewise been shown to reverse DM1 phenotypes in vivo [76, 77].

Fukuyama CMD is caused by a deficiency in fukutin, a ribitol-phosphate transferase required for glycosylating the DAGC member α-dystroglycan [78]. The founder variant of Fukuyama CMD in the Japanese population is an ancestral insertion of the SVA retrotransposon in the 3′-UTR of the *FKTN* mRNA [79, 80]. This initiates a cryptic splicing event that truncates *FKTN* exon 10 and fuses it to a part of the SVA element, creating an abnormal fukutin variant that does not localize to the Golgi apparatus as it should [81, 82]. An AO cocktail against the cryptic donor and acceptor sites, as well as an exonic splicing enhancer site, was able to correct this "exon-trapping" phenomenon in patient-derived cells and humanized knock-in Fukuyama CMD mice [82]. Normal glycosylation of α-dystroglycan was restored post-treatment.

2.3 Gapmers

Because they induce RNase H-mediated degradation of target transcripts, gapmer AOs have been used for muscular dystrophies caused by aberrant gene overexpression or toxic gain-of-function mutations. From the disorders we have covered so far, this would apply to FSHD and DM1. Our group has previously developed locked nucleic acid (LNA) and 2′-*O*-methoxyethyl (2′-MOE) gapmers to knock down *DUX4* mRNA expression [83, 84]. We achieved near-complete *DUX4* inhibition in immortalized patient-derived muscle cells at levels higher than what has been achieved with steric-blocking AOs, with the restoration of DUX4 downstream target gene expression and improvements in in vitro myotube fusion and size. A separate study of one 2′-MOE gapmer revealed that it could improve in vitro muscle membrane repair [85]. Intramuscular administration of these gapmers knocked down *DUX4* expression in an FSHD mouse model as well. Systemic treatment studies are currently being done for these gapmers.

While gapmer studies have been scarce for FSHD, a completely opposite scenario is found for DM1. Gapmers of various chemistries have been designed to target expanded CUG repeats, resulting in preferential elimination of mutant *DMPK* transcripts [75, 86]. A wealth of preclinical work has established the potential of these gapmers for treating DM1, to such an extent that it is the only muscular dystrophy for which a gapmer therapy has reached clinical testing. This was the mixed constrained ethyl (cEt)/2′-MOE gapmer IONIS-DMPK-2.5Rx (ISIS 598769, Ionis Pharmaceuticals). Unfortunately, results from the phase I/IIa trial revealed a lack of therapeutic benefit upon treatment with IONIS-DMPK-2.5Rx, citing issues of inadequate gapmer delivery [75, 87]. Further development of the drug has since been discontinued.

Another interesting application of gapmers would be to deal with dominant-negative mutations, as seen in Ullrich CMD. The collagen VI-related CMDs are muscular dystrophies caused by mutations in one of three genes coding for the α chains of collagen type VI (*COL6A1*, *COL6A2*, *COL6A3*); Ullrich CMD is a severe form of the disorder [88]. In the majority of Ullrich CMD patients, mutant α chains are formed due to in-frame deletions or missense mutations [89, 90]. Such chains cannot be properly incorporated into collagen tetramers during assembly, thereby preventing collagen secretion into the extracellular matrix. Individuals who instead carry null *COL6A* gene mutations in one copy of the gene are asymptomatic, highlighting a dominant-negative mechanism of action [88, 90, 91]. Gapmers against mutant *COL6A3* transcripts in a patient fibroblast cell line selectively degraded these targets over the wild-type version and improved collagen type VI deposition in the extracellular matrix [90].

2.4 Other Oligonucleotide-Based Therapies

Briefly, other oligonucleotide-based therapeutic approaches have been developed for muscular dystrophies besides those using AOs. These include the use of RNA interference, modified U7 small nuclear RNAs, aptamers, and DNA decoys. RNA interference and U7 small nuclear RNAs function in a similar way to AOs and can target transcripts for splice modulation and/or direct their degradation [92, 93]. Aptamers and DNA decoys on the other hand are designed to target disease-associated proteins and exert therapeutic effects by inhibiting their pathological activity [94, 95].

3 CRISPR Genome Editing

The discovery of genome editing tools such as transcription activator-like effector nucleases, zinc finger nucleases, and the clustered regularly interspaced short palindromic repeats (CRISPR) system has ushered in a new era for basic and translational research. Of the three, CRISPR is the most extensively used for muscular dystrophy therapeutic development at present. CRISPR is a programmable, RNA-guided endonuclease derived from the bacterial adaptive immune system [96]. CRISPR has since been adapted and modified to work using two basic components [97–99]. First is the CRISPR-associated endonuclease or Cas enzyme, which is responsible for cleaving target DNA sequences. Target sequences have to contain a nearby protospacer adjacent motif (PAM) for successful recognition by Cas enzymes – numerous Cas variants are available for use, each with their own preferred PAM sequences. The second would be the guide RNA (gRNA), which directs Cas enzyme activity to the DNA sequence of choice and gives CRISPR its programmable nature. Once a double-stranded DNA break has been made, it is repaired through nonhomologous end-joining (NHEJ) or homology-directed repair (HDR). NHEJ randomly creates insertions or deletions (indels) at the site of damage, whereas HDR uses a homologous template to precisely repair the region.

Similar to what we saw with AOs, much of the work done using CRISPR for muscular dystrophies has been on DMD. Conventional CRISPR approaches for DMD are divided into those using single or multiple gRNAs [100]. Single-gRNA CRISPR can reframe mutant exons through indel formation or skip out-of-frame exons by targeting critical sequences for splicing. Using multiple gRNAs, e.g., one for each intron flanking an exon, allows for the deletion of mutant or out-of-frame *DMD* exons. More recently, CRISPR strategies have been developed that use impaired or catalytically inactive versions of the Cas enzyme. The modified Cas endonuclease is typically fused to another protein in these cases to achieve a desired effect, taking advantage of the highly sequence-specific recognition capabilities of CRISPR but without the production of double-stranded DNA breaks [101]. In one strategy, defective Cas enzymes have been fused to cytidine or adenine deaminases, allowing for the editing of DNA sequences at the level of a single base [102, 103]. Base editing has been used to correct *DMD* point mutations or, when targeted to splice sites, skip unwanted *DMD* exons [104–106]. Another strategy would be to fuse transcriptional activators or repressors to defective Cas enzymes, giving CRISPR the ability to modulate gene expression [99, 107–109]. In the context of DMD, this has been used to upregulate the expression of utrophin, a homolog of dystrophin and a known modifier of the DMD phenotype [110]. Interestingly, CRISPR has also been used to directly edit the utrophin gene, eliminating a site used by miRNAs to inhibit utrophin expression [111].

We have previously written a number of reviews on the development of CRISPR therapies for DMD and direct interested readers there for more information [100, 112, 113]. Briefly, preclinical testing revealed that CRISPR editing of the *DMD* gene could successfully restore dystrophin production in patient-derived muscle cells and dystrophic animal models, ameliorating skeletal muscle and cardiac phenotypes to varying degrees. However, the safety of CRISPR therapies for human use remains uncertain and would require further study before clinical trials can be conducted. CRISPR has also been applied to the creation of new human-induced pluripotent stem cell and animal models of DMD [100, 112, 114]. These efforts increase the representation of patient mutations in preclinical testing, enabling the study of therapeutic effects on different genetic backgrounds and phenotypes.

The described strategies have been applied in other muscular dystrophies. In FSHD, CRISPR corrected a mutation of the *SMCHD1* gene, which codes for a chromatin repressing-protein involved in silencing *DUX4* expression [115, 116]. CRISPR has also been used to inhibit *DUX4* expression directly or by modulating the expression of epigenetic regulators [117, 118]. In DM1, various CRISPR therapies have been tested to remove/contract the expanded *DMPK* region or repress the transcription of the pathological allele [119]. Mutations in the dysferlin [120], α-sarcoglycan [120], laminin-α2 [121], and collagen VI genes [53], which all cause a form of muscular dystrophy, have been corrected using CRISPR as well. CRISPR-mediated upregulation of *Lama1*, a modifier gene for LAMA2 MD, has also been attempted [122].

4 Gene Replacement

Approaches to replace defective genes with their functional versions have been developed for several muscular dystrophies, primarily the LGMDs and DMD. Considered a more straightforward form of treatment, gene replacement also overcomes the issue of patient applicability associated with some personalized genetic therapies, e.g., AOs and CRISPR. Of the LGMDs, gene replacement therapies for types 2B (*DYSF*), 2C (*SGCG*), 2D (*SGCA*), and 2E (*SGCB*) have reached clinical trials. In a phase I trial, intramuscular delivery of a human *SGCG* construct via adeno-associated virus serotype 1 (AAV1) vectors into LGMD2C patients was well-tolerated and led to increased but limited *SGCG* expression at higher doses [123]. Western blot failed to detect γ-sarcoglycan protein in all but one treated patient, however. A new clinical trial is being planned based on promising results from a study delivering *SGCG* via AAV8 [124]. On the other hand, a phase I/II trial showed that intramuscular AAV1 delivery of *SGCA* into LGMD2D patients strongly restored α-sarcoglycan protein levels with no adverse effects [125]. Protein rescue persisted for up to 6 months post-treatment in two patients [126]. The same group proceeded to do another phase I/II trial, packaging *SGCA* into AAVrh74 and delivering it via isolated limb perfusion [127]. The treatment was well tolerated but resulted in modest to no functional improvement. Treatment with AAVrh74-*SGCB* in a phase I/II trial revealed sustained production of β-sarcoglycan in LGMD2E patients, at more than 50% of normal levels on average, even 2 years after treatment [128]. Significant improvements in motor function were observed. Two serious adverse events were observed (elevated liver enzymes, dehydration from vomiting) but were resolved upon medical treatment [129, 130].

Gene replacement therapies for calpain-3 and fukutin-related protein are currently in preclinical testing and appear to be promising for treating LGMD2A and 2I, respectively [131–136]. AAV8-myotubularin 1 gene replacement therapy for X-linked myotubular myopathy (AT132, Audentes Therapeutics) has likewise been promising in preclinical studies [137–139] and is being tested in a phase I/II clinical trial. Unfortunately, the FDA put the trial on hold in June 2020 following the deaths of two patients who received high doses of AT132 [140]. This was followed by a third death, that of the last patient in the high-dose cohort, roughly 2 months later [141]. The deaths were caused by progressive liver dysfunction leading to sepsis or gastrointestinal bleeding and were attributed to the high AT132 dose used [142]. This raised concerns over the immunogenicity and safety of AAV treatments, as serious adverse events have been associated with high-dose AAV gene therapy in other clinical trials [142]. The FDA lifted the hold on the trial in December of the same year [143], and results are expected in March 2024.

Gene replacement becomes more challenging in the case of larger genes such as *DYSF* (6.2 kb coding sequence) and *DMD* (11.5 kb coding sequence), as these exceed the typical packaging limit of AAVs at approximately 5 kb. Although delivery of full-length *DYSF* has been attempted previously [144], it has been shown that splitting the cDNA into two with a 1 kb overlap and delivering them via a dual-AAVrh74 vector system yielded higher dysferlin levels in dysferlin-deficient mice [145, 146]. Note that the overlap region allows for reconstitution of full-length *DYSF* via homologous recombination. Dose-dependent improvements in membrane resealing capabilities and diaphragm-specific force were also observed. A phase I trial of this therapy has been completed, but no results have been released at present.

Another solution has been to use abbreviated versions of the genes of concern. This approach has been tested extensively in the case of DMD, using what are now collectively called micro-dystrophins [147]. The large *DMD* gene codes for a 427-kDa full-length protein composed of four major domains: an N-terminal domain for actin binding, a central rod domain for structural flexibility and membrane localization, a cysteine-rich domain for facilitating interactions with the DAGC, and a C-terminal domain also for binding to the DAGC [14]. In contrast, the micro-dystrophin constructs currently used in human clinical trials code for 138–147 kDa proteins and lack majority of the rod domain (having only 4–5 of the 24 spectrin-like repeats) as well as the entire C-terminal domain [147]. Results from various phase I and II trials showed that micro-dystrophin levels were successfully increased in treated patients [148–151], but only one trial so far (PF-06939926, Pfizer) has demonstrated measurable improvement in muscle motor function [150]. One trial (SGT-001, Solid Biosciences) was put on hold twice by the FDA, in March 2018 and November 2019, due to serious adverse events in two patients [147, 152]. Both events were resolved, and the latest hold was lifted in October 2020 [153]. Pfizer has started a phase III trial for their micro-dystrophin therapy in late 2020; Sarepta is planning to begin a phase III trial soon. In partnership with Sarepta, Genethon has also recently launched a phase I/II/III trial for another micro-dystrophin therapy [154]. Replacement therapy with abbreviated dysferlin and laminin-α2 (another large gene, >9 kb cDNA) constructs have been attempted as well in animal models [155, 156].

Efforts to deliver micro-utrophin have also been explored as a surrogate gene therapy to minimize the risk of immune response activation that comes with micro-dystrophin therapy [157]. Indeed, pre-existing immunity to dystrophin was found in 20 out of 91 DMD patients in one study, with regression analyses showing that the likelihood of a T cell response positively correlated with age but negatively correlated with corticosteroid use [158]. A phase I micro-dystrophin trial detected dystrophin-specific T cells by interferon-γ enzyme-linked immune absorbent spot (ELISPOT) assay at least 15 days after treatment in three patients and before treatment in two patients [159]. Side-by-side AAV9 micro-dystrophin and micro-utrophin intramuscular injections in an adult dystrophic canine model with the entire *dystrophin* gene deleted further demonstrated that micro-utrophin was nonimmunogenic compared to micro-dystrophin, which showed a T cell response via ELISPOT 2 weeks post-treatment [160]. Micro-utrophin therapy was shown to ameliorate dystrophic histopathology, normalize serum creatine kinase levels, improve muscle function, and prevent heart failure with no signs of toxicity in dystrophic mice [160–162]. Histopathological improvement and lack of adverse effects were observed in canine DMD models as well [160].

Other surrogate gene therapies have undergone preclinical testing and include muscleblind-like protein 1 for DM1 [163]; mini-agrin for LAMA2 MD [164–166]; SERCA2a for DMD and LGMD2F [167, 168]; follistatin for DMD and FSHD [169–171]; and cytotoxic T cell GalNac transferase (GALGT2) for DMD, LAMA2 MD, LGMD2D, LGMD2I, and potentially other forms of muscular dystrophy [172–178]. Intramuscular follistatin delivery by AAV1 improved performance in the 6-min walk test in some BMD patients as part of a phase I/IIa trial, with no adverse events reported [179]. A phase I/II trial of intramuscular AAV1-follistatin therapy for DMD patients however reported no improvement in functional outcome measures [180]. GALGT2 gene therapy on the other hand is showing more promise and is currently being tested in a phase I/II clinical trial, with results expected last November 2021.

5 Cell Therapy

Another way to deliver functional genes into dystrophic muscle fibers would be by fusing them with healthy muscle cells. A host of preclinical studies have demonstrated the potential of myoblast transplantation for treating muscular dystrophy, leading to clinical trials [181, 182]. Particularly promising results were obtained in a phase I/IIa clinical trial on the safety and efficacy of autologous myoblast transplantation for treating oculopharyngeal muscular dystrophy (OPMD) [183]. OPMD is an adult-onset disease mainly characterized by progressive weakness of the levator palpebrae and pharyngeal muscles, leading to ptosis (droopy eyelid) and swallowing difficulties, respectively [184]. It is caused by a trinucleotide repeat expansion in the coding region of the poly(A) binding protein nuclear 1 gene (*PABPN1*), whose protein product is involved in mRNA polyadenylation [185]. Transplantation of myoblasts from unaffected muscles to the pharyngeal muscles stabilized or significantly improved pharyngeal phenotypes in OPMD patients even up to 2 years post-treatment, with no adverse effects [183]. Dose-dependent improvements in swallowing time were observed. Compared to OPMD where only a few muscle groups are affected, efficacy endpoints were more difficult to achieve for DMD, which presents as a body-wide myopathy [181, 182]. While efficacy was generally poor in earlier clinical trials for myoblast transplantation in DMD, recent human trials using immunosuppression and high-density intramuscular administration of allogeneic myoblasts have seen improvements in the levels of dystrophin and muscle function observed after treatment [181, 182].

Taking it a step further, one could also provide dystrophic muscles with healthy myogenic stem or progenitor cells, potentially improving regenerative/repair capacity. Satellite cells, mesenchymal stem cells, mesoangioblasts, muscle side population cells, $PWI^{+}PAX7^{-}$ interstitial cells, muscle-derived stem cells, $SMALD^{+}$ cells, cardiosphere-derived cells, $CD133^{+}$ cells, $CD34^{+}CD56^{+}CD144^{+}$ myoendothelial cells, induced pluripotent stem cells (iPSCs), embryonic stem cells, and directly-reprogrammed fibroblasts have all been used towards cell therapy development for muscular dystrophies [181, 182]. Of these, mesenchymal stem cells (from the umbilical cord), mesoangioblasts, cardiosphere-derived cells, and $CD133^{+}$ cells have proceeded to clinical trials. Treatment outcomes resulting from these therapies have been relatively modest but are expected to improve with continued optimization, similar to what was seen with myoblast transplantation [181, 182]. It is important to mention here that genome editing strategies such as CRISPR are paving the way for the autologous transplantation of genetically corrected cells, e.g., as has been shown for iPSCs in the context of DMD, LGMD2A, and LGMD2D [186–189].

6 Small Molecules

6.1 Nonsense Suppression

A class of small molecule drugs can treat muscular dystrophies particularly caused by nonsense point mutations. The approach, called nonsense suppression, uses chemical compounds that can alter the conformation of an mRNA transcript or the ribosome itself such that the ribosome reads through premature stop codons [190]. Nonsense suppression promotes the insertion of non-cognate tRNAs at the premature stop codon site in the ribosome, allowing for the incorporation of an amino acid [191, 192]. This suppressive activity is mostly specific to premature stop codons and not natural stop codons [190]. The aminoglycosides were perhaps the earliest compounds used for nonsense suppression therapy; their capacity for stop codon read-through was first demonstrated for streptomycin in prokaryotes in 1964 [193, 194]. One such aminoglycoside, gentamicin, has been used for nonsense suppression in DMD and has reached clinical trials. Despite showing increased dystrophin production with gentamicin treatment in some trials, functional improvement was generally not observed and the potential for clinical benefit remained uncertain [190, 194]. Concerns regarding the toxicity of gentamicin have also been raised [190, 194, 195], prompting the search for other pharmacological agents of nonsense suppression.

A screen of approximately 800,000 small molecule compounds identified the oxadiazole derivative PTC124 as a lead candidate for nonsense suppression [196]. Now more widely known as ataluren (Translarna, PTC Therapeutics), the orally bioavailable drug has received conditional approval from the European Medicines Agency in 2014 for the treatment of DMD [197]. Appreciable dystrophin restoration was demonstrated upon treatment of DMD patients with ataluren in a phase IIa trial [198]. Subsequent phase IIb and phase III trials failed to show strong functional improvement in the 6-min walk test after daily treatment with ataluren for 48 weeks [199, 200]. However, a meta-analysis of these two trials revealed that the improvements observed in the 6-min walk test with ataluren treatment were statistically significant, regardless of how patients were stratified according to their baseline performance in the test [201]. The FDA has previously rejected ataluren but is willing to consider new trial data for future review [202]. Real-world data on ataluren is constantly being collected and monitored through the Strategic Targeting of Registries and International Database of Excellence (STRIDE) registry [203]. Delayed progression of DMD was observed in patients treated with ataluren and the standard of care (mainly corticosteroids), particularly in terms of loss of ambulation and performance in timed function tests [204]. Recently, ataluren has also been tested for nonsense mutations in dysferlin transcripts, with improved grip strength, endurance, and protection against eccentric contractions in vivo as a result of treatment [205]. Moving forward, high-throughput screens will constantly identify new compounds for nonsense suppression in the context of other genetic disorders [206–208] and may have the potential for the treatment of muscular dystrophies.

6.2 Disease Modification

Finally, the majority of pharmacological agents are designed to target the secondary pathologies associated with muscular dystrophies. Perhaps one of the most important and widely used in this list of disease-modifying therapies would be the corticosteroids, which are presently considered the gold standard of care for DMD. Prednisone (off-label use) and deflazacort (FDA-approved for DMD in 2017) are most often used for treatment, which both belong to the glucocorticoid class of corticosteroids [209, 210]. Glucocorticoids recognize and bind the glucocorticoid receptor in the cytoplasm. In its bound state, the receptor translocates to the nucleus where it can act either as a transcriptional activator or as a repressor [211]. It is thought that the glucocorticoid receptor ultimately promotes anti-inflammatory signaling, resulting in the positive effects observed in muscular dystrophy patients [210, 211]. Treatment with prednisone and deflazacort have each led to prolonged survival and ambulation, as well as increased muscle strength and motor function in DMD patients

[210]. Improvements in pulmonary and cardiac function have also been observed. Comparative studies suggest that deflazacort yields better therapeutic outcomes than prednisone [212–216]. Unfortunately, prednisone and deflazacort have both been associated with adverse effects in patients – which include weight gain, behavioral changes, poor bone health, and cataract development – limiting their long-term use for therapy [210, 211].

In response to this, the corticosteroid analog vamorolone (Santhera Pharmaceuticals and ReveraGen BioPharma) has been developed as a safer alternative. Originally called VBP15, vamorolone was identified from a screen of Δ-9,11-modified steroids that have a replacement of the glucocorticoid hydroxyl group in the C-ring with a double bond between carbons 9 and 11 [217, 218]. The Δ-9,11 modification prevents glucocorticoid receptor signaling via glucocorticoid response elements, which are thought to be the major cause behind the adverse effects induced by glucocorticoids [210, 217, 219]. Vamorolone was chosen for further development due to its enhanced capacity to inhibit pro-inflammatory NF-κB signaling in vitro as well as its favorable safety and efficacy profile in vivo [217, 218]. Notably, no glucocorticoid-related adverse effects were observed after treatment of *mdx* mice with vamorolone, unlike what was seen with prednisone [218].

Interim results from a long-term extension phase IIa trial showed that vamorolone treatment for 18 months (2 or 6 mg/kg/day) significantly improved performance in timed function tests and increased North Star Ambulatory Assessment scores in DMD patients compared to baseline, at levels comparable to group-matched corticosteroid-treated patients from the Cooperative International Neuromuscular Research Group Duchenne Natural History Study [220]. Stunting was not observed in vamorolone-treated patients as opposed to corticosteroid-treated patients, but no significant differences in weight gain was found between the two groups. The incidences of corticosteroid-related adverse effects such as Cushingoid features, weight gain, hypertrichosis/hirsutism, and behavioral changes were also lower in vamorolone-treated than in corticosteroid-treated patients. Similar results for safety and efficacy were observed at the end of the 2.5-year phase IIa trial [221]. A phase IIb trial has since been initiated, consisting of patients treated with either low- or high-dose vamorolone (2 or 6 mg/kg/day, respectively), 0.75 mg/kg/day prednisone, or placebo for 48 weeks. Interim 24-week results revealed significant improvements in timed motor function tests from baseline in both vamorolone-treated cohorts, with no significant differences observed in performance between 6 mg/kg/day vamorolone-treated and prednisone-treated patients [222]. The number of treatment emergent adverse events was also lower in vamorolone-treated than prednisone-treated patients, suggesting a more favorable safety profile for vamorolone. The phase IIb study has an expected completion date of August 2021, with more results to follow then.

Aside from corticosteroids and their analogs, a myriad of promising pharmacological agents targeting inflammation, fibrosis, muscle growth and development, cell death, autophagy, metabolism, pathology-associated proteins, genetic modifiers, and other disease-specific features or signaling pathways have been developed for muscular dystrophies. Due to their incredible diversity and number, we cannot possibly cover all such disease-modifying therapies here and so invite readers to consult the literature specific to their muscular dystrophy of interest for further information.

7 Conclusions

Our brief survey has introduced us to a wide array of current muscular dystrophy therapeutics, a sizeable number of which have gone on to clinical trials or have even been approved for patient use. Genetic strategies target the root cause of disease and either correct, eliminate, or replace mutant genes and/or their products. Corrective genetic approaches such as those using AOs, CRISPR, or nonsense suppression additionally offer a personalized, mutation-specific approach to treating certain muscular dystrophies. In particular, four exon skipping AOs and the nonsense-suppressing small molecule drug ataluren have been approved for DMD therapy. Cellular therapies likewise aim to replace mutant genes, with the normal versions of these genes provided through the fusion of dystrophic muscle with healthy or genetically corrected muscle cells. Delivery of muscle progenitor cells is envisioned to be more beneficial, as these could lay the foundation for establishing a regenerative cell population in the recipient muscle. Finally, numerous pharmacological therapies have been developed to target secondary muscle pathologies, and are thus broadly applicable to various forms of muscular dystrophy. Corticosteroid treatment in particular is currently considered the gold standard of care for DMD.

While our progress has certainly been commendable, we should recognize that the development of each of these therapies is not without their respective hurdles. These challenges generally fall into one of two main categories, efficacy or safety; fortunately, solutions to these challenges are constantly being explored. As we review these concerns, it is important to keep in mind that the efficacy and safety of a therapeutic are often linked. For instance, higher doses of a therapeutic may lead to better efficacy but at the risk of inducing more adverse effects. This balance in efficacy and safety must continue to be considered in preclinical and clinical trials. Moreover, it would help if we had a deeper understanding of muscular dystrophy biology as well as the mechanisms of action and behaviors of these therapies in vivo through further basic research. As the development of muscular dystrophy therapeutics continues, it is becoming more and more apparent that a "cure" for this group of disorders may not come from a single approach but rather a combination of approaches [223]. Some examples include employing CRISPR to create genetically corrected myoblasts or myogenic progenitor cells for cell therapy [186–189], performing gene replacement with surrogate gene therapy to improve therapeutic outcomes [224, 225], and using pharmacological agents other than corticosteroids to augment the efficacy of genetic and cellular therapies [181, 226–229]. It would be interesting to see if such holistic approaches to therapy will predominate in succeeding clinical trials. We certainly still have quite a ways to go before we can completely cure any form of muscular dystrophy. But, at the rate the field is going, the future is showing much promise.

Acknowledgments

We would like to thank Dr. Annemieke Aartsma-Rus (Leiden University Medical Center) for providing the hDMDdel52;*mdx* mice used to start our in-house colony and generate the histological image shown in Fig. 1. This work was supported by Women and Children's Health Research Institute (Innovation Grant 2115), the Friends of Garrett Cumming Research & Muscular Dystrophy Canada HM Toupin Neurological Science Research Chair Fund, and the Alberta Innovates Graduate Studentship.

References

1. Mercuri E, Bönnemann CG, Muntoni F (2019) Muscular dystrophies. Lancet 394: 2025–2038. https://doi.org/10.1016/S0140-6736(19)32910-1
2. Mercuri E, Muntoni F (2013) Muscular dystrophies. Lancet. https://doi.org/10.1016/S0140-6736(12)61897-2
3. Waldrop MA, Flanigan KM (2019) Update in Duchenne and Becker muscular dystrophy. Curr Opin Neurol 32:722–727. https://doi.org/10.1097/WCO.0000000000000739
4. Bertini E, D'Amico A, Gualandi F, Petrini S (2011) Congenital muscular dystrophies: a brief review. Semin Pediatr Neurol 18:277–288. https://doi.org/10.1016/j.spen.2011.10.010
5. Lim KRQ, Yokota T (2018) Invention and early history of exon skipping and splice modulation. In: Yokota T, Maruyama R (eds) Exon skipping: methods and protocols. Springer, New York, pp 3–30
6. Lim KRQ, Yokota T (2020) Invention and early history of gapmers. In: Yokota T, Maruyama R (eds) Gapmers methods protocols. Springer, pp 3–19
7. Lee JJA, Yokota T (2013) Antisense therapy in neurology. J Pers Med 3:144–176. https://doi.org/10.3390/jpm3030144
8. Donis-Keller H (1979) Site specific enzymatic cleavage of RNA. Nucleic Acids Res 7:179–192. https://doi.org/10.1093/nar/7.1.179
9. Inoue H, Hayase Y, Iwai S, Ohtsuka E (1987) Sequence-dependent hydrolysis of RNA using modified oligonucleotide splints and RNase H. FEBS Lett 215:327–330
10. Hoffman EP, Brown RH, Kunkel LM (1987) Dystrophin: the protein product of the Duchenne muscular dystrophy locus. Cell 51:919–928. https://doi.org/10.1016/0092-8674(87)90579-4
11. Petrof BJ, Shrager JB, Stedman HH et al (1993) Dystrophin protects the sarcolemma

from stresses developed during muscle contraction. Proc Natl Acad Sci U S A 90:3710–3714. https://doi.org/10.1073/pnas.90.8.3710
12. Ervasti JM, Campbell KP (1993) A role for the dystrophin-glycoprotein complex as a transmembrane linker between laminin and actin. J Cell Biol 122:809–823. https://doi.org/10.1083/jcb.122.4.809
13. Ervasti JM (2007) Dystrophin, its interactions with other proteins, and implications for muscular dystrophy. Biochim Biophys Acta 1772:108–117. https://doi.org/10.1016/j.bbadis.2006.05.010
14. Gao QQ, McNally EM (2015) The dystrophin complex: structure, function, and implications for therapy. Compr Physiol 5:1223–1239. https://doi.org/10.1002/cphy.c140048
15. Monaco AP, Bertelson CJ, Liechti-Gallati S et al (1988) An explanation for the phenotypic differences between patients bearing partial deletions of the DMD locus. Genomics 2:90–95. https://doi.org/10.1016/0888-7543(88)90113-9
16. Lim KRQ, Nguyen Q, Yokota T (2020) Genotype–phenotype correlations in Duchenne and Becker muscular dystrophy patients from the Canadian Neuromuscular Disease Registry. J Pers Med 10:241. https://doi.org/10.3390/jpm10040241
17. Bladen CL, Salgado D, Monges S et al (2015) The TREAT-NMD DMD global database: analysis of more than 7,000 Duchenne muscular dystrophy mutations. Hum Mutat 36:395–402. https://doi.org/10.1002/humu.22758
18. Thanh LT, Nguyen TM, Helliwell TR, Morris GE (1995) Characterization of revertant muscle fibers in Duchenne muscular dystrophy, using exon-specific monoclonal antibodies against dystrophin. Am J Hum Genet 56:725–731
19. Klein CJ, Coovert DD, Bulman DE et al (1992) Somatic reversion/suppression in Duchenne muscular dystrophy (DMD): evidence supporting a frame-restoring mechanism in rare dystrophin-positive fibers. Am J Hum Genet 50:950–959
20. Gurvich OL, Maiti B, Weiss RB et al (2009) DMD exon 1 truncating point mutations: amelioration of phenotype by alternative translation initiation in exon 6. Hum Mutat 30:633–640. https://doi.org/10.1002/humu.20913
21. Wein N, Vulin A, Falzarano MS et al (2014) Translation from a DMD exon 5 IRES results in a functional dystrophin isoform that attenuates dystrophinopathy in humans and mice. Nat Med 20:992–1000. https://doi.org/10.1038/nm.3628
22. Winnard AV, Mendell JR, Prior TW et al (1995) Frameshift deletions of exons 3-7 and revertant fibers in Duchenne muscular dystrophy: mechanisms of dystrophin production. Am J Hum Genet 56:158–166. 7825572
23. Duan D, Goemans N, Takeda S et al (2021) Duchenne muscular dystrophy. Nat Rev Dis Prim 7:13. https://doi.org/10.1038/s41572-021-00248-3
24. Takeshima Y, Nishio H, Sakamoto H et al (1995) Modulation of in vitro splicing of the upstream intron by modifying an intra-exon sequence which is deleted from the dystrophin gene in dystrophin Kobe. J Clin Invest 95:515–520. https://doi.org/10.1172/JCI117693
25. Lim KRQKRQ, Maruyama R, Yokota T (2017) Eteplirsen in the treatment of Duchenne muscular dystrophy. Drug Des Devel Ther 11:533–545. https://doi.org/10.2147/DDDT.S97635
26. Anwar S, Yokota T (2020) Golodirsen for Duchenne muscular dystrophy. Drugs Today 56:491–504. https://doi.org/10.1358/dot.2020.56.8.3159186
27. Roshmi RR, Yokota T (2019) Viltolarsen for the treatment of Duchenne muscular dystrophy. Drugs Today 55:627–639. https://doi.org/10.1358/dot.2019.55.10.3045038
28. Shirley M (2021) Casimersen: first approval. Drugs. https://doi.org/10.1007/s40265-021-01512-2
29. Miyatake S, Mizobe Y, Tsoumpra MK et al (2019) Scavenger receptor class A1 mediates uptake of morpholino antisense oligonucleotide into dystrophic skeletal muscle. Mol Ther – Nucleic Acids. https://doi.org/10.1016/j.omtn.2019.01.008
30. Frank DE, Schnell FJ, Akana C et al (2020) Increased dystrophin production with golodirsen in patients with Duchenne muscular dystrophy. Neurology 94:e2270–e2282. https://doi.org/10.1212/WNL.0000000000009233
31. Wagner KR, Kuntz N, Koenig E et al (2020) Casimersen treatment in patients with duchenne muscular dystrophy: safety, tolerability, and pharmacokinetics over 144 weeks of treatment. https://investorrelations.sarepta.com/static-files/4fd181c5-73ec-4c89-ba29-e25ea2b7c776. Accessed 29 Apr 2021

32. Komaki H, Takeshima Y, Matsumura T et al (2020) Viltolarsen in Japanese Duchenne muscular dystrophy patients: a phase 1/2 study. Ann Clin Transl Neurol 7:2393–2408. https://doi.org/10.1002/acn3.51235
33. Clemens PR, Rao VK, Connolly AM et al (2020) Safety, tolerability, and efficacy of viltolarsen in boys with Duchenne muscular dystrophy amenable to exon 53 skipping. JAMA Neurol 77:982. https://doi.org/10.1001/jamaneurol.2020.1264
34. Harris E, Bladen CL, Mayhew A et al (2016) The clinical outcome study for dysferlinopathy. Neurol Genet 2:e89. https://doi.org/10.1212/NXG.0000000000000089
35. Fanin M, Angelini C (2016) Progress and challenges in diagnosis of dysferlinopathy. Muscle Nerve 54:821–835. https://doi.org/10.1002/mus.25367
36. Bansal D, Miyake K, Vogel SS et al (2003) Defective membrane repair in dysferlin-deficient muscular dystrophy. Nature 423: 168–172. https://doi.org/10.1038/nature01573
37. Lee JJA, Maruyama R, Duddy W et al (2018) Identification of novel antisense-mediated exon skipping targets in DYSF for therapeutic treatment of dysferlinopathy. Mol Ther – Nucleic Acids 13:596–604. https://doi.org/10.1016/j.omtn.2018.10.004
38. Barthélémy F, Blouin C, Wein N et al (2015) Exon 32 skipping of dysferlin rescues membrane repair in patients' cells. J Neuromuscul Dis 2:281–290. https://doi.org/10.3233/JND-150109
39. Aartsma-Rus A, Singh KHK, Fokkema IFAC et al (2010) Therapeutic exon skipping for dysferlinopathies? Eur J Hum Genet 18: 889–894. https://doi.org/10.1038/ejhg.2010.4
40. Verwey N, Gazzoli I, Krause S et al (2020) Antisense-mediated skipping of dysferlin exons in control and dysferlinopathy patient-derived cells. Nucleic Acid Ther 30:71–79. https://doi.org/10.1089/nat.2019.0788
41. Noguchi S, McNally EM, Othmane KB et al (1995) Mutations in the dystrophin-associated protein gamma-sarcoglycan in chromosome 13 muscular dystrophy. Science (80-) 270:819–822. https://doi.org/10.1126/science.270.5237.819
42. McNally EM, Passos-Bueno MR, Bönnemann CG et al (1996) Mild and severe muscular dystrophy caused by a single gamma-sarcoglycan mutation. Am J Hum Genet 59: 1040–1047
43. Gao QQ, Wyatt E, Goldstein JA et al (2015) Reengineering a transmembrane protein to treat muscular dystrophy using exon skipping. J Clin Invest 125:4186–4195. https://doi.org/10.1172/JCI82768
44. Wyatt EJ, Demonbreun AR, Kim EY et al (2018) Efficient exon skipping of SGCG mutations mediated by phosphorodiamidate morpholino oligomers. JCI Insight. https://doi.org/10.1172/jci.insight.99357
45. Demonbreun AR, Wyatt EJ, Fallon KS et al (2019) A gene-edited mouse model of Limb-Girdle muscular dystrophy 2C for testing exon skipping. Dis Model Mech. https://doi.org/10.1242/dmm.040832
46. Aoki Y, Nagata T, Yokota T et al (2013) Highly efficient in vivo delivery of PMO into regenerating myotubes and rescue in laminin-α2 chain-null congenital muscular dystrophy mice. Hum Mol Genet 22:4914–4928. https://doi.org/10.1093/hmg/ddt341
47. Tomé FM, Evangelista T, Leclerc A et al (1994) Congenital muscular dystrophy with merosin deficiency. C R Acad Sci III 317:351–357
48. Helbling-Leclerc A, Zhang X, Topaloglu H et al (1995) Mutations in the laminin alpha 2-chain gene (LAMA2) cause merosin-deficient congenital muscular dystrophy. Nat Genet 11:216–218. https://doi.org/10.1038/ng1095-216
49. Durbeej M (2015) Laminin-α2 chain-deficient congenital muscular dystrophy: pathophysiology and development of treatment. Curr Top Membr 76:31–60. https://doi.org/10.1016/bs.ctm.2015.05.002
50. Geranmayeh F, Clement E, Feng LH et al (2010) Genotype-phenotype correlation in a large population of muscular dystrophy patients with LAMA2 mutations. Neuromuscul Disord 20:241–250. https://doi.org/10.1016/j.nmd.2010.02.001
51. Wheeler TM, Lueck JD, Swanson MS et al (2007) Correction of ClC-1 splicing eliminates chloride channelopathy and myotonia in mouse models of myotonic dystrophy. J Clin Invest. https://doi.org/10.1172/JCI33355
52. Koebis M, Kiyatake T, Yamaura H et al (2013) Ultrasound-enhanced delivery of morpholino with Bubble liposomes ameliorates the myotonia of myotonic dystrophy model mice. Sci Rep 3:2242. https://doi.org/10.1038/srep02242
53. Bolduc V, Foley AR, Solomon-Degefa H et al (2019) A recurrent COL6A1 pseudoexon

insertion causes muscular dystrophy and is effectively targeted by splice-correction therapies. JCI Insight. https://doi.org/10.1172/jci.insight.124403

54. Aguti S, Bolduc V, Ala P et al (2020) Exon-skipping oligonucleotides restore functional collagen VI by correcting a common COL6A1 mutation in Ullrich CMD. Mol Ther – Nucleic Acids 21:205–216. https://doi.org/10.1016/j.omtn.2020.05.029
55. Gualandi F, Manzati E, Sabatelli P et al (2012) Antisense-induced messenger depletion corrects a COL6A2 dominant mutation in Ullrich myopathy. Hum Gene Ther 23:1313–1318. https://doi.org/10.1089/hum.2012.109
56. Lemmers RJLF, van der Vliet PJ, Klooster R et al (2010) A unifying genetic model for facioscapulohumeral muscular dystrophy. Science 329:1650–1653. https://doi.org/10.1126/science.1189044
57. De Iaco A, Planet E, Coluccio A et al (2017) DUX-family transcription factors regulate zygotic genome activation in placental mammals. Nat Genet 49:941–945. https://doi.org/10.1038/ng.3858
58. Hendrickson PG, Doráis JA, Grow EJ et al (2017) Conserved roles of mouse DUX and human DUX4 in activating cleavage-stage genes and MERVL/HERVL retrotransposons. Nat Genet 49:925–934. https://doi.org/10.1038/ng.3844
59. Hewitt JE, Lyle R, Clark LN et al (1994) Analysis of the tandem repeat locus D4Z4 associated with facioscapulohumeral muscular dystropothhy. Hum Mol Genet 3:1287–1295. https://doi.org/10.1093/hmg/3.8.1287
60. Lim KRQKRQ, Nguyen Q, Yokota T (2020) DUX4 signalling in the pathogenesis of facioscapulohumeral muscular dystrophy. Int J Mol Sci 21:729. https://doi.org/10.3390/ijms21030729
61. Wang LH, Tawil R (2016) Facioscapulohumeral dystrophy. Curr Neurol Neurosci Rep 16:66. https://doi.org/10.1007/s11910-016-0667-0
62. Lim KRQ, Yokota T (2021) Genetic approaches for the treatment of facioscapulohumeral muscular dystrophy. Front Pharmacol 12:642858. https://doi.org/10.3389/fphar.2021.642858
63. Vanderplanck C, Ansseau E, Charron S et al (2011) The FSHD atrophic myotube phenotype is caused by DUX4 expression. PLoS One 6:e26820. https://doi.org/10.1371/journal.pone.0026820
64. Marsollier A-C, Ciszewski L, Mariot V et al (2016) Antisense targeting of 3′ end elements involved in DUX4 mRNA processing is an efficient therapeutic strategy for facioscapulohumeral dystrophy: a new gene-silencing approach. Hum Mol Genet 25:1468–1478. https://doi.org/10.1093/hmg/ddw015
65. Chen JC, King OD, Zhang Y et al (2016) Morpholino-mediated knockdown of DUX4 toward facioscapulohumeral muscular dystrophy therapeutics. Mol Ther 24:1405–1411. https://doi.org/10.1038/mt.2016.111
66. Ansseau E, Vanderplanck C, Wauters A et al (2017) Antisense oligonucleotides used to target the DUX4 mRNA as therapeutic approaches in faciosscapulohumeral muscular dystrophy (FSHD). Genes (Basel) 8:93. https://doi.org/10.3390/genes8030093
67. Derenne A, Tassin A, Nguyen TH et al (2020) Induction of a local muscular dystrophy using electroporation in vivo: an easy tool for screening therapeutics. Sci Rep 10:11301. https://doi.org/10.1038/s41598-020-68135-7
68. Lu-Nguyen NB, Malerba A, Dickson G, Popplewell L (2021) Systemic antisense therapeutics inhibiting DUX4 expression improves muscle function in an FSHD mouse model. bioRxiv. https://doi.org/10.1101/2021.01.14.426659
69. Brook JD, McCurrach ME, Harley HG et al (1992) Molecular basis of myotonic dystrophy: expansion of a trinucleotide (CTG) repeat at the 3′ end of a transcript encoding a protein kinase family member. Cell 68:799–808. https://doi.org/10.1016/0092-8674(92)90154-5
70. Fu Y, Pizzuti A, Fenwick R et al (1992) An unstable triplet repeat in a gene related to myotonic muscular dystrophy. Science (80-) 255:1256–1258. https://doi.org/10.1126/science.1546326
71. Mahadevan M, Tsilfidis C, Sabourin L et al (1992) Myotonic dystrophy mutation: an unstable CTG repeat in the 3′ untranslated region of the gene. Science (80-) 255:1253–1255. https://doi.org/10.1126/science.1546325
72. Philips AV (1998) Disruption of splicing regulated by a CUG-binding protein in myotonic dystrophy. Science (80-) 280:737–741. https://doi.org/10.1126/science.280.5364.737
73. Lee JE, Cooper TA (2009) Pathogenic mechanisms of myotonic dystrophy. Biochem Soc Trans 37:1281–1286. https://doi.org/10.1042/BST0371281

74. Kaliman P, Llagostera E (2008) Myotonic dystrophy protein kinase (DMPK) and its role in the pathogenesis of myotonic dystrophy 1. Cell Signal 20:1935–1941. https://doi.org/10.1016/j.cellsig.2008.05.005
75. Overby SJ, Cerro-Herreros E, Llamusi B, Artero R (2018) RNA-mediated therapies in myotonic dystrophy. Drug Discov Today 23: 2013–2022. https://doi.org/10.1016/j.drudis.2018.08.004
76. Cerro-Herreros E, Sabater-Arcis M, Fernandez-Costa JM et al (2018) miR-23b and miR-218 silencing increase Muscleblind-like expression and alleviate myotonic dystrophy phenotypes in mammalian models. Nat Commun 9:2482. https://doi.org/10.1038/s41467-018-04892-4
77. Cerro-Herreros E, González-Martínez I, Moreno-Cervera N et al (2020) Therapeutic potential of antagomiR-23b for treating myotonic dystrophy. Mol Ther – Nucleic Acids 21: 837–849. https://doi.org/10.1016/j.omtn.2020.07.021
78. Saito F, Matsumura K (2011) Fukuyama-type congenital muscular dystrophy and defective glycosylation of α-dystroglycan. Skelet Muscle 1:22. https://doi.org/10.1186/2044-5040-1-22
79. Kobayashi K, Nakahori Y, Miyake M et al (1998) An ancient retrotransposal insertion causes Fukuyama-type congenital muscular dystrophy. Nature 394:388–392. https://doi.org/10.1038/28653
80. Toda T, Segawa M, Nomura Y et al (1993) Localization of a gene for Fukuyama type congenital muscular dystrophy to chromosome 9q31–33. Nat Genet 5:283–286. https://doi.org/10.1038/ng1193-283
81. Tachikawa M, Kanagawa M, Yu C-C et al (2012) Mislocalization of fukutin protein by disease-causing missense mutations can be rescued with treatments directed at folding amelioration. J Biol Chem 287:8398–8406. https://doi.org/10.1074/jbc.M111.300905
82. Taniguchi-Ikeda M, Kobayashi K, Kanagawa M et al (2011) Pathogenic exon-trapping by SVA retrotransposon and rescue in Fukuyama muscular dystrophy. Nature 478:127–131. https://doi.org/10.1038/nature10456
83. Lim KRQ, Maruyama R, Echigoya Y et al (2020) Inhibition of DUX4 expression with antisense LNA gapmers as a therapy for facioscapulohumeral muscular dystrophy. Proc Natl Acad Sci 117:16509–16515. https://doi.org/10.1073/pnas.1909649117
84. Lim KRQ, Bittel A, Maruyama R et al (2020) DUX4 transcript knockdown with antisense 2′-O-methoxyethyl gapmers for the treatment of facioscapulohumeral muscular dystrophy. Mol Ther. https://doi.org/10.1016/j.ymthe.2020.10.010
85. Bittel AJ, Sreetama SC, Bittel DC et al (2020) Membrane repair deficit in facioscapulohumeral muscular dystrophy. Int J Mol Sci 21: 5575. https://doi.org/10.3390/ijms21155575
86. Nguyen Q, Yokota T (2020) Degradation of toxic RNA in myotonic dystrophy using gapmer antisense. Oligonucleotides:99–109
87. Myotonic Dystrophy Foundation (2017) Ionis pharmaceuticals reports on DMPKRx phase 1/2 clinical trial. https://us8.campaign-archive.com/?u=8f5969cac3271759ce78c8354&id=8cc67ae9b8&e=cd1f4d18fe. Accessed 7 June 2021
88. Bönnemann CG (2011) The collagen VI-related myopathies: muscle meets its matrix. Nat Rev Neurol 7:379–390. https://doi.org/10.1038/nrneurol.2011.81
89. Allamand V, Briñas L, Richard P et al (2011) ColVI myopathies: where do we stand, where do we go? Skelet Muscle 1:30. https://doi.org/10.1186/2044-5040-1-30
90. Marrosu E, Ala P, Muntoni F, Zhou H (2017) Gapmer antisense oligonucleotides suppress the mutant allele of COL6A3 and restore functional protein in Ullrich muscular dystrophy. Mol Ther – Nucleic Acids 8:416–427. https://doi.org/10.1016/j.omtn.2017.07.006
91. Foley AR, Hu Y, Zou Y et al (2011) Large genomic deletions: a novel cause of Ullrich congenital muscular dystrophy. Ann Neurol 69:206–211. https://doi.org/10.1002/ana.22283
92. Gadgil A, Raczyńska KD (2021) U7 snRNA: a tool for gene therapy. J Gene Med. https://doi.org/10.1002/jgm.3321
93. Wilson RC, Doudna JA (2013) Molecular mechanisms of RNA interference. Annu Rev Biophys 42:217–239. https://doi.org/10.1146/annurev-biophys-083012-130404
94. Zhou J, Rossi J (2017) Aptamers as targeted therapeutics: current potential and challenges. Nat Rev Drug Discov 16:181–202. https://doi.org/10.1038/nrd.2016.199
95. Osako MK, Nakagami H, Morishita R (2011) Development and modification of decoy oligodeoxynucleotides for clinical application. In: Nucleic acid drugs. Springer, Berlin/Heidelberg, pp 49–59
96. Terns MP, Terns RM (2011) CRISPR-based adaptive immune systems. Curr Opin

Microbiol 14:321–327. https://doi.org/10.1016/j.mib.2011.03.005

97. Mali P, Yang L, Esvelt KM et al (2013) RNA-guided human genome engineering via Cas9. Science 339:823–826. https://doi.org/10.1126/science.1232033

98. Jiang F, Doudna JA (2017) CRISPR-Cas9 structures and mechanisms. Annu Rev Biophys 46:505–529. https://doi.org/10.1146/annurev-biophys-062215-010822

99. Sander JD, Joung JK (2014) CRISPR-Cas systems for editing, regulating and targeting genomes. Nat Biotechnol 32:347–355. https://doi.org/10.1038/nbt.2842

100. Lim KRQ, Yoon C, Yokota T (2018) Applications of CRISPR/Cas9 for the treatment of Duchenne muscular dystrophy. J Pers Med 8:38. https://doi.org/10.3390/jpm8040038

101. Brezgin S, Kostyusheva A, Kostyushev D, Chulanov V (2019) Dead Cas systems: types, principles, and applications. Int J Mol Sci 20:6041. https://doi.org/10.3390/ijms20236041

102. Kim YB, Komor AC, Levy JM et al (2017) Increasing the genome-targeting scope and precision of base editing with engineered Cas9-cytidine deaminase fusions. Nat Biotechnol 35:371–376. https://doi.org/10.1038/nbt.3803

103. Gaudelli NM, Komor AC, Rees HA et al (2017) Programmable base editing of A•T to G•C in genomic DNA without DNA cleavage. Nature 551:464–471. https://doi.org/10.1038/nature24644

104. Yuan J, Ma Y, Huang T et al (2018) Genetic modulation of RNA splicing with a CRISPR-guided cytidine deaminase. Mol Cell 72:380–394.e7. https://doi.org/10.1016/j.molcel.2018.09.002

105. Ryu S-M, Koo T, Kim K et al (2018) Adenine base editing in mouse embryos and an adult mouse model of Duchenne muscular dystrophy. Nat Biotechnol 36:536–539. https://doi.org/10.1038/nbt.4148

106. Chemello F, Chai AC, Li H et al (2021) Precise correction of Duchenne muscular dystrophy exon deletion mutations by base and prime editing. Sci Adv 7:eabg4910. https://doi.org/10.1126/sciadv.abg4910

107. Cheng AW, Wang H, Yang H et al (2013) Multiplexed activation of endogenous genes by CRISPR-on, an RNA-guided transcriptional activator system. Cell Res 23:1163–1171. https://doi.org/10.1038/cr.2013.122

108. Qi LS, Larson MH, Gilbert LA et al (2013) Repurposing CRISPR as an RNA-guided platform for sequence-specific control of gene expression. Cell 152:1173–1183. https://doi.org/10.1016/j.cell.2013.02.022

109. Gilbert LA, Larson MH, Morsut L et al (2013) CRISPR-mediated modular RNA-guided regulation of transcription in eukaryotes. Cell 154:442–451. https://doi.org/10.1016/j.cell.2013.06.044

110. Wojtal D, Kemaladewi DU, Malam Z et al (2016) Spell checking nature: versatility of CRISPR/Cas9 for developing treatments for inherited disorders. Am J Hum Genet 98:90–101. https://doi.org/10.1016/j.ajhg.2015.11.012

111. Sengupta K, Mishra MK, Loro E et al (2020) Genome editing-mediated utrophin upregulation in Duchenne muscular dystrophy stem cells. Mol Ther – Nucleic Acids 22:500–509. https://doi.org/10.1016/j.omtn.2020.08.031

112. Nguyen Q, KRQ L, Yokota T et al (2020) Genome editing for the understanding and treatment of inherited cardiomyopathies. Int J Mol Sci 21:733. https://doi.org/10.3390/ijms21030733

113. Aslesh T, Erkut E, Yokota T (2021) Restoration of dystrophin expression and correction of Duchenne muscular dystrophy by genome editing. Expert Opin Biol Ther 1–13. https://doi.org/10.1080/14712598.2021.1872539

114. Lim KRQ, Nguyen Q, Dzierlega K et al (2020) CRISPR-generated animal models of Duchenne muscular dystrophy. Genes (Basel) 11:342. https://doi.org/10.3390/genes11030342

115. Lemmers RJLF, Tawil R, Petek LM et al (2012) Digenic inheritance of an SMCHD1 mutation and an FSHD-permissive D4Z4 allele causes facioscapulohumeral muscular dystrophy type 2. Nat Genet 44:1370–1374. https://doi.org/10.1038/ng.2454

116. Goossens R, van den Boogaard ML, Lemmers RJLF et al (2019) Intronic SMCHD1 variants in FSHD: testing the potential for CRISPR-Cas9 genome editing. J Med Genet 56:828–837. https://doi.org/10.1136/jmedgenet-2019-106402

117. Himeda CL, Jones TI, Jones PL (2016) CRISPR/dCas9-mediated transcriptional inhibition ameliorates the epigenetic dysregulation at D4Z4 and represses DUX4-fl in FSH muscular dystrophy. Mol Ther 24:527–535. https://doi.org/10.1038/mt.2015.200

118. Himeda CL, Jones TI, Virbasius C-M et al (2018) Identification of epigenetic regulators

of DUX4-fl for targeted therapy of facioscapulohumeral muscular dystrophy. Mol Ther 26:1797–1807. https://doi.org/10.1016/j.ymthe.2018.04.019
119. Marsh S, Hanson B, Wood MJA et al (2020) Application of CRISPR-Cas9-mediated genome editing for the treatment of myotonic dystrophy type 1. Mol Ther 28:2527–2539. https://doi.org/10.1016/j.ymthe.2020.10.005
120. Turan S, Farruggio AP, Srifa W et al (2016) Precise correction of disease mutations in induced pluripotent stem cells derived from patients with limb girdle muscular dystrophy. Mol Ther 24:685–696. https://doi.org/10.1038/mt.2016.40
121. Kemaladewi DU, Maino E, Hyatt E et al (2017) Correction of a splicing defect in a mouse model of congenital muscular dystrophy type 1A using a homology-directed-repair-independent mechanism. Nat Med 23:984–989. https://doi.org/10.1038/nm.4367
122. Kemaladewi DU, Bassi PS, Erwood S et al (2019) A mutation-independent approach for muscular dystrophy via upregulation of a modifier gene. Nature 572:125–130. https://doi.org/10.1038/s41586-019-1430-x
123. Herson S, Hentati F, Rigolet A et al (2012) A phase I trial of adeno-associated virus serotype 1-γ-sarcoglycan gene therapy for limb girdle muscular dystrophy type 2C. Brain 135:483–492. https://doi.org/10.1093/brain/awr342
124. Israeli D, Cosette J, Corre G et al (2019) An AAV-SGCG dose-response study in a γ-sarcoglycanopathy mouse model in the context of mechanical stress. Mol Ther – Methods Clin Dev 13:494–502. https://doi.org/10.1016/j.omtm.2019.04.007
125. Mendell JR, Rodino-Klapac LR, Rosales-Quintero X et al (2009) Limb-girdle muscular dystrophy type 2D gene therapy restores α-sarcoglycan and associated proteins. Ann Neurol 66:290–297. https://doi.org/10.1002/ana.21732
126. Mendell JR, Rodino-Klapac LR, Rosales XQ et al (2010) Sustained alpha-sarcoglycan gene expression after gene transfer in limb-girdle muscular dystrophy, type 2D. Ann Neurol 68:629–638. https://doi.org/10.1002/ana.22251
127. Mendell JR, Chicoine LG, Al-Zaidy SA et al (2019) Gene delivery for limb-girdle muscular dystrophy type 2D by isolated limb infusion. Hum Gene Ther 30:794–801. https://doi.org/10.1089/hum.2019.006
128. Sarepta Therapeutics (2021) Sarepta Therapeutics' investigational gene therapy SRP-9003 for the treatment of limb-girdle muscular dystrophy type 2E shows sustained expression and functional improvements 2 years after administration
129. Gurunathan S (2020) Sarepta therapeutics announces positive results from the SRP-9003 gene therapy trial to treat LGMD2E. https://strongly.mda.org/sarepta-therapeutics-announces-positive-results-from-the-srp-9003-gene-therapy-trial-to-treat-lgmd2e/. Accessed 7 June 2021
130. GeneTherapyLive (2021) SRP-9003 efficacy, tolerability sustained at 2 years in LGMD2E. https://www.genetherapylive.com/view/srp-9003-efficacy-tolerability-sustained-at-2-years-in-lgmd2e. Accessed 7 June 2021
131. Bartoli M, Roudaut C, Martin S et al (2006) Safety and efficacy of AAV-mediated calpain 3 gene transfer in a mouse model of limb-girdle muscular dystrophy type 2A. Mol Ther 13:250–259. https://doi.org/10.1016/j.ymthe.2005.09.017
132. Lostal W, Roudaut C, Faivre M et al (2019) Titin splicing regulates cardiotoxicity associated with calpain 3 gene therapy for limb-girdle muscular dystrophy type 2A. Sci Transl Med 11:eaat6072. https://doi.org/10.1126/scitranslmed.aat6072
133. Vannoy CH, Leroy V, Lu QL (2018) Dose-dependent effects of FKRP gene-replacement therapy on functional rescue and longevity in dystrophic mice. Mol Ther – Methods Clin Dev 11:106–120. https://doi.org/10.1016/j.omtm.2018.10.004
134. Xu L, Lu PJ, Wang C-H et al (2013) Adeno-associated virus 9 mediated FKRP gene therapy restores functional glycosylation of α-dystroglycan and improves muscle functions. Mol Ther 21:1832–1840. https://doi.org/10.1038/mt.2013.156
135. Qiao C, Wang C-H, Zhao C et al (2014) Muscle and heart function restoration in a limb girdle muscular dystrophy 2I (LGMD2I) mouse model by systemic FKRP gene delivery. Mol Ther 22:1890–1899. https://doi.org/10.1038/mt.2014.141
136. Gicquel E, Maizonnier N, Foltz SJ et al (2017) AAV-mediated transfer of FKRP shows therapeutic efficacy in a murine model but requires control of gene expression. Hum Mol Genet 26:1952–1965. https://doi.org/10.1093/hmg/ddx066
137. Childers MK, Joubert R, Poulard K et al (2014) Gene therapy prolongs survival and restores function in murine and canine models of myotubular myopathy. Sci Transl Med

6:220ra10. https://doi.org/10.1126/scitranslmed.3007523
138. Mack DL, Poulard K, Goddard MA et al (2017) Systemic AAV8-mediated gene therapy drives whole-body correction of myotubular myopathy in dogs. Mol Ther 25:839–854. https://doi.org/10.1016/j.ymthe.2017.02.004
139. Shieh PB, Bönnemann CG, Müller-Felber W et al (2020) Re: "moving forward after two deaths in a gene therapy trial of myotubular myopathy" by Wilson and Flotte. Hum Gene Ther 31:787–787. https://doi.org/10.1089/hum.2020.217
140. Audentes Therapeutics (2020) Letter to XLMTM patient community. https://myotubulartrust.org/wp-content/uploads/23JUNE2020-Letter-to-Patient-Community_Sent.pdf. Accessed 7 June 2021
141. Audentes Therapeutics (2020) Audentes therapeutics provides update on the ASPIRO clinical trial evaluating AT132 in patients with X-linked myotubular myopathy. https://www.audentestx.com/press_release/audentes-therapeutics-provides-update-on-the-aspiro-clinical-trial-evaluating-at132-in-patients-with-x-linked-myotubular-myopathy/. Accessed 7 June 2021
142. Nature Biotechnology (2020) High-dose AAV gene therapy deaths. Nat Biotechnol 38:910–910. https://doi.org/10.1038/s41587-020-0642-9
143. Audentes Therapeutics (2020) Hold lifted by FDA on ASPIRO clinical trial of AT132 for treatment of X-linked myotubular myopathy (XLMTM). https://www.astellas.com/en/news/16301. Accessed 7 June 2021
144. Grose WE, Clark KR, Griffin D et al (2012) Homologous recombination mediates functional recovery of dysferlin deficiency following AAV5 gene transfer. PLoS One 7:e39233. https://doi.org/10.1371/journal.pone.0039233
145. Potter RA, Griffin DA, Sondergaard PC et al (2018) Systemic delivery of dysferlin overlap vectors provides long-term gene expression and functional improvement for dysferlinopathy. Hum Gene Ther 29:749–762. https://doi.org/10.1089/hum.2017.062
146. Sondergaard PC, Griffin DA, Pozsgai ER et al (2015) AAV dysferlin overlap vectors restore function in dysferlinopathy animal models. Ann Clin Transl Neurol 2:256–270. https://doi.org/10.1002/acn3.172
147. Duan D (2018) Systemic AAV micro-dystrophin gene therapy for Duchenne muscular dystrophy. Mol Ther 26:2337–2356. https://doi.org/10.1016/j.ymthe.2018.07.011
148. Gonzalez P (2020) SGT-001 microdystrophin gene therapy for Duchenne muscular dystrophy. https://s3.us-east-2.amazonaws.com/solid-bio-assets.investeddigital.com/pubs/Gonzalez_ASGCT_2020_11May2020.pdf. Accessed 29 Apr 2021
149. Pfizer (2021) Pfizer doses first participant in phase 3 study for Duchenne muscular dystrophy investigational gene therapy. https://www.pfizer.com/news/press-release/press-release-detail/pfizer-doses-first-participant-phase-3-study-duchenne. Accessed 25 Apr 2021
150. Pfizer (2020) Pfizer's new phase 1B results of gene therapy in ambulatory boys with Duchenne muscular dystrophy (DMD) support advancement into pivotal phase 3 study. https://www.pfizer.com/news/press-release/press-release-detail/pfizers-new-phase-1b-results-gene-therapy-ambulatory-boys. Accessed 25 Apr 2021
151. Sarepta (2021) Sarepta therapeutics announces top-line results for part 1 of study 102 evaluating SRP-9001, its investigational gene therapy for the treatment of Duchenne muscular dystrophy. https://investorrelations.sarepta.com/news-releases/news-release-details/sarepta-therapeutics-announces-top-line-results-part-1-study-102. Accessed 25 Apr 2021
152. Genetic Engineering & Biotechnology News (2019) FDA places another clinical hold on trial for solid's DMD gene therapy. https://www.genengnews.com/news/fda-places-another-clinical-hold-on-trial-for-solids-dmd-gene-therapy/. Accessed 7 June 2021
153. Solid Biosciences (2020) Solid biosciences announces FDA lifts clinical hold on IGNITE DMD clinical trial. https://www.solidbio.com/about/media/press-releases/solid-biosciences-announces-fda-lifts-clinical-hold-on-ignite-dmd-clinical-trial. Accessed 7 June 2021
154. Genethon (2021) Genethon announces first patient dosed in clinical trial of investigational gene therapy GNT 0004 for Duchenne muscular dystrophy. https://www.genethon.fr/wp-content/uploads/2021/04/PR_GENETHON_DMD-1.pdf. Accessed 25 Apr 2021
155. Krahn M, Wein N, Bartoli M et al (2010) A naturally occurring human minidysferlin protein repairs sarcolemmal lesions in a mouse model of dysferlinopathy. Sci Transl Med 2:50ra69. https://doi.org/10.1126/scitranslmed.3000951

156. Packer D, Martin PT (2021) Micro-laminin gene therapy can function as an inhibitor of muscle disease in the dyW mouse model of MDC1A. Mol Ther – Methods Clin Dev 21: 274–287. https://doi.org/10.1016/j.omtm.2021.02.004
157. Duan D (2019) Micro-utrophin therapy for Duchenne muscular dystrophy. Mol Ther 27: 1872–1874. https://doi.org/10.1016/j.ymthe.2019.10.011
158. Flanigan KM, Campbell K, Viollet L et al (2013) Anti-dystrophin T cell responses in Duchenne muscular dystrophy: prevalence and a glucocorticoid treatment effect. Hum Gene Ther 24:797–806. https://doi.org/10.1089/hum.2013.092
159. Mendell JR, Campbell K, Rodino-Klapac L et al (2010) Dystrophin immunity in Duchenne's muscular dystrophy. N Engl J Med 363:1429–1437. https://doi.org/10.1056/NEJMoa1000228
160. Song Y, Morales L, Malik AS et al (2019) Non-immunogenic utrophin gene therapy for the treatment of muscular dystrophy animal models. Nat Med 25:1505–1511. https://doi.org/10.1038/s41591-019-0594-0
161. Kennedy TL, Guiraud S, Edwards B et al (2018) Micro-utrophin improves cardiac and skeletal muscle function of severely affected D2/mdx mice. Mol Ther – Methods Clin Dev 11:92–105. https://doi.org/10.1016/j.omtm.2018.10.005
162. Banks GB, Chamberlain JS, Odom GL (2020) Microutrophin expression in dystrophic mice displays myofiber type differences in therapeutic effects. PLoS Genet 16: e1009179. https://doi.org/10.1371/journal.pgen.1009179
163. Kanadia RN, Shin J, Yuan Y et al (2006) Reversal of RNA missplicing and myotonia after muscleblind overexpression in a mouse poly(CUG) model for myotonic dystrophy. Proc Natl Acad Sci 103:11748–11753. https://doi.org/10.1073/pnas.0604970103
164. Moll J, Barzaghi P, Lin S et al (2001) An agrin minigene rescues dystrophic symptoms in a mouse model for congenital muscular dystrophy. Nature 413:302–307. https://doi.org/10.1038/35095054
165. Qiao C, Li J, Zhu T et al (2005) Amelioration of laminin-alpha2-deficient congenital muscular dystrophy by somatic gene transfer of miniagrin. Proc Natl Acad Sci U S A 102: 11999–12004. https://doi.org/10.1073/pnas.0502137102
166. Qiao C, Dai Y, Nikolova VD et al (2018) Amelioration of muscle and nerve pathology in LAMA2 muscular dystrophy by AAV9-mini-agrin. Mol Ther – Methods Clin Dev 9:47–56. https://doi.org/10.1016/j.omtm.2018.01.005
167. Wasala NB, Yue Y, Lostal W et al (2020) Single SERCA2a therapy ameliorated dilated cardiomyopathy for 18 months in a mouse model of Duchenne muscular dystrophy. Mol Ther 28:845–854. https://doi.org/10.1016/j.ymthe.2019.12.011
168. Goonasekera SA, Lam CK, Millay DP et al (2011) Mitigation of muscular dystrophy in mice by SERCA overexpression in skeletal muscle. J Clin Invest 121:1044–1052. https://doi.org/10.1172/JCI43844
169. Haidet AM, Rizo L, Handy C et al (2008) Long-term enhancement of skeletal muscle mass and strength by single gene administration of myostatin inhibitors. Proc Natl Acad Sci 105:4318–4322. https://doi.org/10.1073/pnas.0709144105
170. Giesige CR, Wallace LM, Heller KN et al (2018) AAV-mediated follistatin gene therapy improves functional outcomes in the TIC-DUX4 mouse model of FSHD. JCI Insight 3:e123538. https://doi.org/10.1172/jci.insight.123538
171. Kota J, Handy CR, Haidet AM et al (2009) Follistatin gene delivery enhances muscle growth and strength in nonhuman primates. Sci Transl Med 1:6ra15. https://doi.org/10.1126/scitranslmed.3000112
172. Xu R, Camboni M, Martin PT (2007) Postnatal overexpression of the CT GalNAc transferase inhibits muscular dystrophy in mdx mice without altering muscle growth or neuromuscular development: evidence for a utrophin-independent mechanism. Neuromuscul Disord 17:209–220. https://doi.org/10.1016/j.nmd.2006.12.004
173. Martin PT, Xu R, Rodino-Klapac LR et al (2009) Overexpression of Galgt2 in skeletal muscle prevents injury resulting from eccentric contractions in both mdx and wild-type mice. Am J Physiol Physiol 296:C476–C488. https://doi.org/10.1152/ajpcell.00456.2008
174. Xu R, Jia Y, Zygmunt DA, Martin PT (2019) rAAVrh74.MCK.GALGT2 protects against loss of hemodynamic function in the aging mdx mouse heart. Mol Ther 27:636–649. https://doi.org/10.1016/j.ymthe.2019.01.005
175. Xu R, Chandrasekharan K, Yoon JH et al (2007) Overexpression of the cytotoxic T

cell (CT) carbohydrate inhibits muscular dystrophy in the dyW mouse model of congenital muscular dystrophy 1A. Am J Pathol 171: 181–199. https://doi.org/10.2353/ajpath.2007.060927
176. Xu R, DeVries S, Camboni M, Martin PT (2009) Overexpression of Galgt2 reduces dystrophic pathology in the skeletal muscles of alpha sarcoglycan-deficient mice. Am J Pathol 175:235–247. https://doi.org/10.2353/ajpath.2009.080967
177. Thomas PJ, Xu R, Martin PT (2016) B4GALNT2 (GALGT2) gene therapy reduces skeletal muscle pathology in the FKRP P448L mouse model of limb girdle muscular dystrophy 2I. Am J Pathol 186: 2429–2448. https://doi.org/10.1016/j.ajpath.2016.05.021
178. Chicoine LG, Rodino-Klapac LR, Shao G et al (2014) Vascular delivery of rAAVrh74.MCK.GALGT2 to the gastrocnemius muscle of the rhesus macaque stimulates the expression of dystrophin and laminin α2 surrogates. Mol Ther 22:713–724. https://doi.org/10.1038/mt.2013.246
179. Mendell JR, Sahenk Z, Malik V et al (2015) A phase 1/2a follistatin gene therapy trial for Becker muscular dystrophy. Mol Ther 23: 192–201. https://doi.org/10.1038/mt.2014.200
180. Mendell JR (2020) Study results: clinical intramuscular gene transfer of rAAV1.CMV.huFollistatin344 trial to patients with Duchenne muscular dystrophy. https://clinicaltrials.gov/ct2/show/results/NCT02354781. Accessed 7 June 2021
181. Biressi S, Filareto A, Rando TA (2020) Stem cell therapy for muscular dystrophies. J Clin Invest 130:5652–5664. https://doi.org/10.1172/JCI142031
182. Ausems CRM, van Engelen BGM, van Bokhoven H, Wansink DG (2021) Systemic cell therapy for muscular dystrophies. Stem Cell Rev Rep 17:878–899. https://doi.org/10.1007/s12015-020-10100-y
183. Périé S, Trollet C, Mouly V et al (2014) Autologous myoblast transplantation for oculopharyngeal muscular dystrophy: a phase I/Iia clinical study. Mol Ther 22:219–225. https://doi.org/10.1038/mt.2013.155
184. Abu-Baker A, Rouleau GA (2007) Oculopharyngeal muscular dystrophy: recent advances in the understanding of the molecular pathogenic mechanisms and treatment strategies. Biochim Biophys Acta Mol basis Dis 1772:173–185. https://doi.org/10.1016/j.bbadis.2006.10.003
185. Brais B, Bouchard J-P, Xie Y-G et al (1998) Short GCG expansions in the PABP2 gene cause oculopharyngeal muscular dystrophy. Nat Genet 18:164–167. https://doi.org/10.1038/ng0298-164
186. Jin Y, Shen Y, Su X et al (2020) Effective restoration of dystrophin expression in iPSC Mdx-derived muscle progenitor cells using the CRISPR/Cas9 system and homology-directed repair technology. Comput Struct Biotechnol J 18:765–773. https://doi.org/10.1016/j.csbj.2020.03.012
187. Young CS, Hicks MR, Ermolova NV et al (2016) A single CRISPR-Cas9 deletion strategy that targets the majority of DMD patients restores dystrophin function in hiPSC-derived muscle cells. Cell Stem Cell 18:533–540. https://doi.org/10.1016/j.stem.2016.01.021
188. Selvaraj S, Dhoke NR, Kiley J et al (2019) Gene correction of LGMD2A patient-specific iPSCs for the development of targeted autologous cell therapy. Mol Ther 27:2147–2157. https://doi.org/10.1016/j.ymthe.2019.08.011
189. Tedesco FS, Gerli MFM, Perani L et al (2012) Transplantation of genetically corrected human iPSC-derived progenitors in mice with limb-girdle muscular dystrophy. Sci Transl Med 4:140ra89–140ra89. https://doi.org/10.1126/scitranslmed.3003541
190. Finkel RS (2010) Read-through strategies for suppression of nonsense mutations in Duchenne/Becker muscular dystrophy: aminoglycosides and ataluren (PTC124). J Child Neurol 25:1158–1164. https://doi.org/10.1177/0883073810371129
191. Roy B, Leszyk JD, Mangus DA, Jacobson A (2015) Nonsense suppression by near-cognate tRNAs employs alternative base pairing at codon positions 1 and 3. Proc Natl Acad Sci 112:3038–3043. https://doi.org/10.1073/pnas.1424127112
192. Roy B, Friesen WJ, Tomizawa Y et al (2016) Ataluren stimulates ribosomal selection of near-cognate tRNAs to promote nonsense suppression. Proc Natl Acad Sci 113:12508–12513. https://doi.org/10.1073/pnas.1605336113
193. Davies J, Gilbert W, Gorini L (1964) Streptomycin, suppression, and the code. Proc Natl Acad Sci 51:883–890. https://doi.org/10.1073/pnas.51.5.883
194. Malik V, Rodino-Klapac LR, Viollet L, Mendell JR (2010) Aminoglycoside-induced mutation suppression (stop codon read-through) as a therapeutic strategy for

Duchenne muscular dystrophy. Ther Adv Neurol Disord 3:379–389. https://doi.org/10.1177/1756285610388693

195. Bertoni C, Namgoong J (2016) Clinical potential of ataluren in the treatment of Duchenne muscular dystrophy. Degener Neurol Neuromuscul Dis 37. https://doi.org/10.2147/DNND.S71808
196. Welch EM, Barton ER, Zhuo J et al (2007) PTC124 targets genetic disorders caused by nonsense mutations. Nature 447:87–91. https://doi.org/10.1038/nature05756
197. European Medicines Agency (2014) European Medicines Agency recommends first-in-class medicine for treatment of Duchenne muscular dystrophy. https://www.ema.europa.eu/en/news/european-medicines-agency-recommends-first-class-medicine-treatment-duchenne-muscular-dystrophy. Accessed 9 June 2021
198. Finkel RS, Flanigan KM, Wong B et al (2013) Phase 2a study of ataluren-mediated dystrophin production in patients with nonsense mutation Duchenne muscular dystrophy. PLoS One 8:e81302. https://doi.org/10.1371/journal.pone.0081302
199. Bushby K, Finkel R, Wong B et al (2014) Ataluren treatment of patients with nonsense mutation dystrophinopathy. Muscle Nerve 50:477–487. https://doi.org/10.1002/mus.24332
200. McDonald CM, Campbell C, Torricelli RE et al (2017) Ataluren in patients with nonsense mutation Duchenne muscular dystrophy (ACT DMD): a multicentre, randomised, double-blind, placebo-controlled, phase 3 trial. Lancet 390:1489–1498. https://doi.org/10.1016/S0140-6736(17)31611-2
201. Campbell C, Barohn RJ, Bertini E et al (2020) Meta-analyses of ataluren randomized controlled trials in nonsense mutation Duchenne muscular dystrophy. J Comp Eff Res 9:973–984. https://doi.org/10.2217/cer-2020-0095
202. Gardner J (2021) PTC aims again for FDA review of Duchenne drug despite latest miss. https://www.biopharmadive.com/news/ptc-duchenne-data-fourth-fda-review/594642/. Accessed 9 June 2021
203. Muntoni F, Desguerre I, Guglieri M et al (2019) Ataluren use in patients with nonsense mutation Duchenne muscular dystrophy: patient demographics and characteristics from the STRIDE Registry. J Comp Eff Res 8:1187–1200. https://doi.org/10.2217/cer-2019-0086
204. Mercuri E, Muntoni F, Osorio AN et al (2020) Safety and effectiveness of ataluren: comparison of results from the STRIDE Registry and CINRG DMD Natural History Study. J Comp Eff Res 9:341–360. https://doi.org/10.2217/cer-2019-0171
205. Seo K, Kim EK, Choi J et al (2021) Functional recovery of a novel knockin mouse model of dysferlinopathy by readthrough of nonsense mutation. Mol Ther – Methods Clin Dev 21:702–709. https://doi.org/10.1016/j.omtm.2021.04.015
206. Keeling KM, Xue X, Gunn G, Bedwell DM (2014) Therapeutics based on stop codon readthrough. Annu Rev Genomics Hum Genet 15:371–394. https://doi.org/10.1146/annurev-genom-091212-153527
207. Liang F, Shang H, Jordan NJ et al (2017) High-throughput screening for readthrough modulators of CFTR PTC mutations. SLAS Technol Transl Life Sci Innov 22:315–324. https://doi.org/10.1177/2472630317692561
208. Borgatti M, Altamura E, Salvatori F et al (2020) Screening readthrough compounds to suppress nonsense mutations: possible application to β-thalassemia. J Clin Med 9:289. https://doi.org/10.3390/jcm9020289
209. Bylo M, Farewell R, Coppenrath VA, Yogaratnam D (2020) A review of deflazacort for patients with Duchenne muscular dystrophy. Ann Pharmacother 54:788–794. https://doi.org/10.1177/1060028019900500
210. Kourakis S, Timpani CA, Campelj DG et al (2021) Standard of care versus new-wave corticosteroids in the treatment of Duchenne muscular dystrophy: can we do better? Orphanet J Rare Dis 16:117. https://doi.org/10.1186/s13023-021-01758-9
211. Timmermans S, Souffriau J, Libert C (2019) A general introduction to glucocorticoid biology. Front Immunol. https://doi.org/10.3389/fimmu.2019.01545
212. McDonald CM, Sajeev G, Yao Z et al (2020) Deflazacort vs prednisone treatment for Duchenne muscular dystrophy: a meta-analysis of disease progression rates in recent multicenter clinical trials. Muscle Nerve 61:26–35. https://doi.org/10.1002/mus.26736
213. Griggs RC, Miller JP, Greenberg CR et al (2016) Efficacy and safety of deflazacort vs prednisone and placebo for Duchenne muscular dystrophy. Neurology 87:2123–2131. https://doi.org/10.1212/WNL.0000000000003217

214. Marden JR, Freimark J, Yao Z et al (2020) Real-world outcomes of long-term prednisone and deflazacort use in patients with Duchenne muscular dystrophy: experience at a single, large care center. J Comp Eff Res 9: 177–189. https://doi.org/10.2217/cer-2019-0170
215. Shieh PB, Mcintosh J, Jin F et al (2018) Deflazacort versus prednisone/prednisolone for maintaining motor function and delaying loss of ambulation: a post HOC analysis from the ACT DMD trial. Muscle Nerve 58:639–645. https://doi.org/10.1002/mus.26191
216. Bello L, Gordish-Dressman H, Morgenroth LP et al (2015) Prednisone/prednisolone and deflazacort regimens in the CINRG Duchenne Natural History Study. Neurology 85: 1048–1055. https://doi.org/10.1212/WNL.0000000000001950
217. Reeves EKM, Hoffman EP, Nagaraju K et al (2013) VBP15: preclinical characterization of a novel anti-inflammatory delta 9,11 steroid. Bioorg Med Chem 21:2241–2249. https://doi.org/10.1016/j.bmc.2013.02.009
218. Heier CR, Damsker JM, Yu Q et al (2013) VBP15, a novel anti-inflammatory and membrane-stabilizer, improves muscular dystrophy without side effects. EMBO Mol Med 5:1569–1585. https://doi.org/10.1002/emmm.201302621
219. Baudy AR, Reeves EKM, Damsker JM et al (2012) Δ-9,11 modification of glucocorticoids dissociates nuclear factor-κB inhibitory efficacy from glucocorticoid response element-associated side effects. J Pharmacol Exp Ther 343:225–232. https://doi.org/10.1124/jpet.112.194340
220. Smith EC, Conklin LS, Hoffman EP et al (2020) Efficacy and safety of vamorolone in Duchenne muscular dystrophy: an 18-month interim analysis of a non-randomized open-label extension study. PLoS Med 17: e1003222. https://doi.org/10.1371/journal.pmed.1003222
221. Santhera Pharmaceuticals (2021) Santhera and ReveraGen announce new 2.5-year treatment data with vamorolone in Duchenne muscular dystrophy. https://www.globenewswire.com/news-release/2021/04/28/2218239/0/en/Santhera-and-ReveraGen-Announce-New-2-5-year-Treatment-Data-with-Vamorolone-in-Duchenne-Muscular-Dystrophy.html. Accessed 9 June 2021
222. Santhera Pharmaceuticals and ReveraGen BioPharma (2021) Santhera and ReveraGen announce positive and statistically highly significant topline results with vamorolone in pivotal VISION-DMD study. https://www.santhera.com/assets/files/press-releases/2021-06-01_PR_VamTopline_e_final.pdf. Accessed 9 June 2021
223. Cordova G, Negroni E, Cabello-Verrugio C et al (2018) Combined therapies for Duchenne muscular dystrophy to optimize treatment efficacy. Front Genet. https://doi.org/10.3389/fgene.2018.00114
224. Rodino-Klapac LR, Janssen PML, Shontz KM et al (2013) Micro-dystrophin and follistatin co-delivery restores muscle function in aged DMD model. Hum Mol Genet 22: 4929–4937. https://doi.org/10.1093/hmg/ddt342
225. Abmayr S, Gregorevic P, Allen JM, Chamberlain JS (2005) Phenotypic improvement of dystrophic muscles by rAAV/microdystrophin vectors is augmented by Igf1 codelivery. Mol Ther 12:441–450. https://doi.org/10.1016/j.ymthe.2005.04.001
226. Wang DW, Mokhonova EI, Kendall GC et al (2018) Repurposing dantrolene for long-term combination therapy to potentiate antisense-mediated DMD exon skipping in the mdx mouse. Mol Ther – Nucleic Acids 11:180–191. https://doi.org/10.1016/j.omtn.2018.02.002
227. Barthélémy F, Wang RT, Hsu C et al (2019) Targeting RyR activity boosts antisense exon 44 and 45 skipping in human DMD skeletal or cardiac muscle culture models. Mol Ther – Nucleic Acids 18:580–589. https://doi.org/10.1016/j.omtn.2019.09.020
228. O'Leary DA, Sharif O, Anderson P et al (2009) Identification of small molecule and genetic modulators of AON-induced dystrophin exon skipping by high-throughput screening. PLoS One 4:e8348. https://doi.org/10.1371/journal.pone.0008348
229. Quattrocelli M, Cassano M, Crippa S et al (2010) Cell therapy strategies and improvements for muscular dystrophy. Cell Death Differ 17:1222–1229. https://doi.org/10.1038/cdd.2009.160

Chapter 2

Viltolarsen: From Preclinical Studies to FDA Approval

Rohini Roy Roshmi and Toshifumi Yokota

Abstract

Viltolarsen is a phosphorodiamidate morpholino antisense oligonucleotide (PMO) designed to skip exon 53 of the *DMD* gene for the treatment of Duchenne muscular dystrophy (DMD), one of the most common lethal genetic disorders characterized by progressive degeneration of skeletal muscles and cardiomyopathy. It was developed by Nippon Shinyaku in collaboration with the National Center of Neurology and Psychiatry (NCNP) in Japan based on the preclinical studies conducted in the DMD dog model at the NCNP. After showing hopeful results in pre-clinical trials and several clinical trials across North America and Japan, it received US Food and Drug Administration (FDA) approval for DMD in 2020. Viltolarsen restores the reading frame of the *DMD* gene by skipping exon 53 and produces a truncated but functional form of dystrophin. It can treat approximately 8-10% of the DMD patient population. This paper aims to summarize the development of viltolarsen from preclinical trials to clinical trials to, finally, FDA approval, and discusses the challenges that come with fighting DMD using antisense therapy.

Key words Viltolarsen, Duchenne muscular dystrophy, Antisense therapy, Gene therapy, Viltepso, Clinical trials

1 Introduction

Duchenne muscular dystrophy (DMD) is a lethal X-linked recessive disorder, with a prevalence rate of 1:3000–1:6000 boys globally [1]. It is characterized by progressive muscle degeneration, cardiorespiratory complications, and a high level of creatine kinase (CK) [2, 3]. The first symptoms of muscle weakness and difficulty in walking are observed when patients are as young as 1–3 years old. Most patients are wheelchair users by the age of 12. Consequently, death usually occurs in their 20s or 30s due to respiratory and cardiac failure [4–6].

1.1 Mutations in the DMD *Gene*

DMD is caused due to mutations in the *DMD* gene which result in the lack of functional dystrophin protein [7]. The *DMD* gene is the longest gene in humans, consisting of 79 exons and is approximately 2.4 million base pairs [8, 9]. The gene's long length makes it

Rika Maruyama and Toshifumi Yokota (eds.), *Muscular Dystrophy Therapeutics: Methods and Protocols*, Methods in Molecular Biology, vol. 2587, https://doi.org/10.1007/978-1-0716-2772-3_2,

highly vulnerable to mutations. Analysis of genetic data from 7149 patient-specific mutations showed that 80% of DMD-causing mutations are large mutations, of which 86% were deletions [10, 11]. Other mutations such as duplications, point mutations, nonsense mutations, mid-intronic mutations, and missense mutations can lead to a lack of dystrophin protein in muscles as well. Fortunately, exons 45–55 and exons 2–19 of the DMD gene have been identified as are mutation hotspots and are usually targeted while developing therapeutics [12].

Although dystrophin makes up only 0.002% of total muscle protein, its function is crucial for muscle integrity [13]. The dystrophin protein has four domains: an actin-binding amino-terminal domain, a rod domain that provides structural plasticity, a cysteine-rich domain, and a carboxy-terminal domain [14, 15]. This 427 kDa rod-shaped protein connects the actin cytoskeleton to the extracellular matrix (ECM) via the dystrophin–glycoprotein complex (DGC) [16, 17]. Lack of dystrophin disrupts the connection between the cytoskeleton and the ECM, which leads to destabilization of sarcolemma and eventually results in muscle necrosis.

1.2 Currently Approved Therapies for DMD

Perhaps the most clinically advanced therapies in DMD are those that work by exon skipping. Exon skipping therapies for DMD aim to restore the reading frame of the *DMD* gene using DNA-like molecules called antisense oligonucleotides (AOs), which produces a truncated but functional form of dystrophin protein in muscles [18–20]. This strategy is motivated by positive observations made in Becker muscular dystrophy patients (BMD) [21]. BMD patients have in-frame mutations in their *DMD* gene that codes for a truncated but functional form of dystrophin. They usually have milder symptoms and live longer than DMD patients.

AOs called phosphorodiamidate morpholino oligomers (PMOs) or morpholinos are extensively researched as potential therapeutic agents for DMD because of its safety features and high performance in preclinical studies [18]. Currently, there are four PMO drugs approved by the Food and Drug administrations (FDA), namely eteplirsen (brand name Exodys 51), golodirsen (brand name Vyondys 53), viltolarsen (brand name Viltepso), and casimersen (brand name Amondys 45) [22–25]. In this review, we will discuss the journey of viltolarsen from preclinical trials to FDA approval in 2020. Additionally, glucocorticoids such as prednisone and deflazacort have been used for decades to reduce the complications of DMD [26, 27]. Both drugs increase muscle strength and prolong ambulation. They also maintain heart health and improve respiratory function. However, there are many adverse effects of corticosteroids such as weight gain, height reduction and delayed puberty [28].

2 Viltolarsen: An Overview

Viltolarsen, also known as NS-065/NCNP-01, is a PMO drug that restores dystrophin production in DMD patients by skipping exon 53 of the *DMD* gene [29]. It has the potential to treat 8% of DMD patients, including patients with deletions such as exons 43–52, 45–52, 47–52, 48–52, 50–52, or 52 of the *DMD* gene [30, 31]. Viltolarsen hybridizes to its target site on exon 53 of the dystrophin pre-mRNA and hides it from the splicing machinery [32, 33]. Exon 53 is skipped during splicing, and a mature in-frame mRNA is produced. The resultant mature mRNA is translated to form a truncated but functional form of dystrophin.

2.1 Preclinical Studies

In a preclinical trial conducted by Yokota et al., PMOs showed successful exon skipping in the DMD pre-mRNA, restored dystrophin expression in muscles and improved muscle function in a dog model of DMD. Based on the hopeful results from this pre-clinical trial, viltolarsen's development was encouraged [34]. Viltolarsen-Viltolarsen is a 21-nucleotide PMO that hybridizes to exon 53 between positions 36 and 56 (CCTCCGGTTCTGAAGGTGTTC) [31, 35]. In a preclinical study by Watanabe et al., its sequence has been thoroughly optimized to ensure the highest efficiency in exon skipping. Exon skipping efficiency of several sequences targeting exon 53 were investigated in human rhabdomyosarcoma cells and myotubes differentiated from fibroblasts from DMD patients. Viltolarsen successfully skipped exon 53 in both cell types according to reverse transcriptase-polymerase chain reaction (RT-PCR) results. Additionally, western blotting results showed that it induced dystrophin protein expression in MyoD-transduced fibroblasts from DMD patients.

2.2 Timeline of Drug Development

Nippon Shinyaku Pharma (NS Pharma) and the National Center of Neurology and Psychiatry Japan (NCNP) began their collaboration to develop viltolarsen in 2009 [36]. Their joint efforts made it possible for viltolarsen to receive the SAKIGAKE and Orphan Drug designations from the Ministry of Health, Labour and Welfare (MHLW) of Japan in 2015. Not long after, in 2016–2017, viltolarsen was granted Fast Track designation, Orphan Drug status, and Rare Paediatric disease designation in the United States. The New Drug Application (NDA) for viltolarsen was submitted by NS Pharma in 2019. Soon afterward, on March 25, 2020, viltolarsen was approved in Japan, and on August 12, 2020, it received FDA approval for treating eligible DMD patients in the United States [37].

3 Safety Profile

In a phase I clinical trial (NCT02081625) conducted on 10 DMD patients, no severe adverse reactions were observed [33]. Some mild and manageable adverse reactions, such as *N*-acetyl-β-D-glucosaminidase (NAG), proteinuria, albuminuria, anemia, increased interleukin, elevated brain natriuretic peptide, increased diastolic pressures, increased complement component C3, increased β2-microglobulin, decreased ejection fraction, and eczema, occurred among the patients. However, all adverse side effects were manageable, with or without treatment. None of the patients quit the clinical trial because of side effects from the drug. Moreover, antibodies against human dystrophin protein or viltolarsen were undetectable in blood serum analysis.

In a phase II clinical trial (NCT02740972) conducted by Clemens et al. on 16 DMD patients, 94% of the participants experienced adverse side effects [38]. However, the adverse effects did not interfere with the design of the clinical trial, and no participant discontinued the study. No serious adverse effects were observed in the first 4 weeks of the study in more than 1 participant. Between the 5th week until the 24th week of the study, side effects such as cough (4 participants) and nasopharyngitis (4 participants) were the most common in patients. Two patients also suffered from diarrhea and vomiting. All participants, except two with mild adverse reactions, recovered from these treatment-emergent side effects at the end of the study. Additionally, blood serum and urine analysis of participants were normal.

Similarly, in a phase I/II clinical trial (JapicCTI-163291) conducted by Komaki et al., mild adverse reactions were observed, but none of the participants quit the study due to the side effects [39, 40]. Overall, from the observations made in the clinical trials, it is evident that viltolarsen is well-tolerated up to an 80 mg/kg dose. Currently, a 48-week phase 3 clinical trials called RACER53 is being conducted to assess the safety of viltolarsen in 74 patients with DMD [41]. Throughout the duration of the study, routine physical check ups, electrocardiograms, laboratory tests, and records of events of adverse effects will be taken into account to determine the safety of viltolarsen.

4 Pharmacokinetics

In the phase I clinical trial conducted by Komaki et al., they analyzed the pharmacokinetic properties of viltolarsen at different doses: low (1.25 mg/kg), medium (5.0 mg/kg), and high (20.0 mg/kg) [33]. Plasma analysis showed that the maximum drug concentration (C_{max}) was reached in $\leq$1 h and elevated in a dose-dependent manner. Viltolarsen had a short half-life between 1.52 and 1.84 h, and 70–80% of the drug was cleared by the kidneys within 24 h.

This study was followed by a phase I/II clinical trial in Japanese patients with DMD [40]. Participants in this study were divided into two cohorts and were intravenously administered 40 mg/kg or 80 mg/kg of viltolarsen. Analysis of blood samples showed that C_{max} increased in a dose-dependent manner, and urine analysis showed that 92.0–93.1% viltolarsen was excreted in urine within 24 h of administration [30, 42].

Similar results were reproduced by Clemens et al. Pharmacokinetics of viltolarsen was dose-dependent, the drug had a short half-life, and it was excreted in the urine within 24 h of administration [43].

5 Efficacy of Viltolarsen

In the phase I clinical trial conducted by Komaki et al., the amount of exon 53 skipping calculated from RT-PCR results increased in a dose-dependent manner [33]. One patient (NS-07), who received 20 mg/kg of viltolarsen weekly, showed the highest increase in exon skipping of 47.5%. This can be attributed to the elevated amounts viltolarsen participant NS-07 received due to their greater body weight. Also, 7 out of 10 participants showed a significant increase in dystrophin/spectrin ratio (measured by immunofluorescence) in the sarcolemma. Although results from immunofluorescence looked positive, western blotting only detected dystrophin protein in patient NS-07.

In the phase I/II clinical trial following NCT02081625, the patients were divided into two cohorts (40 mg/kg or 80 mg/kg) [40]. An increase in exon 53 skipping was detected in all participants by RT-PCR, during weeks 12 and 24 of treatment. Higher exon skipping was observed in the 80 mg/kg cohort. For 87.5% of the patients, dystrophin protein was detected in the western blot. The increase in dystrophin protein levels from baseline to weeks 12 and 24 was significantly higher in the 80 mg/kg cohort. Exon skipping and dystrophin protein expression were dose-dependent and time-dependent (duration of treatment). However, expectations from immunofluorescence analysis were not met. No significant difference in the number of dystrophin-positive fibers was observed between baseline and week 24 of treatment, except for one patient from 80 mg/kg cohort. The authors explained that the background staining of non-muscle cell regions made it difficult to detect the number of dystrophin-positive fibers.

Consistent with Komaki et al.'s findings, in the phase II clinical trial by Clemen et al., western blotting showed a significant increase in dystrophin expression between baseline and week 25 of treatment in both 40 mg/kg and 80 mg/kg cohorts [38]. On average, 5.7% and 5.9% of normal dystrophin levels were restored in the 40 mg/kg and 80 mg/kg cohorts, respectively. Measurements from RT-PCR showed dose-dependent exon 53 skipping in all participants. Additionally, dystrophin-positive fibers were observed in immunofluorescence in both cohorts.

6 Therapeutic Benefits of Viltolarsen

In the phase 2 trial conducted by Clemens et al., differences in disease progression between the viltolarsen-treatment group and natural history controls from CINRG DNHS were analyzed using timed function tests and muscle strength assessments [38]. Viltolarsen-treated patients showed significant improvements in velocity in the time to run/walk 10-m test (TTRW), 6-min walk test (6MWT), and time to stand from supine (TTSTAND). Improvements were also observed in time to climb 4 stairs test (TTCLIMB), velocity in time to stand from supine test, and North Star Ambulatory Assessment (NSAA), but the improvement was not significant compared to the control. The results were reproduced in clinical trial JapicCTI-163291, where viltolarsen-treated patients showed improvements in TTSTAND, TTRW, and 6MTW tests post-treatment. A major clinical trials called RACER53 (April 2020–December 2024) is currently being conducted to evaluate the efficacy and safety of viltolarsen. It is a phase 3, randomized, double-blind, placebo-controlled study, where 74 DMD patients, with mutation amenable to exon 53 skipping, will be administered 80 mg/kg viltolarsen or placebo weekly for 48 weeks. Various functional tests including TTRW, TTSTAND, 6MWT, TTCLIMB, NSAA and hand held dynamometer will be done on the participants to evaluate efficacy of viltolarsen [41].

7 Clinical Trials

7.1 Phase I Clinical Trial: NCT02081625

A phase I open-label, dose-escalation study was conducted by Komaki et al. to assess the safety and efficacy of viltolarsen [33]. Ten DMD patients, with mutations in the *DMD* gene amenable by exon 53 skipping, were recruited for the study. The participants' age ranged from 5 to 18 years, and 30% of them were ambulant. Viltolarsen was intravenously administered weekly for 12 weeks at three different dose cohorts: low (1.25 mg/kg; $n = 3$), medium (5 mg/kg; $n = 3$), and high (20 mg/kg; $n = 4$). Blood and urine samples were collected on days 0, 7, and 77, and days 0 and 7, respectively. It was used for pharmacokinetic analysis. Muscle biopsies were collected pre-treatment and post-treatment for RT-PCR and Western blot.

7.2 Phase I/II Clinical Trial: JapicCTI-163291

Komaki et al. conducted this phase I/II open-label, dose-finding, exploratory study from April 2016 to November 2017 in Japan [40]. Patients were initially administered 40 mg/kg of viltolarsen weekly. After ensuring the safety of the participants, some of them were allocated to the 80 mg/kg group. The study primarily aimed to study the efficacy of viltolarsen 12- and 24-weeks post-treatment. Additionally, it explored pharmacokinetics and safety of viltolarsen and motor function improvements in response to treatment. Before treatment, muscle biopsies were collected as well as baseline motor function was assessed. At week 12, 50% of the patients ($n = 4$) gave muscle biopsies for RT-PCR, immunohistochemistry, and western blot. The remaining 50% provided muscle biopsies at week 24. For pharmacokinetics, urine samples were collected for 24 h since infusion on day 1 of treatment and week 24. Additionally, a blood sample was collected pre-treatment, 30 min into the infusion, at the end of infusion, and at regular time intervals after infusion on day 1 of week 1 and week 24 (15 min, 1, 2, 4, 8, and 23 h). Motor functional was evaluated again on weeks 12 and 24.

7.3 Phase II Clinical Trial: NCT02740972

This multicenter, two-period, dose-finding, randomized phase II clinical trial was conducted in the United States by Clemens et al. The study primarily aimed to evaluate the safety and efficacy of viltolarsen [38]. The therapeutic benefit of viltolarsen was also assessed through motor function tests. Sixteen ambulatory participants were recruited from 6 different places in North America. Eight participants each were assigned to 40 mg/kg cohort or 80 mg/kg cohort. Muscle biopsies from all participants were collected pre-treatment. The first period of the study (4 weeks) was double-blinded and placebo-controlled. Also, 75% of the participants received viltolarsen, and 25% received a placebo. In the second period of the study (from week 5), all participants received either 40 or 80 mg/kg of viltolarsen for 20 weeks. Muscle biopsies were collected at 24 weeks of treatment.

7.4 Phase II Extension Study: NCT03167255

Participants who completed phase II dose-finding study (NCT02740972) are eligible to participate in this extension study [44]. This phase II multicenter, open-label, extension study will intravenously administer viltolarsen weekly for 192 weeks. Participants will continue in their 40 mg/kg and 80 mg/kg cohorts assigned in NCT02740972. This study aims to evaluate the safety of viltolarsen and the effect this drug intervention has on TTSTAND. Several other motor function tests like TTRW, TTCLIMB, NSAA, and 6MWT will be assessed as well.

8 Challenges

While the progress of exon skipping therapies has certainly been impressive, we have to recognize that exon skipping is not without hurdles. One of the biggest challenges in developing exon skipping therapy is the small number of participants in the clinical trials [45]. Recruiting a homogenous patient population is extremely difficult due to the limited number of eligible patients. Hence, clinical trials of exon-skipping drugs are often conducted with patients of various ages, ambulatory status, and DMD-causing mutations. Consequently, applicability becomes another major challenge. For example, only 8% of DMD patients are eligible for treatment with viltolarsen. Nevertheless, there are other FDA-approved PMO drugs that can treat patients ineligible for treatment with viltolarsen [46]. To make participant recruitment easier and to provide more information to the patient population regarding clinical trials, databases such as Remudy have been developed [47]. To reach a broader DMD patient subpopulation, multiple exon skipping PMO cocktails are being investigated [48]. For example, exons 45-55 skipping can treat about 47% of all DMD patients [49]. Recently Yokota et al, showed successful dystrophin protein restoration of up to 3% of wild-type by skipping exons 45-55 in vivo [49]. Besides its limited applicability to the overall DMD patient population, PMO drugs have limited efficacy in cardiac tissue due to poor uptake [50]. As cardiorespiratory complications cause majority of deaths in the DMD patient population, its low efficacy in heart is a serious concern in exon skipping therapy. A potential solution to this problem is peptide conjugated PMOs (PPMOs), which have shown significant increase in exon skipping efficacy in the heart [51, 52]. Currently Sarepta is developing a PPMO by conjugating a cell-penetration peptide (CPP) to eteplirsen (SRP-5051), an already FDA-approved drug for treating DMD [53]. SRP-5051 is currently being investigated in a phase II dose-expansion study (NCT04004065). Although it has shown improvements in exon skipping efficiency, questions are being raised about its safety due to severe cases of hypomagnesemia in some of the participants. Another CPP knows as DG9 has also shown greater dystrophin restoration in heart muscles (from 2.5% to 7.7% of wild-type) in vivo [49].

9 Conclusion

It has been more than 16 decades since French neurologist Guillaume Duchenne described DMD, but we are still in search of a cure for it [54]. As there are many unique DMD-causing mutations, it has been difficult to curate a drug that can treat all patients.

Nevertheless, researchers around the world are using different approaches to find potential therapies for DMD. Some of these approaches are exon skipping using clustered regularly interspaced short palindromic repeats (CRISPR), peptide-conjugated morpholinos (PPMOs), vivo-morpholinos (vPMOs), and micro-dystrophin therapy [55–59]. Although it is applicable for a small patient population, the FDA approval of viltolarsen is another significant step in curating treatment options for people with DMD.

References

1. Mendell JR, Shilling C, Leslie ND et al (2012) Evidence-based path to newborn screening for duchenne muscular dystrophy. Ann Neurol 71: 304–313. https://doi.org/10.1002/ana.23528
2. Moser H (1984) Duchenne muscular dystrophy: pathogenetic aspects and genetic prevention. Hum Genet 66:17–40. https://doi.org/10.1007/BF00275183
3. Mercuri E, Muntoni F (2013) Muscular dystrophies. Lancet 381:845–860. https://doi.org/10.1016/S0140-6736(12)61897-2
4. Bushby K, Finkel R, Birnkrant DJ et al (2010) Diagnosis and management of Duchenne muscular dystrophy, part 1: diagnosis, and pharmacological and psychosocial management. Lancet Neurol 9:77–93. https://doi.org/10.1016/S1474-4422(09)70271-6
5. Manzur AY, Kinali M, Muntoni F (2008) Update on the management of Duchenne muscular dystrophy. Arch Dis Child 93:986–990. https://doi.org/10.1136/adc.2007.118141
6. Emery AE (2002) The muscular dystrophies. Lancet 359:687–695. https://doi.org/10.1016/S0140-6736(02)07815-7
7. Muntoni F, Torelli S, Ferlini A (2003) Dystrophin and mutations: one gene, several proteins, multiple phenotypes. Lancet Neurol 2: 731–740. https://doi.org/10.1016/S1474-4422(03)00585-4
8. Koenig M, Hoffman EP, Bertelson CJ et al (1987) Complete cloning of the duchenne muscular dystrophy (DMD) cDNA and preliminary genomic organization of the DMD gene in normal and affected individuals. Cell 50:509–517. https://doi.org/10.1016/0092-8674(87)90504-6
9. Roberts RG, Coffey AJ, Bobrow M, Bentley DR (1993) Exon structure of the human dystrophin gene. Genomics 16:536–538. https://doi.org/10.1006/geno.1993.1225
10. Bladen CL, Salgado D, Monges S et al (2015) The TREAT-NMD DMD global database: analysis of more than 7,000 duchenne muscular dystrophy mutations. Hum Mutat 36:395–402. https://doi.org/10.1002/humu.22758
11. Carter JC, Sheehan DW, Prochoroff A, Birnkrant DJ (2018) Muscular dystrophies. Clin Chest Med 39:377–389. https://doi.org/10.1016/j.ccm.2018.01.004
12. Oudet C, Hanauer A, Clemens P et al (1992) Two hot spots of recombination in the DMD gene correlate with the deletion prone regions. Hum Mol Genet 1:599–603. https://doi.org/10.1093/hmg/1.8.599
13. Hoffman EP, Brown RH, Kunkel LM (1987) Dystrophin: the protein product of the duchenne muscular dystrophy locus. Cell 51: 919–928. https://doi.org/10.1016/0092-8674(87)90579-4
14. Gao QQ, McNally EM (2015) The dystrophin complex: structure, function, and implications for therapy. Comprehensive Physiology, Wiley, pp 1223–1239
15. Ervasti JM (2007) Dystrophin, its interactions with other proteins, and implications for muscular dystrophy. Biochim Biophys Acta - Mol Basis Dis 1772:108–117. https://doi.org/10.1016/j.bbadis.2006.05.010
16. Nichols B, Takeda S, Yokota T (2015) Nonmechanical roles of dystrophin and associated proteins in exercise, neuromuscular junctions, and brains. Brain Sci 5:275–298. https://doi.org/10.3390/brainsci5030275
17. Ervasti JM, Campbell KP (1991) Membrane organization of the dystrophin-glycoprotein complex. Cell 66:1121–1131. https://doi.org/10.1016/0092-8674(91)90035-W
18. Yokota T, Lu Q, Partridge T et al (2009) Efficacy of systemic morpholino exon-skipping in duchenne dystrophy dogs. Ann Neurol 65: 667–676. https://doi.org/10.1002/ana.21627
19. Lu QL, Mann CJ, Lou F et al (2003) Functional amounts of dystrophin produced by skipping the mutated exon in the mdx

dystrophic mouse. Nat Med 9:1009–1014. https://doi.org/10.1038/nm897

20. Echigoya Y, Lim KRQ, Melo D et al (2019) Exons 45–55 skipping using mutation-tailored cocktails of antisense morpholinos in the DMD gene. Mol Ther 27:2005–2017. https://doi.org/10.1016/j.ymthe.2019.07.012
21. Koenig M, Beggs AH, Moyer M et al (1989) The molecular basis for duchenne versus becker muscular dystrophy: correlation of severity with type of deletion. Am J Hum Genet 45(4):498
22. Mendell JR, Rodino-Klapac LR, Sahenk Z et al (2013) Eteplirsen for the treatment of Duchenne muscular dystrophy. Ann Neurol. https://doi.org/10.1002/ana.23982
23. O'Keefe L (2020) FDA approves targeted treatment of rare Duchenne muscular dystrophy mutation. FDA news release
24. Aartsma-Rus A, Corey DR (2020) The 10th oligonucleotide therapy approved: golodirsen for duchenne muscular dystrophy. Nucleic Acid Ther 30:67–70. https://doi.org/10.1089/nat.2020.0845
25. Shirley M (2021) Casimersen: first approval. Drugs. https://doi.org/10.1007/s40265-021-01512-2
26. Bylo M, Farewell R, Coppenrath VA et al (2020) A review of deflazacort for patients with duchenne muscular dystrophy. Ann Pharmacother 54(8):788–794. https://doi.org/10.1177/1060028019900500
27. Moxley RT, Ashwal S, Pandya S et al (2005) Practice parameter: corticosteroid treatment of duchenne dystrophy: report of the quality standards subcommittee of the American academy of neurology and the practice committee of the child neurology society. Neurology 64(1): 13–20. https://doi.org/10.1212/01.WNL.0000148485.00049.B7
28. Moxley RT, Pandya S, Ciafaloni E et al (2010) Change in natural history of duchenne muscular dystrophy with longterm corticosteroid treatment: implications for management. J Child Neurol 25(9):1116–1129. https://doi.org/10.1177/0883073810371004
29. Roshmi RR, Yokota T (2019) Viltolarsen for the treatment of Duchenne muscular dystrophy. Drugs Today 55:627. https://doi.org/10.1358/dot.2019.55.10.3045038
30. Dhillon S (2020) Viltolarsen: First Approval. Drugs 80:1027–1031. https://doi.org/10.1007/s40265-020-01339-3
31. Watanabe N, Nagata T, Satou Y et al (2018) NS-065/NCNP-01: an antisense oligonucleotide for potential treatment of exon 53 skipping in duchenne muscular dystrophy. Mol Ther Nucleic Acids 13:442–449. https://doi.org/10.1016/j.omtn.2018.09.017
32. Echevarría L, Aupy P, Goyenvalle A (2018) Exon-skipping advances for Duchenne muscular dystrophy. Hum Mol Genet 27: R163–R172. https://doi.org/10.1093/hmg/ddy171
33. Komaki H, Nagata T, Saito T et al (2018) Systemic administration of the antisense oligonucleotide NS-065/NCNP-01 for skipping of exon 53 in patients with Duchenne muscular dystrophy. Sci Transl Med 10:eaan0713. https://doi.org/10.1126/scitranslmed.aan0713
34. Yokota T, Lu Q, Partridge T et al (2009) Efficacy of systemic morpholino exon-skipping in duchenne dystrophy dogs. Ann Neurol 65(6): 667–676. https://doi.org/10.1002/ana.21627
35. Inxight: drugs. In: U.S. Department of Health and Human Services. https://drugs.ncats.io/drug/SXA7YP6EKX
36. Dhillon S (2020) Viltolarsen: first approval. Drugs 80(10):1027–1031. https://doi.org/10.1007/s40265-01339-3
37. Pharma NS, Fda N, Dystrophy DM, et al (2020) NEWS RELEASE NS Pharma's VILTEPSO™ (viltolarsen) injection Now FDA-approved in the U.S. for the treatment of duchenne muscular dystrophy in patients amenable to exon 53 skipping therapy. 1–4
38. Clemens PR, Rao VK, Connolly AM et al (2020) Safety, tolerability, and efficacy of viltolarsen in boys with duchenne muscular dystrophy amenable to exon 53 skipping. JAMA Neurol 77:982. https://doi.org/10.1001/jamaneurol.2020.1264
39. Komaki H, Takeshima Y, Matsumura T et al (2018) DMD CLINICAL THERAPIES II. Neuromuscul Disord 28:S68. https://doi.org/10.1016/j.nmd.2018.06.157
40. Komaki H, Takeshima Y, Matsumura T et al (2020) Viltolarsen in Japanese Duchenne muscular dystrophy patients: a phase 1/2 study. Ann Clin Transl Neurol 7:2393–2408. https://doi.org/10.1002/acn3.51235
41. Study to Assess the Efficacy and Safety of Viltolarsen in Ambulant Boys With DMD (RACER53), *clinicaltrials.gov*, ClinicalTrials.gov Identifier: NCT04060199 (2019)
42. (2020) Nippon Shinyaku Co. Ltd. Viltepso injection 250 mg: Japanese prescribing information. https://www.pmda.go.jp/
43. Clemens PR, Rao VK, Connolly AM (2018) A phase II, dose-finding study to assess the safety, tolerability, pharmacokinetics, and pharmacodynamics of NS-065/NCNP-01 in boys with

Duchenne muscular dystrophy (DMD). 13th New Directions Biol Dis Skelet Muscle

44. Extension Study of NS-065/NCNP-01 in Boys With Duchenne Muscular Dystrophy (DMD). In: ClinicalTrials.gov. https://clinicaltrials.gov/ct2/show/record/NCT03167255

45. Rodrigues M, Yokota T (2018) An overview of recent advances and clinical applications of exon skipping and splice modulation for muscular dystrophy and various genetic diseases. pp 31–55

46. Himič V, Davies KE (2021) Evaluating the potential of novel genetic approaches for the treatment of Duchenne muscular dystrophy. Eur J Hum Genet. https://doi.org/10.1038/s41431-021-00811-2

47. Nakamura H, Kimura E, Mori-Yoshimura M et al (2013) Characteristics of Japanese Duchenne and Becker muscular dystrophy patients in a novel Japanese national registry of muscular dystrophy (Remudy). Orphanet J Rare Dis 8(1):60. https://doi.org/10.1186/1750-1172-8-60

48. Echigoya Y, Lim KRQ, Melo D et al (2019) Exons 45–55 skipping using mutation-tailored cocktails of antisense morpholinos in the DMD gene. Mol Ther 27(11):2005–2017. S1525001619303259. https://doi.org/10.1016/j.ymthe.2019.07.012

49. Lim KRQ, Woo S, Melo D et al (2022) Development of DG9 peptide-conjugated single- and multi-exon skipping therapies for the treatment of Duchenne muscular dystrophy. Proc Natl Acad Sci 119(9):e2112546119. https://doi.org/10.1073/pnas.2112546119

50. Meyers TA, Townsend D (2019) Cardiac pathophysiology and the future of cardiac therapies in duchenne muscular dystrophy. Int J Mol Sci 20(17):4098. https://doi.org/10.3390/ijms20174098

51. Echigoya Y, Nakamura A, Nagata T et al (2017) Effects of systemic multiexon skipping with peptide-conjugated morpholinos in the heart of a dog model of Duchenne muscular dystrophy. Proc Natl Acad Sci 114(16):4213–4218. https://doi.org/10.1073/pnas.1613203114

52. Jearawiriyapaisarn N, Moulton HM, Buckley B et al (2008) Sustained dystrophin expression induced by peptide-conjugated morpholino oligomers in the muscles of mdx mice. Mol Ther 16(9):1624–1629. S1525001616320615. https://doi.org/10.1038/mt.2008.120

53. Sheikh O, Yokota T (2022) Pharmacology and toxicology of eteplirsen and SRP-5051 for DMD exon 51 skipping: an update. Arch Toxicol 96(1):1–9. https://doi.org/10.1007/s00204-021-03184-z

54. Tyler KL (2003) Origins and early descriptions of ?Duchenne muscular dystrophy? Muscle Nerve 28:402–422. https://doi.org/10.1002/mus.10435

55. Lim KRQ, Echigoya Y, Nagata T et al (2019) Efficacy of multi-exon skipping treatment in duchenne muscular dystrophy dog model neonates. Mol Ther 27:76–86. https://doi.org/10.1016/j.ymthe.2018.10.011

56. Yin H, Moulton HM, Seow Y et al (2008) Cell-penetrating peptide-conjugated antisense oligonucleotides restore systemic muscle and cardiac dystrophin expression and function. Hum Mol Genet 17:3909–3918. https://doi.org/10.1093/hmg/ddn293

57. Yokota T, Nakamura A, Nagata T et al (2012) Extensive and prolonged restoration of dystrophin expression with vivo-morpholino-mediated multiple exon skipping in dystrophic dogs. Nucleic Acid Ther 22:306–315. https://doi.org/10.1089/nat.2012.0368

58. Min Y-L, Bassel-Duby R, Olson EN (2019) CRISPR correction of Duchenne muscular dystrophy. Annu Rev Med 70:239–255. https://doi.org/10.1146/annurev-med-081117-010451

59. Hakim CH, Wasala NB, Pan X et al (2017) A five-repeat micro-dystrophin gene ameliorated dystrophic phenotype in the severe DBA/2J-mdx model of Duchenne muscular dystrophy. Mol Ther Methods Clin Dev 6:216–230. https://doi.org/10.1016/j.omtm.2017.06.006

Part II

Animal Models, Sample Preparation, and Assessment

Chapter 3

Rapid Freezing of Skeletal and Cardiac Muscles Using Isopentane Cooled with Liquid Nitrogen and Tragacanth Gum for Histological, Genetic, and Protein Expression Studies

Saeed Anwar and Toshifumi Yokota

Abstract

Histological and molecular genetic evaluation of skeletal and cardiac muscles is an indispensable part of understanding muscle biology and the pathology of muscle disorders. Proper processing of the muscle tissue is a prerequisite for optimal evaluation. However, the processing of skeletal muscle samples often comes with many challenges. One of the commonly used methods of frozen tissue preparation involves optimal cutting temperature compound (OCT compound) embedding. This method is considered optimal for the processing of most of the routinely studied tissue samples. However, the processing of skeletal muscle samples using this method is often unsuitable as it causes artifacts and low DNA, RNA, and protein yield and quality due to the slow freezing of skeletal muscle tissues that allows ice crystals to form. One of the most suitable methods for skeletal muscle tissue processing for histological, genetic, and molecular studies is rapid freezing of freshly collected tissue samples using isopentane cooled with liquid nitrogen and tragacanth gum, which provides distinct advantages in consuming less time, preserving the cell morphology, and helping higher nucleic acids and protein yields. This chapter describes a protocol for rapid freezing of freshly collected skeletal muscle tissues using isopentane pre-chilled with liquid nitrogen and tragacanth gum. Skeletal muscle tissue samples processed using this protocol can be used for histological and immunological staining investigations and studies requiring DNA, RNA, and proteins from these tissues.

Key words Rapid freezing, Skeletal muscle, Isopentane, Liquid nitrogen, Histology, RNA extraction, Immunostaining, Cryosection, DNA extraction, Protein purification

1 Introduction

Skeletal muscles are voluntary muscles, under the control of the somatic nervous system, attached mainly by tendons to the skeleton. These muscles are composed of a variety of functionally diverse fiber types and feature a striated appearance resulting from the highly organized contractile apparatus that dominates the cellular

Rika Maruyama and Toshifumi Yokota (eds.), *Muscular Dystrophy Therapeutics: Methods and Protocols*,
Methods in Molecular Biology, vol. 2587, https://doi.org/10.1007/978-1-0716-2772-3_3,

and tissue organization [1]. It is one of the most dynamic and plastic tissues in the body that undergo significant changes during physiological processes, e.g., aging, exercise, trauma, and many pathological conditions [2–8]. The molecular and histological evaluation of skeletal muscles is thus a focus of interest in the study of muscle biology in sports sciences, basic myology research, and experimental myopathology. These studies are vital for understanding the cellular and molecular mechanisms involved in normal muscle biology, adaptive responses to tissue injury and physical activities, post-trauma muscle regeneration, and many skeletal muscle diseases, e.g., muscular dystrophies.

Since the skeletal muscle is a unique tissue in terms of structure and function, histological and molecular studies on these tissues present several challenges [9, 10]. For most other tissues that are frequently studied, fixation using aldehydes (e.g., formalin) and embedding using paraffin is considered sufficient and often optimal [11–13]. However, the formalin fixation and paraffin embedding (FFPE) method causes undesired cell shrinkage artifacts when it comes to skeletal muscle tissues. Also, the DNA, RNA, and protein yields from samples prepared by FFPE are not satisfactory.

Freezing is considered the most preferred method for histological, genetic, and molecular studies on skeletal muscles. Freshly frozen muscle tissues are most suitable for histological staining and immunohistochemistry analyses of pathological conditions as they preserve the tissue architecture [14, 15]. In addition, DNA, RNA, and protein studies using frozen tissue samples are also beneficial over traditional FPPE-processed samples. Fresh frozen tissues are also suitable for tissue banking [16]. Therefore, fresh freezing of skeletal muscle samples has become a standard for most research and clinical studies.

However, improperly frozen tissue creates artifacts, which significantly compromise interpretation [15]. A rapid freezing method, also called the snap freezing method, often uses liquid nitrogen directly to achieve quick freezing. However, the use of liquid nitrogen directly has its disadvantages. The direct use of liquid nitrogen sometimes causes cracks on the tissues, thus making cryostat sectioning problematic [15]. Also, the vapor phase of liquid nitrogen sometimes leads to uneven and slow freezing of the tissues, leading to artifacts. An alternative rapid freezing technique using isopentane (2-methylbutane) cooled with liquid nitrogen overcomes the above-mentioned issues. One of the commonly used methods of frozen tissue preparation involves optimal cutting temperature compound (OCT compound) embedding; however, this method leads to slow freezing of skeletal muscle tissues that allows ice crystals to form, often resulting in artifacts.

One of the most suitable methods for skeletal muscle tissue processing for histological, genetic, and molecular studies is rapid freezing of fresh muscle samples using isopentane cooled with liquid nitrogen and tragacanth gum. In this technique, the skeletal muscle tissue is frozen by dipping directly into isopentane pre-cooled with liquid nitrogen. This technique is considered preferable for most applications as it freezes the tissue quickly and helps avoid the formation of ice crystals and ruptured cell membranes while keeping the tissue uncracked.

This chapter describes a protocol for rapid freezing of skeletal muscle tissue samples for histological, genetic, and protein studies. It provides essential guidelines for the appropriate processing of skeletal muscle tissue for a range of histological, molecular, and cellular studies in standard laboratory facilities.

2 Materials

2.1 Harvesting and Processing Skeletal Muscle Tissue Samples

1. Murine models: *mdx* mice and wild-type littermates or other suitable murine models. Canine models: Canine X-linked muscular dystrophy (CXMD) dogs and wild-type littermates or other suitable canine models.
2. Anesthetic to induce euthanasia in murine models: isoflurane.
3. For anesthesia in canine models: thiopental sodium.
4. Anesthesia delivery equipment.
5. Metzenbaum, Iris, and fine surgical scissors.
6. Toothed and non-toothed surgical forceps.
7. Scalpels with blades.
8. Pins.
9. Dissection pans and pads.
10. Kimwipes (Kimberly-Clark Professional, USA).

2.2 Mounting Sample on Disc Corks

1. Cork disk pads or embedding molds.
2. Tragacanth gum: prepared by mixing tragacanth powder in equal amounts of water until it becomes soft and sticky. Put the mixture into 10 mL or 25 mL syringes.
3. Dissection spatula.

2.3 Rapid Freezing Using Isopentane

1. Metal beaker cup.
2. Metal holder or tong to hold the metal beaker cup.
3. Cryo-gloves.
4. Isopentane.
5. Liquid nitrogen in a cryogenic storage Dewar.
6. Metal forceps.
7. Dry ice.

2.4 Cryosectioning the Muscle Samples

1. Cryostat.
2. Poly-L-lysine–coated slides.
3. 1.5 mL tubes.
4. Forceps.
5. Camel hair brushes.
6. Low-profile microtome blades.

3 Methods

3.1 Harvesting and Processing Skeletal Muscle Tissue Samples

1. For works on murine models, euthanize the animal with an overdose of isoflurane or by cervical dislocation or as per a protocol authorized by an institutional review board (*see* **Note 1**). Confirm death by firmly squeezing the footpad (*see* **Note 2**). For works on canine models, inject the animal with thiopental sodium for induction of general anesthesia. Then while maintaining the anesthetic status, euthanize the animal by bleeding from the carotid artery.
2. Immediately following euthanasia, collect the connective tissue-free (e.g., fascia) skeletal muscles of interest (*see* **Note 3**). For murine models of muscle disease, collect an entire muscle. For canine models, trim the muscle sample to 0.1–0.4 inches in cross-sectional diameter and 0.3–0.6 inches in length (*see* **Notes 4** and **5**).
3. Using Kimwipes, gently blot excess liquid off of the tissue sample; however, the tissue sample should not be allowed to be completely dried before it is frozen (*see* **Note 6**).

3.2 Mounting Sample on Disc Corks

1. Put just a sufficient amount of tragacanth gum on the sticky surface of a cork disc (*see* **Note** 7) labeled with animal ID and muscle name on the backside.
2. Place the muscle sample in its thinnest dimension (*see* **Note 8**) directly into the tragacanth gum (*see* Fig. 1a).
3. Use a small dissecting spatula to ensure that the muscle sample is appropriately sealed and fixed to the cork and there are no gaps left (*see* Fig. 1a).
4. Leave the cork with the tissue sample at room temperature until freezing it using isopentane (*see* **Note 9**).

3.3 Rapid Freezing Using Isopentane

1. Add isopentane to a metal cup until it reaches a depth of approximately 1.5–2 inches.
2. Using a holder or tong, put the metal cup in contact with the liquid nitrogen in a cryogenic storage Dewar (*see* **Note 10**). Wait until it becomes cold enough for freezing (*see* **Note 11**).

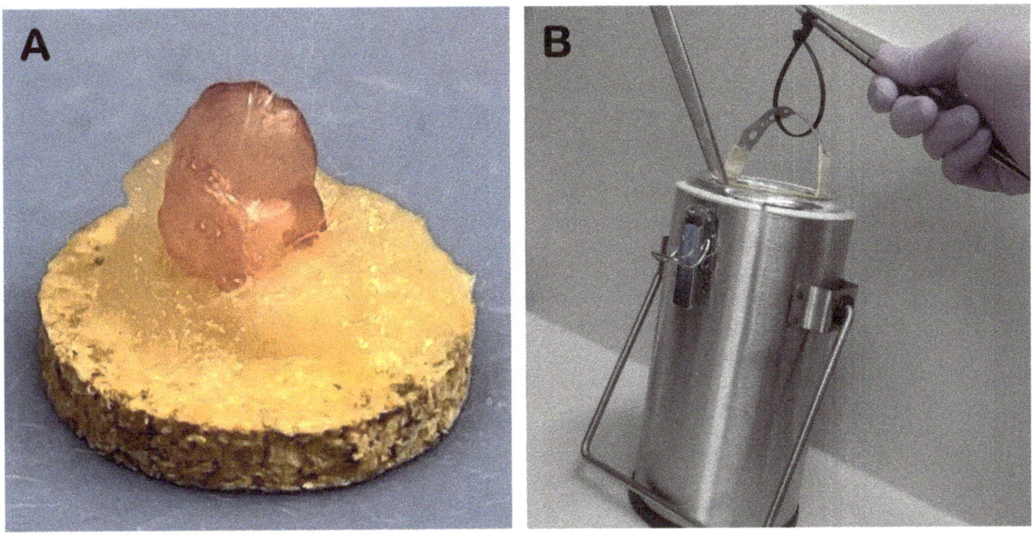

Fig. 1 Preparation of skeletal muscle samples for histological, genetic, and protein studies. (**a**) An appropriately mounted whole quadriceps isolated from 8 weeks old mdx mouse. (**b**) Freezing a skeletal muscle sample on a cork pad using pre-chilled isopentane in a metal cup while keeping the cup in contact with liquid nitrogen in a benchtop cryogenic storage Dewar but preventing the mixture of the two liquids

3. Using a pre-chilled metal tong, put the cork with the sample into the cooled isopentane inside the metal cup and shake vigorously for 30–60 s (*see* **Note 12**) (*see* Fig. 1b).
4. Place the cork with the muscle sample immediately on dry ice until transferred to a −80 °C freezer (*see* **Note 13**).

3.4 Cryosectioning the Muscle Samples

1. Set up cryostat for sectioning. The working temperature should be −25 °C (±2 °C). Set the section thickness at 5–8 μm for immunohistochemistry, 7–10 μm for hematoxylin and eosin staining, and 15–20 μm for procedures involving DNA, RNA, or protein isolation. Put in a microtome blade and leave the machine for 15 min before sectioning can be initiated. Also, keep all the forceps and camel hairbrushes to be used inside the cryostat chamber.
2. Transfer the cork with muscle samples to the cryosection chamber and allow them to equilibrate to the cryostat chamber temperature for 30–40 min (*see* **Note 13**).
3. Label poly-L-lysin-coated glasses slides for histology studies and 1.5 mL tubes for DNA, RNA, or protein isolation. Keep the slides at room temperature and the 1.5-mL tubes at −25 °C (±2 °C).
4. Place a cork with muscle sample on a sample stub using water and then hold the chuck with cork onto the microtome's object holder.

5. Fold out the anti-roll plate and slowly lower down the muscle near the blade stage. Fold the anti-roll plate back and lower the sample onto the blade using the rotating handle to find a clean cut. Continue slicing until a smooth section is produced by each cutting cycle (*see* **Notes 14** and **15**).
6. For histology studies, find a smooth, uniform cryosection ready for collection on the blade stage, fold the anti-roll plate out, and lower a glass slide over the section (*see* **Note 16**). Leave the slide at room temperature to dry for 30–90 min and then store it at −80 °C until further processed for analysis (*see* **Note 17**). For DNA, RNA, or protein isolation, collect 30–40 cryosections in a 1.5 mL tube and keep the tube on dry ice. The sample can be stored at −80 °C until the DNA, RNA, or protein is isolated.

4 Notes

1. This step is recommended to be performed in a fume hood. Use a secondary containment method, e.g., by using a desiccator or a bell jar containing a cotton swab soaked in isoflurane.
2. If euthanized chemically, e.g., by isoflurane overdose, and approved by the institutional review board, a confirmatory secondary euthanasia can be performed. Euthanasia induced by carbon dioxide is not favored as it may drive the muscles to become hypoxic.
3. The choice of muscle samples depends on the type of downstream investigations to be done. We usually collect the following muscles from both limbs of the canine models: tibialis anterior (TA), extensor digitorum longus (EDL), gastrocnemius, soleus, biceps femoris, rectus femoris, biceps brachii, triceps brachii, deltoid, extensor carpi ulnaris (ECU), extensor carpi radialis (ECR), flexor carpi ulnaris (FCU), flexor carpi radialis (FCR), gracilis, intercostal muscles, lateral dorsi, esophagus, and sternocleidomastoid. For works on murine models, we collect the following muscles from both limbs: TA, EDL, gastrocnemius, soleus, quadriceps, triceps brachii, and biceps brachii. We also collect diaphragms from both canine and murine models.
4. We usually use 0.5 inches round cork disks. The suggested tissue diameter works best with this size of cork disks. Larger disks can be used if the object holder of the cryostat allows; however, the diameter of the tissue sample should be optimized accordingly, which should preferably be between 20% and 60% of the cork diameter.

5. Samples collected through core biopsy using a hollow needle or probe to remove tissue from the body of a large animal (e.g., dog model) are sometimes preferred as these do not require euthanasia.
6. This step is very tricky as excessive moisture will cause mounting difficulty and, most importantly, will result in ice crystal artifacts after freezing. If the tissue remains excessively moist after several times of blotting with Kimwipes, it can be further dried by rolling in baby powder. However, powder drying is usually not recommended. Over-drying of the sample can also affect the quality of the downstream studies, especially the protein investigations.
7. This is to provide a foundation for the sample to stand stable on the cork.
8. For example, the proximal end of a TA muscle should go into the tragacanth gum, so the distal end remains to protrude out of the gum for sectioning by a cryostat. For the case of the mouse diaphragm, fold the tissue twice and then mount it on the tragacanth gum. For cardiac muscles, we usually sample the whole heart from small animal models (e.g., mouse) and mount them on tragacanth gum after washing it carefully using PBS followed by 0.1 M potassium chloride solution (for 30 s each) in a way that the ventricular regions protrude out of the gum. To sample cardiac muscles from large animal models (e.g., dog), we separately sample different parts of the heart, e.g., left ventricular free wall, right ventricular free wall, atria, and septum. It is recommended to split the ventricular wall tissues into sub-endocardial, mid-myocardial, and sub-epicardial regions. These samples are then trimmed down to the desired size and then mounted on the tragacanth gum.
9. The tissue should be appropriately frozen within 20–25 min after isolation.
10. Ensure that liquid nitrogen does not enter the metal cup, as it creates a foam insufficiently cold for freezing. The foam is also difficult to work with.
11. In our experience, the isopentane takes around 3–5 min to become sufficiently cold for freezing (less than −140 °C). At appropriate temperature, the container of isopentane becomes a slush., and white pebble-like structures are seen at the bottom of the cup. Carefully observe for these changes after the initial fog disappears. Proceeding with the freezing of the samples before this stage will result in artifacts. If isopentane seems entirely frozen to solid, it can be thawed and chilled again before subsequent use.

12. Perform this step while keeping the metal cup in contact with liquid nitrogen. Note that appropriately cooled isopentane causes minimal fog when the cork with the sample meets it (*see* Fig. 1b). If excessive fog is observed, it indicates that the isopentane was not cooled enough.
13. Take every precaution to prevent accidental thawing after freezing. After freezing, the sample should be placed directly on dry ice until it is transferred to a −80 °C freezer and then transferred to a −80 °C freezer directly from dry ice. Do not let the samples to expose to room temperature. Similar precautions should be taken when transferring the sample from the −80 °C freezer to a cryostat.
14. Stopping or hesitating during the cutting cycle may result in sectioning with varying thickness. It can also dislodge the muscle from the stub.
15. Sweep debris from the blade stage using a fine camel hairbrush until a uniform and flat section is produced. Adjust the anti-roll plate and stub orientation if needed.
16. Static attraction should draw the cryosection to adhere and melt onto the glass slide. Pressing the cryosection too hard may cause it to adhere to the blade stage and break down.
17. 3–5 sections can be drawn on each glass slide. To obtain a representative set of muscle cryosections on each slide, alter the sectioning interval between sampling areas depending on the length of the muscle. For example, place every sixth section on the same slide and cut them at the interval of every 100–200 μm.

Acknowledgments

T.Y. is supported by University of Alberta Faculty of Medicine and Dentistry, the Friends of Garrett Cumming Research and Muscular Dystrophy Canada Endowed Research Chair, Henry M Toupin Neurological Science Research Chair Fund, Muscular Dystrophy Canada, Canada Foundation for Innovation (CFI), Alberta Advanced Education and Technology (AET), Canadian Institutes of Health Research (CIHR), Jesse's Journey—The Foundation for Gene and Cell Therapy, the Women and Children's Health Research Institute (WCHRI), and Alberta Innovates Health Solutions (AIHS). S.A. is supported by scholarships from the Maternal and Child Health (MatCH) Program and the Alberta Innovates Graduate Student Scholarship (AI GSS). The authors thank Tejal Aslesh and Esra Erkut (University of Alberta) for their generous help in capturing photos for this manuscript.

References

1. Frontera WR, Ochala J (2015) Skeletal muscle: a brief review of structure and function. Calcif Tissue Int 96
2. Macaluso F, Isaacs AW, Myburgh KH (2012) Preferential type II muscle fiber damage from plyometric exercise. J Athl Train 47
3. Lieber RL, Fridén J (2000) Functional and clinical significance of skeletal muscle architecture. Muscle Nerve 23:1647–1666
4. Wang J-F, Forst J, Schröder S, Schröder JM (1999) Correlation of muscle fiber type measurements with clinical and molecular genetic data in Duchenne muscular dystrophy. Neuromuscul Disord 9
5. Hopker JG, Coleman DA, Gregson HC, Jobson SA, Von der Haar T, Wiles J et al (2013) The influence of training status, age, and muscle fiber type on cycling efficiency and endurance performance. J Appl Physiol 115:723
6. Song Y, Forsgren S, Liu J-X, Yu J-G, Stål P (2014) Unilateral muscle overuse causes bilateral changes in muscle fiber composition and vascular supply. PLoS One 9
7. Claflin DR, Larkin LM, Cederna PS, Horowitz JF, Alexander NB, Cole NM et al (2011) Effects of high- and low-velocity resistance training on the contractile properties of skeletal muscle fibers from young and older humans. J Appl Physiol. 111
8. Kung TA, Cederna PS, van der Meulen JH, Urbanchek MG, Kuzon WM, Faulkner JA (2014) Motor unit changes seen with skeletal muscle sarcopenia in oldest old rats. J Gerontol Ser A Biol Sci Med Sci. 69:657–665
9. Ben-Ezra J, Johnson DA, Rossi J, Cook N, Wu A (1991) Effect of fixation on the amplification of nucleic acids from paraffin-embedded material by the polymerase chain reaction. J Histochem Cytochem. 39:351
10. Foss RD, Guha-Thakurta N, Conran RM, Gutman P (1994) Effects of fixative and fixation time on the extraction and polymerase chain reaction amplification of RNA from paraffin-embedded tissue. Diagnostic Mol Pathol. 3:148
11. Klopfleisch R, Weiss A, Gruber A (2011) Excavation of a buried treasure – DNA, mRNA, miRNA and protein analysis in formalin fixed, paraffin embedded tissues. Histol Histopathol Cell Mol Biol 26:797–810
12. Maes E, Broeckx V, Mertens I, Sagaert X, Prenen H, Landuyt B et al (2013) Analysis of the formalin-fixed paraffin-embedded tissue proteome: pitfalls, challenges, and future prospectives. Amino Acids 45:205
13. Donczo B, Guttman A (2018) Biomedical analysis of formalin-fixed, paraffin-embedded tissue samples: the Holy Grail for molecular diagnostics. J Pharm Biomed Anal. 155:125
14. van der Loos C (2007) A focus on fixation. Biotech Histochem. 82:141
15. Keith J (2016) Diagnostic skeletal muscle biopsy: quality assurance considerations for the Anatomical Pathology department. Diagnostic Histopathol. 22:345
16. Loken S, Demetrick D (2005) A novel method for freezing and storing research tissue bank specimens. Hum Pathol. 36:977

Chapter 4

Cardiac and Skeletal Muscle Pathology in the D2/*mdx* Mouse Model and Caveats Associated with the Quantification of Utrophin

Tahnee L. Kennedy and Hannah F. Dugdale

Abstract

Duchenne muscular dystrophy (DMD) (the most common form of muscular dystrophy) is caused by a lack of dystrophin protein. Currently, although many therapeutic strategies are under investigation, there is no cure for DMD and unfortunately, patients succumb to respiratory and/or cardiac failure in their second or third decade of life. Preclinical work has focused on the mouse model C57BL/10ScSn-Dmd^{mdx}/J (BL10/*mdx*), which does not exhibit a robust pathophenotype. More recently, the D2.B10-Dmd^{mdx}/J (D2/*mdx*) mouse has been utilized, which presents a more severe pathology and therefore more closely mimics the human pathophenotype, particularly in the heart. Here, we outline important considerations when utilizing the D2/*mdx* model by highlighting the differences between these models in addition to describing histological and immunohistochemical methods utilized in Kennedy et al. (Mol Ther Methods Clin Dev 11:92–105, 2018) for both cardiac and skeletal muscle, which can quantify these differences. These considerations are particularly important when investigating treatment strategies that may be affected by regeneration; such is the case for upregulation of the dystrophin paralogue, utrophin.

Key words d2/*mdx*, Utrophin, Duchenne, Cardiac muscle, Skeletal muscle

1 Introduction

Duchenne muscular dystrophy (DMD) is a severe and progressive muscle-wasting disorder affecting 1:5000 boys [2, 3]. DMD is caused by mutations in the dystrophin gene leading to the loss of dystrophin, a large structural protein located at the sarcolemma. In its absence, the sarcolemma is highly susceptible to contraction-induced injury, causing muscle degeneration and replacement of contractile material with adipose and fibrotic tissue [4, 5]. These processes

The original version of this chapter was revised. The correction to this chapter is available at https://doi.org/10.1007/978-1-0716-2772-3_31

Rika Maruyama and Toshifumi Yokota (eds.), *Muscular Dystrophy Therapeutics: Methods and Protocols*, Methods in Molecular Biology, vol. 2587, https://doi.org/10.1007/978-1-0716-2772-3_4,

typically lead to loss of ambulation between 8 and 12 years, and patients succumb to respiratory and/or cardiac failure in their second or third decade of life [6]. DMD is primarily recognized as a skeletal muscle disorder, and historically, respiratory muscle insufficiencies accounted for the vast majority of patient deaths [7]. However, with improved disease management, cardiomyopathy has emerged as a leading cause of death in patients receiving assisted ventilation [8]. Despite recent progress and extensive research over the last three decades, there is still no effective treatment for DMD. Thus, development of treatments that target skeletal and cardiac muscle is essential for improving both quality of life and longevity of DMD patients.

Utrophin is a structural and functional autosomal paralogue of dystrophin [9] localized to the sarcolemma during fetal development and confined to the neuromuscular junction (NMJ) in mature muscle [10]. Preclinical studies have shown that both transgenic utrophin over-expression and pharmacological modulation using small compounds can prevent pathology in the classical C57BL/10ScSn-Dmd^{mdx}/J (BL10/*mdx*) mouse model [11–13]. Furthermore, there are more than 3000 mutations of the dystrophin gene that can lead to DMD; as such, targeting dystrophin often requires personalized approaches, the development and regulatory approval of which have not been fully fledged. As the utrophin gene is fully intact in all DMD patients and already expressed in adult tissue, utrophin modulation offers a strategy for treating all patients with a reduced risk of immunogenic complications. Additionally, utrophin modulation has the potential to target both cardiac and skeletal muscle. Taken together, these characteristics render strategies targeted at utrophin upregulation highly promising therapies for DMD patients.

Most preclinical studies in the mouse have been carried out in BL10/*mdx* mice that show a relatively mild phenotype, particularly in the heart. In recent years, the D2.B10-Dmd^{mdx}/J (D2/*mdx*) mouse model has been recognized as a potential preclinical model for DMD due to the lesser regenerative capacity of the D2/2J (D2/wild-type [WT]) parent strain. Coley et al. [14] and Vohra et al. [15] showed that pathology in D2/*mdx* mice was more severe than in age-matched BL10/*mdx* mice. D2/*mdx* mice were therefore thought to offer a promising and more acutely affected model for cardiac studies. However, there are some important considerations when utilizing this model in order to capture pathology and correctly interpret results.

1.1 Considerations for Using the D2/mdx Mouse Model for Assessing Heart Pathology

1.1.1 Phenotypic Shift in Cardiac Dysfunction

Cardiomyopathy is a leading cause of death in DMD patients receiving assisted ventilation [8]. Heart defects are evident in nearly all patients by the end of adolescence and are also present in more than 90% of BMD patients [16]. One of the most promising aspects of the D2/*mdx* mouse model when it was first generated was the early onset of a cardiac functional deficit, relative to the BL10/*mdx* mouse. Vohra et al. reported a reduction in LV ejection fraction (LVEF) at 7 months of age using cardiac cine magnetic resonance imaging (MRI) [15]. Coley and colleagues reported similar findings with LVEF depression, assessed using echocardiography, observed at 28 weeks of age in D2/*mdx* mice [14]. Interestingly, Vohra and colleagues saw a recovery in LVEF in D2/*mdx* mice at 52 weeks whereas Coley et al. assessed only one timepoint. In our previous study, we observed an increase in LVEF at 21 weeks of age, assessed via cardiac cine-MRI, which was normalized by 28 weeks of age [1]. We also observed a reduction in LVEF (from its elevated level) with micro-utrophin treatment when the function was assessed at 21 weeks of age, supporting the elevated EF as being pathogenic. Taken together, these observations highlight the D2/*mdx* model as dynamic, and the mechanisms behind the functional shifts are unclear. These observations indicate that phenotypic shift is present across D2/mdx mouse colonies and that timings of functional assessment are paramount to capture deficits.

1.2 Underlying D2/WT Pathology

The more studies assessing the D2/*mdx* phenotype, the more it becomes apparent that the D2/WT mouse also exhibited some basal pathology [1]. Overt calcification of the pericardium has been observed in multiple studies assessing the D2/*mdx* model; furthermore, the JAX website reports skeletal muscles and cardiac calcification as part of a common occurrence in the D2/WT strain. This may cause complications in novel therapies as improvements in pathology cannot be attributed to a reduction in the dystrophic phenotype alone; however, this may be overcome by the addition of a D2/WT control group. Although the presence of a WT pathology complicates D2/*mdx* studies, the model can still be effectively used. However, caution and context are necessary when designing experiments and interpreting data. Correct control groups and the inclusion of calcification measurements as well as standard hematoxylin and eosin (H&E) and fibrosis assessments are paramount when working with D2/*mdx* mice.

1.3 Considerations for Assessing Skeletal Muscle Pathology in D2/mdx Mice

1.3.1 BL10/mdx Mouse vs D2/mdx Mouse Phenotypes

Contrary to DMD patients, BL10/*mdx* mice have a significantly larger body and muscle mass than their control counterparts [14, 15, 17, 18]. Conversely, D2/*mdx* mice (compared to controls) possess a significantly smaller body mass [14, 15, 17, 18]. Differences in muscle size are not merely associated with a reduction in total body weight demonstrated by muscle-to-body weight ratios. Consistent with whole body and muscle mass findings, analysis of individual myofibers revealed that BL10/*mdx* have a high proportion of hypertrophic fibers whereas D2/*mdx* mice possess very few; instead, they have a much larger proportion of small fibers [14, 15]. Taken together, these findings signify the presence of hypertrophy in BL10/*mdx* and atrophy in D2/*mdx* mice; the latter more closely recapitulates the patient atrophic phenotype.

Skeletal muscle atrophy and more severe dystrophic phenotype contribute to the impaired muscle function observed in the D2/*mdx* compared to BL10/6*mdx* mice. Skeletal muscle endurance is reduced compared to control littermates in both BL10/*mdx* and D2/*mdx* strains; however, the latter is far more striking [18]. The diaphragm is the most affected skeletal muscle in the BL10/6*mdx* mouse and as such similar reductions in maximal and specific force are observed in both strains [19]. The extensor digitorum longus (EDL) of the BL10/*mdx* mouse does, however, produce similar [14] or even exceed [19] the maximal force production of its control counterpart, whereas D2/*mdx* mice experience a 40–50% reduction, compared to controls [19]. Preventing decline in muscle function is the primary outcome for most DMD clinical trials and most desirable result for patients; subsequently, being able to recapitulate functional decline in DMD more closely in the D2/*mdx* mouse provides a model whereby potential interventions may be more easily translated to the human disease.

1.4 Assessing Utrophin in a Lower Regenerative Context

As described above, the BL10/*6mdx* mice present with a milder phenotype than the human pathology. One hypothesis for these differences is the higher regenerative capacity these mice seemingly possess. One reliable indicator of muscle regeneration is the number of centrally nucleated fibers (CNFs). BL10/*mdx*, BL6/*mdx*, and D2/*mdx* mice all display increases in CNFs compared to their control counterparts, which in the BL6/*mdx* mice declines with age [20]. However, in the D2/*mdx* mice, the number of CNFs is the lowest of these strains and remains unchanged with age [14, 19, 20]. Similarly, to CNFs, developmental myosin heavy chain (dMHC) provides a robust indicator of muscle regeneration. In D2/*mdx* mice, dMHC is barely detectable at 2, 6, 12, and 18 months, whereas dMHC is observed at much greater levels in BL6/*mdx* at all timepoints (although it does decline with age) [20]. Additionally, following multiple injuries and therefore multiple regeneration cycles, D2/*mdx* mice have a reduced number of myofibers compared to BL10/*mdx* mice [18]. Contrastingly, BL10/*mdx* mice have previously been documented to undergo regeneration following 50 bouts of induced muscle damage, without loss of fibers or increases in fibrosis [21]. These regenerative differences are predominantly attributed to an impaired capacity for self-renew by muscle precursor cells (satellite cells), which becomes exhausted and unable to facilitate adequate muscle repair in the D2/*mdx* mice [18]. Together these findings indicate that D2/*mdx* have an overall lower regenerative capacity across their lifespan.

As previously outlined, in mature muscle, utrophin is limited to the NMJ. During fetal development and muscular regeneration, however, utrophin is present in the sarcolemma as a precursor to dystrophin [22]. Consequently, increases in regeneration can increase utrophin [23]. Quantification of utrophin may therefore be confounded by an increase in regeneration rather than an interventional upregulation of utrophin itself. Subsequently, by utilizing the D2/*mdx* model, changes in utrophin are more likely to present following intervention and not due to increased regeneration.

Here we outline the methodologies used to characterize the D2/*mdx* mouse model phenotype as well as discussing the considerations that need to be made when designing experiments using the D2/*mdx* model. Furthermore, given the preclinical findings pertaining to utrophin attenuation in dystrophic muscles, we also outline how to assess utrophin expression following therapeutic intervention in this model and factors differentiating the D2/*mdx* mouse model from the classical BL10/*mdx* strain.

2 Materials

2.1 Sample Preparation

- Isopentane solution (Merk; PHR1661).
- OCT.
- Liquid nitrogen.
- Dry ice.

2.2 Histology

2.2.1 H&E Staining

- Paraformaldehyde 4% in PBS (Biotium; 22,025).
- Hematoxylin (Mayer's modified; Abcam; ab220365).
- Eosin Y, xanthene dye (Abcam; 146,325).

2.2.2 Alizarin Red Staining

- Alizarin Red S (Signa-Aldrich; A5533).
- Hist-clear II Histology Clearing Agent 1 US Gallon (Scientific Laboratory Supplies; MAT1334).
- 100% ethanol.

2.2.3 Immuno-fluorescence Staining

- Alexa Fluor® 488 goat anti-mouse IgG (Life Technologies; ab150117).
- Fetal bovine serum (FBS).

3 Methods

Considerations for assessing utrophin and dystrophic pathophysiology in D2/*mdx* mice (Fig. 1):

1. As cardiac dysfunction is dynamic across D2/*mdx* mouse colonies, a longitudinal cardiac functional study needs to be performed in order to identify the point of significant cardiac dysfunction in each colony.
2. When designing studies to assess new therapeutics using D2/*mdx* mice, appropriate controls and timing of start and end points are crucial. As D2/WT exhibit pathology, a treated WT cohort should be included to account for the impact the treatment has on the non-dystrophic pathology. Furthermore,

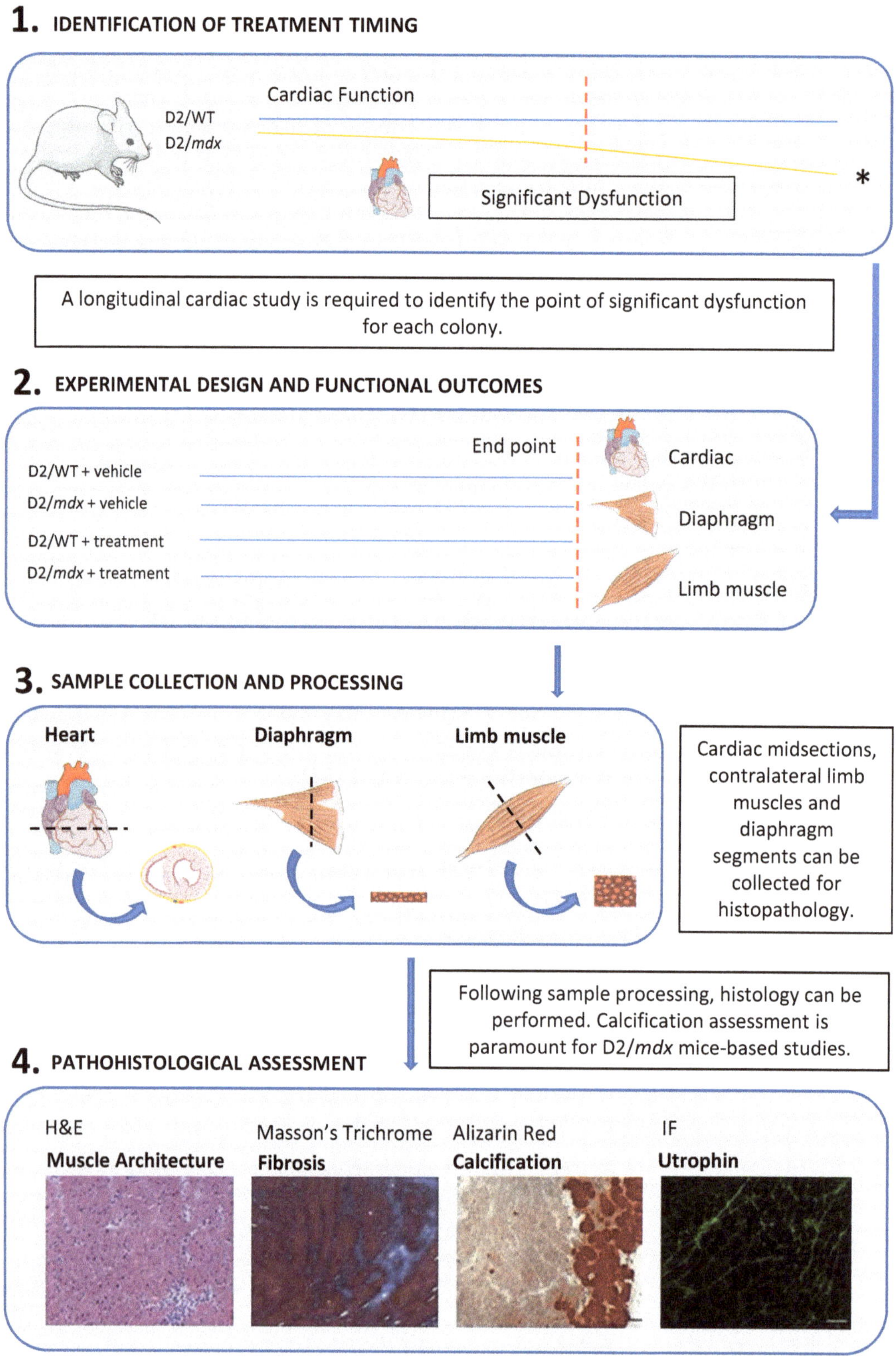

Fig. 1 Considerations for utilising the D2/*mdx* mouse model for assessing new therapeutics. Depiction of the experimental design requirements, samples processing and analysis to correctly interpret the data generated using the D2/*mdx* mouse model. Image adapted from Kennedy et al. [1]

all DMD therapies to date have shown to be prophylactic; as such, earlier treatment initiation if possible should be used. Endpoint analysis (if assessing cardiac function) should be informed by **step 1**. These guidelines will provide the greatest opportunity for capturing and correctly interpreting treatment impact.

3. Calcification assessments must be included in endpoint analysis in order to capture and account for the impact of treatment on the non-dystrophic pathology present in the D2/*mdx* mouse model.

3.1 Functional Outcomes

It should be noted that although we have outlined cardiac and skeletal muscle function, we do not describe these procedures in detail here. Protocols for cardiac, diaphragm, and limb muscle EDL have been previously published [1].

Regarding experimental outline, mice are culled following cardiac assessments and the diaphragm and EDL must be excised and prepared immediately following culling. There should be less than 1 min between in vivo blood cessation to the organs and the muscle preparation being placed into an oxygenated buffer. Two technicians are required to carry out both diaphragm and limb muscle testing on a single cohort of mice. Ideally, a third person is also present to dissect the remaining muscles and any other tissues of interest.

3.2 Sample Collection and Processing

3.2.1 Sample Preparation

1. Prepare isopentane for the freezing of samples by cooling in a beaker over liquid nitrogen or dry ice (*see* **Note 1**). Isopentane protects the OCT from bubbling and fracturing the samples, as it would if directly exposed to liquid nitrogen.
2. Cull mice via CO_2 overdose in accordance with your country's relevant scientific procedures act (*see* **Note 2**).
3. Following heart dissection, trim atria and remove and weigh ventricles (if required). The aorta has a high hydroxyproline content; therefore, if using samples for hydroxyproline assays, take care to fully remove. Excise the middle third of the ventricles and lay lumen side up before coating in OCT.
4. For diaphragm muscle, cut a segment longitudinally along the myofibers and lay flat before coating in OCT. When preparing samples for sectioning, cut through the middle of the OCT-coated muscle segment for myofiber cross-sectional sections (*see* Fig. 1). Keeping track of muscle orientation prior to freezing is vital for correct sample processing (*see* **Note 3**).
5. For EDL muscle, the contralateral EDL muscle (from functional testing) should be fully dissected and coated in OCT. When preparing samples for sectioning, cut through the middle of the OCT-coated muscle for myofiber cross-sectional

sections (*see* Fig. 1). Keeping track of muscle orientation prior to freezing is vital for correct sample processing (*see* **Note 3**).

6. Following coating of samples in OCT, submerge into the thawing isopropanol (from **step 1**) until OCT is solid (OCT will turn from clear to fully white). Tin foil cups can be used to produce a mold in which the tissue can be embedded in OCT.
7. Immediately store the frozen samples at −80 °C.

3.2.2 Cryosectioning

1. Transport tissues to be processed on dry ice to prevent thawing.
2. Set cryostat to 20 °C; prior to cutting, place samples in the cryostat for 10 min to equilibrate. Place all tools (forceps, scalpels, brushes) that will come into contact with the samples in the cryostat to cool to prevent sample thawing.
3. When mounting samples, using OCT make a small disc on the chuck and place it in cryostat until OCT begins to turn white; at this point, the OCT is cold enough so as not to thaw the muscle preparation when mounting (which can cause ice crystal damage). Place muscle in OCT before it fully freezes in the required orientation (see below). Cut serial cross-sections (10 mm) and mount them on anti-static slides.
   (i) Heart: Mount heart preparation flush to the chuck so that when cutting a full cross-section is achieved, the sample should be orientated base-to-apex.
   (ii) Diaphragm: Cut through the middle of the prepared sample to produce a mid/cross-section of the muscle segment. Place half of the sample upright in the OCT so that the mid-section is orientated toward the blade. Store the remaining half for later use. Diaphragm muscles can be difficult to section (*see* **Note 4**).
   (iii) EDL: Cut through the middle of the prepared sample to produce a mid/cross-section of the whole muscle. Place half of the sample upright in the OCT so that the mid-section is orientated toward the blade. Store the remaining half for later use.
4. Following sectioning, label slides and store at −80 °C.

3.3 Pathohistological Assessment

3.3.1 H&E Staining of OCT-Embedded Muscle Sections

It is important to note that solutions used for H&E staining can vary and lose potency with time and number of uses. It is good practice to stain one slide with the working solutions to make sure the resulting stain is the desired intensity. It should also be noted that all samples to be compared must be stained at the same time. If large sample numbers do not allow for all sections to be stained in a single run, divide samples into batches and evenly distribute the experimental cohorts across batches.

1. Allow slides to equilibrate to room temperature for 5–10 min.
2. If slides have been paraffin embedded, slides will need to be incubated in 4% paraformaldehyde solution (2 × 5 min).
3. Submerge slides in room temperature water for 1 min. Subsequently drain the slides on a tissue (slides do not need to be dry).
4. Incubate slides for 5 min in Mayer's hematoxylin solution (*see* **Note 5**). Rinse slide in running tap water for 10 min; this can be done by placing a bell jar containing slides in a larger container to be filled with water or by placing a paper towel over the bell jar to prevent water turbulence from damaging muscle sections.
5. Following hematoxylin staining, exposing the section to a weak alkaline solution turns the nuclei staining from purple to blue in a process called "bluing"; this is not always necessary/desired.
6. Incubate slides for 10 min (incubation time should be tested and optimized before each run) in solution containing 0.5% Eosin and 1:100 acetic acid (*see* **Note 6**). Three one-minute washes must follow the staining to remove excess solution.
7. Air dry samples until excess liquid is no longer present on the slide but before the section begins to dry out.
8. Mounting: Using a compatible mounting medium, apply coverslip ready for the acquisition of digital images.

3.3.2 Calcification Assessments of OCT-Embedded Muscle Sections

Although calcification is sometimes assessed in the BL10/*mdx* mouse model, it is not a staple test for pathology like H&E or fibrosis. As D2/WT mice exhibit skeletal muscle and pericardial calcification, calcification must be checked when testing new therapies in D2/*mdx* mice. Although calcification is exacerbated in D2/*mdx* mice compared to D2/WT mice, a treated D2/WT cohort must be included and assessed for calcification in order to capture and account for any effect treatment is having on non-dystrophic pathology.

Calcification can be assessed by using an Alizarin Red stain. The protocol for Alizarin Red staining is straightforward; however, protocols often dictate dehydration in acetone and using xylene for clearing. These can be substituted for ethanol and Histo-clear, as safer alternatives, without compromising the contrast of the stain.

Like H&E staining, Alizarin Red solution can lose potency with time and number of uses. It is good practice to stain one slide with the working solutions to make sure the resulting stain is the desired intensity. It should also be noted that all samples to be compared must be stained at the same time. If large sample numbers do not allow for all sections to be stained in a single run, divide samples into batches and evenly distribute the experimental cohorts across batches.

1. Allow slides to equilibrate to room temperature for 5–10 min.
2. Make Alizarin Red solution by mixing 2 g of Alizarin Red S with 100 mL of distilled water. Adjust pH to 4.1–4.3 and mix well.
3. De-paraffinize slide (if required) in distilled water, 2 × 1-min incubations.
4. Stain slides in working Alizarin Red solution for 30 s to 5 min (incubation time should be tested and optimized before each run).
5. Dehydrate slides in ethanol, 20 × dips followed by a 1:1 solution of the ethanol–Histo-clear solution, 20 × dips.
6. Clear in Histo-clear and mount using a compatible mounting medium; apply coverslip ready for the acquisition of digital images.

3.3.3 Fibrosis Assessments in OCT-Embedded Muscle Sections

Masson's trichrome is a common fibrosis stain, and commercial kits are available (Sigma-Aldrich; HT15), which include extensive protocols, so the protocol will not be outlined here. It may be helpful to note that fixation of slides with Bouin's solution at 37 °C for 1 h (a standard inclusion for Masson's trichome protocols) can cause deterioration of the muscles tissue (especially when embedded in OCT). Removal of Bouin's solution step from the protocol will result in loss of contrast with sections being entirely blue; fibrosis cannot be assessed. Reducing the time of staining (to 15–30 min) can reduce the damage caused to sections. All samples to be compared need to be stained together, like H&E and Alizarin Red staining.

3.3.4 Utrophin Assessments in Dystrophic Muscle

As mentioned previously, regenerating myofibers express utrophin, and this can convolute the interpretation of utrophin modulating strategies. Co-staining for utrophin with a marker for regeneration, such as dMHC, can help delineate regenerating fibers from mature fibers stabilized by utrophin. For this reason, IF is preferred over western blot. As cardiac muscle does not regenerate, utrophin staining alone is sufficient.

3.3.5 Immunofluorescence

1. Allow slides to equilibrate to room temperature for 5–10 min.
2. Circle section with a hydrophobic pen and permeabilize samples using 0.3% Triton X in PBS.
3. Block sections for 30 min with 10% FBS/PBS.

4. Incubate slides with utrophin primary antibody, developed in-house [24], 1:2000 in 5% FBS and PBS for 2 h at room temperature.
5. Incubate sections with secondary antibody, Alexa Fluor® 488 donkey anti-goat IgG antibody, 1:2000 in 5% FBS and PBS for 1 h at room temperature.
6. Wash sections in PBS and air dry before application of cover slip with fluorescent mounting medium.

4 Notes

1. Isopentane, liquid nitrogen, and dry ice should all be used in a well-ventilated area, and appropriate cold-proof gloves and eye protection should be worn. To check whether the isopentane is sufficiently cold, a piece of dry ice can be submerged; if this fizzes, it is not yet ready for sample preparation.
2. Culling mice via CO_2 prevents diaphragm rupture and cardiac damage that would occur with cervical dislocation.
3. Put a small piece of cardboard in one side of the cup to use as an indicator of tissue orientation.
4. Diaphragm sections are prone to rolling during cryo-sectioning. Cutting the section and leaving it under the roll plate for 10–15 s before lifting can help prevent/minimize this.
5. Hematoxylin stains the nuclei blue. In Duchenne patients, this stain can assist diagnosis due to the presence of centrally nucleated myofibers and can also indicate infiltration of inflammatory cells.
6. Eosin staining stains myofibers red; in Duchenne patients, this displays the typical dystrophic phenotype in which variable sizes of myofibers are observed in addition to connective tissue deposition. In contrast to the BL10/*mdx*, in which the Eosin stain reveals a greater proportion of hypertrophic fibers, the D2/*mdx* display a reduced number of large fibers and an increase in small fibers, leading to an overall smaller average fiber diameter more closely recapitulating the patient histology [15]. Further to this and as previously discussed, the D2/*mdx* mice possess a lower number of centrally nucleated fibers than the BL10/*mdx* due in part to their lower regenerative capacity. These findings can be observed by determining nuclei position by staining with hematoxylin as outlined above, in addition to obtaining fiber outline using Eosin.

Acknowledgement

As Kennedy et al. [1] referenced in Fig. 1, the work flow described there was used in a study performed at the University of Oxford, under Prof. Kay Davies. Original experiments were funded by the MRC.

References

1. Kennedy TL et al (2018) Micro-utrophin improves cardiac and skeletal muscle function of severely affected D2/mdx mice. Mol Ther Methods Clin Dev 11:92–105
2. Mendell JR et al (2012) Evidence-based path to newborn screening for Duchenne muscular dystrophy. Ann Neurol 71(3):304–313
3. Mah JK et al (2014) A systematic review and meta-analysis on the epidemiology of Duchenne and Becker muscular dystrophy. Neuromuscul Disord 24(6):482–491
4. Forcina L et al (2017) Insights into the pathogenic secondary symptoms caused by the primary loss of dystrophin. J Funct Morphol Kinesiol 2(4):44
5. Head S, Williams D, Stephenson G (1994) Increased susceptibility of EDL muscles from mdx mice to damage induced by contraction with stretch. J Muscle Res Cell Motil 15(4): 490–492
6. Bach JR et al (1987) Management of end stage respiratory failure in Duchenne muscular dystrophy. Muscle Nerve 10(2):177–182
7. Rideau Y et al (1983) Prolongation of life in Duchenne's muscular dystrophy. Acta Neurol 5(2):118–124
8. Birnkrant DJ, Ararat E, Mhanna MJ (2016) Cardiac phenotype determines survival in Duchenne muscular dystrophy. Pediatr Pulmonol 51(1):70–76
9. Love DR et al (1989) An autosomal transcript in skeletal muscle with homology to dystrophin. Nature 339(6219):55–58
10. Lin S, Burgunder J-M (2000) Utrophin may be a precursor of dystrophin during skeletal muscle development. Dev Brain Res 119(2): 289–295
11. Tinsley J et al (1998) Expression of full-length utrophin prevents muscular dystrophy in mdx mice. Nat Med 4(12):1441–1444
12. Krag TO et al (2004) Heregulin ameliorates the dystrophic phenotype in mdx mice. Proc Natl Acad Sci 101(38):13856–13860
13. Tinsley JM et al (2011) Daily treatment with SMTC1100, a novel small molecule utrophin upregulator, dramatically reduces the dystrophic symptoms in the mdx mouse. PLoS One 6(5):e19189. https://doi.org/10.1371/journal.pone.0019189. PMID: 21573153; PMCID: PMC3089598
14. Coley WD et al (2016) Effect of genetic background on the dystrophic phenotype in mdx mice. Hum Mol Genet 25(1):130–145
15. Vohra R et al (2017) Magnetic resonance monitoring of disease progression in mdx mice on different genetic backgrounds. Am J Pathol 187(9):2060–2070
16. Finsterer J, Stöllberger C (2008) Cardiac involvement in Becker muscular dystrophy. Can J Cardiol 24(10):786–792
17. van Putten M et al (2019) Natural disease history of the D2-mdx mouse model for Duchenne muscular dystrophy. FASEB J 33(7): 8110–8124
18. Fukada S-I et al (2010) Genetic background affects properties of satellite cells and mdx phenotypes. Am J Pathol 176(5):2414–2424
19. Hammers DW et al (2020) The D2. mdx mouse as a preclinical model of the skeletal muscle pathology associated with Duchenne muscular dystrophy. Sci Rep 10(1):1–12
20. Rodrigues M et al (2016) Impaired regenerative capacity and lower revertant fibre expansion in dystrophin-deficient mdx muscles on DBA/2 background. Sci Rep 6(1):1–9
21. Luz MA, Marques MJ, Santo Neto H (2002) Impaired regeneration of dystrophin-deficient muscle fibers is caused by exhaustion of myogenic cells. Braz J Med Biol Res 35(6): 691–695
22. Clerk A et al (1993) Dystrophin-related protein, utrophin, in normal and dystrophic human fetal skeletal muscle. Histochem J 25(8):554–561
23. Lin S, Gaschen F, Burgunder J-M (1998) Utrophin is a regeneration-associated protein transiently present at the sarcolemma of regenerating skeletal muscle fibers in dystrophin-deficient hypertrophic feline muscular dystrophy. J Neuropathol Exp Neurol 57(8): 780–790
24. Weir AP, Burton EA, Harrod G, Davies KE (2002) A- and B-utrophin have different expression patterns and are differentially up-regulated in mdx muscle*. J Biol Chem 277:45285–45290

Chapter 5

Physiological Assessment of Muscle, Heart, and Whole Body Function in the Canine Model of Duchenne Muscular Dystrophy

Chady H. Hakim, James Teixeira, Stacy B. Leach, and Dongsheng Duan

Abstract

Duchenne muscular dystrophy (DMD) is a lethal muscle disease caused by dystrophin deficiency. Patients gradually lose motor function, become wheelchair-bound, and die from respiratory and/or cardiac muscle failure. Dystrophin-null dogs have been used as a large animal model for DMD since 1988 and are considered an excellent bridge between rodent models and human patients. While numerous protocols have been published for studying muscle and heart physiology in mice, few such protocols exist for studying skeletal muscle contractility, heart function, and whole-body activity in dogs. Over the last 20 years, we have developed and adapted an array of assays to evaluate whole-body movement, gait, single muscle force, whole limb torque, cardiac electrophysiology, and hemodynamic function in normal and dystrophic dogs. In this chapter, we present detailed working protocols for these assays and lessons we learned during the development and use of these protocols.

Key words Duchenne muscular dystrophy, DMD, Canine model, Skeletal muscle, Heart, Muscle contraction, ECG, Echo, Gait, Force, Dog

1 Introduction

Duchenne muscular dystrophy (DMD) is the most severe form of muscular dystrophy affecting 1 in every 5000 male births [1]. DMD is caused by the loss of a cytosolic structural protein known as dystrophin [2, 3]. Dystrophin plays an essential role in lateral and longitudinal force transmission during contraction [4–6]. Dystrophin-deficient muscle becomes sensitive to contraction-induced injury, undergoes degeneration, and loses force production capacity. DMD patients become immobilized to a wheelchair in their early teens and die before 30 years of age from cardiac and/or respiratory muscle failure.

Rika Maruyama and Toshifumi Yokota (eds.), *Muscular Dystrophy Therapeutics: Methods and Protocols*, Methods in Molecular Biology, vol. 2587, https://doi.org/10.1007/978-1-0716-2772-3_5,

Loss of muscle and heart function are physiological hallmarks of DMD. Since the cloning of the dystrophin gene in 1987 [3], numerous studies have been conducted to characterize the physiological consequences of dystrophin deficiency. Most of these studies are performed in mdx mice, the first mammalian DMD model discovered in 1984 and validated in 1989 [7, 8]. While results obtained from mdx mice have provided important insights into our understanding of DMD pathology and physiology, it has also been recognized that the mdx model has important limitations. Clinical presentations are largely attenuated in mdx mice due to various compensatory mechanisms such as utrophin upregulation and enhanced regeneration. Force reduction is a primary physiological phenotype in DMD patients. However, the absolute muscle force is not reduced in mdx mice.

Dystrophin deficiency was confirmed in a spontaneous canine myopathy model in 1988 [9]. Despite the fact that the canine model more faithfully reproduces the dystrophic phenotype seen in human patients, few studies have utilized the dog model. The discrepancy in the use of the mouse versus the dog model is attributed to many factors. Among these is the lack of physiological methods to study muscle function in canines. Over the last three decades, a series of protocols have been developed by our lab and many other labs. In this chapter, we present protocols to evaluate bodywide motor function (whole-body activity and gait), skeletal muscle function (single muscle force and hindlimb torque), and cardiac function (electrocardiography, Holter monitoring, and echocardiography) in canines.

2 Materials

2.1 Evaluation of Bodywide Motor Function

2.1.1 Overnight Activity Recording (See Fig. 1)

1. Personal protective equipment (PPE), including disposable gowns and/or dedicated scrubs, a head cover, shoe covers, a surgical mask, and gloves.
2. A low-light charge-coupled device (CCD) camera equipped with an infrared lens (Supercircuits Inc., Austin, TX, USA).
3. Near-infrared (NIR) light source (Clover electronics USA, Cerritos, CA, USA).
4. Custom-made mounting bracket for the CCD camera and the NIR light source (University of Missouri Physics Machine Shop, Columbia, MO, USA) (*see* **Note 1**).
5. Custom-made aluminum kennel (University of Missouri Physics Machine Shop, Columbia, MO, USA) (*see* Fig. 1) (*see* **Note 2**).
6. Digital thermometer (ReliOn Inc., Spokane Valley, WA, USA).

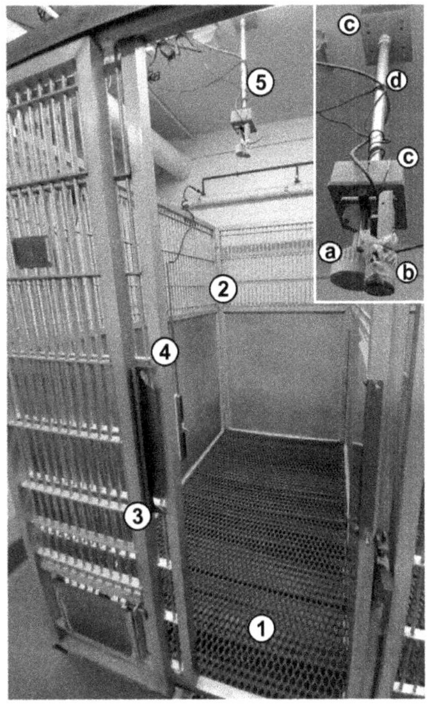

Fig. 1 Custom-made aluminum kennel to record dog activity overnight. (1) Kennel floor made of thick high-density plastic mesh and is raised off the ground, (2) back and side walls are made of a flat aluminum sheet (bottom portion) and aluminum grid (upper portion), (3) front wall made of the aluminum grid, (4) sliding door made of the aluminum grid, and (5) custom-made mounting bracket to suspend the CCD camera and the NIR light source from the ceiling, enlarged in the insert. (a) NIR light source, (b) CCD camera, (c) 4 inch × 4 inch electrical PVC junction boxes, and (d) 1-inch steel galvanized steel conduit

7. Stethoscope (3M™ Littmann® Stethoscopes, St. Paul, MN, USA).
8. Overnight Dog Recording (ODR) software (*see* **Note 3**).
9. Overnight Dog-Video Analyzer (ODVA) software (*see* **Note 3**).

2.1.2 Gait Analysis

1. Personal protective equipment (PPE), including disposable gowns and/or dedicated scrubs, a head cover, shoe covers, a surgical mask, and gloves.
2. GAIT4 Dog® walkway system (CIR Systems Inc., Sparta, NJ, USA). The system consists of a portable walkway mat with encapsulated sensors, a USB cable, a webcam, and a personal computer (PC) laptop (*see* Fig. 2) (*see* **Note 7**).
3. GAITFour software (CIR Systems Inc., Sparta, NJ, USA).
4. Harness (Petco Animal Supplies, Inc., San Diego, CA, USA).
5. Leash (Petco Animal Supplies, Inc., San Diego, CA, USA).

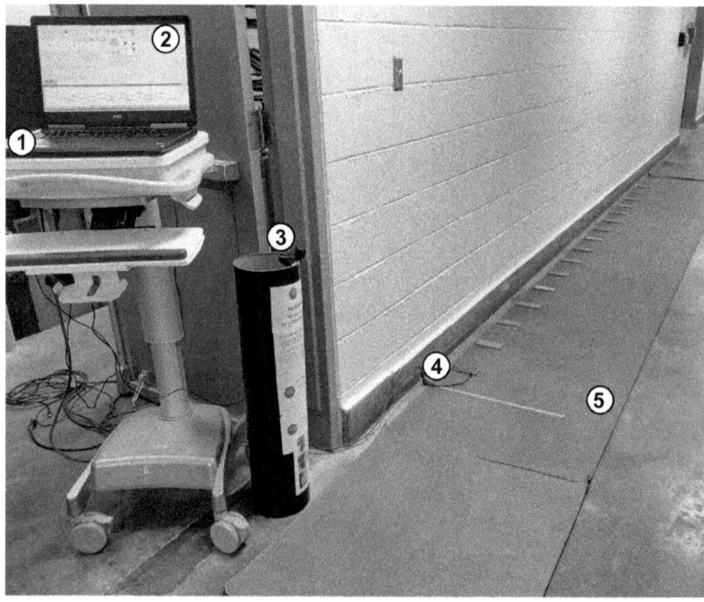

Fig. 2 GAIT4 Dog® walkway system. (1) PC laptop, (2) GAITFour software, (3) webcam, (4) USB cable, and (5) portable walkway mat with encapsulated sensors

2.2 Evaluation of Skeletal Muscle Function

2.2.1 Single Muscle Force Measurement (See Fig. 3a)

1. PPE including disposable gowns and/or dedicated scrubs, a head cover, shoe covers, a surgical mask, and gloves.
2. Stethoscope (3M™ Littmann® Stethoscopes, St. Paul, MN, USA).
3. Digital thermometer (ReliOn Inc., Spokane Valley, WA, USA).
4. Cerenia (Maropitant citrate) (Zoetis Inc., Kalamazoo, MI, USA).
5. Rimadyl (Carprofen) (Zoetis Inc., Kalamazoo, MI, USA).
6. Disposable sterile syringe (Becton, Dickinson and Company, Franklin Lakes, NJ, USA).
7. Custom-made muscle force assay platform (University of Missouri Physics Machine Shop, Columbia, MO, USA). The platform consists of the main platform, a tri-axis mount, a dual-axis mount, and two bone pins (*see* **Note 13**) (*see* Fig. 3b, c).
8. In situ muscle function assay system (Aurora Scientific Inc., Aurora, ON, Canada). The system consists of a 310C-LR Dual-Mode transducer that measures/controls the muscle force and length, a 310C-LR force transducer controller module, a 701A stimulator, and a 604A signal interface (*see* **Note 14**) (*see* Fig. 3b).

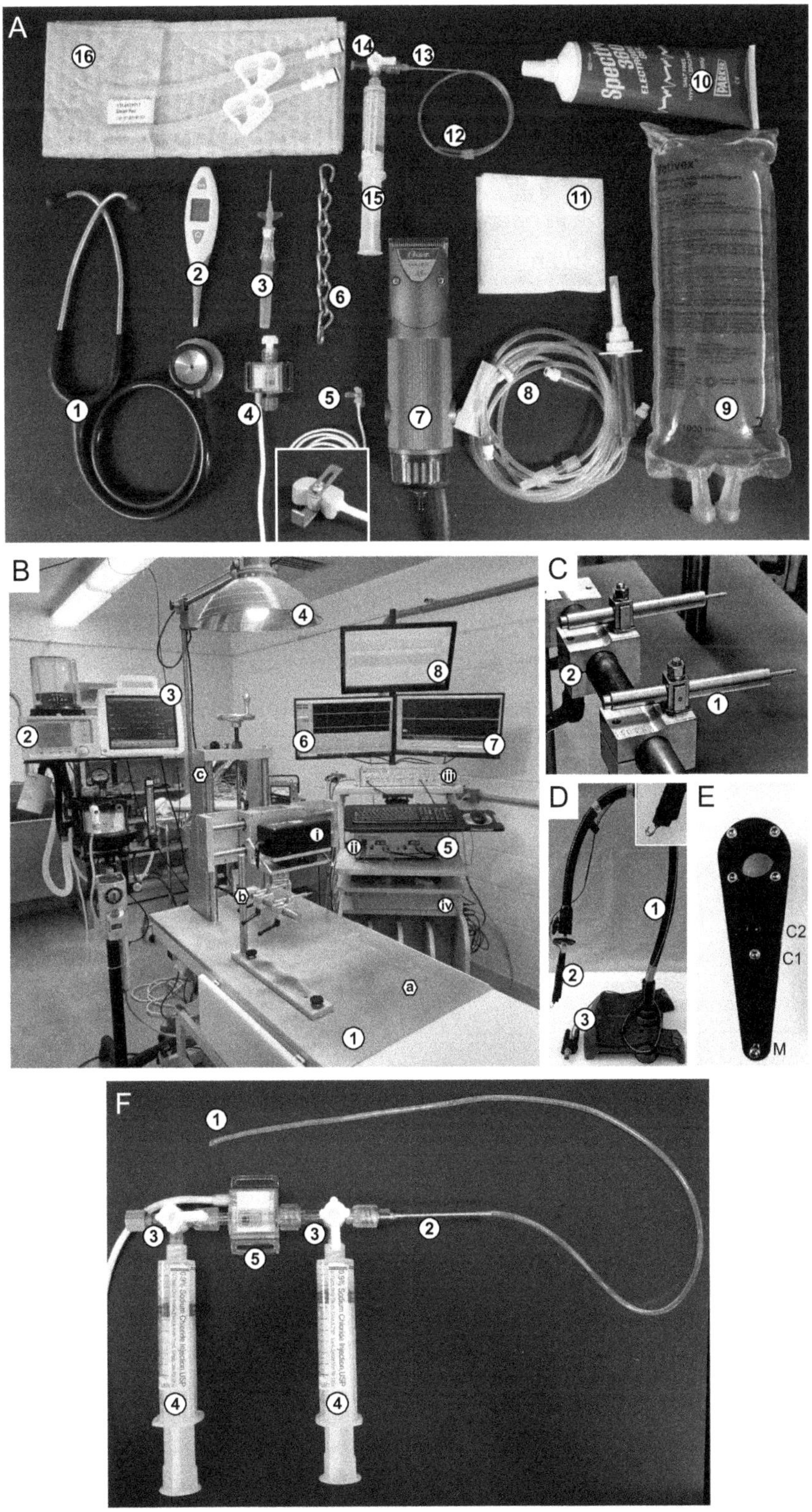

Fig. 3 Experiment setting to measure single muscle force. (**a**) Materials for the single muscle force assay. (1) Stethoscope, (2) digital thermometer, (3) shielded I.V. catheter, (4) MLT0669 blood pressure transducer,

9. Dynamic Muscle Control (DMC) software (Aurora Scientific Inc., Aurora, ON, Canada). The software controls the force-length transducer and stimulator and allows for real-time acquisition of the force and length data.
10. Dynamic muscle control data analysis (DMA) software (Aurora Scientific Inc., Aurora, ON, Canada). The software analyzes the contractile data acquired by the DMC software.
11. 3PS transonic flow probe (Transonic Systems Inc., Ithaca, NY, USA).
12. TS420 perivascular flowmeter module (Transonic Systems Inc., Ithaca, NY, USA).
13. MLT0669 blood pressure transducer (ADInstruments Inc., Colorado Springs, CO, USA).
14. FE221 blood pressure amplifier module (ADInstruments Inc., Colorado Springs, CO, USA).
15. PowerLab (ADInstruments Inc., Colorado Springs, CO, USA).
16. Lab Chart software (ADInstruments Inc., Colorado Springs, CO, USA).
17. Animal hair clipper with a #50 blade (Oster Professional Products, Sunbeam Products, Inc., Boca Raton, FL, USA).
18. Acepromazine (acepromazine maleate) (MWI, Boise, ID, USA).
19. Torbugesic (butorphanol tartrate) (Zoetis Inc., Kalamazoo, MI, USA).
20. Atropine sulfate (MWI, Boise, ID, USA).

Fig. 3 (continued) (5) 3PS transonic flow probe, enlarged in the insert, (6) stainless steel chain, (7) animal hair clipper with a #50 blade, (8) SmartSite gravity set, (9) lactated Ringer's buffer, (10) electrode gel, (11) non-woven gauze sponges, (12) medical-grade silicone tube, (13) stainless steel blunt tip Luer lock needle, (14) three-way stopcock with a male Luer lock, (15) BD PosiFlush pre-filed saline, and (16) conductive blanket. (**b**) Custom-made muscle force assay platform. (1) Custom-made muscle force assay platform, consisting of (a) the main platform, (b) dual-axis mount, and (c) tri-axis mount; (2) ventilator; (3) veterinary vital sign monitor; (4) heat lamp; (5) in situ muscle function assay system, consisting of (i) 310C-LR dual-mode transducer that measures/controls the muscle force and length, (ii) 310C-LR force transducer controller module, (iii) 701A stimulator, and (vi) 604A signal interface; (6) Dynamic Muscle Control software; (7) dynamic muscle control data analysis software; and (8) Lab Chart software. (**c**) A close view of the dual-axis mounts. (1) Custom-made stainless-steel bone pins and (2) stainless-steel bone pin mounts. (**d**) Custom-made bipolar platinum electrodes. (1) Gooseneck holder, (2) bipolar electrode, enlarged in the insert, and (3) BNC connector. (**e**) Custom-made lever arm for the force transducer with three muscle attachment sites: M, C1, and C2. (**f**) Carotid artery catheter assembly. (1) Medical-grade silicone tube, (2) stainless steel blunt tip Luer lock needle, (3) three-way stopcock with male Luer lock, (4) BD PosiFlush pre-filed saline, and (5) MLT0669 blood pressure transducer

21. Shielded I.V. catheter (20G × 1.00 in. needle) (Becton, Dickinson and Company, Franklin Lakes, NJ, USA).
22. Diprivan (propofol) (Fresenius Kabi, Lake Zurich, IL, USA).
23. Conductive blankets (Adroit Medical Systems Inc., Loudon, TN, USA) connected to a heated circulating water bath (ThermoFisher Scientific, Hampton, NH, USA).
24. Fluriso (Isoflurane) (MWI, Boise, ID, USA).
25. Ventilator (DRE Veterinary, Louisville, KY, USA).
26. Veterinary vital sign monitor (DRE Veterinary, Louisville, KY, USA).
27. Lactated Ringer's buffer (Baxter, Deerfield, IL, USA).
28. SmartSite gravity set (CareFusion, San Diego, CA, USA).
29. 70% isopropyl alcohol (Walmart Stores Inc., Bentonville, AR, USA).
30. Non-woven gauze sponges (4 inch × 4 inch) (Medline Industries Inc., Northfield, IL, USA).
31. Surgical instruments (Fine science tools, Inc., Foster City, CA, USA).
32. Medical-grade silicone tube (OD: 1.95 mm, and ID: 1.4 mm) (Dow Corning Corp., Midland, MI, USA).
33. Stainless Steel Blunt Tip Luer Lock needle (15 g) (ThermoFisher Scientific, Hampton, NH, USA).
34. Three-Way Stopcock with Male Luer Lock (ThermoFisher Scientific, Hampton, NH, USA).
35. BD PosiFlush pre-filed saline (0.9% sodium chloride) syringe for intravenous (IV) injection (Becton, Dickinson and Company, Franklin Lakes, NJ, USA).
36. Braided silk suture (Surgical Specialties Corp., Wyomissing, PA, USA).
37. Electrode gel (Parker Laboratories Inc., Fairfield, NJ, USA).
38. Soft tape measure (Walmart Stores Inc., Bentonville, AR, USA).
39. Stainless steel chain (Lowes, Mooresville, NC, USA).
40. Temperature probe (Yellow Springs Instrument Inc., Yellow Springs, OH, USA).
41. Saran wrap (Walmart Stores Inc., Bentonville, AR, USA).
42. Heating lamp (Tensor Lighting Company, Boston, MA, USA).
43. Custom-made bipolar platinum electrodes (*see* Fig. 3d) (*see* **Note 15**).

2.2.2 Hindlimb Muscle Torque Measurement (See Fig. 4A)

1. PPE including disposable gowns and/or dedicated scrubs, a head cover, shoe covers, a surgical mask, and gloves.
2. Custom-made muscle force assay platform (University of Missouri Physics Machine Shop, Columbia, MO, USA). The platform consists of the main platform and a tri-axis mount (*see* Fig. 3b) (*see* **Note 26**).
3. Custom-made forelimb stereotactic frame. The frame consists of three components: a base, a hip joint mount, and a stifle joint mount (*see* Fig. 4B) (*see* **Note 27**).
4. In situ muscle function assay system (Aurora Scientific Inc., Aurora, ON, Canada). The system consists of a 310C-LR Dual-Mode transducer that measures/controls the muscle force and length, a 310C-LR force transducer controller module, a 701A stimulator, and a 604A signal interface.
5. Dynamic Muscle Control (DMC) software (Aurora Scientific Inc., Aurora, ON, Canada). The software controls the force-length transducer and stimulator and allows for real-time acquisition of the force and length data.
6. Dynamic muscle control data analysis (DMA) software (Aurora Scientific Inc., Aurora, ON, Canada). The software analyzes the contractile data acquired by the DMC software.
7. Stethoscope (3M™ Littmann® Stethoscopes, St. Paul, MN, USA).
8. Digital thermometer (ReliOn Inc., Spokane Valley, WA, USA).
9. Animal hair clipper with a #50 blade (Oster Professional Products, Sunbeam Products, Inc., Boca Raton, FL, USA).
10. Acepromazine (acepromazine maleate) (MWI, Boise, ID, USA).
11. Torbugesic (butorphanol tartrate) (Zoetis Inc., Kalamazoo, MI, USA).
12. Atropine sulfate (MWI, Boise, ID, USA).
13. Disposable sterile syringe (Becton, Dickinson and Company, Franklin Lakes, NJ, USA).
14. Shielded I.V. catheter (20G × 1.00 in. needle) (Becton, Dickinson and Company, Franklin Lakes, NJ, USA).
15. Diprivan (propofol) (Fresenius Kabi, Lake Zurich, IL, USA).
16. Conductive blankets (Adroit Medical Systems Inc, Loudon, TN, USA) connected to a heated circulating water bath (ThermoFisher Scientific, Hampton, NH, USA).
17. Syringe infusion pump (Harvard Apparatus, Holliston, MA, USA).
18. Ventilator (DRE Veterinary, Louisville, KY, USA).

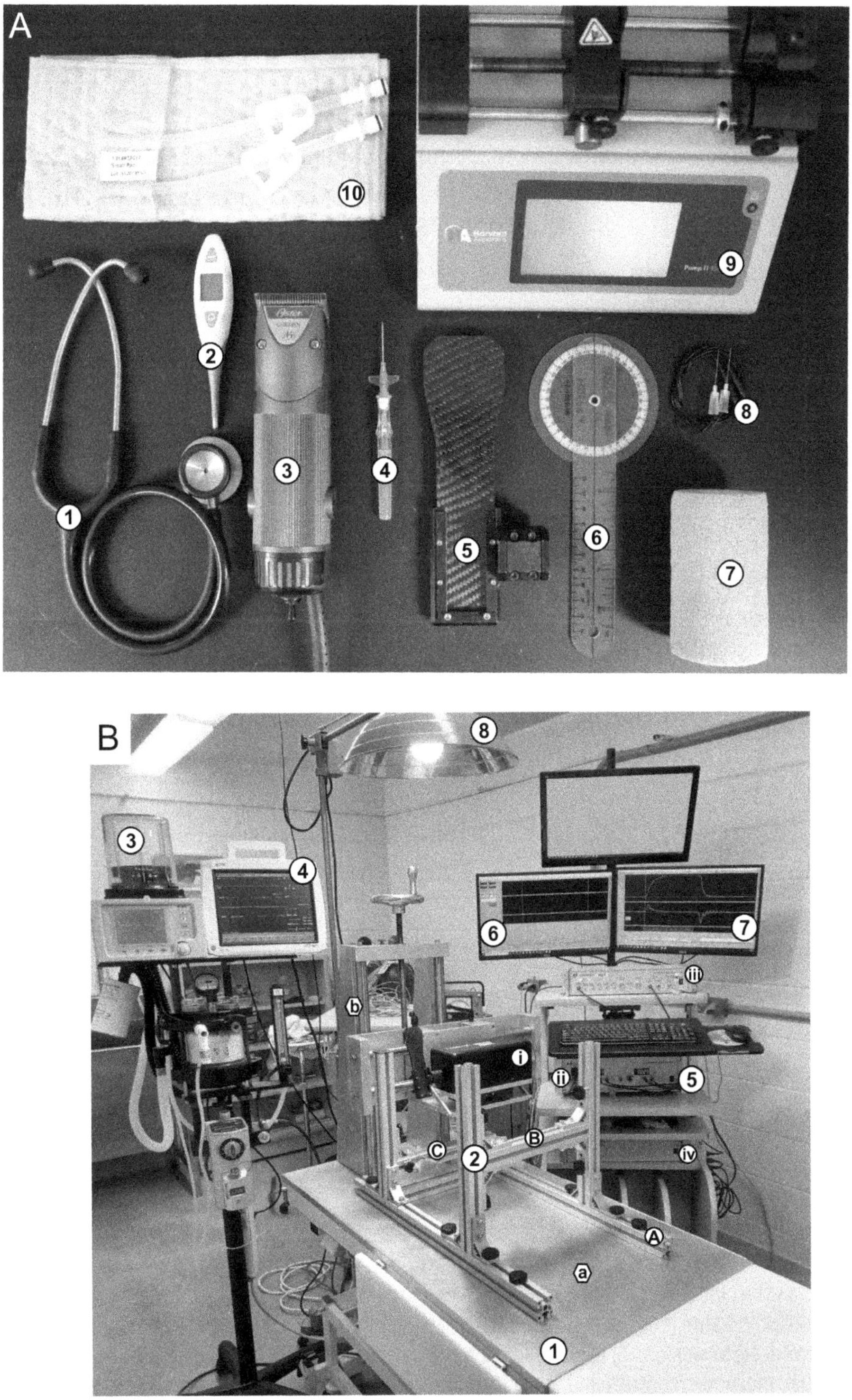

Fig. 4 Experiment setting for hindlimb torque measurement. **A.** Materials needed. (1) Stethoscope, (2) digital thermometer, (3) animal hair clipper with a #50 blade, (4) shielded I.V. catheter, (5) custom-made hindlimb paw pedal, (6) goniometer, (7) 3M Vetrap^{TM}, (8) monopolar needle electrodes, (9) syringe infusion pump, and

19. Veterinary vital sign monitor (DRE Veterinary, Louisville, KY, USA).
20. Soft tape measure (Walmart Stores Inc., Bentonville, AR, USA).
21. Custom-made hindlimb paw pedal (*see* **Note 28**).
22. 3M Vetrap™ (2 inch × 5 yard) (3M™ Animal Care Products, St. Paul, MN, USA).
23. Goniometer (Medline Industries Inc., Northfield, IL, USA).
24. 70% isopropyl alcohol (Walmart Stores Inc., Bentonville, AR, USA).
25. Non-woven gauze sponges (4 inch × 4 inch) (Medline Industries Inc., Northfield, IL, USA).
26. Monopolar needle electrodes 25 mm × 27 ga (Consolidated Neuro Supply, Inc., Milford, OH, USA).
27. Medical adhesive tape (ThermoFisher Scientific, Hampton, NH, USA).

2.3 Evaluation of Heart Function

2.3.1 Electrocardiography (ECG) (See Fig. 6)

1. PPE including disposable gowns and/or dedicated scrubs, a head cover, shoe covers, a surgical mask, and gloves.
2. Stethoscope (3M™ Littmann® Stethoscopes, St. Paul, MN, USA).
3. Digital thermometer (ReliOn Inc., Spokane Valley, WA, USA).
4. CardioTouch 3000 electrocardiography machine (Bionet Company, Seoul, South Korea).
5. ECG color-coded wires (Bionet Company, Seoul, South Korea).
6. Electrocardiography paper (Medigraphics, Lexington, KY, USA).
7. Padded blanket (Standard Textile, Cincinnati, OH, USA).
8. 70% isopropyl alcohol (Walmart Stores Inc., Bentonville, AR, USA).
9. Spray bottle (Walmart Stores Inc., Bentonville, AR, USA).

Fig. 4 (continued) (10) conductive blankets. **B.** Custom-made muscle force assay platform. (1) Custom-made muscle force assay platform, consisting of (a) the main platform and (b) tri-axis mount; (2) custom-made forelimb stereotactic frame, consisting of (A) base, (B) hip joint mount, and (C) stifle joint mount; (3) ventilator; (4) veterinary vital sign monitor; (5) in situ muscle function assay system, consisting of (i) 310C-LR dual-mode transducer that measures/controls the muscle force and length, (ii) 310C-LR force transducer controller module, (iii) 701A stimulator, and (vi) 604A signal interface; (6) Dynamic Muscle Control software; (7) dynamic muscle control data analysis software; and (8) heat lamp

10. Alligator clips (Mueller Electric Company, Cleveland, OH, USA).
11. Bic Pen (150 mm) (Staples, Framingham, MA, USA).
12. ECG calipers (Elite Medical Instruments, Orange County, CA, USA).

2.3.2 Holter Monitoring (See Fig. 7)

1. Forest Medical Trillium 5900 Holter monitor (Forest Medical LLC., Syracuse, NY, USA).
2. Five ECG color-coded electrode cable (Forest Medical LLC., Syracuse, NY, USA).
3. Animal hair clipper with a #50 blade (Oster Professional Products, Sunbeam Products, Inc., Boca Raton, FL, USA).
4. 70% isopropyl alcohol (Walmart Stores Inc., Bentonville, AR, USA).
5. Non-woven gauze sponges (4 inch × 4 inch) (Medline Industries Inc., Northfield, IL, USA).
6. ECG Electrode Pads (Medline Industries Inc., Northfield, IL, USA).
7. Spectra 360 Electrode Gel (Parker Laboratories Inc., Fairfield, NJ, USA).
8. Cast padding wrap (4 inches × 4 yards) (BSN Medical Inc., Rutherford College, NC, USA).
9. 3M Vetrap™ (4 inch × 5 yard) (3M™ Animal Care Products, St. Paul, MN, USA).
10. Holter vest (Nightrider.com performance products, division of S&P Mullen Enterprises, Inc., Oldsmar, FL, USA).
11. Trillium Platinum Vet software (Forest Medical LLC., Syracuse, NY, USA).
12. Data transfer cable (Forest Medical LLC., Syracuse, NY, USA).

2.3.3 Echocardiography (See Fig. 8)

1. Custom-made echocardiographic table and pad or padded tabletop ultrasound exam table (Avante Animal Health, Louisville, KY, USA).
2. Mobile lift table (Avante Animal Health, Louisville, KY, USA).
3. Aquasonic 100 ultrasound transmission gel (Parker Laboratories Inc., Fairfield, NJ, USA).
4. M5Sc-D and 6S-D sector (phased) array transducer probes (GE Medical Systems, Waukesha, WI, USA).
5. GE Vivid E90 ultrasonographic system (GE Medical Systems, Waukesha, WI, USA).
6. GE Echo picture archiving and communication system (PACS) software (GE Medical Systems, Waukesha, WI, USA).

3 Methods

3.1 Evaluation of Bodywide Motor Function

3.1.1 Overnight Activity Recording

The following protocol describes a video capturing/processing system to quantify dog mobility at night.

1. Follow the institutional guidelines for animal care and handling. Obtain approval from the institutional animal care and use committee (ACUC). Wear all PPE before working or handling animals.
2. Secure the CCD camera and the NIR light source to the custom-made mounting bracket and attach the bracket to the ceiling above the custom-made aluminum kennel. Position the camera and light source at the center of the kennel at a height to acquire a full view and provide even illumination of the cage floor (*see* Fig. 1) (*see* **Note 1**).
3. Perform a physical exam, and record the vital signs (heart rate, respiratory rate and effort, and rectal body temperature), capillary refill time (CRT), and mucous membrane color.
4. Place the dog in the kennel 8 h before the recording time to acclimate the dog to the housing environment (*see* **Note 4**).
5. Turn on the camera, light source, and computer. Load the ODR software and set the recording time for 12 h from 6 p.m. to 6 a.m. the next day (*see* **Note 5**).
6. Return the dog to the housing kennel and analyze the data using the ODVA software (*see* **Note 6**).

3.1.2 Gait Analysis

1. Follow the institutional guidelines for animal care and handling. Obtain approval from the institutional ACUC. Wear all PPE before working or handling animals.
2. Set up the GAIT4 Dog® portable walkway mat on flat ground and connect it to the laptop computer through a USB cable (*see* Fig. 2). Load the GAITFour software and confirm that the walkway is linked to the software. Set the webcam to overlook the walkway (*see* Fig. 2). Enter the dog information into the software (*see* **Note 8**).
3. Perform a physical exam, and record the vital signs (heart rate, respiratory rate and effort, and rectal body temperature), CRT, and mucous membrane color.
4. Measure the dog's height (from the ground to the withers) and body weight. Record both measurements into the software. Set the scan rate and the sensor size, according to the body weight, in the software (*see* **Note 9**).
5. Select the appropriate harness size according to the dog's body size and put it on. Connect the leash to the harness. Allow the dog to walk back and forth on the walkway mat to get acclimated with the test (*see* **Note 10**).

6. Record the dog's gait while allowing the dog to walk from one end to the other end of the mat at a steady gait speed. Repeat the gait measurement until achieving five constant recordings. If needed, allow the dog to rest between recordings (*see* **Note 11**).
7. Perform a second physical exam, and record the vital signs, CRT, and mucous membrane color before returning the dog to the kennel.
8. Load the recorded gait into the analysis module of the GAIT-Four software, select the analysis parameters of interest, and analyze the recording (*see* **Note 12**).

3.2 Muscle Force Assessment

The following protocol describes measuring the contractile force of the left extensor carpi ulnaris (ECU) muscle (*see* **Note 16**).

3.2.1 Single Muscle Force Measurement

1. Follow the institutional guidelines for animal care and handling. Obtain approval from the institutional ACUC. Wear all PPE before working or handling animals. Use aseptic techniques while performing the surgery.
2. Fast the dog the night before the procedure.
3. Perform a physical exam, and record the vital signs (heart rate, respiratory rate and effort, and rectal body temperature), CRT, and mucous membrane color.
4. Prepare the first pre-anesthetic cocktail according to the body weight. First, withdraw the calculated dosage of Cerenia (1–5 mg/kg) into a syringe, then withdraw the calculated dosage of Rimadyl (4.4 mg/kg) into a separate syringe. Next, add the withdrawn volume of Rimadyl into the Cerenia syringe and gently invert to mix both agents. Inject the cocktail subcutaneously on the dorsal aspect of the dog's back 2 h before starting the procedure.
5. Check all the components of the all-in-one muscle force system and calibrate the force transducer (*see* **Note 17**).
6. Connect the 3PS transonic flow probe to the TS420 perivascular flowmeter module and the MLT0669 blood pressure transducer to the FE221 blood pressure amplifier module. Then, connect both modules to a PowerLab and calibrate the flow probe and blood pressure transducer.
7. Perform a second physical exam 2 h after the first pre-anesthetic cocktail injection and record the vital signs, CRT, and mucous membrane color (*see* **Note 18**).
8. Restrain the animal gently and shave the body hair from the medial to the lateral side of the right neck and the lateral and medial sides of the right forelimb.

9. Prepare the second pre-anesthetic cocktail according to the body weight. First, withdraw the calculated dosage of acepromazine (0.02 mg/kg), butorphanol (0.4 mg/kg), and atropine (0.04 mg/kg) into individual syringes. Mix all the withdrawn agents into an appropriately sized syringe and invert gently to mix. Inject the cocktail intramuscularly into the epaxial muscle (*see* **Note 19**).
10. Place an intravenous (IV) catheter into the saphenous vein. Induce the dog through intravenous injection of propofol (6 mg/kg) and intubate immediately. Position the dog in dorsal recumbency on the main platform of a custom-made muscle force assay platform. Place a conductive blanket underneath the animal and another on top of the animal to maintain body temperature at 37 °C throughout the experiment.
11. Maintain the anesthesia with 2–4% isoflurane while mechanically ventilating the animal throughout the experiment. Monitor the heart rate, electrocardiograph, oxygen saturation (SpO_2), CO_2, blood pressure, and body temperature using the veterinary vital sign monitor throughout the experiment. Record vital signs, CRT, mucous membrane color, the palpebral reflex, and the pedal reflex every 15 min. Infuse Lactated Ringer's buffer intravenously at 4 mL/kg/h using the Saphenous vein catheter.
12. Disinfect the neck skin with a 70% isopropanol soaked gauze and surgically expose a 4–6 cm segment of the right carotid artery. Insert a silicone tube in the artery and advance it to the thoracic aorta for measuring the central blood pressure (*see* Fig. 3f). Secure the tube to the artery using a 1-0 braided silk suture and close the skin incision site with a 2-0 braided silk suture (*see* **Note 20**).
13. Position the dog on the left lateral side. Disinfect the medial skin above the elbow at the left forelimb with 70% isopropanol-soaked gauze and surgically expose the brachial artery. Place the 3PS transonic flow probe around the brachial artery to measure blood flow. Fill the space between the artery and the probe with electrode gel, and close the incision site with a 2-0 braided silk suture.
14. Position the dog on the right lateral side and extend the left forelimb. Disinfect the lateral skin of the left forelimb with 70% isopropanol-soaked gauze. Make an incision from the elbow joint to the carpal joint to expose the ECU muscle. Measure the total length of the entire ECU muscle, including the tendon, from the proximal tendon insertion at the medial epicondyle of the humerus to the distal tendon insertion at the carpus. Determine the optimal fiber length (Lf) using the following equation ($Lf = \text{total length} - (\text{total length} \times 16\%)$) (*see* **Note 21**).

15. Extend the forearm incision proximately to expose the triceps brachii lateral muscle. Retract the triceps brachii lateral muscle and the triceps brachii accessory caudally. Dissect the radial nerve between the triceps brachii accessory and the brachialis muscle. Tie a 1-0 braided silk suture around the radial nerve and then cut the nerve close to the collateral radial artery.
16. Attach two stainless-steel bone pins to the forelimb, one in the olecranon and the other in the radius bone close to the styloid process.
17. Position the dog in the dorsal recumbency position. Secure the bone pins in the dual-axis horizontal mounts and adjust the height of the mounts vertically to match the animal's height. Cut the distal ECU tendon at the carpus bone insertion and attach a stainless-steel chain to the tendon using a size 0 Vicryl suture. Secure the chain to the lever arm and align the force transducer according to the forelimb position using the tri-axes mount (*see* Fig. 3e) (*see* **Note 22**).
18. Place a temperature probe between the ECU muscle and the radius bone to monitor the muscle temperature throughout the experiment. Cover the exposed ECU muscle and tendon with a warm wet saline gauze and then with Saran wrap to avoid moisture evaporation. Position a heat lamp 30–38 cm above the forelimb to maintain the ECU muscle temperature at 37 °C throughout the experiment. Secure the distal end of the radial nerve on a bipolar electrode (*see* **Note 23**).
19. Set the resting tension of the ECU muscle to 200 g using the translational knob of the X-axis sub-component in the tri-axis mount. Connect the blood pressure transducer to the carotid artery tube, and then connect the transducer to the blood pressure amplifier module. Connect the flow probe to the perivascular flowmeter module. Connect the blood pressure amplifier and the perivascular flowmeter modules to the Power lab. Monitor the blood flow and arterial blood pressure with the PowerLab data acquisition software (LabChart) throughout the experiment.
20. Once the ECU muscle temperature reaches 37 °C, stimulate the ECU muscle three times at 150 Hz and 10 mA for 200 msec with a 1-min rest between each stimulation to warm up the muscle [10, 11]. Load the DMA software and analyze the muscle force (*see* **Note 24**).
21. Rest the muscle for 2 min. Stimulate the ECU muscle at 60 Hz and 10 mA for 500 msec, and then analyze the muscle force. Keep repeating the muscle stimulation while gradually increasing the resting tension by 100 g until the muscle force reaches the highest value and any additional increase in the resting

tension results in a force reduction. Rest the muscle for 1 min between each stimulation. The resting tension that yielded the maximal force is considered the optimal resting tension.

22. Rest the ECU muscle for 2 min. Set the muscle at optimal resting tension and stimulate the muscle at 60 Hz and 10 mA for 500 msec, and then analyze the muscle force. Keep repeating the muscle stimulation while gradually increasing the current by 10 mA until the muscle force reaches the highest value and any additional increase in the current results in a force reduction. Rest the muscle for 1 min between each stimulation. The current that yielded the maximal force is the optimal current.
23. Rest the ECU muscle for 2 min. Set the muscle at optimal resting tension and stimulate the muscle at optimal current and 100 Hz for 500 msec. Then, analyze the force. Keep repeating the muscle stimulation while gradually increasing the stimulation duration by 100 msec until the muscle force reaches the highest value and any additional increase in the stimulation time results in a force reduction. Rest the muscle for 1 min between each stimulation. The time that yielded the maximal force is the optimal stimulation duration.
24. Rest the ECU muscle for 2 min. Set the muscle at optimal resting tension and stimulate the muscle at optimal current and stimulation duration at different frequencies (5, 20, 40, 60, 80, 100, and 120 Hz). Rest the muscle for 1 min between each stimulation. Analyze the muscle force. The frequency that yielded the maximal force is the optimal stimulation. Record the maximal muscle force as the absolute tetanic force (Po).
25. Rest the muscle for 2 min. Set the muscle at optimal resting tension and stimulate the muscle with 1 Hz. Analyze the muscle force and record the force as the absolute twitch force (Pt).
26. Rest the muscle for 2 min. Set the muscle at optimal resting tension and apply 10 repetitive cycles of eccentric contraction. At each cycle, stimulate the muscle at optimal current and frequency for the optimal stimulation time plus 1 s. During the last second of stimulation, stretch the muscle by 5% of muscle length at a speed of 5% muscle length per second. Rest the muscle for 1 min between each cycle and adjust the resting tension to optimal. Calculate the force drop after eccentric contraction with the formula: Force drop % = (F1 – Fn)/F1. F1 is the tetanic force obtained during the first cycle. Fn is the tetanic force obtained during the nth cycle.
27. At the end of the experiment, euthanize the animal and carefully dissect the ECU muscle. Measure the muscle weight and

calculate the PCSA according to the following equation: (muscle weight in gram × cos10.03°)/(1.056 g/cm^3 × Lf in cm) (*see* **Note 25**).

28. The force value obtained from the DMA program represents the absolute force. Calculate the specific muscle force by dividing the absolute force with the PCSA.
29. Determine the time to peak tension (TPT), half relaxation time (1/2 RT), maximum rate of muscle contraction (max df/dt), maximum rate of muscle relaxation (max -df/dt), time to max df/dt, and time to max -df/dt for the Pt and Po using the DMA software.

3.2.2 Hindlimb Torque Measurement

The following protocol describes measuring the contractile force of the flexor and extensor muscle group of the right hindlimb (*see* **Note 29**).

1. Follow the institutional guidelines for animal care and handling. Obtain approval from the institutional ACUC. Wear all PPE before working or handling animals. Use aseptic techniques while performing the surgery.
2. Fast the dog the night before the procedure.
3. Secure the custom-made hindlimb stereotactic frame to the main platform of the custom-made muscle force assay platform. Turn on all the equipment for the in situ muscle function assay system and calibrate the force transducer.
4. Perform a physical exam, and record the vital signs (heart rate, respiratory rate and effort, and rectal body temperature), CRT, and mucous membrane color.
5. Restrain the animal gently and shave the body hair from the stifle joint to the tibiotarsal joint at the lateral and medial sides of the right hindlimb. Measure the body weight and lower limb volume (*see* **Note 30**).
6. Prepare the pre-anesthetic cocktail according to the body weight. First, withdraw the calculated dosage of acepromazine (0.02 mg/kg), butorphanol (0.4 mg/kg), and atropine (0.04 mg/kg) into individual syringes. Mix all the withdrawn agents into an appropriately sized syringe and invert gently to mix. Inject the cocktail intramuscularly into the epaxial muscle (*see* **Note 19**).
7. Place an IV catheter into the saphenous vein. Induce the dog through intravenous injection of propofol (6 mg/kg) and intubate immediately. Position the dog in dorsal recumbency on the main platform of a custom-made muscle force assay platform. Place a conductive blanket underneath the animal and another on top of the animal to maintain body temperature at 37 °C throughout the experiment.

8. Maintain the anesthesia with a constant rate infusion of 0.1–0.4 mg/kg/min propofol using an infusion pump while mechanically ventilating the animal throughout the experiment. Monitor the heart rate, electrocardiograph, oxygen saturation (SpO_2), CO_2, blood pressure, and body temperature using the veterinary vital sign monitor throughout the experiment. Record vital signs, CRT, mucous membrane color, the palpebral reflex, and the pedal reflex every 15 min (*see* **Note 31**).
9. Measure and record the hindlimb paw length from the center axis of rotation of the tarsal joint to the distal tip of the longest digit.
10. Align the center axis of rotation of the tarsal joint with the axis of rotation of the custom-made hindlimb paw pedal and secure the metatarsal and the digits to the pedal plate using Vetrap™. Insert the axis of rotation of the pedal into the force transducer shaft. Position and secure the stifle and hip joint of the right hindlimb to the stereotactic frame such that the stifle joint is less than 90° angle (*see* Fig. 5B). Secure the pedal to the force transducer shaft such that the tarsal joint is at a 100° angle (*see* Fig. 5B) (*see* **Note 32**).
11. Disinfect the lateral skin at the stifle joint with 70% isopropanol-soaked gauze. Palpate and identify the common peroneal nerve site and then carefully place a pair of monopolar needle electrodes subcutaneously and position them to be laid on top of the common peroneal nerve (*see* Fig. 5B).
12. Stimulate the hindlimb tibiotarsal flexion muscle group repeatedly at 1 Hz using 10 mA. Adjust the position of the needles to obtain the maximum twitch force. Continue the stimulation while gradually increasing the current by 10 mA until the muscle force reaches the highest value, and any additional increase in the current results in a force reduction (*see* **Note 33**).
13. Rest for 1 min. Stimulate the hindlimb tibiotarsal flexion muscle group three times at 60 Hz and optimal current for 500 msec with a 1-min rest between each stimulation to warm up the muscle. Load the DMA software and analyze the muscle force.
14. Rest for 2 min. Stimulate the hindlimb tibiotarsal flexion muscle group at 80 Hz and optimal current for 500 msec. Analyze the force. Repeat the stimulation while gradually increasing the stimulation duration by 100 msec until the force reaches the highest value, and any additional increase in the stimulation time results in a force reduction. Rest the muscle for 1 min between each stimulation. The time that yielded the maximal force is the optimal stimulation duration.

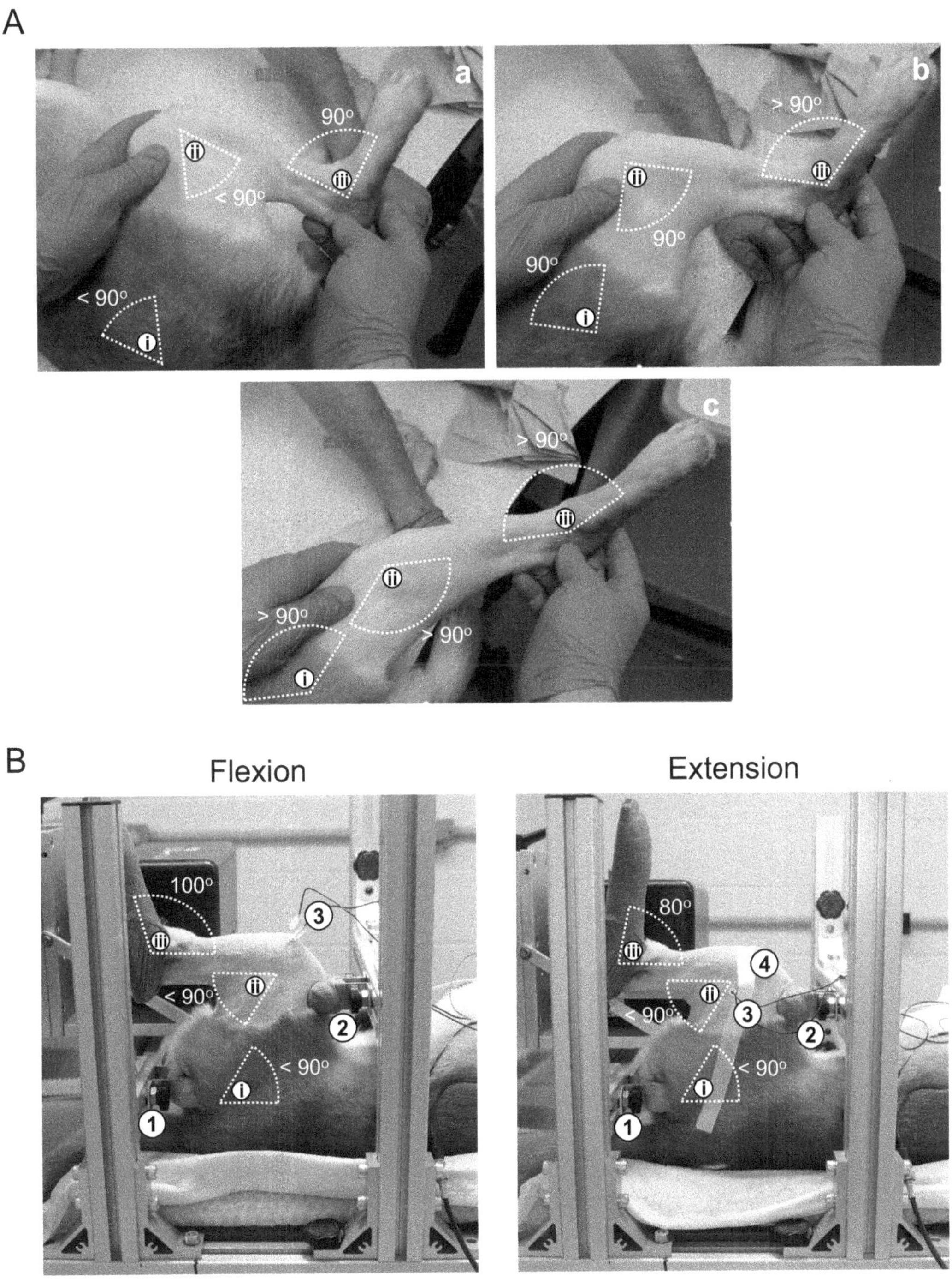

Fig. 5 Positioning the dog for hindlimb torque measurement. **A.** a–c, Representative images show how the hip and stifle angle change influence the tibiotarsal angle. (i) hip angle, (ii) stifle angle, and (iii) tibiotarsal angle. a. When the hip and stifle joints are set at less than a 90° angle, the tibiotarsal joint is extended to a 90° angle. b. When the hip and stifle joints are set at a 90° angle, the tibiotarsal joint is extended to more than a 90° angle. c. When the hip and stifle joints are fully extended (>90° angle), the tibiotarsal joint is fully extended. **B**. Experimental setup for the flexion and extension torque measurement. (1) Hip joint mount, (2) stifle joint mount, (3) monopolar needle electrodes, and (4) medical adhesive tape. (i) Hip angle, (ii) stifle angle, and (iii) tibiotarsal angle

15. Rest for 2 min. Stimulate the hindlimb tibiotarsal flexion muscle group at optimal current and stimulation duration using different frequencies (10, 20, 40, 60, 80, 100, and 120 Hz). Rest the muscle for 1 min between each stimulation. Analyze the force. The frequency that yielded the maximal force is the optimal stimulation. Record the maximal muscle force as the absolute tetanic force (Po).
16. Rest for 2 min. Stimulate the hindlimb tibiotarsal flexion muscle group with 1 Hz at optimal current. Analyze the force and record it as the absolute twitch torque.
17. Rest the muscle for 2 min. Apply 10 repetitive cycles of eccentric contraction with 5 s of rest between each cycle. At each cycle, Stimulate the forelimb muscle group at optimal current and frequency for the optimal stimulation time plus 200 msec. During the last 200 msec of stimulation, stretch the muscle group by rotating the pedal 29° to extend the paw away from the body. Repeat the 10 eccentric contraction cycles two times with 4 min of rest between each group of cycles. Calculate the force drop after eccentric contraction with the formula: Force drop % = (F1 − Fn)/F1. F1 is the first tetanic force obtained during the first cycle of the first eccentric contraction group. Fn is the tetanic force obtained during the nth cycle.
18. Remove the monopolar needle electrodes and change the tarsal joint angle to 80° (*see* Fig. 5) (*see* **Note 34**).
19. Insert the monopolar needle electrodes at the distal aspect of the gastrocnemius muscle bellies to stimulate the tibial nerve (*see* Fig. 5B).
20. Stimulate the hindlimb tibiotarsal extension muscle group repeatedly at 1 Hz using 10 mA. Adjust the position of the needles to obtain the maximum twitch force. Continue the stimulation while gradually increasing the current by 10 mA until the muscle force reaches the highest value, and any additional increase in the current results in a force reduction (*see* **Note 35**).
21. Rest for 1 min. Stimulate the hindlimb tibiotarsal extension muscle group three times at 60 Hz and optimal current for 500 msec with a 1-min rest between each stimulation to warm up the muscle. Load the DMA software and analyze the muscle force.
22. Rest for 2 min. Stimulate the hindlimb tibiotarsal extension muscle group at 80 Hz and optimal current for 500 msec. Analyze the force. Repeat the stimulation while gradually increasing the stimulation duration by 100 msec until the force reaches the highest value, and any additional increase in the stimulation time results in a force reduction. Rest the muscle for 1 min between each stimulation. The time that yielded the maximal force is the optimal stimulation duration.

23. Rest for 2 min. Stimulate the hindlimb tibiotarsal flexion muscle group at optimal current and stimulation duration using different frequencies (10, 20, 40, 60, 80, 100, and 120 Hz). Rest the muscle for 1 min between each stimulation. Analyze the force. The frequency that yielded the maximal force is the optimal stimulation. Record the maximal muscle force as the absolute tetanic torque.
24. At the end of the experiment, recover the dog from anesthesia and monitor the dog until it can maintain an unassisted sternal position. Return the dog to the kennel and follow with a post-operative assessment for three days to confirm no adverse effects from the procedure.
25. Calculate the absolute twitch and tetanic muscle force (Pt and Po) by dividing the absolute twitch and tetanic torque force over the paw length, respectively. Normalize the torque and the absolute force by the body weight and the hindlimb volume.

3.3 Heart Function Assessment

3.3.1 ECG

1. Follow the institutional guidelines for animal care and handling. Obtain approval from the institutional ACUC. Wear all PPE before working or handling animals.
2. Perform a physical exam, and record the vital signs (heart rate, respiratory rate, and rectal body temperature), CRT, and mucous membrane color.
3. Turn on the ECG machine and enter the dog information.
4. Place the dog on top of a padded blanket on the exam table (*see* **Note 36**).
5. Gently restrain the dog in right lateral recumbency with the head and neck in line with the body and the limbs perpendicular to the body (*see* Fig. 6a) (*see* **Note 37**).
6. Spray the proximal aspect of both the elbow and stifle joint on both sides with 70% isopropyl alcohol.
7. Connect the alligator clips to the end of each of the four ECG color-coded wires.
8. Clamp the alligator clip on the skin at the designated joint according to the color of each wire on the ECG cable ("white wire" in the right elbow, "green wire" in the right stifle, "black wire" in the left elbow, and "red wire" in the left stifle in Fig. 6a) (*see* **Note 38**).
9. Check the ECG trace on the system. Once the trace is stabilized with a clear baseline and without any artifact from movement, record a 10-s ECG trace at a paper speed of 25 mm/s with a 10 mm/mv sensitivity, and save the recording. Repeat the recording at a paper speed of 50 mm/s for 10 s and save it. Print the ECG trace for both recordings. The printout includes ECG traces from leads I, II, III, aVR, aVL, and aVF.

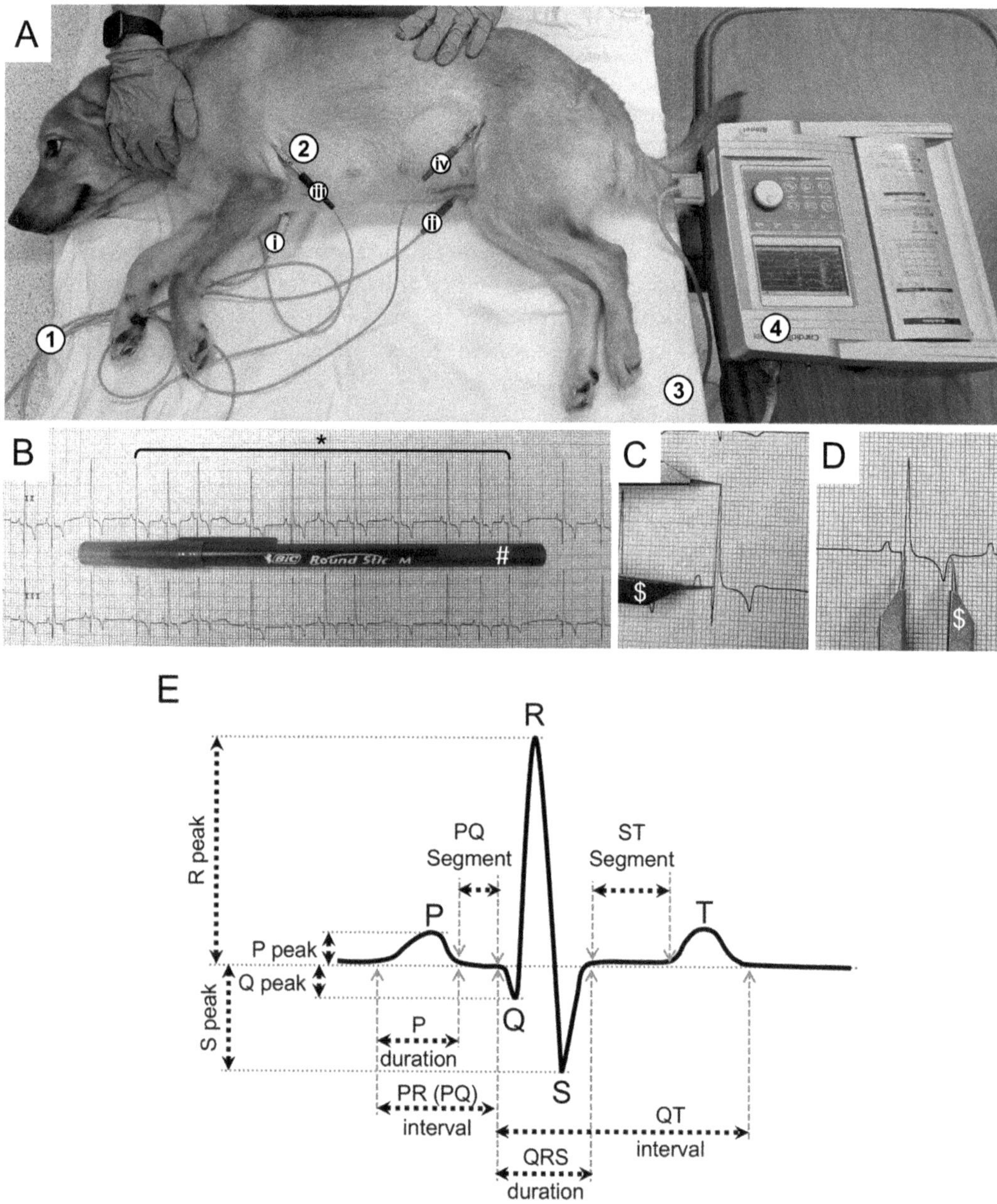

Fig. 6 ECG assay setup. (**a**) Dog position (a right lateral recumbency) and ECG electrode locations. (1) Color-coded ECG wires: (i) "white wire" clipped in the right elbow, (ii) "green wire" clipped in the right stifle, (iii) "black wire" clipped in the left elbow, and (iv) "red wire" clipped in the left stifle; (2) alligator clips; (3) padded blanket; and (4) CardioTouch 3000 electrocardiography machine. (**b**) Determination of the heart rate from lead II at the 25 mm/sec trace using a standard Bic pen. *, counted peaks to determine the heart rate. #, Standard Bic pen (150 mm). (**c**) An example of how to measure the amplitude of a peak using an ECG caliper. $, ECG caliper. (**d**) An example of how to measure the interval between two waves using an ECG caliper. $, ECG caliper. (**e**) A cartoon illustration of the ECG tracing showing how to determine the amplitude of the P, Q, R, S, and T peaks, the duration of the P and QRS waves, and the PR and QT intervals

10. Remove the alligator clips and return the dog to the kennel.
11. Lay a standard Bic pen (150 mm of length) on top of lead II at the 25 mm/s trace paper. Align the left end of the pen to any R peak on the trace. Count the number of R peaks within the length of the pen. Do not include the peak used for aligning the pen (*see* Fig. 6b). Multiply the total counted peaks by ten to determine the heart rate.
12. Measure the amplitude of the P, Q, R, S, and T peaks for all the leads (*see* Fig. 6c, e) and the wavelength of the P, P-R, QRS, and QT intervals (*see* Fig. 6d, e) for lead II from the 50 mm/s trace using an ECG caliper and according to the diagram presented in Fig. 6e (*see* **Note 39**).

3.3.2 Holter Monitoring

1. Connect the five ECG color-coded electrode cables to the Trillium Holter monitor. Power on the device and enter the dog information. Set the device to record for 24 h at a sampling rate of 256 Hz (*see* **Note 40**).
2. Shave the skin hair (2 inches × 2 inches) between the shoulder blades but slightly caudal (area #1), halfway between the spine and the sternum on the right lateral side (area #2), on the chest behind the elbow on the right ventral side (area#3), halfway between the spine and the sternum on the left lateral side (area #4), and on the chest behind the elbow on the left ventral side (area#5) (*see* Fig. 7B).
3. Clean the skin with 70% isopropanol-soaked gauze by wiping in a circular motion using a center-out pattern. Allow the skin to air dry or pat with dry clean gauze.
4. Disconnect the ECG electrode cable from the Holter, and attach an electrode pad to each end of the color-coded cable wires. Next, remove adhesive backing and add a pea-sized drop of the electrode gel to the center of the pad. Then, stick the pad to the designated shaved area on the skin according to the cable color (white wire in area #1, green wire in area #2, black wire in area #3, red wire in area #4, and brown wire in area #5) (*see* Fig. 7B) (*see* **Note 41**).
5. Extend the electrode cable to the left lateral aspect of the neck. Using cast padding, wrap around the circumference of the thorax twice to cover all the electrode pads (*see* Fig. 7B) (*see* **Note 42**).
6. Repeat the same cast pad wrapping pattern using Vetrap (*see* Fig. 7B) (*see* **Note 43**).
7. Apply the appropriately sized Holter vest onto the dog and feed the electrode cable through the pocket hole on the vest. Next, connect the cable to the Holter monitor and wrap any excess wires around the monitor. Then, place them into the vest pocket (*see* Fig. 7B) (*see* **Note 44**).

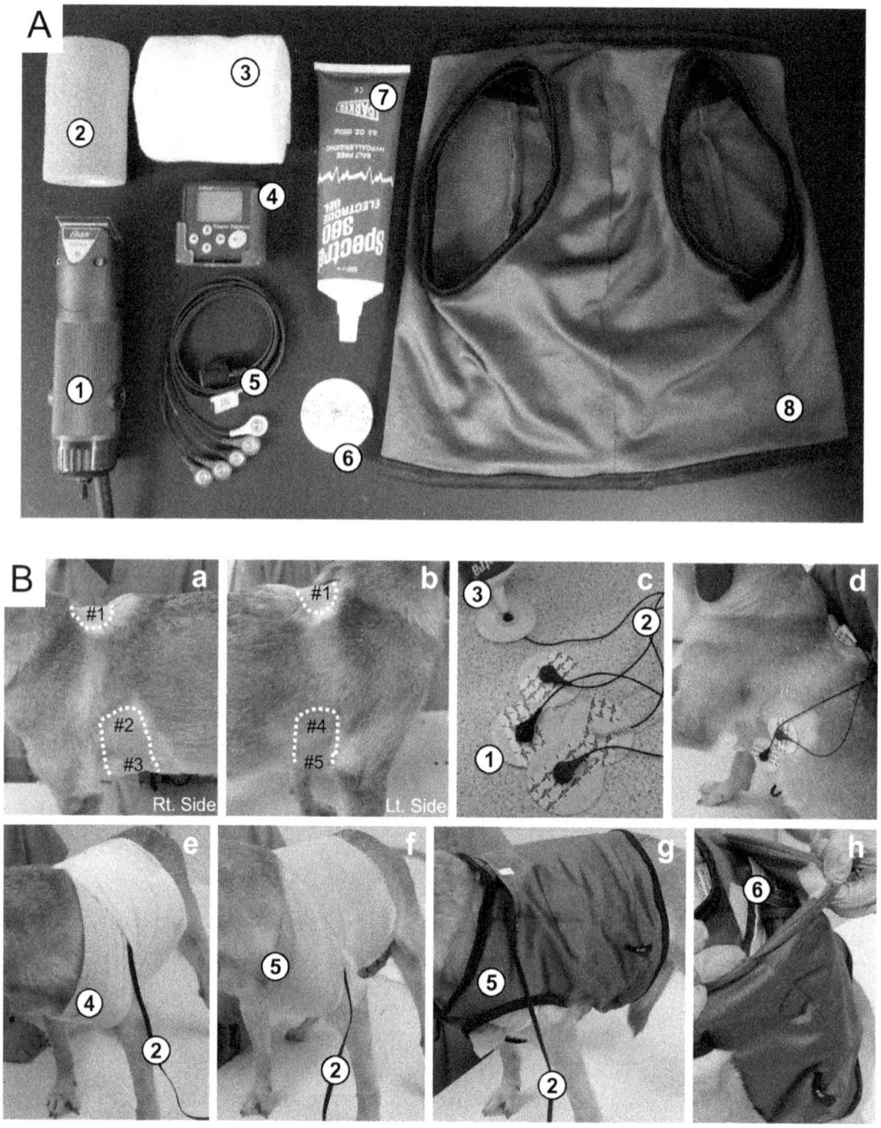

Fig. 7 Experimental materials and setup for Holter monitoring. **A**. Materials needed. (1) Animal hair clipper with a #50 blade, (2) 3M Vetrap^{TM}, (3) cast padding wrap, (4) Forest Medical Trillium 5900 Holter monitor, (5) five color-coded ECG electrode cables, (6) ECG electrode pad, (7) electrode gel, and (8) Holter vest. **B**. Step-by-step Holter setup. a, Shave the skin hair between the shoulder blades but slightly caudal (area #1), and halfway between the spine and the sternum on the right lateral side (area #2), and on the chest behind the elbow on the right ventral side (area #3). b, Shave the skin hair halfway between the spine and the sternum on the left lateral side (area #4) and on the chest behind the elbow on the left ventral side (area #5). c, Attach an electrode pad to each end of the color-coded cable wires. Remove the adhesive backing and add a pea-sized drop of the electrode gel to the center of the pad. d, Stick the pad to the designated shaved area on the skin according to the cable color (white wire in area #1, green wire in area #2, black wire in area #3, red wire in area #4, and brown wire in area #5). e, Wrap around the thorax circumference twice to cover all the electrode pads using cast padding. f, Repeat the same cast pad wrapping pattern using Vetrap. g, Apply the appropriately sized Holter vest onto the dog and feed the electrode cable through the pocket hole on the vest. h, Connect the cable to the Holter monitor, wrap any excess wires around the monitor, and place them into the vest pocket. (1) ECG electrode pads, (2) five color-coded ECG electrode cable, (3) electrode gel, (4) cast padding wrap, (5) 3M Vetrap^{TM}, and (6) Holter vest

8. Following 24 h of recording, remove the Holter monitor in the reverse order of how it was applied (*see* **Note 45**).
9. Using the transfer cable, upload the recorded data from the Holter monitor to the Trillium Platinum Vet software, and analyze the data according to company instructions.

3.3.3 Echocardiography

1. Gently place the dog in lateral recumbency on a padded ultrasonographic exam table with cutouts to allow the transducer probes to be placed from underneath (*see* Fig. 8A) (*see* **Note 46**).
2. Wet the fur coat near the apex beat with alcohol and apply a generous amount of a water-soluble ultrasound transmission gel (*see* **Note 47**).
3. Determine the best phased array transducer probe suitable for the patient size (*see* **Note 48**).
4. Acquire a series of standard two-dimensional imaging views for measurement of left ventricular and left atrial chamber diameters and volumes (*see* **Note 49**) (*see* Fig. 8B). Left ventricular internal chamber diameters can be indexed or normalized to the dog's body weight and compared to previously established reference intervals established by Visser et al.[12]. Color and spectral Doppler interrogation of each valve and cardiac chamber should be performed in a systematic approach to evaluate for valvular regurgitation, stenosis, and shunts. Pulse wave tissue Doppler imaging with the sample volume placed on the ventricular myocardium directly adjacent to the lateral mitral annulus using the left apical five-chamber view can be used to measure peak early diastolic velocity (E'), peak atrial systolic velocity (A'), and peak systolic velocity (S') to assess left ventricular diastolic and systolic function, respectively. It is recommended to store a minimum of 3–5 cardiac cycles for each cine loop using simultaneous lead II ECG monitoring.
5. Many ultrasound systems have workstations and PACS systems to facilitate long-term storage and offline analysis and reporting.

4 Notes

1. Both the camera and light source were suspended from the ceiling through a custom-made mounting bracket. The bracket was made of two 4 inch × 4 inch electrical polyvinyl chloride (PVC) junction boxes and a 1-inch steel galvanized steel conduit. The conduit was cut at the desired length. Each end of the conduit was secured to the PVC box using conduit fitting connectors. One box was attached to the ceiling, and the other was suspended to the kennel through the conduit.

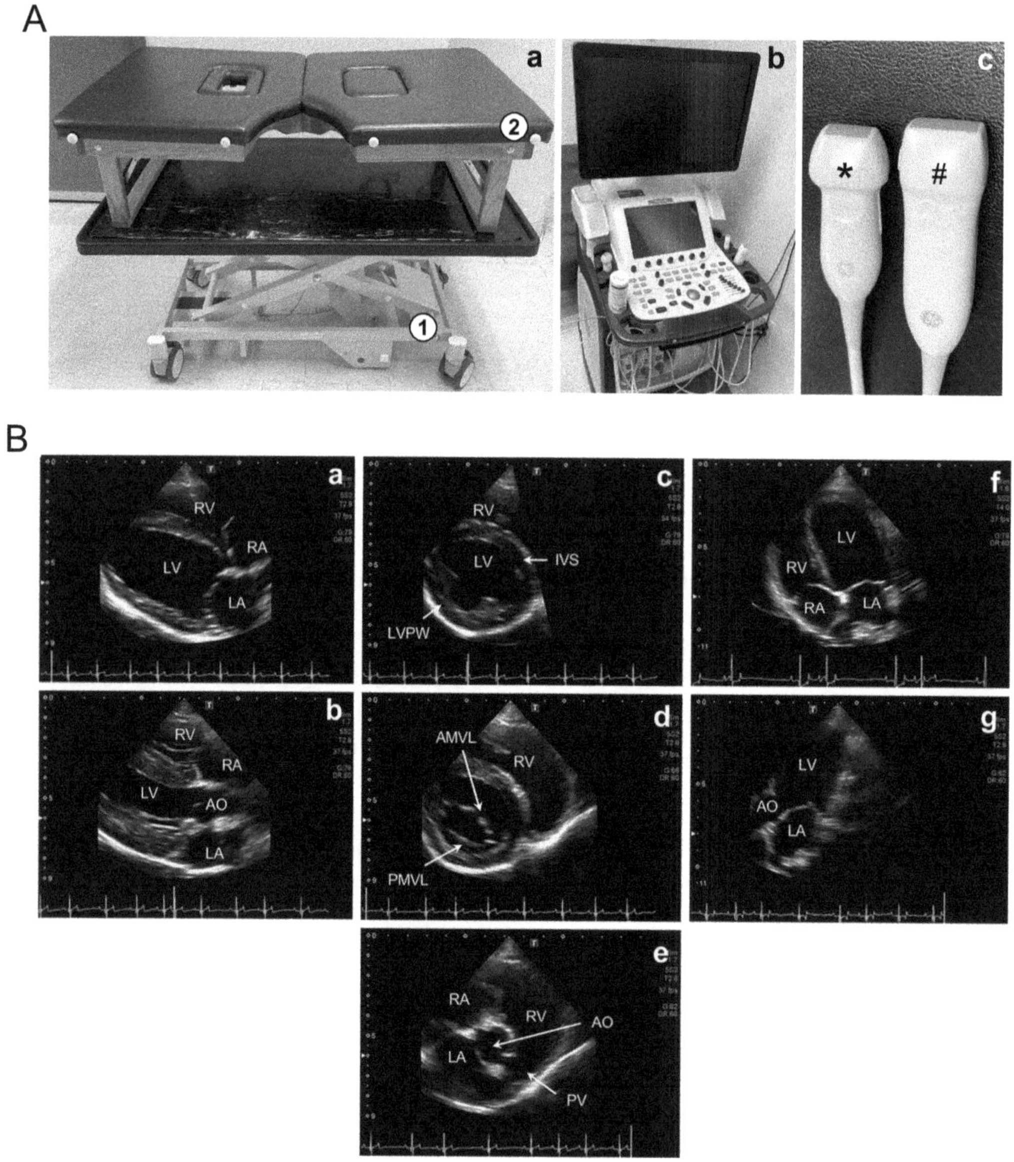

Fig. 8 Experimental setup for echocardiography. **A.** Materials needed. a, Exam table. (1) Mobile lift table and (2) custom-made echocardiographic padded table. b, GE Vivid E90 ultrasonographic system. c, Transducer probes. (*), 6S-D probe, and (#), M5Sc-D probe. **B.** Representative standard two-dimensional imaging views. **a–b**, Right parasternal long-axis images. a, Right parasternal four-chamber long-axis view. b, Right parasternal five-chamber long-axis view. **c–e**, Short-axis images. c, Right parasternal short-axis view at the level of the left ventricle. d, Right parasternal short-axis view at the level of the mitral valve. e, Right parasternal short-axis view at the level of the aorta and heart base. **f–g**, Left apical images. f, Left apical four-chamber view. g, Left apical five-chamber view. *RA* right atrium, *RV* right ventricle, *LA* left atrium, *LV* left ventricle, *AO* aorta, *IVS* interventricular septum, *LVPW* left ventricular posterior wall, *AMVL* anterior mitral valve leaflet, *PMVL* posterior mitral valve leaflet, *PV* pulmonic valve

Rubber dampers were added between the box and ceiling to isolate any vibration from the ceiling to the camera. The CCD camera and NIR light source were attached to the suspended box and were set to ensure full visibility and even illumination of the kennel floor. The CCD camera was wrapped in plastic to protect from water splashes when washing the kennel (*see* Fig. 1).

2. The main structure of the custom-made kennel is made of the heavy-duty galvanized aluminum square tube welded together to form a cubic shape with a dimension of 181 cm width, 122 cm length, and 198 cm height. The floor is made of thick high-density plastic mesh and is raised off the ground to allow water and urine to flow through without pooling. Therefore, it allows the dog to lay down and walk on a clean and dry surface. The back and side walls are made of a flat aluminum sheet reinforced with galvanized aluminum square tubes. The front wall contains a sliding door and is made of welded wire panels to provide clear visibility of the dog. The kennel ceiling is open without any obstruction (*see* Fig. 1).
3. The development of the ODR and the ODVA software is detailed in our published manuscript [13].
4. We prefer recording the dog during the night to eliminate potential interferences from the environment or animal caregivers.
5. We recommend using an electronic outdoor timer to turn on the camera and light source to eliminate human interferences before the recording. Also, we recommend turning on the camera and light source half an hour before the recording to warm up the hardware. The start and end of the recording time can be pre-set on the recording software.
6. The outcome parameters include the percentage of time in motion, number of movements, number of short movements, number of long movements, amplitude of movements, duration of a single movement, and duration of a single rest. The reference values for these outcome measurements can be found in our published manuscript [13]. We have previously published a manual method to analyze the recorded video [14]. In the manual method, we first converted the video to images (one frame every 3 s). We then determined dog movement by manually quantifying differences between neighboring frames. The manual approach is labor-intensive and time-consuming. We recommend not to use the manual method.
7. In this chapter, we describe the method of using GAIT4 walkway system to study dog gait. Several different methods have also been developed and utilized to study normal and DMD dog gait. These include video recording-based methods [14, 15] and wearable inertial sensor-based methods [16–18].

8. Use caution when unpacking and setting the GAIT4 Dog® portable walkway mat on the ground. Avoid tossing, pulling, and dragging the mat on the ground to prevent damaging the encapsulated sensors. Use a floor wire cover to run the USB cable across the open floor.
9. We found that 180 Hz is the optimal scanning rate for recording the gait. We recommend setting the sensor size to ½ inch for dogs more than 10 kg.
10. The dog should be accustomed to walking on a leash and harness. In addition, the dog should be well trained on walking in a straight line at the center of the mat while being leashed. We recommend training the dog with a trained handler once a day for one week before the test day.
11. To avoid distracting the dog while walking in a straight line on the mat, we set the mat next to a wall (*see* Fig. 2). We also like to place custom-made rails on the other side of the mat (not showing in the figure). The custom-made rail is made of 1 inch PVC pipe to form a rectangle shape with two supporting plates at the base. The inner space of the rectangle is covered with mesh fabric wrapped around the PVC pipe and secured with zip ties. To obtain a steady gait speed, we let the dog start walking 6–10 ft before the beginning of the mat and terminate the walk 6–10 ft after the end of the mat. In addition, the walking speed of the handler should not influence the walking speed of the dog. The handler should be walking one to two steps behind the dog and without applying any tension on the leash. We recommend developing a scored scale based on criteria such as leash pull, distraction, change in gait, stopping, and motions to keep or eliminate passes. Use caution not to overexert the dog when repeating the gait recording multiple times to achieve perfect passes. Rests and breaks should be given as needed based on the dog's physical parameter assessment, such as heart rate, respiratory rate and effort, mucous membrane color and consistency, fatigue, and energy level.
12. The GAITFour software provides over 70 analysis parameters. Parameters can be selected based on the research need. We recommend normalizing the total scaled pressure of both front and hind limbs to the body weight; and the step length, stride length, and speed of both front and hind limbs to the dog height (the distance from the ground to the highest point of the dog's shoulder blades, the withers).
13. The custom-made muscle force assay platform was designed by the author CHH to accommodate canines with different body weights and sizes (*see* Fig. 3b). The platform is manufactured using aluminum and stainless steel materials, and it includes the main platform, a tri-axis mount, and a dual-axis mount (*see*

Fig. 3b). The main platform (25 × 48 inches) is made of a ¾ inch aluminum slab, and it provides support for the entire system. The dual-axis mount is used to align the forelimb with the force transducer through a linear motion along with the X-axis and Y-axis using custom-made sliding mechanisms (*see* Fig. 3b). The forelimb is secured to the dual-axis using two custom-made stainless-steel bone pins (*see* Fig. 3c). The tri-axis mount is used to align the force transducer with the forelimb through linear motion along the X-axis, Y-axis, and Z-axis using custom-made sliding mechanisms (*see* Fig. 3b).

14. The current in situ ECU muscle force assay system is based on the force/length dual-mode transducer made by Aurora Scientific Inc [19]. Aurora Scientific also provides the stimulator, interface, DMC, and DMA software for data acquisition and analysis. We have previously developed a primitive in situ ECU force assay protocol based on the force transducer from a different supplier (SM-250–38, Interface, Scottsdale, AZ) [20]. Although this primitive protocol has allowed us to study functional ischemia in the canine model [21], and to establish the proof-of-principle for single adeno-associated virus (AAV)-mediated micro-dystrophin gene therapy and dual AAV-mediated larger mini-dystrophin gene therapy in the canine model [22, 23], we also noticed many limitations. For example, we had to assemble the assay system using components from different companies. We also had to control the muscle length manually during eccentric contraction because there was no suitable lever system. Furthermore, the software used to analyze muscle force was primitive and only yielded a few contractile parameters. The new all-in-one automated assay system/protocol described here does not have the aforementioned limitations and it has allowed us to generate more accurate and consistent force data and to comprehensively study the contractile kinetics [19].

15. To make the bipolar electrode, solder two 0.635 mm thick platinum wires (Surepure chemicals LLC., FlorhamPark, NJ, USA) to two 22G electric wires (Jameco Electronics, Belmont, CA, USA), respectively. Cover each solder connection with an electric heat shrink wrap (Jameco Electronics) and connect the electrodes together using electrical tape (Jameco Electronics). Bend the tip of the platinum wire into a hook shape. Attach the assembled electrode to the gooseneck holder (Walmart Stores Inc., Bentonville, AR, USA). Connect the other ends of the 22G electric wire to the stimulator via a BNC connector (Jameco Electronics) (*see* Fig. 3d).

16. The surgical procedure described can be adapted to study other extensor muscles on both sides of the forelimb. The surgical procedure is composed of five steps: (i) placement of the artery

blood flow probe, (ii) exposure of the extensor muscle, (iii) determination of the ECU muscle length, (iv) exposure of the radial nerve, and (v) fixation of the forelimb with bone pins.

17. The all-in-one assay system consists of a main supporting platform, two muscle-mounting apparatuses, and a modified Aurora Scientific 310C-LR dual-mode lever system, which includes a force/length transducer, a stimulator, force/length control software, and data acquisition/analysis software. The system can be potentially adopted to study the function of other dog muscles or to study muscle function in other large animal models. The force transducer is calibrated using certified standard weights (ThermoFisher Scientific, Hampton, NH, USA).
18. The second physical exam is performed to identify any adverse reaction from the pre-anesthetic cocktail injection.
19. Avoid injecting more than 2 mL of the second pre-anesthetic agent intramuscularly as it causes muscle damage. If the total calculated volume for all three agents is more than 2 mL, combine the acepromazine and butorphanol into one syringe and inject intramuscularly into the epaxial muscle. The atropine is then given subcutaneously on the dorsal side of the dog's body.
20. Assemble the carotid artery catheter by first inserting a stainless steel blunt tip Luer lock needle into one end of a silicone tube, then attaching a three-way stopcock with male Luer lock to the needle (*see* Fig. 3f). Connect a BD PosiFlush pre-filed saline to one end of the stopcock and fill the silicone tube with saline. Once the tube is inserted in the carotid artery and secured, connect the MLT0669 blood pressure transducer and adjust the stopcock valve as presented in Fig. 4E.
21. We have previously reported that the tendon length is 16% of the length of the entire ECU preparation. The experimental ECU muscle length was then calculated by subtracting the tendon length from the length of the entire ECU preparation (Yang et al., 2012).
22. The muscle attachment site designed by the manufacturer (M) was positioned 80 mm away from the center of the rotation axis at the bottom end of the lever arm and can only withstand up to 100N of force (*see* Fig. 3e). To measure force greater than 100N, we added two muscle attachment sites to the lever arm. The first site (C1) was positioned at half the distance (40 mm) between the M site and the rotation axis of the lever arm (*see* Fig. 3e). The second site (C2) was positioned 30 mm away from the rotation axis center of the lever arm (*see* Fig. 3e). These modifications allowed us to measure muscle force up to 200 N at the C1 and 300 N at the C2 sites. The Aurora force transducer was designed to study isometric,

concentric, and eccentric contraction. The lever arm rotates along its axis during concentric and eccentric contraction. The range of the movement is regulated by the position of muscle attachment. When the muscle was attached at the M site in a standard 310C-LR Aurora force transducer, the range of the movement (lever arm excursion length) was set at 40 mm (i.e., ±20 mm). To meet the needs of studying eccentric contraction in the canine muscle, the excursion length of the lever arm at the M site was set to 56 mm (i.e., ± 28 mm) by the manufacturer in our force transducer. When the muscle was attached at the C1 and C2 sites, the range of the movement was ±14 mm (50% reduction) and ±10.5 mm (62.5% reduction), respectively. During an eccentric contraction, the contracting muscle was stretched to 105% of the muscle length by force. For the muscle that was attached to the M site, the stretch distance was calculated according to the formula: (muscle length) × 5%. For the muscle attached to the C1 site, the stretch distance was calculated according to the formula: (muscle length) × 5% × 2, where 2 was the correction factor. For the muscle attached to the C2 site, the stretch distance was calculated according to the formula: (muscle length) × 5% × 2.66, where 2.66 was the correction factor. For example, if an ECU muscle has a length of 100 mm and is attached to the M site, this muscle will be stretched 5 mm (100 mm × 5%). In other words, the total muscle length at the end of the stretch will be 105 mm. If this ECU muscle is attached to the C1 site, it will be stretched 10 mm (100 mm × 5% × 2). If this ECU muscle is attached to the C2 site, it will be stretched 13.3 mm (100 mm × 5% × 2.66).

23. The radial nerve stimulation results in paw extension due to the contraction of the forelimb extensor muscles. To prevent paw extension from moving the chain that links the muscle with the lever arm, the paw can be gently secured to the main platform using a nylon rope.

24. Maintain the muscle temperature at 37 °C by turning on or off the heat lamp.

25. 10.03° is the average pennation angle of the ECU muscle [20]. 1.0597 g/cm^3 is the muscle density [24]. Lf is the optimal fiber length. Lf was determined by multiplying the measured muscle length by the fiber length/muscle length ratio. This ratio is 0.0448 for the ECU muscle [20].

26. The custom-made muscle force assay platform was designed by the author CHH to accommodate canines with different body weights and sizes (*see* Fig. 3b). The platform is manufactured using aluminum and stainless steel materials, and it includes the main platform, a tri-axis mount, and a dual-axis mount (*see*

Fig. 3b). For the hindlimb assay, the dual-axis mount is removed and replaced with the custom-made hindlimb stereotactic frame (*see* Fig. 4B).

27. The custom-made hindlimb stereotactic frame was designed and constructed by the author CHH to accommodate canines with different body weights and sizes; and with consideration to be adopted to the custom-design muscle force assay platform, previously described in **Note 13**. The frame was made from T-slotted aluminum extrusion rails (1.5 inches × 1.5 inches). The frame consists of three components: a base, a hip joint mount, and a stifle joint mount (*see* Fig. 4B).

    The base was constructed from four aluminum rails (front, back, and two sides). The rails were secured together using aluminum brackets to form a rectangle shape with a dimension of 16 inches in width and 36 inches in length. The base was secured to the main platform of the custom-made muscle force assay platform using stainless steel screws (*see* Fig. 4B). The inside space of the rectangle base was filled with composite decking boards to make it flush with the aluminum rail thickness and provide a flat level surface for the dog's body.

    The hip joint mount consisted of a hip bracket, horizontal and vertical aluminum rails, and four aluminum 90° brackets (*see* Fig. 4B). The hip bracket was made of a 1-inch-wide and 0.2-mm-thick aluminum plate and was shaped to an arc-like shape to fit the humerus head at the hip joint. The arc-like shaped hip bracket was mounted to the horizontal aluminum rail through the T-slot and was allowed to slide horizontally for a proper alignment with the hip joint. Two 90° brackets were secured at opposite sides on one end of the horizontal aluminum rail and were positioned to be perpendicular to the hip bracket. The brackets were allowed to slide vertically through the T-slot of the vertical aluminum rail. The vertical sliding movement allows the proper alignment of the horizontal rail/hip bracket with the animal hip height. The vertical aluminum rail was connected to the base using two 90° brackets. The brackets were secured at the bottom end of the vertical aluminum rail on two opposite sides and were set to be perpendicular to the horizontal aluminum rail. The brackets were allowed to slide horizontally on one of the base side aluminum rail through the T-slot. The horizontal movement allows for proper alignment of the hip bracket with the animal body (*see* Figs. 4B and 5).

    The stifle joint mount consisted of an stifle bracket, horizontal and two vertical aluminum rails, and eight aluminum 90° brackets (*see* Fig. 4B). The stifle bracket was made of a 1-inch-wide and 0.2-mm-thick aluminum plate and was shaped to an arc-like shape to fit the olecranon at the stifle joint. The arc-like shaped stifle bracket was mounted to the horizontal aluminum rail through the T-slot and was allowed to slide horizontally for a proper alignment with the stifle joint. Two 90° brackets were secured at opposite sides on each end of the horizontal aluminum rail and were positioned to be perpendicular to the stifle bracket. The brackets on each end were mounted on a vertical aluminum rail and were allowed to slide simultaneously vertically through the T-slot of the vertical aluminum rail. The vertical sliding movement of the horizontal rail through the brackets allows for proper alignment of the stifle bracket with the animal stifle height. Each of the two vertical aluminum rails was connected to the base using two 90° brackets. The brackets were secured at the bottom end of each of the vertical aluminum rails on two opposite sides and were set to be perpendicular to the horizontal aluminum rail. Each of the two brackets was allowed to slide horizontally on one of the base side aluminum rails through the T-slot. The simultaneously horizontal movement of the two vertical aluminum rails allows for proper alignment of the stifle bracket with the animal body (*see* Figs. 4B and 5).

28. The custom-made hindlimb pedal consists of two components: an axis of rotation frame and a paw support plate. The axis of the rotation frame is manufactured using aluminum materials. The paw support plate is made of two carbon fiber sheets with honeycomb wood lattice in between. The honeycomb is set with phenolic resin between the carbon sheets. This structural assembly makes the plate very lightweight while providing a higher level of stiffness. The plate is secured to the frame using stainless steel screws (*see* Fig. 4A).

29. The protocol described here is adapted from a protocol developed by the Korneray lab [25–27]. We have made several modifications. First, we have designed a novel custom-made stereotactic frame to secure the hindlimb. Our custom-made frame is made of commercially available, highly durable, and sturdy materials (*see* **Note 27**). We provide details on how to manufacture and assemble the custom-made frame. Second, we implemented a hip joint mount. We found that securing the hip and stifle joint stabilizes the hindlimb and reduces background noises from unwanted movement of the upper part of the hindlimb while contracting the lower part of the hindlimb. Third, we found that securing the hip, stifle, and tibiotarsal joints at 90° angles in the stereotactic frame are not ideal for obtaining the maximal torque. As a matter of fact, we found that the hip and stifle angles influence the tibiotarsal joint angle (*see* Fig. 5A). For example, when setting the hip and stifle joints at a 90° angle, the tibiotarsal joint is naturally extended to more than a 90° angle (*see* Fig. 5A-a). Fixing the stifle angle at a 90° angle imposes tension on the force transducer shaft. We

recommend adjusting the hip and/or stifle joint angle accordingly to obtain a resting tension of 0 N/cm for the desired angle of the tibiotarsal joint. Fourth, we recommend testing a range of the tibiotarsal joint angle to determine the best angle for the maximum torque production. In our hands, we found that the optimal angle for the flexor and extensor muscle groups is approximately 100° and 80°, respectively. Fifth, we recommend implementing a detailed protocol to identify the optimal stimulation parameters for maximum torque production.

30. We use the fluid displacement method to determine the volume of the lower hindlimb by submerging it up to the stifle joint.
31. Depending on the animal's hydration and health status, infuse physiological saline (0.9% sodium chloride injection) intravenously at 4 ml/kg/hr using a different intravenous catheter.
32. We found that the maximum tetanic force of the tibiotarsal flexion muscle group is achieved when the tibiotarsal joint angle is 100° (*see* Fig. 5B). However, we recommend that the user test a range of angles (60–120°) to determine which angle provides the maximum tetanic force.
33. The placement of the electrodes is a critical step in the assay and should be performed with great attention to assure optimal nerve stimulation. This step can be achieved by testing multiple placements of the electrodes around the common peroneal nerve until the maximum twitch torque is reached.
34. We found that the maximum tetanic torque of the tibiotarsal extension muscle group is usually achieved when the tibiotarsal joint angle is 80° (*see* Fig. 5B). To get the best results, we recommend testing a range of angles (60–120°) to determine which angle provides the maximum tetanic force.
35. The placement of the electrodes is a critical step in the assay and should be performed with great attention to assure optimal nerve stimulation. This step can be achieved by testing multiple placements of the electrodes around the tibial nerve until the maximum twitch torque is reached. Once the correct placement is achieved, the monopolar needles can be secured to the skin using a medical tape to avoid displacement while contracting the extensor muscle group (*see* Fig. 5B).
36. The padded blanket provides comfort to the dog while being restrained. In addition, it reduces signal interference and artifact caused by the metal conductive surface of the exam table.
37. The dog position described in the method is essential to acquiring a proper ECG. Minimal restraint allows a representative ECG to be recorded while minimizing movement artifacts from the restrainer or the patient.

38. Blunt the teeth of the alligator clips to reduce the discomfort from the clip grip.
39. Consult with a veterinary cardiologist for data interpretation.
40. The Holter monitor operates using one AAA battery. Use a new battery with every recording. In addition, the recorded data is stored in a memory card. Confirm that the card is inserted correctly into the monitor.
41. Stick the electrode pad firmly on the dog's skin to ensure good contact. If needed, apply a strip of medical tape at the top and bottom of the electrode pad. Before wrapping the cast pad, check the conductivity of the electrodes. Connect the electrode cable to the monitor and cycle through the three channels to ensure no artifact in the readings.
42. To stabilize the cast padding, continue the wrapping from the thorax around one of the shoulders and then back to the thorax next to the contralateral elbow.
43. Unroll the 3M Vetrap™ completely to relieve the tension from the roll, and then rewrap the roll loosely back onto itself. Avoid wrapping the Vetrap too tight. In addition, the Vetrap should completely cover the cast padding.
44. Before placing the monitor in the vest pocket, cycle through the three channels to ensure no artifact in the readings from the wrapping process. Putting an e-collar around the dog's neck limits access to the Holter or the electrode cables.
45. Clean any residual electrode gel from the dog skin using 2% chlorhexidine wet gauze to prevent skin infection or irritation from the gel electrodes or the shaving.
46. Echocardiography is usually accomplished in a conscious dog with gentle restraint. Sedation is not typically required. A variety of sedatives used in veterinary medicine can have effects on cardiovascular function, and this should be taken into account whenever performing echocardiographic studies.
47. For dogs with medium-to-long or thick fur coats, shaving the hair with electronic clippers may help facilitate good contact with the ultrasonographic gel and transducer probe.
48. Lower frequency probes provide better tissue penetration but with less image resolution, whereas higher frequency probes provide higher image resolution but provide less tissue penetration. Narrowing the sector window and reducing the depth can provide higher frame rates and better resolution.
49. Specific acoustic windows on the right and left hemithorax have been described in dogs and most commonly include the right parasternal and left apical acoustic windows. The right parasternal window is typically near the right 4th or 5th

intercostal space and between the costochondral junction and sternum, or where the apex beat is palpable. This position allows for the acquisition of the following standard views or tomographic planes: right parasternal four-chamber long-axis view, right parasternal five-chamber long-axis view, and the right parasternal short-axis views at the level of the left ventricular papillary muscles, the mitral valve, and aortic root (*see* Fig. 8B) [28]. The left parasternal apical window is typically between the left 4th to 6th intercostal spaces adjacent to the sternum. This position allows for the acquisition of the left apical four-chamber and five-chamber views (*see* Fig. 8B).

Acknowledgments

The research on the canine DMD model in the Duan lab is currently supported by the National Institutes of Health (NS-90634 and AR-70517), Jesse Davidson Foundation-Defeat Duchenne Canada, Hope for Javier, Jackson Freel DMD Research Fund, Parent Project Muscular Dystrophy, Jett Foundation, Michael's Cause, Ryan's Quest, Solid Biosciences Inc., and the University of Missouri.

Disclosure DD is a member of the scientific advisory board for Solid Biosciences and equity holders of Solid Biosciences. DD is a member of the scientific advisory board for Sardocor Corp. In the last three years, the Duan lab has received research supports unrelated to this project from Solid Biosciences and Edgewise.

References

1. Duan D, Goemans N, Takeda S et al (2021) Duchenne muscular dystrophy. Nat Rev Dis Primers 7:13
2. Hoffman EP, Brown RH Jr, Kunkel LM (1987) Dystrophin: the protein product of the Duchenne muscular dystrophy locus. Cell 51(6):919–928
3. Kunkel LM (2005) 2004 William Allan award address. Cloning of the DMD gene. Am J Hum Genet 76(2):205–214
4. Peter AK, Cheng H, Ross RS et al (2011) The costamere bridges sarcomeres to the sarcolemma in striated muscle. Prog Pediatr Cardiol 31(2):83–88. https://doi.org/10.1016/j.ppedcard.2011.02.003
5. Hughes DC, Wallace MA, Baar K (2015) Effects of aging, exercise, and disease on force transfer in skeletal muscle. Am J Physiol Endocrinol Metab 309(1):E1–E10. https://doi.org/10.1152/ajpendo.00095.2015
6. Gumerson JD, Michele DE (2011) The dystrophin-glycoprotein complex in the prevention of muscle damage. J Biomed Biotechnol 2011:210797. https://doi.org/10.1155/2011/210797
7. Bulfield G, Siller WG, Wight PA et al (1984) X chromosome-linked muscular dystrophy (mdx) in the mouse. Proc Natl Acad Sci U S A 81(4):1189–1192
8. Sicinski P, Geng Y, Ryder-Cook AS et al (1989) The molecular basis of muscular dystrophy in the mdx mouse: a point mutation. Science 244(4912):1578–1580
9. Cooper BJ, Winand NJ, Stedman H et al (1988) The homologue of the Duchenne

locus is defective in X-linked muscular dystrophy of dogs. Nature 334(6178):154–156
10. Hakim CH, Li D, Duan D (2011) Monitoring murine skeletal muscle function for muscle gene therapy. Methods Mol Biol 709:75–89
11. Sheard P, Paul A, Duxson M (2002) Intramuscular force transmission. In: Sensorimotor control of movement and posture, vol 508, pp 495–499
12. Visser LC, Ciccozzi MM, Sintov DJ et al (2019) Echocardiographic quantitation of left heart size and function in 122 healthy dogs: A prospective study proposing reference intervals and assessing repeatability. J Vet Intern Med 33(5):1909–1920. https://doi.org/10.1111/jvim.15562
13. Hakim CH, Peters AA, Feng F et al (2015) Night activity reduction is a signature physiological biomarker for Duchenne muscular dystroophy dogs. J Neuromuscul Dis 2(4): 397–407. https://doi.org/10.3233/JND-150114
14. Shin J-H, Greer B, Hakim CH et al (2013) Quantitative phenotyping of Duchenne muscular dystrophy dogs by comprehensive gait analysis and overnight activity monitoring. PLoS One 8(3):e59875
15. Marsh AP, Eggebeen JD, Kornegay JN et al (2010) Kinematics of gait in golden retriever muscular dystrophy. Neuromuscul Disord 20 (1):16–20. https://doi.org/10.1016/j.nmd.2009.10.007
16. Barthelemy I, Barrey E, Thibaud JL et al (2009) Gait analysis using accelerometry in dystrophin-deficient dogs. Neuromuscul Disord 19(11):788–796. https://doi.org/10.1016/j.nmd.2009.07.014
17. Jenkins GJ, Hakim CH, Yang NN et al (2018) Automatic characterization of stride parameters in canines with a single wearable inertial sensor. PLoS One 13(6):e0198893. https://doi.org/10.1371/journal.pone.0198893
18. Zhang X, Jenkins GJ, Hakim CH et al (2022) Four-limb wireless IMU sensor system for automatic gait detection in canines. Sci Rep 12(1):4788. https://doi.org/10.1038/s41598-022-08676-1
19. Hakim CH, Yang HT, Teixeira J et al (2021) Extensor carpi ulnaris muscle shows unexpected slow-to-fast fiber type switch in Duchenne muscular dystrophy dogs. 14: dmm049006 Dis Model Mech. https://journals.biologists.com/dmm/article/doi/10.1242/dmm.049006/272625/Extensor-carpi-ulnaris-muscle-shows-unexpected
20. Yang HT, Shin J-H, Hakim CH et al (2012) Dystrophin deficiency compromises force production of the extensor carpi ulnaris muscle in the canine model of Duchenne muscular dystrophy. PLoS ONE 7(9):e44438
21. Kodippili K, Hakim CH, Yang HT et al (2018) Nitric oxide dependent attenuation of norepinephrine-induced vasoconstriction is impaired in the canine model of Duchenne muscular dystrophy. J Physiol 596(21): 5199–5266. https://doi.org/10.1113/JP275672
22. Shin J-H, Pan X, Hakim CH et al (2013) Microdystrophin ameliorates muscular dystrophy in the canine model of Duchenne muscular dystrophy. Mol Ther 21(4):750–757. https://doi.org/10.1038/mt.2012.283
23. Kodippili K, Hakim CH, Pan X et al (2018) Dual AAV gene therapy for Duchenne muscular dystrophy with a 7-kb mini-dystrophin gene in the canine model. Hum Gene Ther 29(3): 299–311. https://doi.org/10.1089/hum.2017.095
24. Mendez J, Keys A (1960) Density and composition of mammalian muscle. Metabolism 9: 184–188
25. Kornegay JN, Bogan DJ, Bogan JR et al (1999) Contraction force generated by tarsal joint flexion and extension in dogs with golden retriever muscular dystrophy. J Neurol Sci 166(2): 115–121
26. Kornegay JN, Bogan JR, Bogan DJ et al (2011) Golden retriever muscular dystrophy (GRMD): developing and maintaining a colony and physiological functional measurements. Methods Mol Biol 709:105–123. https://doi.org/10.1007/978-1-61737-982-6_7
27. Childers MK, Grange RW, Kornegay JN (2011) In vivo canine muscle function assay. J Vis Exp (50):pii 2623. https://doi.org/10.3791/2623
28. Thomas WP, Gaber CE, Jacobs GJ et al (1993) Recommendations for standards in transthoracic two-dimensional echocardiography in the dog and cat. Echocardiography Committee of the Specialty of Cardiology, American College of Veterinary Internal Medicine. J Vet Intern Med 7(4):247–252

Part III

Antisense Oligonucleotides and Gene Replacement Therapies

Chapter 6

Restoring Dystrophin Expression by Skipping Exons 6 and 8 in Neonatal Dystrophic Dogs

Md Nur Ahad Shah and Toshifumi Yokota

Abstract

Duchenne muscular dystrophy (DMD) is caused by the mutations in the DMD gene resulting in no dystrophin production. Skipping DMD exons using phosphorodiamidate morpholino oligomers (PMOs) is an emerging treatment strategy that can restore the reading frame of the mutated gene and produce truncated but functional dystrophin protein. To date, four PMOs, including eteplirsen, casimersen, viltolarsen, and golodirsen, have been conditionally approved by the FDA for the treatment of DMD. Since degeneration of muscle fibers and irreversible fibrosis occur from childhood, the earlier treatment is preferred. The canine X-linked muscular dystrophy in Japan (CXMDj), a dog model of DMD, produces no dystrophin and exhibits a severe phenotype similar to human patients from early childhood. As such, CXMDj, which harbors a splice site mutation in intron 6, is a useful model for examining the long-term effects of early PMO treatment. In this chapter, we describe the systemic delivery of a cocktail of four PMOs that can successfully induce multiple exon skipping (exons 6–9) in neonatal dystrophic dogs. We also describe the procedures to evaluate the efficacy and toxicity, including clinical grading of dystrophic dogs, ELISA-based quantification of PMOs, histology, RT-PCR, and western blotting.

Key words Phosphorodiamidate Morpholino Oligomers (PMO), Antisense oligonucleotides (AONs), Dystrophic dogs, Canine X-linked muscular dystrophy (CXMD), Duchenne muscular dystrophies, Multiple exon skipping, Dystrophin

1 Introduction

Duchenne muscular dystrophy (DMD) is a genetic disorder characterized by progressive muscle degeneration and weakness [1]. It is the most common form of muscular dystrophy affecting 1 in every 3000–6000 males at birth [2]. This disease is caused by the lack of the protein dystrophin. Most out-of-frame mutations in the *DMD* gene, which encodes for the dystrophin protein, lead to a premature stop codon that ultimately results in a lack of dystrophin production [3, 4]. Becker muscular dystrophy (BMD) is a milder form of muscular dystrophy, which is mostly caused by in-frame

Rika Maruyama and Toshifumi Yokota (eds.), *Muscular Dystrophy Therapeutics: Methods and Protocols*,
Methods in Molecular Biology, vol. 2587, https://doi.org/10.1007/978-1-0716-2772-3_6,

mutations in the *DMD* gene [5]. In this case, the dystrophin protein produced is shorter in length and is partially functional. Studies in animal models have shown that antisense oligonucleotide-mediated exon skipping can restore the reading frame and turn DMD-like syndromes into BMD-like symptoms [6, 7]. Skipping individual exons can theoretically treat approximately 70% of the DMD patients with deletions, and skipping multiple exons can increase its coverage up to 90% of this population in total [8].

Phosphorodiamidate morpholino oligomers (PMOs) are among the most commonly used antisense oligonucleotides for exon skipping. PMOs have charge-neutral phosphorodiamidate linkages and six-membered morpholine rings [9]. These compounds have been widely used to study nonmammalian animal models and have shown to be capable of modulating the splicing of the pre-mRNA effectively [7, 10]. Because of this, antisense drugs based on PMOs are being extensively studied, and four PMO drugs targeting different exons, including exons 45, 51, and 53, to treat DMD patients have been approved by the FDA since 2016 [11, 12].

Although there are several mouse models available for studying DMD, most of them exhibit a milder phenotype compared to DMD patients [13]. In contrast, the canine X-linked muscular dystrophy in Japan (CXMDj), a dog model that has a point mutation in the *DMD* intron 6 splice acceptor site that leads to the out-of-frame skipping of exon 7 and no dystrophin production [14], causes the dogs to be dystrophic and exhibit a phenotype that is close to DMD patients from early childhood [15]. To restore the reading frame and produce dystrophin, at least two exons (exons 6 and 8) need to be skipped [16–19] (Fig. 1). As such, CXMD is a useful model to test the effects of multiple exon skipping from early childhood.

Although the efficacy of PMO-mediated exon skipping is well studied in clinical trials and young adult animals in preclinical settings, the efficacy can be improved if they are treated earlier. Since degeneration of muscle fibers and irreversible fibrosis occur from childhood, early treatment with PMOs might improve efficacy. The CXMDj dog is a useful model for studying the long-term effects of PMOs after neonatal treatment, as the dog model shows extensive fibrosis and muscle wasting over time [15].

.This chapter describes the method of intravenous injection of a cocktail of four PMOs to induce multiple exon skipping (exons 6–8) in neonatal dystrophic dogs. A cocktail containing four different PMOs was designed to target two different sites in exons 6 and 8, respectively. This cocktail was administered systemically in vivo and has been shown to induce body-wide multiple exon skipping (exons 6–9) in skeletal muscles. Exon 9 is known as an

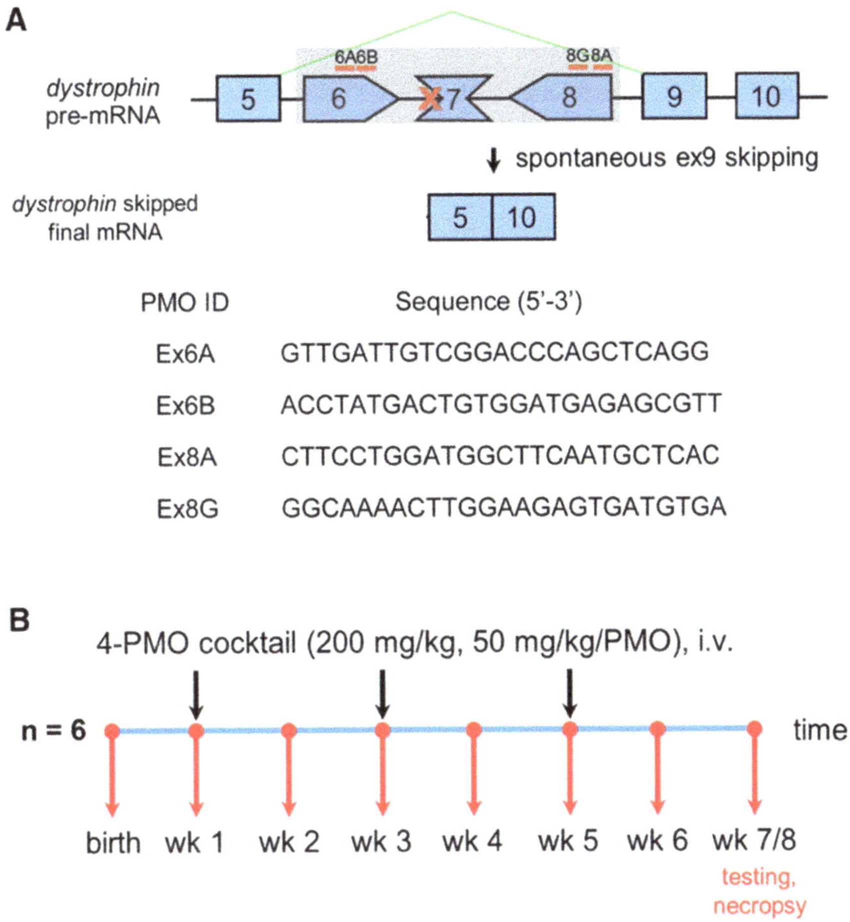

Fig. 1 (**a**) Targeting two exons with a four-PMO-cocktail to restore the reading frame. The CXMD dog model has a point mutation in the *DMD* intron 6 splice acceptor site that leads to the out-of-frame skipping of exon 7. Skipping exons 6 and 8 can restore this reading frame. The PMOs were designed to target two sites of exon 6 and two sites of exon 8 for increased efficacy. (**b**) Treatment schedule for the study. Injection times are marked in black; blood collection times are in red. (Figure reproduced from Ref. [16] with permission from ASGCT)

alternatively spliced exon and was spontaneously skipped along with exons 6-8 upon treatment with PMOs. The treatment increased the level of dystrophin and improved body-wide muscle function. Multiple exon skipping offers a prospect of treating a wide range of populations with out-of-frame mutations in different regions. For example, skipping DMD exons 45–55, which is known to be a mutation hotspot for DMD, could treat approximately 47% of all patients [20]. Hence, PMO-based multiple exon skipping represents a promising therapeutic strategy for treating many patients using a single cocktail drug.

2 Materials

2.1 Animal Model

1. Neonatal dogs with CXMD (canine X-linked muscular dystrophy) mutation.
2. Wild-type neonatal dogs.

2.2 Designing the Antisense Morpholinos

1. Online tools to determine the exonic enhancer site motifs, e.g., ESE finder, Rescue-ESE.

2.3 Injecting the Morpholinos

1. Antisense morpholino cocktail. We used a cocktail of four morpholinos (Ex6A: 5′-GTTGATTGTCGGACCC AGCTC AGG-3′, Ex6B: 5′-ACCTATGACTGTGGATGAGAG CG TT-3′, Ex8A: 5′-CTTCCTGGATGGCTTCAATGCTCAC-3′, Ex8G: 5′-GGCAAAACTTGGAAGAGTGATGTGA-3′) synthesized by Gene Tools, LLC to skip the *dystrophin* exons 6–9.
2. Saline.
3. Syringe infusion pump.
4. 22 G indwelling needles.
5. 50 mL syringe.
6. Serum and blood tests.

2.4 Clinical Grading (Behavior and Gait Test)

1. Video camera.
2. Stopwatch.

2.5 Muscle Sampling/Necropsy

1. Thiopental sodium.
2. Isoflurane.
3. Veterinary surgical instruments: forceps, scalpels, and scissors.
4. Tragacanth gum.
5. Cork disks.
6. Liquid nitrogen.
7. Isopentane.
8. Metal container.
9. Dry ice.
10. 25 mL syringe.
11. Glass vials.
12. Poly-L-lysine-coated slides.

2.6 Hematoxylin and Eosin (H&E) Staining

1. Hematoxylin.
2. Eosin.
3. Ethanol.

4. Shandon Xylene Substitute (Thermo Fisher).
5. Coverslips.
6. Kimwipes.
7. Permount Mounting Medium (Fisher Chemical).
8. Nail polish.

2.7 RNA Extraction and RT-PCR

1. Trizol (Thermo Fisher).
2. Chloroform.
3. Isopropanol.
4. 75% ethanol.
5. RNase-free glycogen.
6. SuperScript IV Reverse Transcriptase kit (Invitrogen).
7. 50 μm random hexamers (Invitrogen).
8. RNaseOUT (Invitrogen).
9. GoTaq Green Master Mix, 2× (Promega).
10. Forward primer in exon 5: 5′-CTGACTCTTGGTTTGATTT GGA-3′.
11. Reverse primer in exon 10: 5′-TGCTTCGGTCTCTGTC AATG-3′.
12. RNase-free water.
13. 2% agarose gel.
14. GelRed Nucleic Acid Gel Stain (Biotium).
15. Gel extraction kit.

2.8 Protein Extraction and Western Blot

1. Lysis buffer: 10% SDS, 70 mM Tris–HCl at pH 6.7, 5 mM EDTA-2Na at pH 8.0, proteinase inhibitor cocktail (Roche, Germany), 5% beta-mercaptoethanol.
2. Coomassie (Bradford) Protein Assay Kit (Thermo Fisher).
3. NuPAGE® LDS Sample Buffer 4× (Invitrogen).
4. NuPAGE® Sample Reducing Agent 10× (Invitrogen).
5. NuPAGE™ Novex™ 3–8% Tris–Acetate Protein Gels (Thermo Fisher).
6. NuPAGE® Tris–Acetate SDS Running Buffer (20×) (Thermo Fisher).
7. NuPAGE™ Antioxidant (Invitrogen).
8. HiMark™ Pre-stained Protein Standard (Thermo Fisher).
9. Polyvinylidene difluoride (PVDF) membrane, pore size 0.45 μm (Millipore, Billerica, MA, USA).
10. Extra thick paper, 2.5 mm thickness, 8.0 cm × 13.5 cm (Thermo Fisher).

11. Methanol.
12. Concentrated Anode Buffer: 0.3 M Tris–HCl, 20% methanol.
13. Anode Buffer: 0.03 M Tris–HCl, 20% methanol.
14. Cathode Buffer: 25 mM Tris–HCl, 20% methanol, 40 mM 6-amino-n-hexanoic acid, 0.01% SDS.
15. Phosphate-buffered saline (PBS) containing 0.05% Tween 20 (PBST).
16. Amersham ECL Prime Blocking Reagent (GE Healthcare, Little Chalfont, UK).
17. NCL-DYS1 anti-dystrophin antibody (Leica Biosystems).
18. Anti-desmin antibody (Abcam).
19. HRP-conjugated anti-mouse immunoglobulin G2a [IgG2a] secondary antibody (Bio-Rad, Hercules, CA, USA).
20. HRP-conjugated goat anti-rabbit (or mouse) IgG H+L secondary antibody (Bio-Rad, Hercules, CA, USA).
21. Amersham ECL Select Western Blotting Detection Reagent (GE Healthcare).

2.9 Immunohistochemistry

1. Cover glasses.
2. Mouse monoclonal anti-dystrophin antibody NCL-DYS1 (Leica Biosystems; targets rod domain).
3. Alexa 594 goat anti-mouse IgG2a highly cross-absorbed (Thermo Fisher).
4. VECTASHIELD HardSet Antifade Mounting Medium with 4′,6-Diamidino-2- Phenylindole Dihydrochloride (DAPI) (Vector Laboratories).
5. Goat serum.
6. Moisture chamber.
7. Phosphate-buffered saline (PBS).
8. Hydrophobic barrier pen.
9. Nail polish.
10. Triton X-100 (Sigma-Aldrich).

2.10 ELISA

1. 40–60 muscle sections (20 μM thickness).
2. Distilled water.
3. Lysis buffer: 1 protease inhibitor tablet (Roche) in 7 mL of Ripa buffer (Thermo Fisher).
4. PMOs used in injection for preparing a standard curve.
5. Hybridization buffer: 1× TE buffer; 1 M final NaCl solution; 0.1% v/v Triton X-100.

6. DNA/phosphorothioate probe: 5′-CTCCTACTCAGACTG TTACTCTGGTGACAC-3′ (first and last seven base pairs are phosphorothioated; the 3′ end is tagged to biotin and the 5′ end to digoxigenin).
7. Avidin-coated plate.
8. Wash buffer: 50 mM Tris–HCl, 150 mM sodium chloride pH 7.6, 0.1% v/v Tween-20.
9. Micrococcal Nuclease (MNase).
10. Reaction Buffer: pH 7.9; 50 mM Tris–HCl, 100 mM NaCl, 5 mM $CaCl_2$, 0.1 mg/mL bovine serum albumin.
11. Anti-digoxigenin antibody conjugated with alkaline phosphatase enzyme.
12. SuperBlock (TBS) Blocking Buffer (Thermo Fisher).
13. AttoPhos Substrate.

3 Methods

3.1 Designing the Antisense Morpholinos

1. Identify ESE sites around the genomic sequence of interest using ESE finder and Rescue-ESE.
2. Design 25–30 mer PMOs that are the antisense sequences of the target sites (*see* **Note 1**).

3.2 Injecting the Morpholino Drugs

1. Restrain the dogs manually in accordance with the animal use guidelines.
2. Inject the PMOs using a venous indwelling needle into a limb vein (cephalic or saphenous vein). The treatment dosage can be 120–240 mg/kg body weight (bw). We used 50 mg/kg for each PMO (total 200 mg/kg) dissolved in saline to make 50 mL solution. Use an infusion pump or syringe driver to administer the 50 mL of drug-containing solution at a rate of 2.5 mL/minute for 20 min (*see* **Note 2**).
3. To examine toxicity, perform weekly blood tests—collect 3 mL of blood from one of the subcutaneous veins of the fore or hindlimbs. Include complete blood count (CBC), gamma-glutamyl transferase (GGT), aspartate aminotransferase (AST), blood urea nitrogen (BUN), alanine aminotransferase (ALT), creatine kinase (CK), and creatinine assessments when doing toxicity analysis.

3.3 Behavior and Gait Test

1. Use a video camera to record the dog's behavior and gait. Use the grading criteria to calculate the total grade (Table 1).
2. Calculate the total grade by adding the scores of each category.
3. Measure how long it takes for the dog to run 15 meters.

Table 1
Grading criteria for behavior and gait tests

Grade	Gait	Mobility disturbance	Muscle atrophy of the limb	Drooling	Hypertrophy of the tongue	Ability to swallow
1	None	None	None	None	None	None
2	Sitting with hind limb extended	Frequent lying down	Suspect hardness	Occasional dribbles of saliva when sitting	Slightly enlarged	Takes time and effort in eating food
3	Bunny-hops with hind limbs	Unable to jump on hind limbs	Can feel hardness or appears thin	Some drool when eating and drinking	Extended outside dentition	Difficulty in eating food from a plate
4	Shuffling walk	Increasing difficulty moving around	Between grades 3 and 5	Strings of drool when eating or drinking	Enlarged and slightly thickened	Difficulty in chewing, swallowing, or drinking
5	Unable to walk	Unable to get up and move around	Extremely thin or hard	Continuous drool	Enlarged and thickened	Unable to eat

3.4 Muscle Sampling/Necropsy

1. To prepare the mounting gum for tissue collection, mix 20 gm of tragacanth gum with an equal amount of water to make them soft and sticky. Store them in 25 mL syringes at −20 °C.
2. Anesthetize the dogs by injecting 20 mg/kg of thiopental sodium using the intravenous route and maintain the isoflurane inhalation rate to 2–3%. Check muscle reflex and monitor heart and respiration rate. The normal range for respiration rate (RR) is 10–20 breaths/min; heart rate (HR), 80–120 bpm; and oxygen saturation (SpO_2) level, 95–100%.
3. Euthanize the dogs by exsanguination from the carotid artery under general anesthesia (*see* **Note 3**).
4. Dissect the following muscles: tibialis anterior (TA), extensor digitorum longus (EDL), gastrocnemius, soleus, biceps femoris, rectus femoris, biceps brachii, triceps brachii, deltoid, extensor carpi ulnaris (ECU), extensor carpi radialis (ECR), flexor carpi ulnaris (FCU), flexor carpi radialis (FCR), gracilis, intercostal, abdominal muscles, diaphragm, lateral dorsi, esophagus, sternocleidomastoid, and the heart. Collect kidney and liver samples to examine toxicity.
5. Cut the muscle samples in small sections (1–1.5 cm in length).

6. Label the cork discs with the animal's ID and muscle name. On the opposite side, put tragacanth gum maintaining a thickness of 0.5–1 cm.
7. Place the muscles in the tragacanth gum such that one-third of the sample is securely embedded within the gum. The longitudinal axis of the samples should be perpendicular to the cork. Apply additional gum around the bottom if necessary.
8. Place the muscle on the cork into the cooled isopentane using a tong and shake vigorously for about 1 min for the penetration of isopentane so that the samples are quickly frozen without any artifacts. Keep them on dry ice, or store them in glass vials at −80 °C.
9. For cryosectioning, place the samples in a cooled cryostat at −25 °C. Mount the cork with the muscle sample on a disc and trim the muscles to get tissue sections. The section thickness is 20 μm for RNA (RT-PCR) and protein (western blot) samples, and 7 μm for hematoxylin and eosin (H&E) staining and immunohistochemistry (IHC). Collect 25–30 sections for RT-PCR and western blot analysis, and 3 sections per slide for H&E staining and IHC. Roughly perform 20–30 sections beforehand to get finer muscle sections for histology and IHC slides.

3.5 Hematoxylin and Eosin (H&E) Staining

1. Place the slides in each of these solutions in the following order and duration:

   Hematoxylin: 15 min

   Running cold water: 15 min

   Eosin: 10 min

   70% Ethanol: 5 s

   90% Ethanol: 5 s

   99% Ethanol: 30 s

   99% Ethanol: 30 s

   Xylene substitute: 5 min

   Xylene substitute: 5 min
2. Tap the slide tray containing the slides on a dry napkin between each step.
3. Air-dry the slides for 15 min after the final xylene wash. Wipe the edges of the slides with a Kimwipe if necessary.
4. Put 2–3 drops of mounting medium over the muscle sections and gently put the coverslips. Avoid air bubbles when placing the coverslips. Let the slides dry for 1 day. Use nail polish around the edges of the coverslips to seal the slides.
5. Visualize using a microscope.

3.6 RNA Extraction and RT-PCR

1. Add 1 mL of cold Trizol to the frozen samples and vortex for 30 s. Incubate for 5 min at room temperature.
2. Add 200 μL of chloroform and vortex. Incubate for 2 min at room temperature.
3. Centrifuge the sample at 12,000× *g* for 15 min at 4 °C.
4. Take the supernatant (clear portion) carefully and transfer it to a new tube (*see* **Note 4**).
5. Add 1 μL of RNase-free glycogen.
6. Add 500 μL of isopropanol and vortex. Incubate for 10 min at room temperature.
7. Centrifuge at 12,000× *g* for 10 min at 4 °C. RNA pellet will be on the bottom of the tube.
8. Decant supernatant and pipette out the remaining isopropanol.
9. Add 1 mL of 75% ethanol.
10. Vortex briefly until the pellet can be seen floating.
11. Centrifuge at 7500× *g* for 5 min at 4 °C.
12. Decant supernatant and centrifuge at 7500× *g* for 5 min at 4 °C.
13. Pipette out the remaining ethanol and air-dry for 5–10 min at room temperature by opening up the tube lid and covering it with a Kimwipe.
14. Add 20 μL of RNase-free water and warm at 65 °C for 10 min.
15. Quantify total RNA concentration using NanoDrop and prepare aliquots of appropriate concentration with RNase-free water. Store at −80 °C.
16. Prepare cDNA using RT-PCR (total 20 μL):

Reagent	Volume
10 mM dNTP	1 μL
Random hexamers	1 μL
RNA sample	200 ng
RNase-free water	Add to make it 13 μL

Incubate at 65 °C for 5 min, then at 4 °C for 5 min. This can be done in the PCR thermocycler.

Add the following reagents to the solution:

Reagent	Volume
5× RT buffer	4 μL
0.1M DTT	1 μL

(continued)

Reagent	Volume
RNaseOUT	1 μL
SuperScript IV RT	1 μL

Use the following program:

(a) 23 °C, 10 min
(b) 52 °C, 10 min
(c) 80 °C, 10 min
(d) 4 °C, 5 min
(e) 4 °C, hold

17. cDNA amplification using PCR (total 25 μL):

Reagent	Volume
GoTaq	12.5 μL
10 μm forward primer	1.65 μL
10 μm reverse primers	1.65 μL
cDNA sample	8 μL
RNase-free water	1.2 μL

Use the following program:

(a) 95 °C, 2 min
(b) 40 cycles of 95 °C, 30 s; 60 °C, 30 s; 72 °C, 42 s
(c) 72 °C, 5 min
(d) 4 °C, hold

18. Prepare 2% agarose gel and add gel red before cooling it down (1:20,000).
19. Run 5 μL of the PCR product on 2% agarose gel and visualize (Fig. 2).
20. Cut the band of interest and collect it into a 1.5 mL tube (*see* **Note 5**). Use the Qiagen gel extraction kit to purify the PCR product. Store the PCR product at 4 °C.
21. Sequence the purified PCR product to confirm the exon skipping. The primers for sequencing are the same as the RT-PCR.

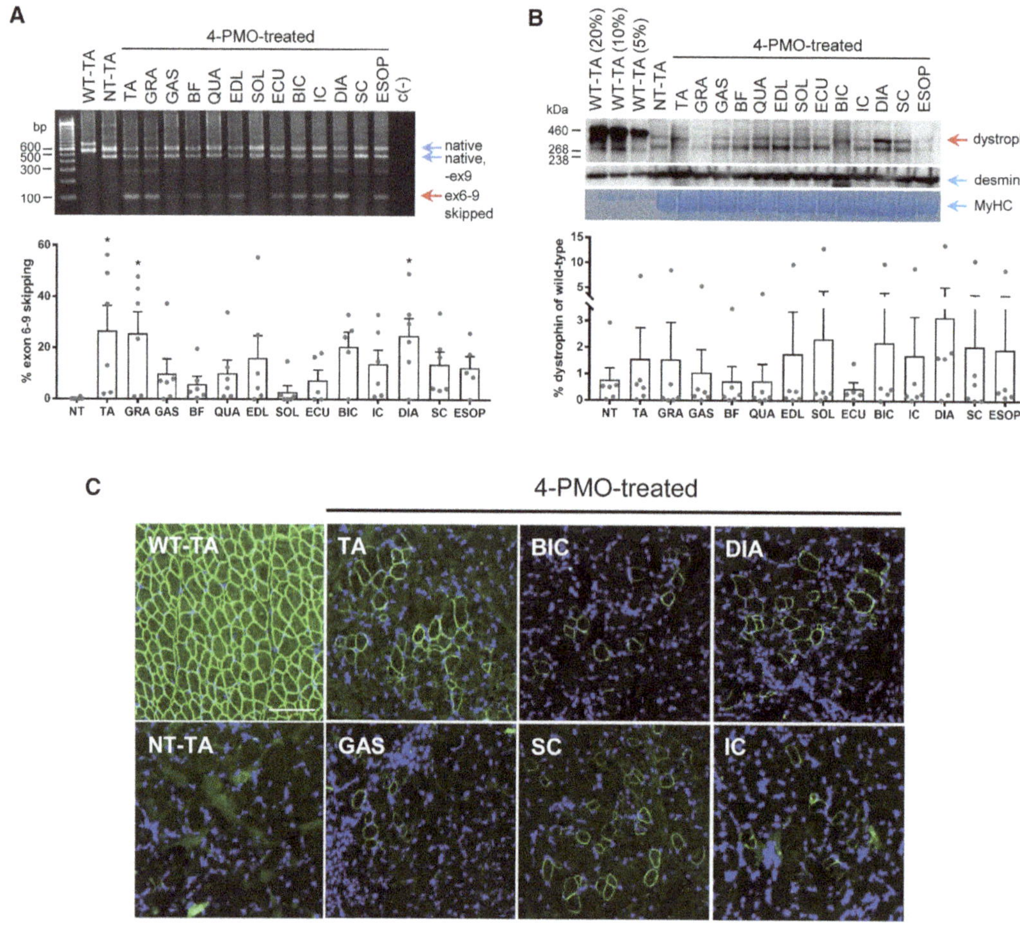

Fig. 2 (**a**) Efficacy of multiple exon skipping by RT-PCR. The top bands represent the native (non-skipped) bands with and without exon 9 (spontaneous skipping). Bottom band represents the skipped mRNA. Skipping percentages have also been quantified. (**b**) Western blot showing dystrophin rescue in treated CXMD skeletal muscles. Desmin and myosin heavy chain (MyHC) were used as control. Level of dystrophin rescue has also been quantified. (**c**) Representative immunohistochemistry images to indicate restored dystrophin. Green, DYS1; blue, nuclei. Abbreviations: *TA* tibialis anterior, *GRA* gracilis major, *GAS* gastrocnemius, *BF* biceps femoris, *QUA* quadriceps, *EDL* extensor digitorum longus, *SOL* soleus, *ECU* extensor carpi ulnaris, *BIC* biceps brachii, *IC* intercostal muscles, *DIA* diaphragm, *SC* sternocleidomastoid, *ESOP* esophagus. (Figure reproduced from Ref. [16] with permission from ASGCT)

3.7 Protein Extraction and Western Blot

1. Add 80 μL of lysis buffer and vortex for 30 s.
2. Incubate at 37 °C for 5 min and centrifuge at max speed for 30 min.
3. Collect the supernatant in a fresh tube and measure the volume.
4. Dilute the samples to 50-fold with water.

5. Measure the protein concentration by Bradford Protein Assay Kit and adjust the protein concentration to 5 μg/μL with the lysis buffer. Store at −80 °C.
6. Make 2× LDS sample buffer:

Reagent	Volume
4× LDS sample buffer	2.5 μL
10× Reducing agent	1 μL
Water	1.5 μL

7. Mix 12 μL of the 5 μg/μL protein sample and 12 μLof 2× LDS sample buffer. Heat the mixed samples at 70 °C for 10 min and spin them down.
8. Load 20 μL of the sample (50 μg protein) per well of the gel.
9. Load 100%, 20%, 10%, and 5% protein samples from wild-type dogs to use as standard.
10. Add NuPAGE antioxidant to the running buffer in the smaller well (1:400).
11. Run the SDS-PAGE at 150 V for 75 min.
12. Soak PVDF membrane for 20 s in methanol and then in the anode buffer for 6 min.
13. Soak blotting papers in each transfer buffer for 6 min.
14. Place the blotting papers, the PVDF membrane, and the gel on a semi-dry transfer machine (Fig. 3).
15. Transfer at 20 V for 70 min.

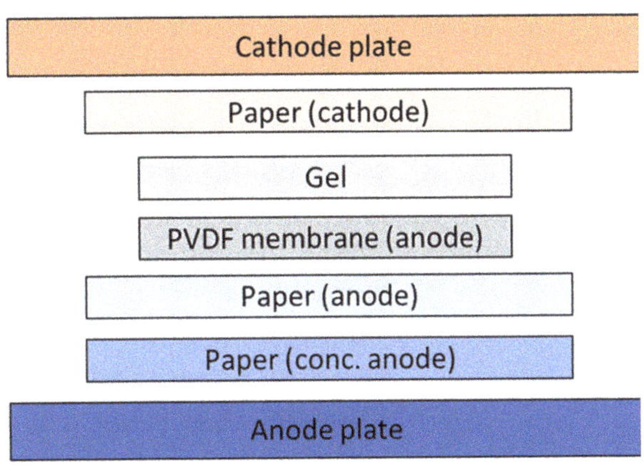

Fig. 3 Semi-dry transfer method for western blotting. Conc. anode buffer, anode buffer, and cathode buffer soaked extra-thick papers (2.5 mm thickness). PVDF membrane soaked in anode, and the gel soaked in cathode buffer. The order is presented in the figure

16. Rinse the membrane by PBST twice.
17. Soak the membrane in 2% blocking reagent in PBST and shake gently for 1 h at room temperature or overnight at 4 °C for blocking.
18. Incubate the membrane with NCL-DYS1 antibody or anti-desmin antibody in blocking solution (1:200 for NCL-DYS1 or 1:4000 for anti-desmin) for 1 h at room temperature or overnight at 4 °C.
19. Wash the membrane three times for 10 min by PBST at room temperature.
20. Incubate the membrane with the HRP-conjugated anti-mouse IgG2a secondary antibody for DYS1 in PBST (1:10,000) or, with the HRP-conjugated goat anti-rabbit IgG secondary antibody for desmin in PBST (1:10,000) for 1 h at room temperature.
21. Wash the membrane three times for 10 min by PBST at room temperature.
22. Detect the signal by the ECL select kit and a chemiluminescent imager (Fig. 2b).

3.8 Immunohistochemistry

1. Air-dry the muscle sections at room temperature for at least 30 min.
2. Prepare PBSTX using PBS with 0.1% Triton X-100.
3. Prepare blocking reagent: 10% goat serum in PBSTX.
4. Draw a square enclosing the muscle sample on the slide using a hydrophobic barrier pen.
5. Add blocking reagent to the samples and incubate them for 10 min in the moisture chamber at room temperature.
6. Add NCL-DYS1 anti-dystrophin antibody diluted in blocking reagent (1:50 dilution) to the samples and incubate in the moisture chamber at room temperature for 1 h.
7. Wash the slides with PBS three times on a shaker, 5 min for each wash. Cover the slides during this step.
8. Add Alexa 594 goat anti-mouse IgG2a antibody diluted in PBSTX (1:1000 dilution) to the samples and incubate in the moisture chamber at room temperature for 1 h.
9. Wash the slides with PBS three times on a shaker, 5 min for each wash. Cover the slides during this step.
10. Wipe off excess liquid. Use the DAPI containing mounting medium to mount the coverslips.
11. Let the slides air-dry for 15 min before storing them at 4 °C.
12. Use nail polish to seal the slides.
13. Visualize using a confocal microscope (Fig. 2c).

3.9 ELISA

1. Collect sera from blood samples every week to check PMO clearance. Use tissue samples to examine PMO uptake.
2. To prepare the serum sample, add 22 μL of serum to 198 μL of distilled water to get 10% serum concentration (*see* **Note 6**).
3. To prepare tissue samples, take 30–40 slices of the muscle section at 20 μM in a 1.5 mL tube and add 150 μL of lysis buffer and vortex. Incubate overnight at 55 °C. Centrifuge the sample at maximum speed for 15 min and collect the supernatant. Quantify and adjust the protein concentration to 2 mg/mL using lysis buffer. Add 22 μL of the protein lysate to 198 μL distilled water. Store at −20 °C.
4. Prepare PMO standard:
   (a) Prepare 1 μM, and serially dilute to make 10 nM, and 1 nM stock:

   For the 1 μM stock, add 1 μL of 1 mM PMO with 1498.5 μL of distilled water.

   For the 10 nM stock, add 5 μL of 1 μM PMO with 495 μL of distilled water.

   For the 1 nM stock, add 150 μL of 10 nM PMO, with 150 μL of serum from wild-type untreated animal (or lysis buffer when using protein), and 1.2 mL of distilled water.

   (b) Serially dilute the PMOs to the following concentrations: 250, 125, 62.5, 31.3, 15.6, 7.8, and 0 pM: Prepare the solvent containing 2.7 mL of distilled water with 300 μL of lysis buffer/serum. Start with the 1 nM stock and combine 250 μL PMO with 750 μL solvent to get 250 pM solution and so on. Transfer 220 μL of each standard into a fresh tube.
5. Prepare the probe: Add 1.5 μL of 100 μM probe with 1498.5 μL of hybridization buffer to make 100 nM stock. Add 100 μL of the 100 nM stock with 19.9 mL of hybridization buffer to make a 0.5 nM working solution.
6. Add 200 μL of 0.5 nM probe to the standard PMO tubes and the sample tubes.
7. Warm at 37 °C for 30 min.
8. Wash the avidin-coated plate with 200 μL of wash buffer three times.
9. Transfer 200 μL of the mixture from **step 6** to each well in the plate. Do this in duplicates.
10. Incubate at 37 °C for 30 min followed by washing the plate with 200 μL of wash buffer three times (*see* **Note 7**).

11. Add 1.5 μL of MNase with 30 mL reaction buffer to make a working concentration of 0.1 U/μL.
12. Transfer 200 μL of 0.1 U/μL MNase solution to each well.
13. Incubate at 37 °C for 1 h followed by washing the plate with 200 μL of wash buffer three times (*see* **Note 7**).
14. Prepare 1:5000 anti-digoxigenin antibody in SuperBlock buffer with 0.25% Tween 20 and add 200 μL to each well (*see* **Note 8**).
15. Incubate at 37 °C for 30 min followed by washing the plate with 200 μL wash buffer three times (*see* **Note 7**).
16. Add 150 μL of AttoPhos substrate per well (*see* **Note 8**).
17. Incubate at 37 °C for 30 min.
18. Use SpectraMax microplate reader to read fluorescence with 444 nM excitation and 555 nm emission (cutoff 530 nm) (*see* **Note 9**).

4 Notes

1. PMOs can be designed to target an exon/intron boundary as well as ESE motifs. Consider the secondary structure and GC contents to create stable PMOs. A sequence with ~60% GC content is ideal. Avoid heterodimers, self-complementary sequences, and sequences containing >3 consecutive guanine residues. Design at least one PMO that targets an ESE site.
2. Depending on the experiment design, the concentration of PMO and the frequency of dose administration may vary. The doses can be administered at a weekly or bi-weekly interval period. The number of total injections also depends on the experiment, but we typically inject the animal 3–5 times since repeated injections result in dystrophin accumulation. For this experiment, we administered the dosage every 2 weeks on neonatal dogs at the age of 1, 3, and 5 weeks (Fig. 1b).
3. Euthanasia and muscle sample collection should be done 1–2 weeks after the last injection.
4. Do not include the colored layers to avoid contamination.
5. The size of exon 6–9 skipped product is 101 bp.
6. This is applicable with a PMO dosage of 12.5 mg/kg. If the dosage is higher (e.g., 50 mg/kg), more dilution is needed (e.g., 5% or 2.5%).
7. Avoid pipetting out the wash buffer. Remove the wash buffer by turning over the plate in one swift motion.

8. When using a 96-well plate, add 4 μL of the anti-digoxigenin antibody in 20 mL of SuperBlock blocking buffer with 0.25% Tween 20.
9. Prepare 15 mL of stock for a 96-well plate.
10. PMO amounts were calculated in reference to the standard curve constructed from fluorescence values given by the respective PMO standards.

Acknowledgments

This work was supported by the Friends of Garrett Cumming Research Chair Fund, HM Toupin Neurological Science Research Chair Fund, Muscular Dystrophy Canada, Alberta Innovates, the University of Alberta Faculty of Medicine and Dentistry, and the Women and Children's Health Research Institute (WCHRI). We also would like to thank Kenji Rowel Q Lim for his assistance.

References

1. Bushby K, Finkel R, Birnkrant DJ et al (2010) Diagnosis and management of Duchenne muscular dystrophy, part 1: diagnosis, and pharmacological and psychosocial management. Lancet Neurol 9:77–93. https://doi.org/10.1016/S1474-4422(09)70271-6
2. Mendell JR, Shilling C, Leslie ND et al (2012) Evidence-based path to newborn screening for Duchenne muscular dystrophy. Ann Neurol 71:304–313
3. Van Deutekom JC, Van Ommen G-JB (2003) Advances in Duchenne muscular dystrophy gene therapy. Nat Rev Genet 4:774–783
4. Ervasti JM (2007) Dystrophin, its interactions with other proteins, and implications for muscular dystrophy. Biochim Biophys Acta 1772:108–117
5. Flanigan KM (2014) Duchenne and Becker muscular dystrophies. Neurol Clin 32:671–688
6. Lu QL, Rabinowitz A, Chen YC et al (2005) Systemic delivery of antisense oligoribonucleotide restores dystrophin expression in body-wide skeletal muscles. Proc Natl Acad Sci 102:198–203
7. Yokota T, Lu Q, Partridge T et al (2009) Efficacy of systemic morpholino exon-skipping in Duchenne dystrophy dogs. Ann Neurol 65:667–676
8. Yokota T, Duddy W, Partridge T (2007) Optimizing exon skipping therapies for DMD. Acta Myol 26:179
9. Arora V, Devi GR, Iversen PL (2004) Neutrally charged phosphorodiamidate morpholino antisense oligomers: uptake, efficacy and pharmacokinetics. Curr Pharm Biotechnol 5:431–439
10. Guncay A, Yokota T (2015) Antisense oligonucleotide drugs for Duchenne muscular dystrophy: how far have we come and what does the future hold? Future Med Chem 7(13):1631–1635. https://doi.org/10.4155/fmc.15.116
11. Hanson B, Wood MJA, Roberts TC (2021) Molecular correction of Duchenne muscular dystrophy by splice modulation and gene editing. RNA Biol 18:1048–1062. https://doi.org/10.1080/15476286.2021.1874161
12. Shirley M (2021) Casimersen: first approval. Drugs 81:1–5
13. Rodrigues M, Echigoya Y, Fukada S-I, Yokota T (2016) Current translational research and murine models for Duchenne muscular dystrophy. J Neuromuscul Dis 3:29–48. https://doi.org/10.3233/JND-150113
14. Sharp NJ, Kornegay JN, Van Camp SD et al (1992) An error in dystrophin mRNA processing in golden retriever muscular dystrophy, an animal homologue of Duchenne muscular dystrophy. Genomics 13:115–121. https://doi.org/10.1016/0888-7543(92)90210-j
15. Yu X, Bao B, Echigoya Y, Yokota T (2015) Dystrophin-deficient large animal models: translational research and exon skipping. Am J Transl Res 7:1314–1331

16. Lim KRQ, Echigoya Y, Nagata T et al (2019) Efficacy of multi-exon skipping treatment in Duchenne muscular dystrophy dog model neonates. Mol Ther 27:76–86. https://doi.org/10.1016/j.ymthe.2018.10.011
17. Maruyama R, Echigoya Y, Caluseriu O et al (2017) Systemic delivery of morpholinos to skip multiple exons in a dog model of Duchenne muscular dystrophy. In: Morpholino oligomers. Springer, pp 201–213
18. Melo D, Maruyama R, Yokota T (2018) Systemic injection of peptide-PMOs into humanized DMD mice and evaluation by RT-PCR and ELISA. In: Exon skipping and inclusion therapies. Springer, pp 263–273
19. Miskew Nichols B, Aoki Y, Kuraoka M et al (2016) Multi-exon skipping using cocktail antisense oligonucleotides in the canine X-linked muscular dystrophy. J Vis Exp:53776. https://doi.org/10.3791/53776
20. Aoki Y, Yokota T, Nagata T et al (2012) Bodywide skipping of exons 45–55 in dystrophic mdx52 mice by systemic antisense delivery. Proc Natl Acad Sci 109:13763–13768

Chapter 7

Restoring Dystrophin Expression with Exon 44 and 53 Skipping in the *DMD* Gene in Immortalized Myotubes

Yusuke Echigoya and Toshifumi Yokota

Abstract

Phosphorodiamidate morpholino oligomer (PMO)-mediated exon skipping is a therapeutic approach that applies to many Duchenne muscular dystrophy (DMD) patients harboring out-of-frame deletion mutations in the *DMD* gene. In particular, PMOs for skipping exon 44 have been developing in clinical trials, such as the drug NS-089/NCNP-02. Two exon 53 skipping PMOs, golodirsen and viltolarsen, have received conditional approval for treating patients due to their ability to restore dystrophin protein expression. Although promising, further development of exon-skipping technology is needed for patients to have more therapeutic benefit. This chapter describes evaluation methods of exon 44 and 53 skipping PMOs in immortalized DMD patient-derived skeletal muscle cells. We introduce how to quantify exon-skipping efficiencies and dystrophin rescue levels represented by RT-PCR and western blotting, respectively. The screening methods using immortalized patient myotubes can serve to find exon-skipping PMO drug candidates.

Key words Duchenne muscular dystrophy, Dystrophin, Exon skipping, Antisense oligonucleotide, Morpholino, PMO, Exon 44, Exon 53, Golodirsen, Viltolarsen, Immortalized cells

1 Introduction

Exon skipping using antisense oligonucleotides, phosphorodiamidate morpholino oligomers (PMOs), is one of the most promising approaches to treating an X-linked recessive disorder, Duchenne muscular dystrophy (DMD) [1]. DMD is caused by reading-frame disrupting mutations in the *dystrophin* (*DMD*) gene and subsequent dystrophin deficiency in skeletal and cardiac muscles. PMO-mediated exon skipping converts out-of-frame *DMD* mRNA to in-frame, translatable one by removing exon(s) neighboring missing mutant exon(s) (Fig. 1). The splicing modulation allows for short but functional dystrophin protein expression as seen in a milder dystrophinopathy, Becker muscular dystrophy [2]. Exon skipping can treat approximately 80% of patients with deletion mutations in theory [3]. The approach to exclude particular

Rika Maruyama and Toshifumi Yokota (eds.), *Muscular Dystrophy Therapeutics: Methods and Protocols*,
Methods in Molecular Biology, vol. 2587, https://doi.org/10.1007/978-1-0716-2772-3_7,

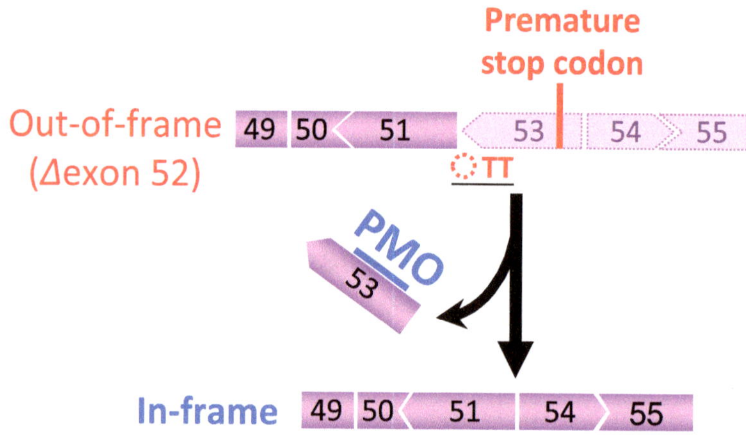

Fig. 1 Schematic mechanism of exon 53 skipping for correcting the exon 52-deleted *DMD* mRNA. Phosphorodiamidate morpholino oligomer (PMO) binds *DMD* pre-mRNA in a sequence-specific manner. It excludes an exon 53 that generates the out-of-frame, i.e., premature stop codons, due to the lack of exon 52 to be paired for correct codons. The reading frame is corrected to an in-frame that allows for the translation of the gene into short but functional dystrophin as seen in a milder phenotype of dystrophinopathy, Becker muscular dystrophy. Boxes and numbers indicate exons and their order, respectively

exons is applicable to the large groups of patients with mutations found in the *DMD* mutation hot spot ranging from exons 43 to 55 [4]; approx. 20% of the patients with a deletion mutation can be treated by skipping exon 51, approx. 15% by exon 53, 14% by exon 45, and 13% by exon 44 [4]. So far, the US Food and Drug Administration (FDA) has conditionally approved four *DMD* exon-skipping drugs composed of PMOs. They target particular exons: exon 51 (eteplirsen, Exondys 51™ [Sarepta Therapeutics]), exon 53 (golodirsen, Vyondys 53™ [Sarepta Therapeutics]; viltolarsen, Viltepso™ [NS Pharma]), and exon 45 (casimersen, Amondys45™ [Sarepta Therapeutics]) [1]. In terms of efficacy and safety, other exon-skipping PMOs have also been in clinical trials, such as the drug NS-089/NCNP-02 for exon 44 skipping (ClinicalTrials.gov Identifier: NCT04129294).

Numerous studies to develop exon-skipping drugs have been conducted using various cell and animal models of DMD. Of them, as demonstrated in our previous reports [5–8], immortalized cell lines derived from patient skeletal muscles provide great opportunities in screening many PMO drug candidates because they have excellent proliferation and differentiation abilities, unlike primary skeletal muscle cells [9]. Although the FDA-approved PMOs are promising, further development of exon-skipping technology is required for patients to more benefit from the therapy [10]. PMOs can induce greater efficiency at skipping a target exon when more appropriate sequences of target pre-mRNA and

PMOs are discovered [6, 11]. Immortalized patient muscle cell lines have a significant advantage in finding such optimal PMO sequences because they allow for the quantitative evaluation of exon-skipping efficiency and dystrophin rescue under appropriate experimental conditions [5–7, 12]. This chapter describes a methodology for evaluating the ability of PMOs to skip exon 44 or 53 in immortalized DMD patient-derived skeletal muscle cells.

2 Materials

2.1 Immortalized Skeletal Muscle Cell Culture

1. Immortalized healthy or DMD patient-derived skeletal muscle cells [5]. The immortalized cells with exonic deletion mutations amenable to exon 44 or 53 skipping can be obtained by request from the MRC Centre for Neuromuscular Diseases Biobank.
2. Growth media: Dulbecco's modified Eagle's medium/nutrient mixture F-12 (DMEM/F-12 with 2.5 mM L-glutamine and 15 mM HEPES, GIBCO) containing skeletal muscle growth medium supplement mix (Promocell, #C-39365), 50 units/mL penicillin–50 μg/mL streptomycin (GIBCO), and 20% fetal bovine serum (*see* **Note 1**).
3. Differentiation media: DMEM/F-12 with 2.5 mM L-glutamine and 15 mM HEPES containing 50 units/mL penicillin–50 μg/mL streptomycin (GIBCO) and 2% horse serum (heat inactivated) (*see* **Note 2**).

2.2 PMO-Based Antisense Oligonucleotides

1. PMO sequences designed according to a guideline, such as one from our previous report, of which the algorisms provide the predictive efficiency of exon skipping [5]. Sample sequences are available in Table 1.
2. Designed and mock PMOs purchased from Gene Tools LLC (*see* **Note 3**).
3. Transfection reagent, Endo-Porter (Gene Tools LLC).
4. Collagen type I-coated multi-well plates (Corning).

2.3 RT-PCR for Evaluating Exon-Skipping Efficiency

1. TRIzol reagent (Invitrogen).
2. Chloroform.
3. Isopropanol.
4. Glycogen 20 mg/mL, RNA grade (Thermo Scientific).
5. 75% ethanol.
6. RNase-free water.
7. One-Step RT-PCR kit, e.g., SuperScript™ III One-Step RT-PCR system with Platinum™ Taq (Invitrogen).

Table 1
Sample 30-mer PMO sequences to test exon 44 or 53 skipping efficiency [5]

Target	Name	Oligo sequence (5′ to 3′)	Distance from ASS	Predicted skip %
Exon 44	hEx44_Ac7-30mer	AAAACGCCGCCATTTCTCAACAGATCT GTC	7	88.2
	hEx44_Ac2-30mer	GCCGCCATTTCTCAACAGATCTGTCAA ATC	2	87.1
	hEx44_Ac14-30mer	CATAATGAAAACGCCGCCATTTC TCAACAG	14	83.0
	hEx44_Ac0-30mer	CGCCATTTCTCAACAGATCTGTCAAA TCGC	0	80.1
	hEx44_Ac18-30mer	ATATCATAATGAAAACGCCGCCATTTC TCA	18	63.3
	hEx44_Ac56-30mer	ACTGTTCAGCTTCTGTTAGCCACTGA TTAA	56	52.0
	hEx44_Ac118-30mer	CTTAAGATACCATTTGTATTTAGCATG TTC	118	16.8
	hEx44_Ac85-30mer	ATTCTCAGGAATTTGTGTCTTTC TGAGAAA	85	7.8
Exon 53	hEx53_Ac9-30mer	GTTCTTGTACTTCATCCCACTGATTC TGAA	9	73.9
	hEx53_Ac18-30mer	TCTGAAGGTGTTCTTGTACTTCATCCC ACT	18	59.0
	hEx53_Ac26-30mer	CCTCCGGTTCTGAAGGTGTTCTTGTAC TTC	26	75.2
	hEx53_Ac30-30mer	GTTGCCTCCGGTTCTGAAGGTGTTCTT GTA	30	69.0
	hEx53_Ac40-30mer	TCATTCAACTGTTGCCTCCGGTTC TGAAGG	40	50.7
	hEx53_Ac64-30mer	CATTGTGTTGAATCCTTTAACATTTCA TTC	64	2.1
	hEx53_Ac115-30mer	TTGGCTCTGGCCTGTCCTAAGACCTGC TCA	115	10.9
	hEx53_Ac198-30mer	TATCTTTGATACTAACCTTGGTTTCTG TGA	198	−14.4

ASS acceptor splice site

8. Primers designed to amplify a region of interest. Sample primers are available in Table 2.
9. 1.5% agarose gel for electrophoresis.
10. SYBR™ Safe DNA Gel Stain (Invitrogen).
11. Equipment: cooled centrifuge, PCR thermal cycler, gel electrophoresis machine, and UV transilluminator or similar imaging system.
12. ImageJ software (NIH).

Table 2
Sample RT-PCR primers to detect mRNA transcripts following exon 44 or 53 skipping [5]

Detection	Name	Sequence 5′ to 3′	Amplicon size
DMD exon 44 skipping in exons 42–43-deleted mRNA	Ex41_56-75_hDMD_Fwd Ex45_105-124_hDMD_Rv	AAGCTGAGGGCTTGTCTGAG GGCTTCCCAATTTTTCCTGT	252 bp in skipping
DMD exon 53 skipping in exon 52-deleted mRNA	Ex51_188-207_hDMD_Fwd Ex54_125-144_hDMD_Rv	GGTGGGTGACCTTGAGGATA GCTTCTCCAAGAGGCATTGA	190 bp in skipping
GAPDH as an internal control	hGAPDH_662-81_Fwd hGAPDH_860-79_Rv	TCCCTGAGCTGAACGGGAAG GGAGGAGTGGGTGTCGCTGT	

2.4 Western Blotting for Quantifying Dystrophin Rescue

1. Lysis buffer: RIPA buffer containing a proteinase inhibitor, e.g., cOmplete, Mini, EDTA-free protease inhibitor cocktail (Roche).
2. Cell scrapers.
3. 21G needles and 1-mL tuberculin syringes.
4. Pierce™ Coomassie (Bradford) Protein Assay Kit (Thermo Scientific).
5. Two times SDS-PAGE sample loading buffer formulated as follows: 10% SDS (w/v), 70 mM Tris–HCl (pH 6.7), 5 mM EDTA–2Na (pH 8.0), 20% glycerol, 0.004% w/v bromophenol blue, and 5% β-mercaptoethanol (*see* **Note 4**).
6. NuPAGE™ Novex™ 3–8% Tris–Acetate Midi gel (Invitrogen).
7. NuPAGE™ Tris–Acetate SDS running buffer (20×) (Invitrogen).
8. NuPAGE™ Antioxidant (Invitrogen).
9. Appropriate protein ladders for high-molecular-weight proteins, e.g., HiMark™ Pre-stained protein standard (Invitrogen).
10. Blotting Filter Papers, 2.5 mm-thick, 8.6 × 13.5 cm (Thermo Fisher).
11. Methanol.
12. Discontinuous buffer system:
    - Concentrated anode buffer: 0.3 M Tris–HCl, 20% methanol.

- Anode buffer: 0.03 M Tris–HCl, 20% methanol.
- Cathode buffer: 25 mM Tris–HCl, 40 mM 6-amino-n-hexanoic acid, 20% methanol, 0.01% SDS.

13. Immobilon polyvinylidene difluoride (PVDF) membrane, 0.45 μm pore size (Millipore).
14. Phosphate-buffered saline (PBS) with 0.05% Tween 20 (PBST).
15. Coomassie Brilliant Blue G-250 Dye, e.g., PageBlue™ Protein Staining Solution (Thermo Scientific).
16. Amersham™ ECL Prime Blocking Agent (Cytiva).
17. Anti-dystrophin antibodies: rabbit polyclonal antibody against the C-terminal domain of dystrophin (ab15277, Abcam) or mouse monoclonal antibody against the rod domain of dystrophin (NCL-DYS1, Leica Biosystems).
18. Secondary antibodies: horseradish peroxidase (HRP)-conjugated anti-rabbit or anti-mouse IgG H + L antibodies (Bio-Rad).
19. Amersham™ ECL Select Western Blotting Detection Reagent (Cytiva).
20. Equipment: heat block/water bath, XCell4 SureLock™ Midi-Cell SDS-PAGE tank (Thermo Fisher), Novex™ Semi-Dry Blotter (Life Technologies), power supply, chemiluminescence detection system.
21. ImageJ software (NIH).

3 Methods

3.1 Culture of Immortalized Healthy or Patient-Derived Skeletal Muscle Cells

1. Prepare and warm 30 mL of growth media per a frozen cell vial in a water bath at 37 °C.
2. Thaw the frozen vial containing approximately 2–3 × 10^6 cells and take it out, leaving a small piece of ice in the vial.
3. Gently transfer cells into the growth media along a tube wall, and mix it using a 25-mL serological pipette a few times but not make bubbles. Put all the cell solution into a 225-cm^2 plastic flask. An immortalized skeletal muscle cell line derived from healthy donors serves as a control toward patient-derived cell lines.
4. Grow cells for 2–3 days at 37 °C with 5% CO_2 until reaching approximately 80–90% confluency. The cellular morphology will change from round to spindle shape during the proliferation. Note that do not make the cells fully confluent in the flask, which may affect the proliferation ability of cells.
5. Adjust the number of cells at 1 × 10^5 cells/mL with GM.

6. Seed cells at approximately 1.7×10^4 cells/cm^2 in a 12- or 24-well plate coated with collagen type I.
7. Culture for approximately 48 h after the seeding to allow cell density to reach approximately ~90% confluency.
8. Replace the growth media with 400 and 800 μL of differentiation media in the 12- and 24-well plates, respectively. Culture for 3 days to differentiate cells to myotubes with multiple nuclei for use in PMO transfection.

3.2 PMO Transfection

1. Incubate stock PMOs at 1 mM in a water bath at 65 °C for 10 min just before making a working solution (*see* **Note 5**).
2. Make solvent mixture containing 6 μM transfection reagent, Endo-Porter in differentiation media.
3. Adjust the concentration of PMOs (~10 μM) with the solvent mixture.
4. Replace the spent media with the working PMO solution, and incubate for 48 h.
5. Aspirate the PMO solution and put a fresh differentiation media.
6. In 72 h of incubation, harvest myotubes for subsequent experiments as follows.

3.3 Total RNA Extraction from the Immortalized Cell-Derived Myotubes

1. Aspirate the media from wells as much as possible, and put 1 mL of TRIzol reagent into the cells/well. Wash out the well-bottom with the reagent roughly by pipetting for lysing cells, and then collect the cell suspension into a 1.7 mL microtube. The suspension can also be stored in a − 80 °C freezer until use.
2. Vortex the tubes for 15 s.
3. Add 200 μL of chloroform and shake tubes vigorously for 30 s according to the manufacture's instruction, and then incubate for 5 min at room temperature.
4. Centrifuge the tubes at 12,000 × g for 15 min at 4 °C. The cooling should be in advance.
5. Gently take the tube out of the centrifuge and transfer the upper aqueous RNA phase into a fresh tube.
6. Add 10 μg RNase-free glycogen and 0.5 mL of 100% isopropanol, followed by vortex briefly. Incubate for 10 min at room temperature.
7. Centrifuge the tubes at 12,000 × g for 15 min at 4 °C.
8. Make sure the precipitation of RNA as a pellet on the tube bottom. Decant the supernatant; at this procedure, the pellet does not flow out due to surface tension. Spin down and aspirate the residual supernatant using a micropipette. Be careful not to disrupt and discard the RNA pellet.

9. Add 1 mL of 75% ethanol into and vortex briefly the tube for washing the pellet.
10. Centrifuge samples at 7500 × g for 5 min at 4 °C.
11. Repeat **step 8** to remove the supernatant.
12. Air-dry the pellet opening the cap and covering with paper for 5 min at room temperature. Ensure no ethanol in the tubes because ethanol carry-over can interrupt downstream RT-PCR.
13. Resuspend the pellet with 30–40 μL of RNase-free water by pipetting with a filtered tip several times to dissolve. Incubate the suspension in a heat block device for 10–15 min at 60 °C. Immediately after the warming, keep it on ice.
14. Measure and adjust RNA concentration (e.g., 40 ng/μL) with nuclease-free water. The RNA samples are stored at −80 °C until use.

3.4 RT-PCR and Evaluation of Exon-Skipping Efficiency

1. Prepare the one-step RT-PCR mixture for a 25 μL reaction using a one-step RT-PCR system, such as SuperScript™ III One-Step RT-PCR System with Platinum™Taq DNA Polymerase (Invitrogen), as follows.
2. Mix 12.5 μL two × reaction buffer, 0.5 μL of each forward and reverse primer at 10 μM, 1 μL of SuperScript™ III RT/Platinum™ Taq Mix, 200 ng total RNA, and a necessary amount of nuclease-free water to fill up to 25 μL reaction mixture.

   Sample primers designed to detect target *DMD* mRNAs and an internal control mRNA are available in Table 2.
3. Set and run a cycle reaction according to the manufacturer's instruction while modifying the condition depending on primers and amplicon sizes: e.g., 50 °C for 5 min; 94 °C for 2 min; 35 cycles at 94 °C for 15 s, 60 °C for 30 s, and 68 °C for 35 s; 68 °C for 5 min; 4 °C for 5 min, and 15 °C min to hold.
4. Electrophorese PCR products on a 1.5% agarose gel for 30 min to separate native, intermediate, and exon-skipped *DMD* mRNA bands.
5. Stain the gel with SYBR Safe DNA Gel Stain (Invitrogen) for 30 min with shaking, and then wash it shortly in distilled water to reduce the background signal.
6. Visualize the PCR bands with a UV transilluminator. Set appropriate exposure time not to saturate the band signal intensities. Once the condition is confirmed, take an image for analysis. Sample results are shown in Fig. 2a.
7. Quantify the signal intensities of both native, intermediate, and skipped bands using ImageJ software (*see* **Note 6**).

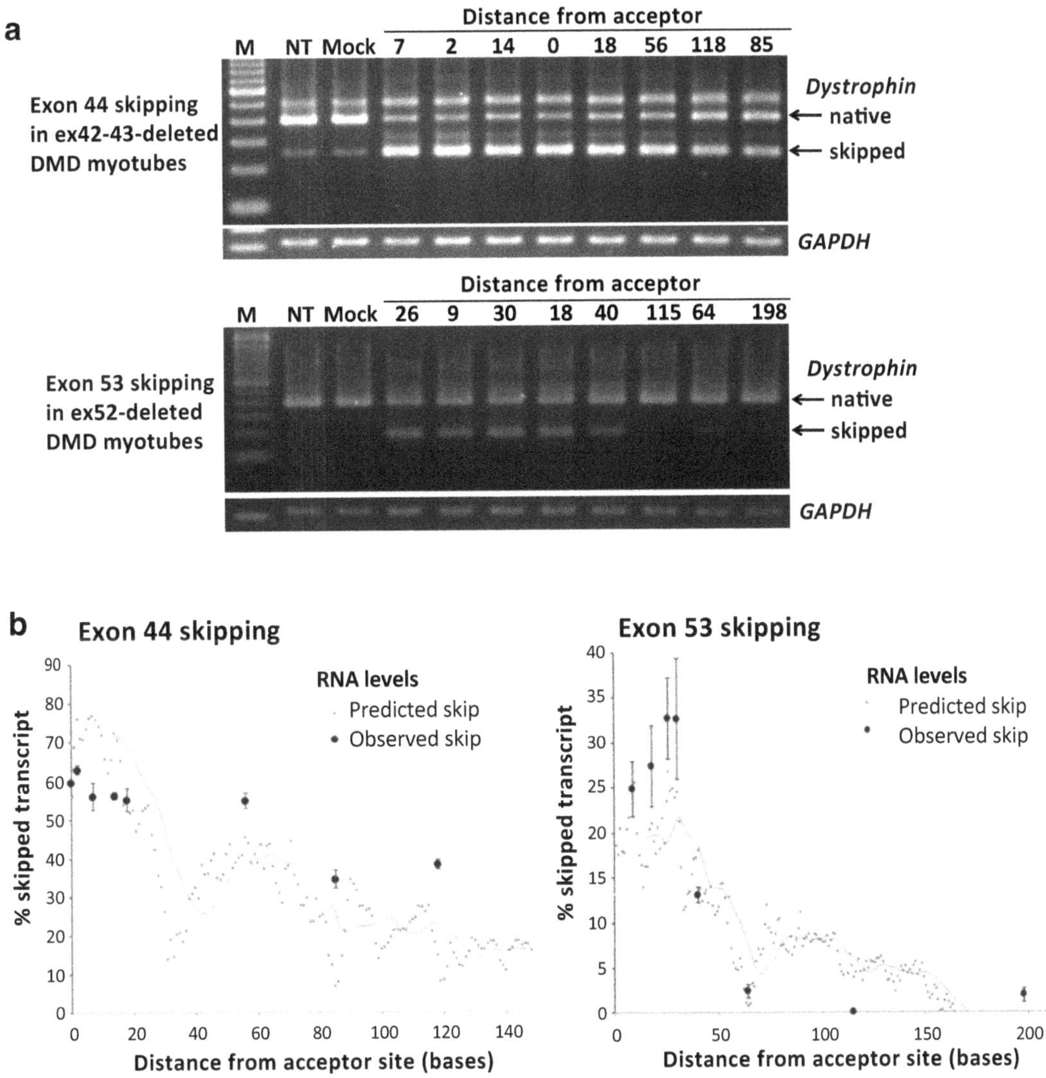

Fig. 2 Sample RT-PCR results of exon 44 and 53 skipping using ten μM PMOs in DMD patient-derived immortalized myotubes with the deletion mutations of exons 42–43 and exon 52, respectively. (**a**) Representative gel images showing the bands of the native and exon-44-skipped (252 bp) or exon-53-skipped (190 bp) transcripts in the immortalized DMD cell lines following exon-skipping treatments. M: 100 bp ladder, NT: non-treated, Mock: random 31-mer; test PMOs are numbered according to their distance from the acceptor splice site. As a loading control, *GAPDH* is shown. (**b**) Actual exon skipping efficiencies (%; mean, standard deviation) calculated in three independent experiments. The actual skipping levels were plotted compared with the predicted exon-skipping efficiencies against distance from the exon acceptor splice site (see Ref. [5] for the predictive algorisms). Images are licensed with permission under a CC BY 4.0. If you would like to know the contents of the license, please check https:/creativecommons.org/licenses/by/4.0/. (Slightly modified from Echigoya et al. [5])

8. Calculate exon-skipping efficiency using the following formula: exon-skipped band intensity/(native + intermediate + exon-skipped band intensities) × 100 (%). Sample results of the efficiency are shown in Fig. 2b.

3.5 Preparation of Protein Samples from Immortalized Myotubes

1. Aspirate the media from a 12-well-plate as much as possible, and add 100 μL/well of cold lysis buffer containing a proteinase inhibitor.
2. Detach cells and collect the debris to one side using a cell scraper.
3. Transfer the debris into a fresh 1.7 mL microtube, and keep it on ice for 30 min to promote the lysis.
4. Homogenize the cells by passing through a 21G needle with a 1-mL tuberculin syringe ten times. Avoid forming bubbles that can degrade protein.
5. Centrifuge the samples at maximum speed (>14,000 × g) for 15 min at 4 °C.
6. Transfer the supernatant into a fresh tube and keep on ice or store it at −80 °C until use.
7. Quantify the concentration of protein extracts using a Coomassie (Bradford) Kit and adjust the concentration with the lysis buffer (e.g., 1 μg/μL).
8. Mix the protein extracts with two × SDS sample buffer prepared according to the formula shown in the material section. Heat the mixture at 70 °C for 10 min. Spin down the tubes and keep them at room temperature until used for SDS-PAGE (*see* **Note 7**).

3.6 SDS-PAGE and Western Blotting for Quantifying Dystrophin Rescue

1. Prepare one × Tris–Acetate SDS-Running buffer with and without an antioxidant agent, and assemble the SDS-PAGE tank with a precast 3–8% Tris–Acetate Midi gel.
2. Fill the running buffer with and without the antioxidant agent to inner and outer buffer chambers, respectively. Ensure that the wells of the gel are submerged with the running buffer.
3. Flush each well of the gel with the running buffer by pipetting several times.
4. Load the protein samples from DMD myotubes (e.g., 12 μg/well in 24 μL) and those from healthy-donor-derived myotubes with the desired amount to serve as calibration curve and positive control (e.g., 10 to 1% of the protein amount of DMD myotubes).
5. Load a protein ladder for high molecular weight, such as HiMark™ Pre-stained protein standard (Invitrogen), to determine 427 kDa of the full-length dystrophin protein (*see* **Note 8**).
6. Run the gel at 150 V for 75 min.
7. Meanwhile, presoak a PVDF membrane in methanol for 20 s; ensure that the membrane is submerged. Then, transfer into and equilibrate it with the anode buffer until use, for at least 10 min on an orbital shaker.

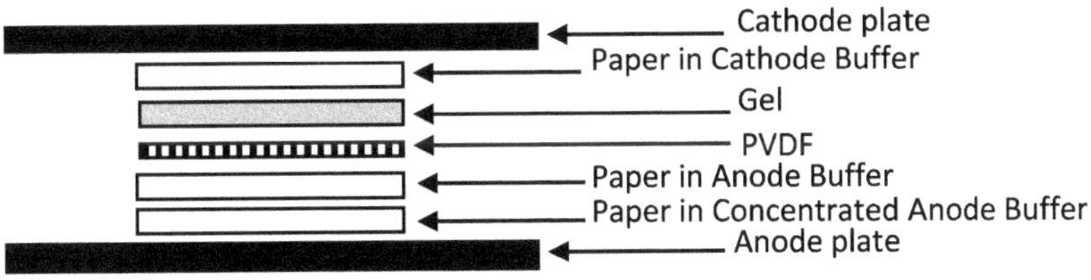

Fig. 3 Semi-dry western blotting assembly for dystrophin protein detection. A gel after SDS-PAGE separation is stacked on two thick papers soaked in anode buffers at different concentrations and PVDF membrane and then sandwiched with a thick paper soaked in cathode buffer

8. Prewet thick blotting papers in the anode, concentrated anode, and cathode buffers, respectively, for 30 min on an orbital shaker.
9. Transfer and rinse the gel after the SDS-PAGE into the cathode buffer.
10. Stack the blotting papers, PVDF membrane, and the gel as shown in Fig. 3.
11. Transfer at 20 V for 70 min.
12. Rinse the membrane with PBST for 5 min with shaking (*see* **Note 9**).
13. Block the membrane with 2% ECL Prime Blocking Reagent in PBST overnight at 4 °C or 1 h at room temperature with shaking.
14. Incubate the membrane in the selected anti-dystrophin antibody diluted in the blocking reagent for 1 h at room temperature with shaking; a dilution ratio of 1:2500 for the rabbit polyclonal anti-dystrophin C-terminal antibody [cat# ab15277, Abcam], and that of 1:400 for the mouse monoclonal anti-dystrophin rod-domain antibody (cat# NCL-DYS1, Leica Biosystems).

    An internal control antibody, such as anti-alpha Tubulin antibody (cat# ab7291, Abcam), is detected on the same membrane; we usually cut the membrane according to the molecular weight difference of the targets and use a single antibody for respectively separated membranes.
15. Wash the membrane three times with PBST for 10 min room temperature with shaking.
16. Incubate with HRP-conjugated secondary antibody diluted at 1:10,000 ratio in PBST, appropriate to the primary antibodies used, for 1 h at room temperature with shaking.
17. Wash the membrane three times with PBST for 10 min at room temperature with shaking.
18. Meanwhile, prepare an enhanced chemiluminescence reagent to detect target protein signals, such as the ECL Select Western

Blotting Detection kit (Cytiva); mix solutions A (luminol) and B (peroxide) of the kit in a 1:1 ratio.

19. React the entire membrane with the detection mixture, and incubate for 5 min at room temperature in the dark.
20. Visualize and save the blot image using a chemiluminescent imaging system while avoiding the saturation of the band intensities. Sample blotting results of immortalized DMD patient-derived myotubes treated with exon 44 and 53 skipping PMOs are shown in Fig. 4a.
21. Using ImageJ, quantify the intensities of the observed dystrophin bands. Construct a calibration curve using the intensities for the healthy control protein standards, and use the equation calculated from the standards to quantify the levels of dystrophin rescue in PMO-treated myotube samples. Sample results are shown in Fig. 4b.

4 Notes

1. DMEM/F-12, skeletal muscle growth medium-supplement mix (Promocell), and antibiotics are pre-mixed as a basal medium. Adding fetal bovine serum (FBS) to the basal medium, prepare a complete growth medium with 20% FBS for only the quantity needed just before use. This preparation method may help prevent or delay the deterioration of the medium.
2. For the same reason as the growth media, horse serum, heat-inactivated is added to DMEM/F-12 immediately before use.
3. Some PMO sequences, such as ones with consecutive guanines (G) or high G content, may be unable to be synthesized. Ask the company for the order of such sequences. PMO as a stock solution is adjusted to 1 mM with nuclease-free water for molecular biology and store at room temperature tightly sealed with parafilm.
4. Prepare concentrated stock component reagents of the SDS-PAGE sample loading buffer: 20% SDS, 0.5 M Tris–HCl, and 0.5 M EDTA-2Na, for making the working solution.
5. Pre-heating the stock PMOs at 65 °C for 10 min just before use can help dissociate the duplex or aggregation of PMO sequences, which may restore PMO activity and may allow for the proper evaluation. Another option to restore the activity may be autoclaving PMOs, which have been shown not to affect the exon-skipping effect, based on our experience.
6. The PCR products should be direct-sequenced to confirm if PMO-mediated exon skipping generates the proper junction of the exons without a target exon.

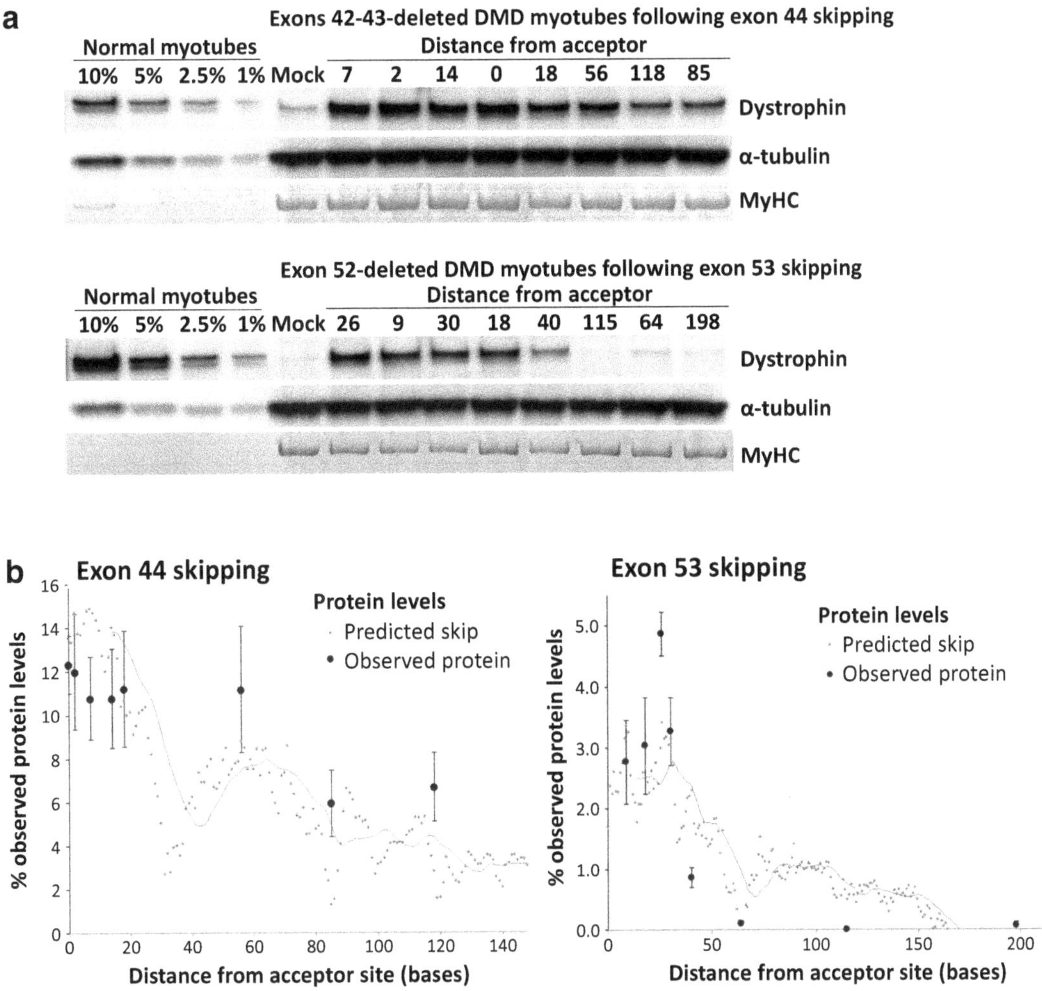

Fig. 4 Sample western blotting results of dystrophin rescue quantification. Immortalized DMD myotubes with the exons 42–43 and exon 52 deletion were treated with exon 44 and 53 skipping PMOs at 10 μM, respectively. (**a**) Western blots that present rescued truncated dystrophin protein by using an anti-dystrophin C-terminal antibody. A calibration curve of full-length dystrophin from immortalized normal myotubes derived from a healthy subject was loaded for comparison; Mock: random 31-mer PMO; test PMOs are numbered according to their distance from the acceptor splice site; MyHC: Myosin Heavy Chain as a loading control and a myogenic marker. (**b**) Actual levels of dystrophin rescue calculated using the standard curve from the healthy myotube bands. The dystrophin protein levels observed in three independent experiments (%; mean, standard deviation) were plotted against the predicted exon-skipping efficiencies and distance from the exon acceptor splice site. Predicted skipped dots shown here are a reference with % of the predicted skipped transcript, of which the scale is the same as that of Fig. 2b. Images are licensed with permission under a CC BY 4.0. If you would like to know the contents of the license, please check https:/creativecommons.org/licenses/by/4.0/. (Slightly modified from Echigoya et al. [5])

7. Prepare two times SDS sample buffer just before use. After heating, keep the heated protein mixture at room temperature, not on ice; cooling the mixtures forms the precipitation of the SDS at a higher concentration in the sample tubes.

8. Load SDS-sample buffer into the remaining wells for the protein samples to enter the gel evenly.
9. Stain the post-transferred gel with a Coomassie Brilliant Blue dye, such as PageBlue™ Protein Staining Solution (Thermo Scientific), for 1 h at room temperature.

Destain the gel in a tray with distilled water and Kimwipes overnight at room temperature with shaking, which allows for the visualization of myosin heavy chain as a loading and differentiation control.

Acknowledgments

This work was partially supported by the Intramural Research Grant (2–6) for Neurological and Psychiatric Disorders of the National Center of Neurology and Psychiatry (NCNP), JSPS KAKENHI (Grant Numbers 18 K14595 and 21 K06976), Nihon University College of Bioresource Sciences, and Nihon University Veterinary Research Center (NUVERC) to Y.E.; Muscular Dystrophy Canada, the Friends of Garrett Cumming Research Fund, the HM Toupin Neurological Science Research Fund, University of Alberta, Canadian Institutes of Health Research (CIHR), Alberta Innovates: Health Solutions (AIHS), Jesse's Journey, and the Women and Children's Health Research Institute (WCHRI) to T.Y.

References

1. Duan D, Goemans N, Takeda S, Mercuri E, Aartsma-Rus A (2021) Duchenne muscular dystrophy. Nat Rev Dis Primers 7(1):13. https://doi.org/10.1038/s41572-021-00248-3
2. Nakamura A, Shiba N, Miyazaki D, Nishizawa H, Inaba Y, Fueki N, Maruyama R, Echigoya Y, Yokota T (2016) Comparison of the phenotypes of patients harboring in-frame deletions starting at exon 45 in the Duchenne muscular dystrophy gene indicates potential for the development of exon skipping therapy. J Hum Genet. https://doi.org/10.1038/jhg.2016.152
3. Bladen CL, Salgado D, Monges S, Foncuberta ME, Kekou K, Kosma K, Dawkins H, Lamont L, Roy AJ, Chamova T, Guergueltcheva V, Chan S, Korngut L, Campbell C, Dai Y, Wang J, Barisic N, Brabec P, Lahdetie J, Walter MC, Schreiber-Katz O, Karcagi V, Garami M, Viswanathan V, Bayat F, Buccella F, Kimura E, Koeks Z, van den Bergen JC, Rodrigues M, Roxburgh R, Lusakowska A, Kostera-Pruszczyk A, Zimowski J, Santos R, Neagu E, Artemieva S, Rasic VM, Vojinovic D, Posada M, Bloetzer C, Jeannet PY, Joncourt F, Diaz-Manera J, Gallardo E, Karaduman AA, Topaloglu H, El Sherif R, Stringer A, Shatillo AV, Martin AS, Peay HL, Bellgard MI, Kirschner J, Flanigan KM, Straub V, Bushby K, Verschuuren J, Aartsma-Rus A, Beroud C, Lochmuller H (2015) The TREAT-NMD DMD global database: analysis of more than 7,000 Duchenne muscular dystrophy mutations. Hum Mutat 36(4):395–402. https://doi.org/10.1002/humu.22758
4. Echigoya Y, Lim KRQ, Nakamura A, Yokota T (2018) Multiple exon skipping in the Duchenne muscular dystrophy hot spots: prospects and challenges. J Pers Med 8(4):41. https://doi.org/10.3390/jpm8040041
5. Echigoya Y, Mouly V, Garcia L, Yokota T, Duddy W (2015) In silico screening based on predictive algorithms as a design tool for exon skipping oligonucleotides in duchenne muscular dystrophy. PLoS One 10(3):e0120058.

https://doi.org/10.1371/journal.pone.0120058

6. Echigoya Y, Lim KRQ, Trieu N, Bao B, Miskew Nichols B, Vila MC, Novak JS, Hara Y, Lee J, Touznik A, Mamchaoui K, Aoki Y, Takeda S, Nagaraju K, Mouly V, Maruyama R, Duddy W, Yokota T (2017) Quantitative antisense screening and optimization for exon 51 skipping in Duchenne muscular dystrophy. Mol Ther 25(11):2561–2572. https://doi.org/10.1016/j.ymthe.2017.07.014
7. Echigoya Y, Lim KRQ, Melo D, Bao B, Trieu N, Mizobe Y, Maruyama R, Mamchaoui K, Tanihata J, Aoki Y, Takeda S, Mouly V, Duddy W, Yokota T (2019) Exons 45-55 skipping using mutation-tailored cocktails of antisense Morpholinos in the DMD gene. Mol Ther. https://doi.org/10.1016/j.ymthe.2019.07.012
8. Lim KRQ, Maruyama R, Echigoya Y, Nguyen Q, Zhang A, Khawaja H, Sen Chandra S, Jones T, Jones P, Chen YW, Yokota T (2020) Inhibition of DUX4 expression with antisense LNA gapmers as a therapy for facioscapulohumeral muscular dystrophy. Proc Natl Acad Sci U S A 117(28):16509–16515. https://doi.org/10.1073/pnas.1909649117
9. Mamchaoui K, Trollet C, Bigot A, Negroni E, Chaouch S, Wolff A, Kandalla PK, Marie S, Di Santo J, St Guily JL, Muntoni F, Kim J, Philippi S, Spuler S, Levy N, Blumen SC, Voit T, Wright WE, Aamiri A, Butler-Browne-G, Mouly V (2011) Immortalized pathological human myoblasts: towards a universal tool for the study of neuromuscular disorders. Skelet Muscle 1:34. https://doi.org/10.1186/2044-5040-1-34
10. Sheikh O, Yokota T (2021) Developing DMD therapeutics: a review of the effectiveness of small molecules, stop-codon readthrough, dystrophin gene replacement, and exon-skipping therapies. Expert Opin Investig Drugs 30(2):167–176. https://doi.org/10.1080/13543784.2021.1868434
11. Echigoya Y, Nakamura A, Nagata T, Urasawa N, Lim KR, Trieu N, Panesar D, Kuraoka M, Moulton HM, Saito T, Aoki Y, Iversen P, Sazani P, Kole R, Maruyama R, Partridge T, Takeda S, Yokota T (2017) Effects of systemic multiexon skipping with peptide-conjugated morpholinos in the heart of a dog model of Duchenne muscular dystrophy. Proc Natl Acad Sci U S A 114(16):4213–4218. https://doi.org/10.1073/pnas.1613203114
12. Lim KRQ, Yokota T (2018) Quantitative evaluation of exon skipping in immortalized muscle cells in vitro. Methods Mol Biol 1828:127–139. https://doi.org/10.1007/978-1-4939-8651-4_7

Chapter 8

Restoring Dystrophin Expression with Duchenne Muscular Dystrophy Exon 45 Skipping in Induced Pluripotent Stem Cell-Derived Cardiomyocytes

Mitsuto Sato, Naoko Shiba, Daigo Miyazaki, Yuji Shiba, and Akinori Nakamura

Abstract

Induced pluripotent stem cell (iPSC)-based disease model is a useful tool that can represent the pathophysiology of patient organs that are inaccessible due to invasiveness. Here, we present a method to induce differentiation of Duchenne muscular dystrophy (DMD) patient-derived iPSCs into cardiomyocytes and restore dystrophin expression by exon skipping using antisense nucleic acids. This involves a 20-day multistep culture process for differentiation to cardiomyocytes, followed by exon-skipping experiments. Additionally, RT-PCR, western blotting, and immunocytochemistry are used to confirm the restoration of dystrophin expression.

Key words Duchenne muscular dystrophy (DMD), Induced pluripotent stem cells (iPSCs), Exon skip, Antisense oligonucleotide (ASO), Cell culture, Cardiomyocytes, RT-PCR, Western blotting, Immunocytochemistry

1 Introduction

Duchenne muscular dystrophy (DMD) is a lethal X-linked muscle disease that develops into whole-body skeletal muscle atrophy and weakness, resulting in patients losing their ability to walk by the age of 13 [1]. Cardiac muscle dysfunction is also observed after the age of 18 [2]. Previously, the most common cause of death in DMD was respiratory failure. However, with the improvement of respiratory support and the introduction of steroid therapy, DMD cardiomyopathy has become the greatest prognostic factor, resulting in heart failure and ventricular arrhythmias, leading to sudden death [3, 4].

In recent years, the US Food and Drug Administration has approved four exon-skipping antisense oligonucleotides (ASOs) for the treatment of DMD. In fact, recovery of dystrophin in

Rika Maruyama and Toshifumi Yokota (eds.), *Muscular Dystrophy Therapeutics: Methods and Protocols*, Methods in Molecular Biology, vol. 2587, https://doi.org/10.1007/978-1-0716-2772-3_8,

skeletal muscle has been confirmed, and maintenance and improvement of motor function are also expected [5, 6]. However, the therapeutic effect on the myocardium has not yet been elucidated. It has been shown that ASOs have lesser effect on dystrophin recovery in cardiac muscles than in skeletal muscles [7]. The pathogenesis of DMD cardiomyopathy is not as well understood as that of skeletal muscles, and the efficacy of therapeutic agents for DMD myocardium, including antisense nucleic acids, remains to be determined.

Human induced pluripotent stem cells (iPSCs) represent an attractive platform for overcoming the current limitations of animal models for disease modeling and drug discovery. Patient-derived iPSCs may be able to represent pathologies of organs that are too invasive to approach, such as the heart and central nervous system. Cardiomyocytes derived from iPSCs (iPSC-CMs) with patient-specific mutations can be used to model the molecular mechanisms underlying the pathophysiology of the disease and to screen for responses to various therapeutic agents [8].

In this protocol, we describe how to differentiate DMD patient-derived iPSCs into cardiomyocytes and restore dystrophin expression by exon skipping using ASOs. We also present methods for evaluating exon-skipping efficiency and dystrophin recovery.

2 Materials

2.1 iPSC Culture

1. 6-well flat-bottom culture plates.
2. Essential 8™ Medium (E8) (Gibco).
3. EDTA solution: Versene Solution (Gibco).
4. Dulbecco's formula without magnesium and calcium (PBS).
5. Matrigel matrix (Corning): Dissolve in RPMI 1640 medium at a dilution of 1:60 for thin coating of culture plates (*see* **Note 1**).
6. Rho-associated protein kinase (ROCK) inhibitor Y27632 dihydrochloride: Dissolve in DMEM to make 5 mM stock solution and store at −20 °C.
7. 5% CO_2 incubator.
8. Phase contrast microscope.

2.2 Cardiomyocytes Differentiation

1. CHIR99021: Dissolve in DMSO to make a 1 mM stock solution. Filter sterilize, aliquot, and store at −20 °C.
2. RPMI1640 medium with 300 mg/L L-glutamine.
3. B-27™ Supplement, minus insulin (Gibco).
4. Recombinant Human/Mouse/Rat Activin A Protein: Dissolve in 4 mM HCl to make 100 μg/mL stock solution.

5. Matrigel matrix (Corning): Dissolve in RPMI 1640 medium at a dilution of 1:3 for Matrigel overlay (*see* **Note 2**).
6. B-27™ Supplement (50X), serum free (Gibco).
7. Recombinant Human BMP-4 Protein: Dissolve in 4 mM HCl + 0.1% BSA to make a 50 μg/mL stock solution.
8. XAV 939: Dissolve in DMSO to make 1 mM stock solution.
9. 12-well flat-bottom culture plates.

2.3 Transfection of ASOs for Exon 45 Skipping

1. Antisense oligonucleotide designed to skip exon 45 (Gene Tools).
2. Standard control oligonucleotide (Gene Tools).
3. Endo-Porter (Gene Tools).
4. Sterile water.

2.4 RT-PCR for Evaluation of Exon-Skipping Efficiency

1. RNA extraction kit.
2. Spectrophotometer.
3. Reverse Transcription Kit.
4. Taq DNA polymerase.
5. 1X Polymerase chain reaction buffer.
6. Oligonucleotide primers.
7. Thin-walled polycarbonate PCR tubes.
8. Thermal cycler.
9. Agarose.
10. Electrophoresis chamber.
11. 5X DNA-loading dye.
12. DNA molecular-weight ladder.
13. Image J software (http://rsbweb.nih.gov/ij/).

2.5 Western Blotting

1. Radioimmunoprecipitation assay (RIPA) buffer.
2. Protease inhibitors.
3. Probe sonicator.
4. BCA protein assay kit.
5. NuPAGE® Bis-Tris Electrophoresis System (Invitrogen): NuPAGE Antioxidant, NuPAGE LDS Sample Buffer, NuPAGE Sample Reducing Agent, and NuPAGE Tris–Acetate SDS Running Buffer.
6. NuPAGE 3–8% Tris–Acetate Protein Gels (Invitrogen).
7. XCell SureLock Mini-Cell (Invitrogen).
8. Semi-dry Transfer Apparatus.
9. Western blot transfer buffer: EzFastBlot HMW (Atto).

10. Hydrophilic polyvinylidene fluoride (PVDF) membranes (pore size: 0.45 μm).
11. Methanol.
12. iBind Western System (Thermo Fisher Scientific): iBind Western Device, iBind cards, and iBind Solution kit.
13. Rabbit anti-dystrophin antibody: ab15277 (abcam).
14. HRP-conjugated anti-rabbit antibody.
15. Western Blotting Detection Reagent.
16. Charge-coupled device (CCD) camera-based imager.

2.6 Immunocytochemistry

1. 4-well chamber slides.
2. 4% paraformaldehyde in PBS.
3. 0.2% Triton X-100 substitute in PBS.
4. 5% goat serum in PBS.
5. Rabbit anti-dystrophin antibody: ab15277 (abcam).
6. Mouse anti-cardiac troponin T: sc-20025 (Santa Cruz Biotechnology).
7. Alexa Fluor 568-conjugated goat anti-rabbit IgG (H + L).
8. Alexa Fluor 488-conjugated goat anti-mouse IgG (H + L).
9. Permanent mounting medium with DAPI.
10. Glass coverslips.

3 Methods

We present an example of a patient with a genetic mutation that is in-framed by exon 45 skipping. Representative deletion mutation patterns that have been shown to be in-framed by exon 45 skipping are 44, 46–47, 46–48, 46–49, 46–51, and 46–53. It is estimated that approximately 9% of patients with DMD are eligible for exon 45 skipping therapy. In this study, the results of experiments using iPSCs from DMD patients with the DMD-exon 46–55 deletion are shown in the figures. The exon 45–55 deletion pattern, which is an in-frame mutation after 45 skipping for exons 46–55 deletion, shows a mild clinical phenotype.

3.1 Preparation of iPSCs

It requires the establishment of iPSC lines from patients with DMD gene mutations that are in-frame after exon 45 skipping. There are some protocols for iPSC generation; we established iPSC lines from T lymphocytes as previously described [9]. The iPSCs were routinely cultured on Matrigel (1:60)-coated 6-well plates in Essential 8 (E8) medium in a humidified 5% CO_2 incubator at 37 °C. The medium was refreshed daily, and cells were routinely passaged at a

Day	-12h	-1	0	1	2	3	4	5	6	7	8	9	10
Phase	Harvest	Differentiation and Maturation											
Molecules	Y-27632	CHIR	ActivinA Matrigel	BMP-4 CHIR		XAV939							
Medium	E8		RPMI/B27-ins							BPMI/B27			

Fig. 1 Protocol for the myocardial differentiation of human iPSCs. Undifferentiated iPSC lines are plated on Matrigel-coated culture dishes and cultured in Essential 8 medium. Before differentiation, the medium is supplemented with 1 mM CHIR99021. On day 0, the medium is changed to RPMI 1640 medium containing B27 (RPMI/B27) without insulin (RPMI/B27-ins) with 100 ng/mL activin A and Matrigel (1:60). The medium is replaced on day 1 with RPMI/B27-ins containing 10 ng/mL BMP-4 and 1 mM CHIR99021. On days 3–4, the medium is replaced with RPMI/B27-ins containing 1 mM XAV939. On day 7, the cells are grown in RPMI/B27 medium, and the medium is changed every 2 days

split ratio of 1:7 using EDTA solution when the colonies reached 80% confluence. The addition of ROCK inhibitor is essential for the effective survival and recovery of singularized iPSCs following seeding. For each passage, 5 μM ROCK inhibitor was added to the medium and incubated for 24 h.

3.2 Differentiation into Cardiomyocytes and Their Enrichment

iPSCs were differentiated into cardiomyocytes, as previously described [10] with modifications (Fig. 1). The optimized procedure is given below.

1. When the undifferentiated iPSC culture becomes 80 – 90% confluent, remove the medium and wash cells with PBS, split all the cells using EDTA solution, and seed at $2.0 \times 10^5/cm^2$ onto a new Matrigel-coated 6-well flat bottom plate and culture in E8.
2. Twelve hours after cell seeding (Differentiation Day −1), the cells should be close to confluence. Add 1 μM of CHIR99021 to the medium.
3. Differentiation day 0: Replace the medium with RPMI 1640 medium containing B27 (RPMI/B27) without insulin (RPMI/B27-ins) with 100 ng/mL activin A and Matrigel (at a 1:60 dilution) (*see* **Note 4**).
4. Differentiation day 1: Replace the medium with RPMI/B27-ins containing 10 ng/mL BMP-4 and 1 μM CHIR99021.
5. Differentiation 3–4: Replace the medium with RPMI/B27-ins containing 1 μM XAV939.
6. Differentiation day 5: Change the medium to RPMI/B27-ins.
7. Differentiation day 7: Replace the medium to RPMI/B27, and change the medium every 2 days. When myocardial differentiation is successful, cell morphology and coloration change (Fig. 2), and spontaneous beating of cardiomyocytes is observed from days 7 to 10 post-iPSC differentiation (*see* **Note 5**).

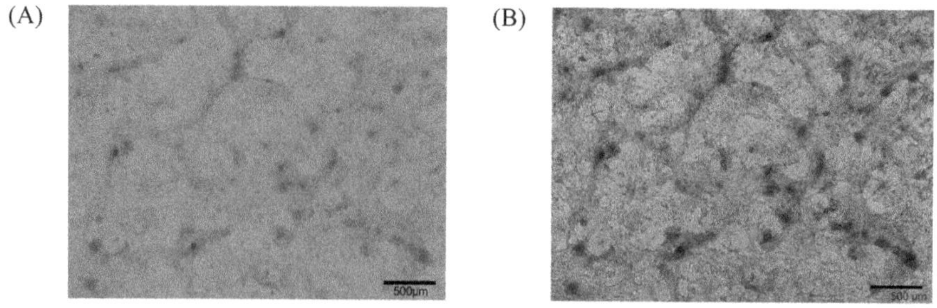

Fig. 2 Representative images of the iPSC-CMs. (**a**) Bright-field image and (**b**) phase-contrast image of iPSC-CMs on differentiation day 10. Scale bar = 200 μm

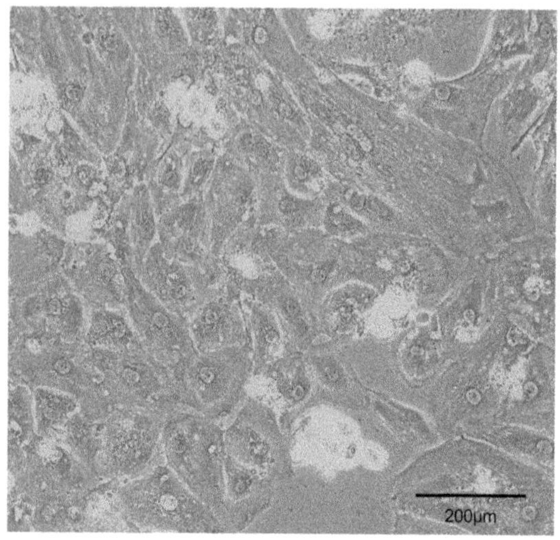

Fig. 3 Representative images of the iPSC-CMs after cardiomyocyte enrichment. Phase-contrast microscope image of iPSC-CMs on differentiation day 20. Scale bar = 100 μm

8. Change the medium every 2 days and wait for the cardiomyocytes' differentiation and maturation (Fig. 3).
9. After 20 days of differentiation, exon-skipping experiments using iPSC-CMs can be performed starting.

3.3 Exon 45 Skipping for DMD iPSC-CMs Using Antisense Nucleic Acid

The sequence design of antisense oligos for exon skipping is an important factor that affects the experimental results. The efficacy of ASO is often unpredictable because of the numerous factors involved in exon skipping, making it difficult to select the optimal sequence. Recently, eSkip-Finder (https://eskip-finder.org) has been released as the first web-based resource to help researchers identify effective exon-skipping ASOs [11]. eSkip-Finder is not only a database of exon-skipping ASOs, but also a rapid analysis program that works by specifying the exon/intron sequence and the length of the ASO to identify effective ASOs for exon skipping based on a machine learning model trained on experimental data. It would be very beneficial to use this resource for the sequence design of antisense nucleic acids.

1. Make 1 mM or less of Morpholino stock solution in sterile water (*see* **Note 7**).
2. Mix 1–10 μM ASO and 6 μM Endo-Porter in RPMI/B27 medium (*see* **Note 8**).
3. Replace the medium with a medium containing the transfection reagents and culture humidified at 37 °C and 5% CO_2.
4. After 48 h of incubation with ASO, change the medium to fresh RPMI/B27.
5. From the seventh day after ASO transfection, evaluate the exon-skipping efficiency (*see* Subheading 3.4) (*see* **Note 9**).
6. From the 14th day after ASO transfection, to confirm the recovery of dystrophin expression, collect cell lysates for western blotting, and seed cells onto chamber slides for immunocytochemistry (*see* Subheadings 3.5 and 3.6) (*see* **Note 10**).

3.4 Evaluation of Exon-Skipping Efficiency by RT-PCR

1. From the seventh day after ASO transfection, remove the differentiation medium and wash with 1x PBS. Add cell lysis buffer, lyse the iPSC-CMs and harvest the total RNA using an RNA extraction kit according to the manufacturer's instructions.
2. Measure the RNA concentration with a spectrophotometer.
3. Synthesize complementary DNA using a reverse transcription kit and perform RT-PCR.
4. Perform 1.0% agarose gel electrophoresis to quantify PCR bands and calculate skipping efficiency by densitometry analysis of the RT-PCR product using ImageJ software (Fig. 4) (*see* **Note 11**).

   Exon-skipping efficiency calculation is as below:

   Exon-skipping efficiency (%) = [skipped band/(skipped band + non-skipped band)] × 100.
5. Store the PCR product at 4 °C for a few days or at −20 °C for long-term storage.

3.5 Western Blot Analysis of Dystrophin Expression After Exon Skipping

1. Extract the total protein from the iPSC-CMs using RIPA buffer with protease inhibitors.
2. Sonicate the cell lysates on ice and centrifuge at 15,000× *g* for 15 min at 4 °C.
3. Collect the supernatant and measure protein concentrations using a BCA protein assay kit.

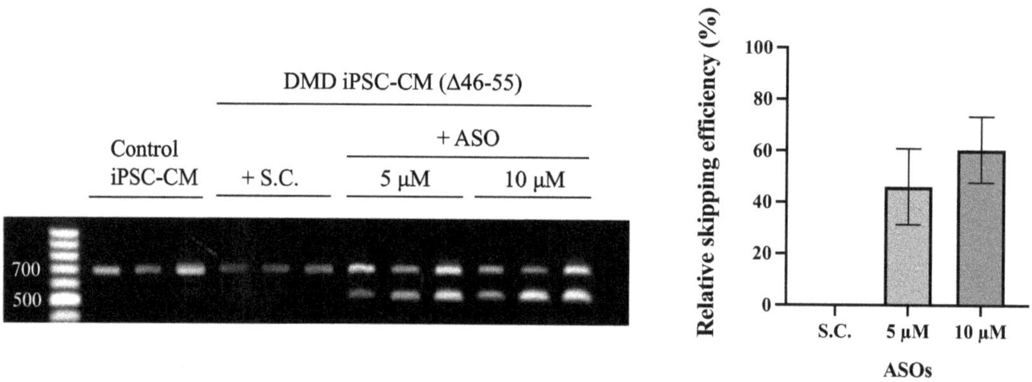

Fig. 4 RT-PCR analysis of dystrophin-mRNA expression in control- and DMD-iPSC-CMs. The forward primer targeted exon 44 and the reverse primer targeted exon 57. (**a**) The lower bands show a 514-bp PCR band, reflecting a 176-bp deletion after exon 45 skipping. (**b**) Quantitation of the RT-PCR-band intensity after exon 45 skipping with 5 or 10 mM PMO. The ratio of the skipping-product density to the total density

4. Add 15 μg of total protein in a microtube and dilute by adding RIPA buffer containing protease inhibitors to a total volume of 10 μL.
5. Prepare the NuPAGE® Bis-Tris electrophoresis system. Mix 5 μL of 4x sample buffer, 2 μL of 10× sample reducing agent, and 3 μL of deionized water.
6. Mix the diluted protein and sample buffer to a total volume of 20 μL. Denature the samples at 70 °C for 10 min.
7. Set the Tris–acetate 3–8% gel in the electrophoresis tank. Load 20 μL of each sample and perform electrophoresis at 150 V for 75 min.
8. Soak a PVDF membrane for 20 s in methanol and then in the blotting buffer until use (at least 10 min).
9. After electrophoresis, cut excess gel and soak in distilled water.
10. Place a blotting paper soaked with blotting buffer, gel, and PVDF membrane on the semi-dry transfer device.
11. Roll out air bubbles between the gel and membrane with a roller.
12. Place the blotting paper soaked with the blotting buffer on top. Transfer at 4 mA/cm^2 for 30 min.
13. After transferring, rinse the membrane twice with distilled water.
14. Prepare the iBind Automated Western System and 1× iBind flex solution (*see* **Note 12**).
15. Place the membrane in 1× iBind flex solution and incubate the membrane for 10 min at RT on a shaker.
16. Prepare rabbit anti-dystrophin antibody (1:500) as the primary antibody in 1× iBind flex solution.

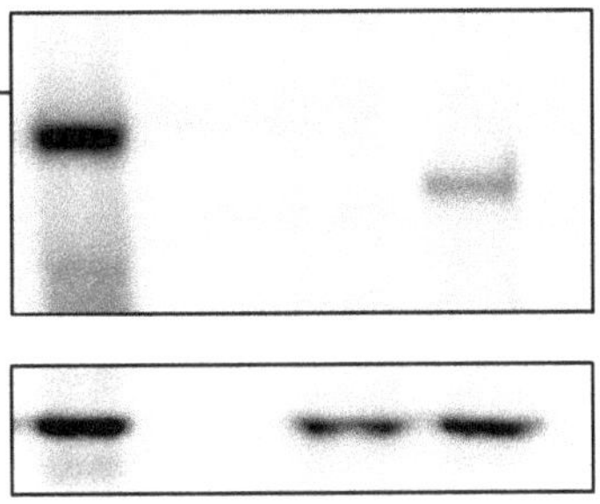

Fig. 5 Western blot analysis of dystrophin in control iPSC-CMs and DMD iPSC-CMs. GAPDH was detected as the loading control. Western blotting analysis revealed restored dystrophin expression, in-frame exon 45 skipping generated truncated dystrophin with a deletion of exons 45–55

17. Prepare HRP-conjugated anti-rabbit antibody (1:2000) as a secondary antibody in 1× iBind flex solution.
18. Open the well cover and add solutions sequentially to the wells according to the manufacturer's instructions.
19. Incubate the membranes with antibodies using iBind Western Device at room temperature for 2.5 h or overnight.
20. Verify that all wells are completely empty, pick up the membrane, and rinse the membrane twice with distilled water.
21. Detect the proteins using a chemiluminescent detection reagent and a CCD camera-based imager (Fig. 5).

3.6 Immunofluorescence Study of Cultured Cells

1. Wash the cells with PBS and detach the cells from the plate using an EDTA solution.
2. Seed the cells onto Matrigel-coated chamber slides.
3. After 3 days, replace the medium and observe the presence of beating live cardiomyocytes under a microscope.
4. Wash once with PBS and fix the iPSC-CMs with 4% paraformaldehyde in PBS for 10 min at room temperature.
5. Permeabilize with 0.2% Triton X-100 substitute in PBS for 5 min.
6. Rinse with PBS twice for 10 min.
7. Block the cells with 5% goat serum for 1 h at room temperature.
8. Incubate with primary antibodies diluted in blocking reagents overnight at 4 °C.
9. Wash the cells with PBS and incubate with secondary antibodies diluted in blocking reagents for 1 h at room temperature.
10. Mount chamber slide with permanent mounting medium with DAPI.
11. Observe the samples under a fluorescence microscope (Fig. 6).

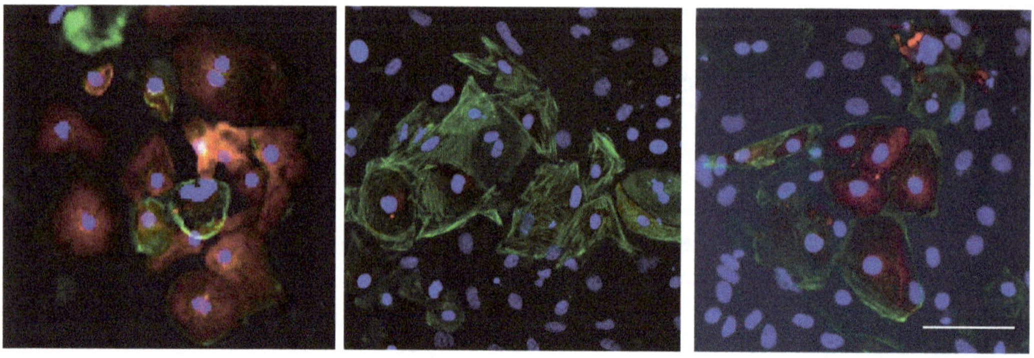

Fig. 6 Immunocytochemistry for dystrophin after ASO treatment. Immunofluorescence of iPSC-CMs with anti-dystrophin (red) and anti-troponin T-C (green) antibodies. Scale bar: 100 mm. Immunocytochemistry revealed restored dystrophin expression

4 Notes

1. Thaw Matrigel in cold room overnight on ice or in a cold storage unit maintained at 4 °C with minimal temperature variation, and store on ice until use. Using a pre-chilled pipette or tube, mix Matrigel uniformly and dilute 1:60 in chilled RPMI medium. Dispense immediately and store at −20 °C. Do not repeat freeze–thaw cycles.
2. In this protocol, a high concentration of Matrigel is coated on the cultured cells during the induction of myogenic differentiation. Thaw Matrigel on ice as described above and dilute 1:3 with RPMI medium using a chilled pipette. Dispense and store at −30 °C.
3. By sandwiching Matrigel on top of cells, it is expected to inhibit cell detachment and promote the induction of differentiation.
4. Empirically, if no beating of the cells is observed after 10 days of cardiomyocyte differentiation, it is assumed that differentiation induction has failed.
5. Cells that have not differentiated into cardiomyocytes are suspended in the medium after 2–3 days of metabolic selection.
6. In most cases, ASOs dissolve easily in water. If the oligo does not dissolve completely in water, warm it at 65 °C, or autoclave it.
7. When transfecting antisense nucleic acids into cardiomyocytes, Endo-Porter can deliver ASOs into cytosol efficiently and uniformly by releasing oligos from endosomes.

8. Exon-skipping efficiency can be assessed at the RNA level after 2–3 days of ASO administration. In our validation, there was no difference in skipping efficiency even 10 days after exon skipping (data not shown).
9. It takes 1–2 weeks to restore dystrophin protein after exon skipping in cardiomyocytes. Therefore, the recovery of dystrophin expression should be analyzed 14 days after transfection. The optimal timing of exon skipping should be considered according to the efficiency of cell differentiation induction, the degree of maturation, and the purpose of the experiment.
10. Or microchip electrophoresis can calculate the exon-skipping efficiency using the molar concentration. It can calculate the skip efficiency more accurately.
11. We used Thermo Fisher Laboratory's iBind Automated Western System because this system not only saves the time and effort of western blotting procedures but also reduces the amount of the antibody used. In addition, the blotting image can also clearly detect bands with a low background. Of course, it is possible to detect dystrophin by western blotting using classical methods.

References

1. Duchenne (1867) The pathology of paralysis with muscular degeneration (Paralysie Myosclerotique), or paralysis with apparent hypertrophy. Br Med J 2(363):541–542
2. Nigro G, Comi LI, Politano L et al (1990) The incidence and evolution of cardiomyopathy in Duchenne muscular dystrophy. Int J Cardiol 26(3):271–277
3. McNally EM (2007) New approaches in the therapy of cardiomyopathy in muscular dystrophy. Annu Rev Med 58:75–88
4. Eagle M, Baudouin SV, Chandler C et al (2002) Survival in Duchenne muscular dystrophy: improvements in life expectancy since 1967 and the impact of home nocturnal ventilation. Neuromuscul Disord 12(10):926–929
5. Mendell JR, Goemans N, Lowes LP et al (2016) Longitudinal effect of eteplirsen versus historical control on ambulation in Duchenne muscular dystrophy. Ann Neurol 79(2): 257–271
6. Komaki H, Nagata T, Saito T et al (2018) Systemic administration of the antisense oligonucleotide NS-065/NCNP-01 for skipping of exon 53 in patients with Duchenne muscular dystrophy. Sci Transl Med 10(437). https://doi.org/10.1126/scitranslmed.aan0713
7. Lu QL, Rabinowitz A, Chen YC et al (2005) Systemic delivery of antisense oligoribonucleotide restores dystrophin expression in bodywide skeletal muscles. Proc Natl Acad Sci U S A 102(1):198–203
8. Rovina D, Castiglioni E, Niro F et al (2020) "Betwixt mine eye and heart a league is took": the progress of induced pluripotent stem-cell-based models of dystrophin-associated cardiomyopathy. Int J Mol Sci 21(19). https://doi.org/10.3390/ijms21196997
9. Okita K, Yamakawa T, Matsumura Y et al (2013) An efficient nonviral method to generate integration-free human-induced pluripotent stem cells from cord blood and peripheral blood cells. Stem Cells 31(3):458–466
10. Rodriguez ML, Graham BT, Pabon LM et al (2014) Measuring the contractile forces of human induced pluripotent stem cell-derived cardiomyocytes with arrays of microposts. J Biomech Eng 136(5):051005
11. Chiba S, Lim KR, Sheri N et al (2021) eSkip-Finder: a machine learning-based web application and database to identify the optimal sequences of antisense oligonucleotides for exon skipping. Nucleic Acids Res. https://doi.org/10.1093/nar/gkab442

Chapter 9

Quantitative Evaluation of Exon Skipping in Urine-Derived Cells for Duchenne Muscular Dystrophy

Katsuhiko Kunitake, Chaitra Sathyaprakash, Norio Motohashi, and Yoshitsugu Aoki

Abstract

Antisense oligonucleotide (ASO)-based exon skipping therapy is thought to be promising for Duchenne muscular dystrophy (DMD). For the screening or assessing patient eligibility before administering ASO to patients, in vitro testing using myoblasts derived from each DMD patient is considered crucial. We previously reported state-of-the-art technology to obtain patient primary myoblasts from *MYOD1*-induced urine-derived cells (UDCs) as a model of DMD. We hypothesize that the myoblasts may potentially reflect specific pathological phenotypes, leading to a path for precision medicine in DMD patients. Here, we describe a detailed protocol for both acquiring *MYOD1*-induced myoblasts from UDCs and evaluating the correction of *DMD* mRNA and protein levels after exon-skipping in the cells.

Key words Duchenne muscular dystrophy (DMD), Urine-derived cells (UDCs), Exon skipping, MYOD1, RT-PCR, Western blotting, Immunocytochemistry

1 Introduction

Duchenne muscular dystrophy (DMD), severe and progressive muscular disease, is caused by frame-shift mutations in the *DMD* gene that result in the lack of dystrophin protein [1]. Antisense oligonucleotide (ASO)-based exon skipping is considered a promising therapy for DMD. This therapy has the capability of altering pre-mRNA splicing to restore the *DMD* reading frame [2]. We have demonstrated an excellent safety profile, promising efficacy, and acceptable pharmacokinetic parameters of exon 53 skipping ASO-drug, viltolarsen of the phosphorodiamidate morpholino oligomer (PMO), by repeated intravenous administration in a first-in-human study (registered as UMIN: 000010964 and ClinicalTrials.gov: NCT02081625) [3]. To develop cost-effective and efficient treatments for DMD, in vitro testing should be performed using patient-derived primary muscle cells to verify

Rika Maruyama and Toshifumi Yokota (eds.), *Muscular Dystrophy Therapeutics: Methods and Protocols*, Methods in Molecular Biology, vol. 2587, https://doi.org/10.1007/978-1-0716-2772-3_9,

the patient's eligibility before starting clinical trials [4]. Previously, we reported a novel technology to apply patient-specific *MYOD1*-induced urine-derived cells (UDCs) [5, 6] as a primary myoblast model of DMD [7]. The strength of our technology is that noninvasively collected urine cells from the patients could be applicable as their own primary myoblasts. Here, we describe a detailed protocol for generating myoblasts via *MYOD1*-induction of UDCs, as a model to evaluate exon skipping efficiency and dystrophin restoration level for DMD.

2 Materials

2.1 Isolation and Primary Culture of UDCs

1. Gelatin-Coated Microplate 6 Well with Lid (IWAKI) for culturing UDCs.
2. Sterile plastic bottles for collecting urine.
3. Sterile PBS without calcium and magnesium.
4. Primary medium (PM): Used first 3 days during isolation: 1:1 mixture of high-glucose Dulbecco's modified Eagle's medium (DMEM) without sodium pyruvate (GE Healthcare) and Ham's F-12 nutrient mix (Thermo Fisher Scientific) supplemented with REGM SingleQuots (Lonza), 10% tetracycline-free fetal bovine serum (FBS; Clontech), 1% penicillin/streptomycin (P/S), and 0.5 μg/mL amphotericin B.
5. Growth medium (GM): Switched from PM 3 days after the beginning of isolation: 1:1 mixture of REGM Bullet Kit (Lonza) and high-glucose DMEM supplemented with 15% tetracycline-free FBS (Clontech), 0.5% Glutamax (Thermo Fisher Scientific), 0.5% nonessential amino acids (Thermo Fisher Scientific), 2.5 ng/mL basic fibroblast growth factor (bFGF; Sigma-Aldrich), 2.5 ng/mL Human PDGF-AB (Peprotech), 2.5 ng/mL EGF (Peprotech), and 1% P/S.
6. 0.25% trypsin–EDTA (Gibco).
7. Gelatin-coated 60-mm dish (IWAKI) for culturing the UDCs after passage.

2.2 Retroviral Construct

1. *MYOD1* (NM_002478) Human Tagged ORF Clone (ORIGENE).
2. CloneAmp HiFi PCR premix (Clontech).
3. Forward and reverse primers targeted respectively on both sides across the inserted *MYOD1* sequence.
4. Sterilized distilled water.
5. 0.7% agarose gel.
6. NucleoSpin Gel and PCR Clean-up Kit (Clontech).
7. NanoDrop (Thermo Fisher) as a spectrophotometer.

8. 500 ng/μL pRetroX-TetOne-Puro Vector (Clontech) containing a Tet-on system and *puromycin-resistant gene*.
9. 10× K buffer (TAKARA) to digest retroviral vector described in Subheading 2.2, **item** 7.
10. 2-15 U restriction enzymes: BamH1 and EcoR1.
11. In-Fusion HD Enzyme Premix (Clontech).
12. *E. coli* DH5α Competent Cells (TAKARA).
13. LB culture plate consisting of 10 g/L bacto tryptone, 5 g/L bacto yeast extract, 5 g/L NaCl, 15 g/L bacto agar, and 50 mg/L ampicillin.
14. LB culture medium without bacto agar.
15. HiSpeed Plasmid Purification Kit (Qiagen).
16. GP2-293 packaging cells (Clontech).
17. 100 mm/Collagen-Coated dish (IWAKI) for culturing packaging cells stated above.
18. DMEM with 10% FBS.
19. pVZV-G capsid vectors.
20. Xfect transfection reagent (Clontech) containing cell-penetrating peptide.
21. Retro-X Concentrator (Clontech).
22. MILTEX HV 0.45 μm filter (MERCK) for filtering the retrovirus supernatant.
23. One Step TB Green PrimeScript RT-PCR Kit (TAKARA) for titration of the retroviral vector.
24. Retrovirus Titer Set (TAKARA) for titration of the retroviral vector.

2.3 Infection with MYOD1-Retroviral Vector in UDCs

1. Gelatin-coated 60-mm dish (IWAKI) for culturing undifferentiated UDCs.
2. Hexadimethrine bromide.
3. 1 μg/mL puromycin.
4. GM described in Subheading 2.1.

2.4 Myogenic Differentiation of MYOD1-Induced UDCs Treated with 3-Deazaneplanocin A Hydrochloride (DZNep)

1. Differentiation medium (DM): High-glucose DMEM with 1% L-alanine–L-glutamine, 5% horse serum, and 1% ITS Liquid Media Supplement (Sigma-Aldrich).
2. 1 μg/mL doxycycline.
3. DZNep.
4. Collagen-Coated Microplate 12 Well/Flat Bottom (IWAKI) for differentiated UDCs.
5. Collagen-Coated Microplate 96 Well/Flat Bottom (IWAKI) for differentiated UDCs.

2.5 Transfection of ASO into MYOD1-Induced UDCs

1. 1 mM ASO (25-mer, PMO).
2. 1 mM Endo-Porter transfection reagent (Gene Tools).
3. 1 mM control antisense (Gene Tools).

2.6 Evaluation of Exon Skipping Efficiency by RT-PCR

1. RNase-free water.
2. QIAGEN OneStep RT-PCR Kit (Qiagen).
3. 5 mM primer pairs designed corresponding to target exon and each patient's mutation.
4. RNeasy kit (Qiagen) for RNA extraction.
5. NanoDrop (Thermo Fisher) as a spectrophotometer.

2.7 Detection of Dystrophin by Western Blotting

1. Radioimmunoprecipitation assay (RIPA) buffer (Thermo Fisher Scientific) containing cOmplete Protease Inhibitor Cocktail (Roche).
2. Pierce BCA Protein Assay Kit (Thermo Fisher Scientific) for measuring the concentration of protein.
3. NuPAGE LDS Sample Buffer (4×) (Invitrogen).
4. NuPAGE Sample Reducing Agent (10×) (Invitrogen).
5. Deionized water.
6. NuPAGE Tris–Acetate SDS Running Buffer (Invitrogen) containing 8.95 g/L tricine, 6.06 g/L tris base, and 1.0 g/L sodium dodecyl sulfate (SDS).
7. NuPAGE 3–8% Tris–Acetate Protein Gels (Invitrogen).
8. XCell SureLock Mini-Cell (Invitrogen) for electrophoresis.
9. EzFastBlot HMW (Atto) as a blotting buffer during semidry.
10. Immobilon-P Transfer Membrane (PVDF) (MERCK).
11. Extra Thick Blot Filter Paper (Bio-Rad).
12. Methanol.
13. Semidry transfer apparatus.
14. Rabbit anti-dystrophin (Abcam; ab15277) as a primary antibody.
15. Mouse anti-α-tubulin (Sigma-Aldrich; T6199) as a primary antibody.
16. Histofine Simple Stain MAX-PO (NICHIREI BIOSCIENCE INC.; 424,151) as a secondary antibody.
17. iBind Flex Western Device (Thermo Fisher Scientific) as an automated Western-processing device.
18. iBind Flex Solution (Thermo Fisher Scientific) mixed with antibodies.

19. ECL Prime Western Blotting Detection Reagent (GE healthcare) as a chemiluminescent detection reagent.
20. ChemiDoc MP Imaging System (Bio-Rad) as a charged-coupled device (CCD) camera-based imager.

2.8 Detection of Dystrophin by Immunocytochemistry

1. Sterile PBS without calcium and magnesium.
2. 4% paraformaldehyde.
3. 0.1% Triton-X (MP Biomedicals) as a nonionic detergent.
4. 10% goat serum diluted by PBS (10% GS) for blocking.
5. Mouse anti-dystrophin (Leica, NCL-DYS1) as a primary antibody.
6. Alexa Fluor 546 goat anti-mouse IgG (H + L; Invitrogen) as a secondary antibody.
7. DAPI solution (1:2000, DOJINDO) for nuclear staining.
8. BZ-9000 or BZ-X800 (KEYENCE) as a fluorescent microscope.
9. BZ-X Analyzer or BZ-800 Analyzer (KEYENCE) for analyzing the fluorescence signal intensity.

3 Methods

3.1 Isolation and Primary Culture of UDCs (See Note 1 and Fig. 1)

1. Collect 50–250 mL of urine samples during spontaneous micturition in a sterilized plastic bottle on day 0 (*see* **Note 2**).
2. Divide the entire urine sample into the required number of 50-mL tubes.
3. Centrifuge tubes at 400× *g* for 10 min at room temperature.
4. Aspirate the supernatant, leaving 3 mL in each tube.
5. Resuspend the pellets in the remaining 3 mL of urine for each tube and then collect cell suspension from different tubes into a single 50 mL tube.
6. Add an equal amount of PBS without calcium and magnesium to the cell suspension and centrifuge it at 200× *g* for 10 min at room temperature.
7. Aspirate the supernatant, leaving 3 mL in the tube.
8. Resuspend the cell pellets in the remaining 3 mL of liquid above and add 6 mL of PM.
9. Seed the total cells in a single well of a gelatin-coated six-well plate (total volume of the well, 9 mL). Culture incubated at 37 °C and 5% CO_2 for 72 h.
10. On day 3, replace the medium with 2 mL of GM.
11. Change GM every other day.

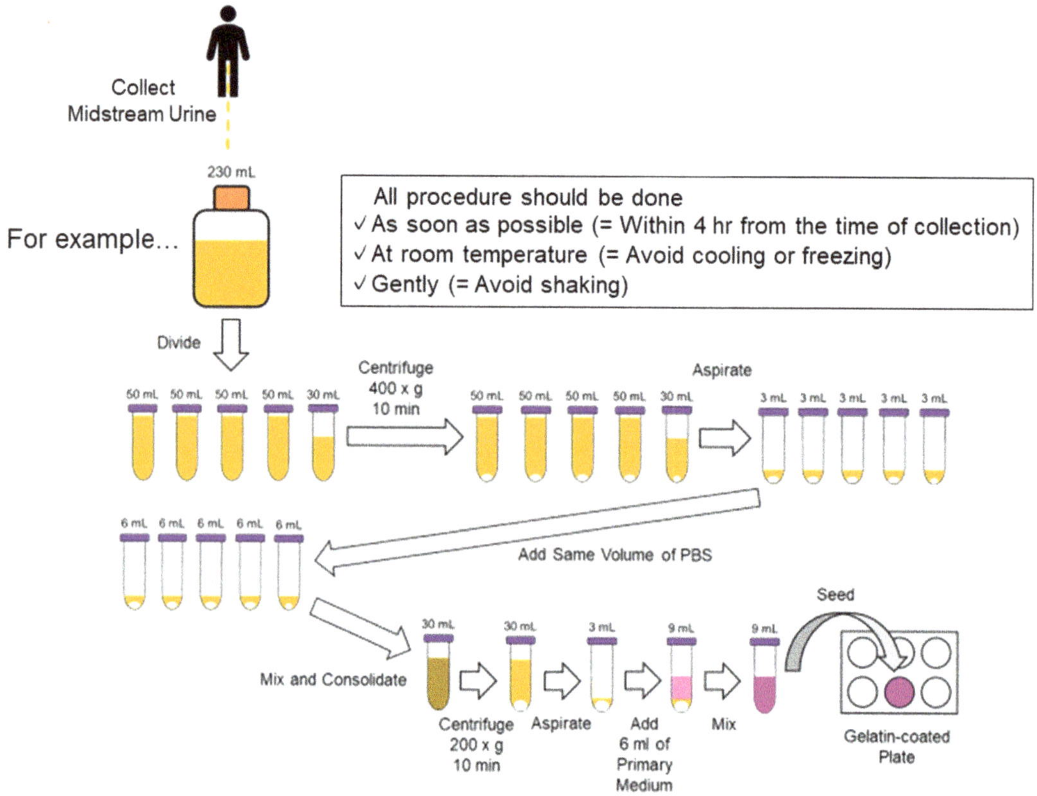

Fig. 1 Isolation and primary culture of UDCs. The process of extracting UDCs-composition from collected urine is illustrated for each step

12. When the UDC culture becomes 80–90% confluent, remove the medium and wash cells with PBS, split all the cells using 0.25% trypsin-EDTA, and seed at 3000–5000 cells/cm^2 onto a new gelatin-coated 60 mm dish (*see* **Note 3**).

3.2 Retroviral Construct

3.2.1 Insertion of MYOD1 into Retroviral Vector

1. Amplify the coding region of *MYOD1* (NM_002478) by PCR. In detail, 80 ng of *MYOD1* is used as a template for PCR with 5 pmol forward and reverse primers, 12.5 μL of 2× CloneAmp HiFi PCR premix, and sterilized water up to 25 μL. The cycling conditions are 35 cycles of 98 °C for 10 s, 55 °C for 10 s, and 72 °C for 10 s.
2. Detect a single band of 963 bp size by 0.7% agarose gel electrophoresis using 1 μL of the amplified PCR product to confirm that the *MYOD1* sequence is amplified successfully.
3. Clean the PCR product using NucleoSpin Gel and PCR Clean-up Kit and determine its concentration with NanoDrop.
4. Incubate the mixture containing 2 μL of 10× K buffer, 2 μL of 500 ng/μL pRetroX-TetOne-Puro Vector retroviral vector, 1 μL each of BamH1 and EcoR1, and 14 μL of sterilized water at 37 °C overnight.

5. Detect a single band by 0.7% agarose gel electrophoresis using 1 μL of the digested product to confirm that the retroviral vector was digested successfully.
6. Clean the digested product using NucleoSpin Gel and PCR Clean-up Kit and determine its concentration by NanoDrop.
7. To clone the amplified *MYOD1* fragment into the digested retroviral vector, perform an in-fusion cloning reaction. Set up the reaction as 100 ng of purified *MYOD1* fragment, 100 ng of digested retroviral vector, 4 μL of 5× In-Fusion HD Enzyme Premix, and sterilized water up to 20 μL, incubate the reaction mixture for 15 min at 50 °C, and then place it on ice.
8. Perform transformation using *E. coli* DH5α Competent Cells according to the manufacturer's instructions.
9. Pick up the selected colony on LB culture plates and culture those in LB culture medium at 200 rpm at 37 °C overnight.
10. Purify the *MYOD1*-inserted retroviral vector using the HiSpeed Plasmid Purification Kit and quantify its amount using NanoDrop.
11. Confirm that *MYOD1* is correctly inserted into the retroviral vector by direct sequencing of the PCR product amplified by the primer pairs.

3.2.2 Production of Retrovirus

1. For retroviral production, seed GP2-293 packaging cells at 50,000 cells/cm^2 on 100 mm/Collagen-Coated dish and culture in DMEM with 10% FBS incubated at 37 °C and 5% CO_2 for 24 h. (*see* **Note 4**).
2. When GP2-293 packaging cells reach 80% confluency, mix 30 μg of *MYOD1*-inserted retroviral vectors, 30 μg of pVZV-G capsid vectors, and Xfect transfection reagent by vortexing, and incubate those for 10 min at room temperature.
3. Add the mixture to the medium of GP2-293 packaging cells and incubate the culture at 37 °C and 5% CO_2.
4. Collect the viral supernatant and replace it with fresh medium at 24 and 48 h after co-transfection.
5. To concentrate the viral supernatant, mix it with Retro-X Concentrator, incubate it at 37 °C overnight, and then centrifuge it at 1500× *g* for 45 min at 4 °C.
6. Filter the retrovirus supernatant through a MILTEX HV 0.45-μm filter.
7. Check the titer of the retroviral vector using both One Step TB Green PrimeScript RT-PCR Kit and Retrovirus Titer Set with thermal cycler system according to the manufacturer's instructions.
8. Aliquot the viral supernatant into small aliquots and store them at −80 °C.

3.3 Infection with MYOD1-Retroviral Vector in UDCs (See Fig. 2)

1. Seed the UDCs at 3000 cells/cm^2 on a gelatin-coated 60-mm dish.
2. After 24 h of seeding, infect the thawed retrovirus at a multiplicity of infection (MOI) of 200 (*see* **Note 5**) with 8 μg/mL hexadimethrine bromide.
3. After a 24 h incubation at 37 °C and 5% CO_2, replace the culture medium with fresh GM containing 1 μg/mL puromycin to select the successful *MYOD1*-induced cells. Change the medium every other day.
4. *MYOD1*-positive cells should be selected within 7 days after adding puromycin.

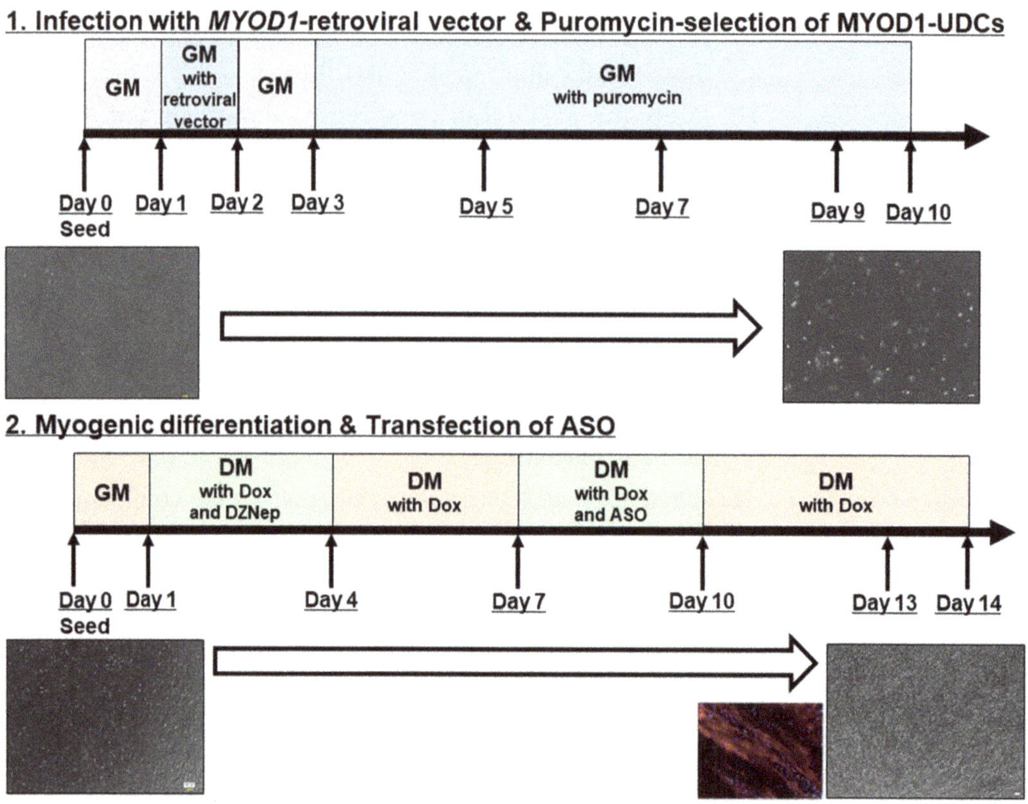

Fig. 2 Time course of infection with *MYOD1*-retroviral vector, puromycin-selection of MYOD1-UDCs, myogenic differentiation, and transfection of ASO. In the upper line (1, infection with *MYOD1*-retroviral vector and puromycin selection of MYOD1-UDCs), medium is replaced on Days 1, 2, 3, 5, 7, and 9. Representative bright-field images are shown on Days 1 and 10. In the lower line (2, myogenic differentiation and transfection of ASO), medium is replaced on Days 1, 4, 7, 10, and 13. Representative bright-field images are shown on Days 1 and 14. Representative image of immunocytochemistry for dystrophin is shown on Day 14. Scale bars indicate 100 μm. *GM* growth medium, *DM* differentiation medium, *Dox* doxycycline

3.4 Myogenic Differentiation of MYOD1-Induced UDCs Treated with 3-Deazaneplanocin A Hydrochloride (DZNep) (See Fig. 2)

1. Plate *MYOD1*-positive UDCs in the collagen-coated wells at a density of 3.5×10^4 cells/cm^2. Culture in GM humidified at 37 °C and 5% CO_2.
2. After 24 h, change GM to DM added 1% doxycycline (Dox) and 5 μM DZNep.
3. After 3 days, replace DM with fresh DM added only 1% Dox, but no DZNep. Then, change the medium every 3 days (*see* **Notes 6–9**).

3.5 Transfection of ASO into MYOD1-Induced UDCs (See Fig. 2)

1. Mix ASO with 6 μM Endo-Porter solution and DM to a final concentration of 1–10 μM.
2. Transfect the mixture above into *MYOD1*-induced UDCs on day 7 after switching GM to DM. Culture incubated at 37 °C and 5% CO_2.
3. After 72 h of incubation with ASO, change the medium to fresh DM containing 1% doxycycline without ASO.
4. 72 h later, change the medium to fresh DM containing 1% Dox, again.
5. 24 h later, proceed to evaluation (*see* **Note 10**).

3.6 Evaluation of Exon Skipping Efficiency by RT-PCR

1. Remove DM and wash 1× with PBS. Add cell lysis buffer, lyse the *MYOD1*-induced myoblasts from UDCs, and harvest the total RNA using the RNeasy kit.
2. Measure the RNA concentration with NanoDrop.
3. Mix the required reagents for one-step RT-PCR reaction in PCR tubes according to the manufacturer's instructions.
4. Place the PCR tubes with the mixture on a thermocycler. The cycling conditions are 1 cycle 50 °C for 30 min, 1 cycle 95 °C for 15 min, 35 cycles 94 °C for 1 min, 60 °C for 1 min, 72 °C for 1 min, 1 cycle 72 °C for 7 min, and finally hold at 4 °C.
5. Perform microchip electrophoresis and calculate the exon skipping efficiency using the equation below:

   Exon skipping efficiency (%) = [skipped band/(skipped band + nonskipped band)] × 100.

3.7 Detection of Dystrophin by Western Blotting

1. Extract the total protein of the *MYOD1*-induced myoblasts from UDCs using RIPA buffer containing cOmplete Protease Inhibitor Cocktail.
2. Sonicate the lysates on ice and centrifuge at 14,000× *g* for 15 min at 4 °C.
3. Collect the supernatant and determine protein concentrations using Pierce BCA Protein Assay Kit.
4. Add 15 μg of total protein in a 0.5 mL tube and dilute by adding RIPA buffer containing cOmplete Protease Inhibitor Cocktail to a total volume of 10 μL.

5. Add 5 μL of NuPAGE LDS Sample Buffer (4×), 2 μL of NuPAGE Sample Reducing Agent (10×), and 3 μL of deionized water. Denature the cell lysates at 70 °C for 10 min.
6. Prepare NuPAGE Tris–Acetate SDS Running Buffer.
7. Load 20 μL of the sample per lane onto NuPAGE 3–8% Tris–Acetate Protein Gel and perform electrophoresis at 150 V for 75 min with XCell SureLock Mini-Cell.
8. During electrophoresis, soak the PVDF for 20 s in methanol and then in the EzFastBlot HMW until use (at least 10 min).
9. Cut Extra Thick Blot Filter Paper to the same size as that of PVDF and soak those in EzFastBlot HMW until use.
10. After electrophoresis, cut the gel to the same size as that of the PVDF membrane and soak the gel in distilled water.
11. Place Extra Thick Blot Filter Paper, PVDF, and gel on the semidry transfer apparatus. Transfer at 4 mA/cm 2 for 30 min.
12. Rinse transferred PVDF with distilled water.
13. Using iBind Flex Western Device at room temperature, incubate rinsed PVDF with Rabbit anti-dystrophin (1:500) and Mouse anti-α-tubulin (1:1200) antibodies diluted by iBind Flex Solution simultaneously, wash with wash buffer, and then incubate with Histofine Simple Stain MAX-PO (1:100) diluted by iBind Flex Solution.
14. Rinse the reacted PVDF in distilled water.
15. Detect the proteins using ECL Prime Western Blotting Detection Reagent and ChemiDoc MP Imaging System.

3.8 Detection of Dystrophin by Immunocytochemistry

1. Wash the *MYOD1*-induced myoblasts from UDCs with PBS and fix them in 4% paraformaldehyde for 10 min at 4 °C.
2. Permeabilize those cells in 0.1% nonionic detergent for 10 min at room temperature and block those with 10% GS for 15 min at 37 °C.
3. Incubate those cells with Mouse anti-dystrophin (1:30) diluted by 10% GS overnight at 4 °C.
4. Wash the cells with PBS and incubate those with Alexa Fluor 546 goat anti-mouse IgG (1:300) diluted by 10% GS for 30 min at room temperature.
5. Wash the cells with PBS and incubate those with DAPI (1: 2000) diluted by PBS for 10 min at room temperature.
6. Image the cells using BZ-9000 or BZ-X800 and use BZ-X Analyzer or BZ-800 Analyzer to semi-quantify the fluorescence signal detected at each well of the 96-well plate.

4 Notes

1. UDCs were isolated according to previously published protocols [8–11] with some modifications.
2. Midstream urine is desirable to reduce the risk of viral contamination. It is recommended that collected urine is used for isolation within 4 h and stored at room temperature. Even keeping the cells at 4 °C might reduce the cell viability.
3. Colonies of UDCs appear within 1–2 weeks.
4. Collagen coating is necessary for culturing HEK293 packaging cells. Transfection may be better at 24 h or more after seeding because packaging cells easily detach from the culture plate.
5. We usually calculate MOI by the equation below:

   MOI = the copy number of retrovirus/the number of UDCs.

   In this protocol, MOI is 200, and the number of UDCs is 6×10^4 cells/60-mm dish.

   Therefore, the required copy number of retrovirus is 1.2×10^7 copies.
6. The *MYOD1*-induced UDCs could be stored at −80 °C or liquid nitrogen before induction of myodifferentiation.
7. We usually add Dox to DM just before each experiment.
8. We have reported that DZNep, a histone methyltransferase inhibitor, could significantly promote the expression of *MYOGENIN*, one of the late muscle regulatory factors, and also accelerate myotube differentiation [7].
9. 5 mM DZNep solution could be stored at −80 °C for 3 months. Use a defrost freezer and avoid repeated freeze-thaw cycles.
10. UDCs fuse and form myotubes within 2–3 weeks after switching GM to DM.

Acknowledgments

This research was supported by the Japan Society for the Promotion of Science Grant-in-Aid for Scientific Research (B) [grant no. 21H02843], Grants-in-Aid for Research on Nervous and Mental Disorders [grant no. 2-6], and the Japan Agency for Medical Research and Development [grant no. 21lm0203086h0003].

References

1. Hoffman EP et al (1987) Dystrophin: the protein product of the Duchenne muscular dystrophy locus. Cell 51:919–928
2. Cirak S et al (2011) Exon skipping and dystrophin restoration in patients with Duchenne muscular dystrophy after systemic phosphorodiamidate morpholino oligomer treatment: an open-label, phase 2, dose-escalation study. Lancet 378:595–605
3. Komaki H et al (2018) Systemic administration of the antisense oligonucleotide NS-065/NCNP-01 for skipping of exon 53 in patients with Duchenne muscular dystrophy. Sci Transl Med 10(437):eaan0713
4. Antoury L et al (2018) Analysis of extracellular mRNA in human urine reveals splice variant biomarkers of muscular dystrophies. Nat Commun 9:3906
5. Rahmoune H et al (2005) Glucose transporters in human renal proximal tubular cells isolated from the urine of patients with non-insulin-dependent diabetes. Diabetes 54:3427–3434
6. Zhang Y et al (2008) Urine derived cells are a potential source for urological tissue reconstruction. J Urol 180:2226–2233
7. Takizawa H et al (2019) Modelling Duchenne muscular dystrophy in MYOD1-converted urine-derived cells treated with 3-deazaneplanocin A hydrochloride. Sci Rep 9:3807
8. Zhou T et al (2012) Generation of human induced pluripotent stem cells from urine samples. Nat Protoc 7:2080–2089
9. Chen W et al (2017) Skeletal myogenic differentiation of human urine-derived cells as a potential source for skeletal muscle regeneration. J Tissue Eng Regen Med 11:334–341
10. Kim EY et al (2016) Direct reprogramming of urine-derived cells with inducible MyoD for modeling human muscle disease. Skelet Muscle 6:32
11. Takizawa H et al (2020) Exon skipping in directly reprogrammed myotubes obtained from human urine-derived cells. J Vis Exp e60840. https://doi.org/10.3791/60840

Chapter 10

Use of Glycine to Augment Exon Skipping and Cell Therapies for Duchenne Muscular Dystrophy

Gang Han, Caorui Lin, and HaiFang Yin

Abstract

Antisense oligonucleotide (AO)-based exon-skipping and cell therapies are the main therapeutic approaches for Duchenne muscular dystrophy (DMD). Insufficient systemic delivery leading to low therapeutic efficacy limits the former; low transplantation efficiency hampers the latter. Here we describe how glycine can address these issues by augmenting satellite proliferation and muscle regeneration, resulting in enhanced AO uptake in regenerating myofibers and cell transplantation efficiency in dystrophic mice. The dual functionality of glycine demonstrated in AO-based exon-skipping and cell therapies presents a simple and efficient method to augment AO potency and cell transplantation efficacy in DMD and other muscle diseases.

Key words Duchenne muscular dystrophy, glycine, oligonucleotides, exon-skipping, cell therapy, satellite cells, regenerating myofibers

1 Introduction

Duchenne muscular dystrophy (DMD) is a devastating muscle-wasting disease caused by the loss of functional dystrophin due to frame-shift mutations in the *dystrophin* gene [1]. Four antisense oligonucleotide (AO)-mediated exon-skipping drugs have been approved by the regulatory authority to treat DMD [2]; however, the therapeutic efficacy is limited due to inadequate systemic delivery [3, 4]. An alternative therapeutic modality is cell therapy, with numerous trials employing different types of cells including myoblasts, satellite cells, bone marrow cells, mesoangioblasts, and CD133+ cells in dystrophic mice or DMD patients [5], but low transplantation efficiency hinders cell therapy's clinical implementation. Thus, approaches to enhance the efficacy of exon-skipping and cell therapies are key to unlock the full potential of these therapeutics.

Rika Maruyama and Toshifumi Yokota (eds.), *Muscular Dystrophy Therapeutics: Methods and Protocols*,
Methods in Molecular Biology, vol. 2587, https://doi.org/10.1007/978-1-0716-2772-3_10,

To enhance systemic delivery of AOs, different groups attempted cell-penetrating, muscle-specific, and chimeric peptides in dystrophic mice with encouraging therapeutic efficacy, though safety remains a concern [6–12]. Polymers and nanoparticles, which have been shown to significantly enhance drug delivery to tumors [13, 14], had limited effect in muscle when investigated in dystrophic mice [15–18]. Exosomes, nano-sized vesicles, are being developed as natural delivery tools for DMD AOs and the results are promising, but it is still at an early stage of development [19]. Previously, we demonstrated that the mixture of glucose/fructose (1:1) can significantly enhance the uptake and activity of AOs in peripheral muscles of adult *mdx* mice, but the results were less ideal in aged *mdx* mice [20, 21]. Recently, we have shown that glycine can dramatically augment the potency of AOs and transplanted myoblasts in peripheral muscles of dystrophic mice by promoting satellite proliferation and muscle regeneration, thus resulting in enhanced AO uptake in regenerating myofibers [22]. Our study provides a simple method to enhance the efficiency of exon-skipping and cell therapies without the requirement for complicated conjugation and thus will accelerate the clinical implementation of these approaches in DMD and other muscle diseases.

Here we describe the method and protocol for the preparation, local intramuscular and systemic intravenous injection of glycine with AOs, transplantation of primary myoblasts, and the evaluation of efficacy in dystrophin-deficient *mdx* mice.

2 Materials

Ultrapure water used to make solutions should be autoclaved prior to use.

2.1 HEK 293T Cell Culture

1. Dulbecco's modified Eagle's medium with high glucose (DMEM-GlutaMaxTM).
2. Fetal bovine serum (FBS).
3. Penicillin–streptomycin.
4. Trypsin.
5. Gelatin from porcine skin, type A.
6. Trypan blue.
7. 1x phosphate-buffered saline, pH 7.4 (PBS).
8. RIPA lysis buffer.
9. 96-well plate reader and 24-well plate.
10. Culture medium: DMEM supplemented with 10% FBS, and 1% penicillin–streptomycin.

2.2 Oligonucleotides

1. PMO 5′-ggccaaacctcggcttacctgaaat-3′.
2. Lissamine-labeled PMO (*see* **Note 1**).
3. FITC-labeled PMO (*see* **Note 1**).

2.3 Chemicals for Mechanistic Studies

1. Tetrahydrofolate (THF).
2. Formate.
3. Bafilomycin A1 (BAF).
4. Methotrexate (MTX).
5. Bitopertin (BP).
6. Torkinib (PP242).

2.4 Glycine

1. Glycine.
2. Autoclaved 15% glycine stock solution followed by 0.22 μm filter sterilization.
3. 5% glycine solution.

2.5 Animal Experiments

1. 6–8-week-old *mdx* mice.
2. 6–8-week-old *C57BL/6* mice.
3. Isoflurane anesthetics (*see* **Note 2**).
4. 0.3 mL Insulin needles.
5. Heat pad.

2.6 Immuno-histochemical Staining for Dystrophin, eMyHC, and Laminin

1. Microscope cover slips: Borosilicate glass, 22 × 40 × 0.15 mm.
2. Microscope slides.
3. Normal goat serum (NGS).
4. Fetal bovine serum (FBS).
5. PBS.
6. Cryostat Microsystems CM1950.
7. Dystrophin antibody-rabbit polyclonal antibody P7 against the dystrophin carboxyl-terminal region (the antibody was kindly provided by Dr. Qilong Lu, Carolinas Medical Center, USA).
8. eMyHC antibody-mouse monoclonal antibody.
9. Laminin antibody-rabbit monoclonal antibody.
10. Alexa 594 donkey anti-rabbit IgG.
11. Alexa 488 rabbit anti-mouse IgG.
12. DAPI containing mounting agent.
13. DAKO fluorescent mounting medium.
14. 4,6-Diamidino-2-phenylindole (DAPI) (*see* **Note 4**).
15. Moisture chamber.
16. Blocking solution: 20% FBS and 20% NGS in PBS.

17. Primary antibody solution: 1/50 dilution of dystrophin in primary diluents (20% NGS in PBS), 1/30 dilution of eMyHC in blocking solution, and 1/400 dilution of laminin in blocking solution.

2.7 Immunohistochemical Staining for FITC-Labeled PMO

1. 4% paraformaldehyde (PFA).
2. Acetone (*see* **Note 2**).
3. Normal goat serum (NGS).
4. Bovine serum albumin (BSA).
5. Blocking solution: 20% goat serum and 10 mg/mL BSA in PBS.
6. Fluorescein antibody-rabbit polyclonal antibody.
7. Wheat germ agglutinin: Alexa Fluor 647 (1/200 dilution).
8. Alexa488 donkey anti-rabbit IgG.
9. Primary antibody solution: 1/200 dilution of fluorescein in blocking solution.

2.8 RNA Extraction and RT-PCR

1. TRIZOL Reagent (*see* **Note 3**).
2. Chloroform, analytical reagent grade (*see* **Note 3**).
3. 2-Propanol, for molecular biology, minimum 99% (*see* **Note 2**).
4. Ethanol, for molecular biology (*see* **Note 2**).
5. DEPC treated RNAase-free water.
6. RT-PCR: QIAGEN One Step RT-PCR Kit (*see* **Note 5**).
7. Nested PCR: Hot Star Taq DNA Polymerase kit.
8. Agarose gel.
9. Ethidium bromide (*see* **Note 6**).

2.9 Preparation of Protein Samples

1. Tris–base.
2. Sodium dodecyl sulfate (SDS).
3. 2-Mercaptoethanol.
4. Lysis buffer: 75 mM Tris–HCl, pH 6.5; 10% SDS, add 5% 2-mercaptoethanol.
5. Bradford kit.

2.10 SDS–Polyacrylamide Gel Electrophoresis (SDS-PAGE)

1. Tris–base.
2. 30% acrylamide/bis-acrylamide solution.
3. N,N,N,N′-Tetramethyl-ethylenediamine (TEMED) (*see* **Notes 2** and **7**).
4. Glycerol.
5. 10% ammonium persulfate (APS).
6. Bromophenol blue.

7. Resolving gel buffer: 1.5 M Tris–HCl, pH 8.8.
8. Stacking gel buffer: 1.0 M Tris–HCl, pH 6.8.
9. 20% SDS.
10. Running Buffer (10×): 247.7 mM Tris–base, 1.91 M glycine, 1% SDS.
11. Laemmli SDS-Sample Buffer (5×): 0.1 M Tris–HCl, pH 6.6; 2% SDS; 2% 2-mercaptoethanol; 20% glycerol; 0.01% bromophenol blue.

2.11 Western Blotting for Dystrophin Protein

1. Transfer buffer (10×): 247.7 mM Tris, 1.91 M glycine. Store at room temperature.
2. Transfer buffer (1×): 10% 10× transfer buffer; 10% methanol, analytical reagent grade (*see* **Note 2**) and 0.01% SDS (*see* **Note 8**).
3. Immobilon-P membrane, PVDF, 0.45 μm.
4. Filter paper.
5. Tween 20.
6. Non-fat dry milk.
7. Washing Solution: Prepare phosphate-buffered saline (PBS) with 0.1% (v/v) Tween 20 to make PBST; 5% 5 M NaCl.
8. Dystrophin antibody: NCL DYS-1 (*see* **Note 9**).
9. Secondary antibody: goat-anti-mouse IgG whole monoclonal peroxidase.
10. Chemi luminescent HRP substrate, Immobilon Western.
11. Fuji medical X-ray film (*see* **Note 10**).

2.12 Immuno-histochemical Staining for Dystrophin-Associated Protein Complex (DAPC)

1. Microscope cover slips: Borosilicate glass, 22 × 40 × 0.15 mm.
2. Microscope slides: Superfrost (R) Plus.
3. Mouse on Mouse (M.O.M.) Detection Kit.
4. MOM blocking solution: Dilute 2 drops of MOM Blocking Solution in 2.5 mL of PBS.
5. MOM diluent solution: Dilute MOM Protein Concentrate (Vector) 1:125 in PBS. Store at 2–8 °C.
6. Primary antibodies: β-Sarcoglycan: NCL-L-β-SARC; α-Sarcoglycan: NCL-l-α- SARC; β-Dystroglycan: NCL-β-DG (*see* **Note 8**).
7. DAPI containing mounting agent.
8. DAKO fluorescent mounting medium.
9. 4,6-Diamidino-2-phenylindole (DAPI) (*see* **Note 5**).
10. Moisture chamber.

2.13 Co-immunohistochemical Staining for PAX7 and Ki67

1. Triton-100.
2. Paired box transcription factor (PAX7) antibody-mouse monoclonal antibody.
3. Ki67 antibody-rabbit monoclonal antibody.
4. PBSTR-1/100 Triton-100 in PBS.
5. pH 6.0 citrate buffer.
6. Affinipure Fab goat anti mouse IgG (AFG).
7. Blocking buffer: 1% Triton 100, 2% BSA, 5% NGS and PBS.
8. Primary antibodies: 1/300 dilution of PAX7 and 1/1000 dilution of Ki67 antibody in primary diluents.

2.14 Tissue Distribution

1. 50 mL syringe and 5 mL syringe needles.
2. Filter paper and black paper.
3. IVIS imaging system.

2.15 Histology

1. Bouin's solution.
2. Xylene (*see* **Note 2**).
3. Ethanol (*see* **Note 2**).
4. Paraffin.
5. Hematoxylin.
6. Eosin.
7. Neutral balsam.
8. Ultra-thin semiautomatic microtome.

2.16 Isolation of Primary Myoblasts

1. D′ Hanks buffer.
2. Type II collagenase.
3. 70 mm cell strainer.
4. 40 mm cell strainer.
5. Ham's F-10.
6. Bovine fibroblast growth factor (bFGF).
7. Digestion buffer: 400 U type II collagenase/mL D-Hanks buffer.
8. Primary myoblast culture medium: Ham's F-10 supplemented with 20% FBS, 10% horse serum, 1% penicillin–streptomycin, and 2.5 ng/mL bFGF.
9. Matrigel: 1/100 dilution in DMEM.

2.17 Myoblast Transplantation in mdx Mice

1. Cardiotoxin (CTX).
2. Tacrolimus (FK506).

2.18 Generation of RagB Knockout HEK293T Cells

1. CRISPR/Cas9 lentivirus package plasmid.
2. Single guide RNA (sgRNA) lentivirus package plasmid.
3. Blasticidin.
4. Puromycin.
5. The sgRNA targeting exon 3 of human RagB is 5′-CCACCAC TAGGGGAACCGGA-3′.

3 Methods

3.1 In Vivo Experiments

3.1.1 Local Intramuscular Injection

1. Prepare 2 μg of PMO in saline or 5% glycine solution.
2. Anaesthetize 6–8-week-old *mdx* mice while monitoring body temperature, respiration rate, and rhythm.
3. Ensure the mouse is sufficiently anesthetized by tail pinch or pedal with drawl reflex techniques. Keep mouse on a heated pad while anesthetized.
4. Inject the solution into tibialis anterior (TA) muscle.
5. Remove the mouse from anesthesia and observe until full recovery.
6. 10 days after administration, the mouse may be sacrificed by cervical dislocation and TA muscles are snap-frozen in liquid nitrogen and stored at −80 °C.

3.1.2 Local Intramuscular Injection for Exon-Skipping Frequency Studies

1. Prepare 2 μg of PMO in saline.
2. Anesthetize 6–8-week-old *mdx* mice while monitoring body temperature, respiration rate, and rhythm.
3. Ensure the mouse is sufficiently anesthetized by tail pinch or pedal with drawl reflex techniques. Keep mouse on a heated pad while anesthetized.
4. Inject PMO solution into TA muscles of *mdx* mice prior to administration of glycine 16 h later.
5. Remove the mouse from anesthesia and observe until full recovery.
6. 10 days after administration, the mouse may be sacrificed by cervical dislocation and TA muscles are snap-frozen in liquid nitrogen and stored at −80 °C.

3.1.3 Local Intramuscular Injection for Mechanistic Studies

1. Prepare 2 μg of PMO in 5% glycine with 0.2 μg of THF, 0.2 μg of MTX, or 3 μg of BP or 5 μg of PP242, or 2 μg of PMO in 50 mg/mL formate or 0.4 μg of THF.
2. Anesthetize 6–8-week-old *mdx* mice while monitoring body temperature, respiration rate, and rhythm.

3. Ensure the mouse is sufficiently anesthetized by tail pinch or pedal with drawl reflex techniques. Keep mouse on a heated pad while anesthetized.
4. Inject the solution into TA muscles.
5. Remove the mouse from anesthesia and observe until full recovery.
6. 10 days after administration, mice may be sacrificed by cervical dislocation and TA muscles are snap-frozen in liquid nitrogen and stored at −80 °C.

3.1.4 Systemic Intravenous Injection

1. Prepare the desired concentration of PMO in saline or 5% glycine solution.
2. Anesthetize 6–8-week-old *mdx* mice while monitoring body temperature, respiration rate, and rhythm.
3. Ensure the mouse is sufficiently anesthetized by tail pinch or pedal with drawl reflex techniques. Keep mouse on a heated pad while anesthetized.
4. Gently inject the solution into the tail vein.
5. Remove the mouse from anesthesia and observe until full recovery.
6. Two weeks after administration, mice may be sacrificed by cervical dislocation and desired tissue samples are snap-frozen in liquid nitrogen and stored at −80 °C.

3.2 Immunohistochemical Staining for Dystrophin, eMyHC, and Laminin

1. Cut 8-μm-thick cryosections from at least two-thirds deep into the muscle at 100-μm intervals. Slides may be kept at −80 °C until ready to stain.
2. Allow the slides to air-dry and circle the sections with a blocking pen so that less reagent is required and the sections do not lift from the slide. Soak the slides in PBS for 10 min at room temperature.
3. Prepare the blocking solution (20% FBS, 20% NGS and PBS) and pipette onto slides and incubate for 1 h at room temperature.
4. Pipette blocking solution onto slides and incubate for 1 h at room temperature.
5. Throw away blocking buffer and pipette primary antibody solution for 1.5 h at room temperature.
6. Throw away the primary antibody and wash the slides 3 times with PBS buffer.
7. Incubate with secondary antibody at a dilution of 1/200 in PBS for 1 h.

8. Throw away the secondary antibody and wash 3 times with PBS buffer.
9. Mount the slides with DAKO mounting media containing DAPI nuclear stain and coverslip.
10. View the slides under the fluorescence microscope.

3.3 Detection of FITC-Labeled PMO in Muscle Tissues

1. Cut 8-μm-thick cryosections from at least two-thirds deep into the muscle at 100 μm intervals. Slides may be kept at −80 °C until ready to stain.
2. Allow the slides to air-dry and circle the sections with a blocking pen so that less reagent is required and the sections do not lift from the slide. Soak the slides in 4% PFA for 15 min at room temperature.
3. Soak the slides in ice-cold acetone for 10 min.
4. Prepare the blocking solution (20% NGS and 10 mg/mL BSA and PBS) and pipette onto slides and incubate for 1 h at room temperature.
5. Throw away blocking buffer and pipette primary antibody solution overnight at 4 °C.
6. Throw away the primary antibody and wash the slides 3 times with PBS buffer.
7. Incubate with secondary antibody at a dilution of 1/200 in PBS for 1 h.
8. Throw away the secondary antibody and wash 3 times with PBS buffer.
9. Prepare the wheat germ agglutinin-Alexa Fluor 647 solution (1:500) and pipette onto slides and incubate for 10 min at room temperature.
10. Throw away the wheat germ agglutinin-Alexa Fluor 647 solution and wash 3 times with PBS buffer.
11. Mount the slides with DAKO mounting media containing DAPI nuclear stain and coverslip.
12. View the slides under the fluorescence microscope.

3.4 RNA Extraction and RT-PCR

1. Muscle tissues are harvested and snap-frozen in liquid nitrogen or dry ice-cold isopentane and stored at −80 °C.
2. Collect 20–30 muscle cryosections of 15-μm thickness into sterile 1.5-mL Eppendorf tubes placed on dry ice (*see* **Note 11**).
3. Put 250 μL of Trizol into the Eppendorf and carry out the subsequent steps according to the manufacturer's instructions for Trizol (*see* **Notes 12** and **13**).
4. Measure the total RNA concentration by spectrophotometer (Nanodrop).

5. Use 200 ng of RNA template for a 10-μL RT-PCR with the OneStep RT-PCR kit. The primer sequences for the initial RT-PCR are Exon20F0 5′-CAGAATTCT GCCAATTGCT-GAG-3′ and Ex26R0 5′-TTCTTCAGCTTGTGTCATCC-3′, which allow amplification of messenger RNA from exons 20 to 26. The cycle conditions are 95 °C for 30 s, 55 °C for 1 min, and 72 °C for 2 min for 25 cycles. RT-PCR product (1 μL) is then used as the template for secondary PCR performed in 25 μL with 0.5 U Taq DNA polymerase. The primer sequences for the second round are Ex20F1 5′-CCCAGTCTACCACCC TATCAGAGC-3′ and Ex24R1 5′-CAGCCATCCAT TTC TGTAAGG-3′. The cycle conditions are 95 °C for 1 min, 57 °C for 1 min, and 72 °C for 2 min for 25 cycles (*see* **Notes 14** and **15**).
6. The PCR products are examined by electrophoresis on a 2% agarose gel.

3.5 Preparation of Protein Samples

1. Collect 40–60 cryosections of 15 μm thickness from each sample into cooled sterile 1.5 mL Eppendorf tubes placed on dry ice (*see* **Note 11**).
2. Prepare fresh lysis buffer with 5% β-mercaptoethanol and place 100 μL in each tube and homogenize mechanically with a pipette (*see* **Note 16**).
3. Place tubes on a preheated heat block (100 °C) for 5 min. Puncture holes in the lids of the tubes to prevent them popping open from the generation of gases.
4. Spin the samples for 10 min at 13000 rpm and then remove the supernatant and place in a new tube. Keep these tubes on ice.
5. Quantify the concentration of protein by Bradford assay according to the supplier's instructions.
6. Protein samples should be aliquoted in 200 μL Eppendorf tubes and stored at −80 °C.

3.6 SDS-Polyacrylamide Gel Electrophoresis (SDS-PAGE)

1. These instructions are based on the Bio-Rad mini-gel system with 0.75-mm-thick gel plates.
2. Wash the glass plates and wipe the glass plates with methanol. It is important to ensure that all the glass plates are clean.
3. Assemble the gel plates in a green plastic gel holder. Ensure that the glass plates are correctly aligned and level with the bench.
4. Place the gel holder and plates into the transparent rack. Press down on the gel holder to ensure that there is a tight seal with the rubber pad underneath so that the gel mixture does not leak out.
5. Prepare 6% resolving gel layer, containing 0.8 mL of 1.5 mol/L Tris–HCl buffer, 0.7 mL of 30% acryl amide/bis-acrylamide

solution, 1.6 mL of water, 166.7 mL of glycerol, 28.3 μL of 20% SDS, 33.3 μL of 10% APS, and 3.3 μL of TEMED and mix thoroughly. Pipette the gel mixture, leaving space for the stacking layer and overlay with water. The resolving layer should take 30 min to polymerize.

6. Pour off the water.
7. Prepare 4% stacking gel, containing 0.2 mL of 1 mol/L Tris–HCl buffer, 0.217 mL of 30% acryl amide/bis-acrylamide solution, 1.1 mL of water, 83.3 μL of glycerol, 16.7 μL of 20% SDS, 16.7 μL of 10% APS, and 1.7 μL of TEMED and mix thoroughly. Pour the stacking gel mixture on top of the resolving gel layer and insert the comb. The stacking layer should take 30 min to polymerize (*see* **Note 17**).
8. Prepare the running buffer by diluting 100 mL of the 10× running buffer with 900 mL of ddH_2O and mix thoroughly.
9. Once the stacking layer has set, place the gels in the electrophoresis cassette, which is in turn placed in the gel unit. Fill the upper and lower chambers of the gel unit with running buffer. Gently remove the comb from the top of the gel by pulling directly upward. Be careful not to deform the wells. Using a P200 micro-pipette wash the wells with Running Buffer.
10. Prepare the samples (10–20 μL depending on the size of comb used) in Laemmli's buffer, and make up the rest of the volume with ddH_2O. Load the pre-stained protein molecular weight marker in the first lane and the samples in the subsequent lanes.
11. Complete the assembly of the gel unit and connect to a power supply. Run the gel at 80 V for 30 min and 120 V for 3 h on ice to prevent the system from overheating.

3.7 Western Blotting for Dystrophin Protein

1. The gel with the samples separated by SDS-PAGE is transferred to a PVDF membrane by electrophoresis. These directions are for the use of the BioRad Mini-PROTEAN(R) system; however, they may be easily adapted for other formats.
2. The gel unit is disconnected and disassembled. The glass plates are carefully peeled apart, and the stacking layer is cut away from the resolving layer using a sharp blade or the sharp edge of a spatula.
3. Prepare the transfer buffer as directed and soak two sponges and two filter papers in the transfer buffer until they are fully saturated.
4. Cut a PVDF membrane to the size of the gel and cut one corner for identification. Soak the membrane in methanol for approximately 1 min (*see* **Note 18**) and rinse in the transfer buffer.

5. Assemble the components in the cassette as follows: sponge, filter paper, gel, membrane, filter paper, sponge and place the sealed cassette in the transfer tank such that the PVDF membrane is between the gel and the anode. It is imperative that this orientation is correct or the proteins will be lost from the gel into the buffer rather than transferred onto the PVDF membrane.
6. Fill the tank with transfer buffer, seal the tank, and transfer overnight at 110 mA at 4 °C.
7. Once the transfer is complete, disassemble the cassette and place the membrane in 5% milk for 1–2 h to block.
8. Following the blocking step, incubate the membrane with dystrophin primary antibody diluted 1:200 in blocking solution overnight in a cold room (shaker).
9. Wash the membrane with blocking solution containing 250 mM NaCl for 10 min, followed by 3 × 5 min washes (*see* **Note 19**).
10. Incubate the membrane with secondary antibody diluted to 1: 5000 in blocking solution for 1 h at room temperature (shaker).
11. Wash the membrane 3 times with PBST containing 250 mM NaCl for 15 min.
12. Pour off the last wash and place 500 μL of each ECL reagent (1:1 ratio) onto the membrane and shake gently to ensure the reagent mixes and contacts the membrane. Maintain for 2 min.
13. Place the membrane on a piece of cling film and carefully fold the cling film over the top to ensure there are no bubbles between the plastic and the membrane.
14. Place the membrane into the cassette and proceed to the dark room under safe light conditions.
15. Place an appropriately sized piece of X-ray film on top of the membrane and seal the cassette. Useful exposure times range between 1 and 15 min. When the decided exposure time has lapsed, remove the film from the cassette and place it in the Kodak medical X-ray processor 102 for developing (*see* **Note 20**).

3.8 Immunohistochemistry for Dystrophin-Associated Protein Complex (DAPC)

1. Cut 8-μm-thick cryosections from at least two-thirds deep into the muscle at 100-μm intervals.
2. Allow the slides to air-dry and circle the sections with a blocking pen so that less reagent is required and the sections do not lift from the slide.
3. Block the slides with the avidin solution for 15 min at room temperature. Wash briefly with PBST.

4. Incubate with Biotin solution for 15 min at room temperature. Wash briefly with PBST.
5. Prepare the MOM blocking solution and pipette onto the sections for 1 h.
6. Discard the blocking solution and wash the slides 3 times with PBST.
7. Prepare the MOM diluent solution and pipette onto the sections for 5 min.
8. Tip off the solution and incubate with the primary antibodies (1/100 in MOM diluent) α-sarcoglycan, β-sarcoglycan, and β-dystroglycan monoclonal mouse antibodies for 80 min (*see* **Note 21**).
9. Wash the slides 3 times for 3 min with PBST.
10. Dilute the anti-mouse IgG with the diluent buffer and incubate for 10 min.
11. Wash the slides 3 times for 3 min with PBST.
12. Dilute the Texas-red fluorescent dye in MOM diluent and incubate for 15 min.
13. Wash the slides 3 times for 3 min with PBST.
14. Mount the slides with DAKO mounting medium containing DAPI nuclear stain and coverslip. Wrap the slides in foil as the DAPI and goat-anti rabbit secondary antibody are light sensitive.
15. View the slides under the fluorescence microscope—red emission for the dystrophin protein; while blue emission for the nuclei.

3.9 Co-Immunohistochemical Staining for PAX7 andKi67

1. Cut 8 μm-thick cryosections from at least two-thirds deep into the muscle at 100 μm intervals.
2. Allow the slides to air-dry and circle the sections with a blocking pen so that less reagent is required and the sections do not lift from the slide.
3. Add 100 μL of 4% PFA on section. Fix the fiber at RT for 15 min.
4. Wash the slides 3 times for 5 min with PBSTR.
5. Heat the slides in citrate buffer for 10 min at 100 °C (*see* **Note 22**).
6. Wash the slides 3 times for 5 min with PBSTR.
7. Prepare the 10% AFG and pipette onto the sections for 1 h at RT.
8. Wash the slides 3 times for 5 min with PBSTR.
9. Prepare the blocking buffer and pipette onto the sections for 30 min at RT.

10. Tip off the solution and incubate with the primary antibodies PAX7 (a gift from Prof. Hongbo Zhang) and Ki67 for overnight at 4 °C.
11. Wash the slides 3 times for 5 min with PBSTR.
12. Incubate with donkey-anti-rabbit and donkey-anti-mouse secondary antibodies at a dilution of 1/200 in PBSTR for 1 hat RT.
13. Throw away the secondary antibody and wash the slides 3 times for 5 min with PBSTR.
14. Mount the slides with DAKO mounting media containing DAPI nuclear stain and coverslip.
15. View the slides under the fluorescence microscope.

3.10 Tissue Distribution

1. Inject 5% glycine into the tail vein of *mdx* mice every other day for 1 week.
2. Prepare 25 mg/kg Lissamine-labeled or 50 mg/kg FITC-labeled PMO in 5% glycine or saline solution with a final volume of 120 μL.
3. Inject the solution into the tail vein of *mdx* mice.
4. Perfuse mice with a large amount of pre-chilled PBS via the hepatic portal vein (*see* **Note 23**).
5. Harvest body-wide tissues and wash tissues with pre-chilled PBS. Dry the tissues with filter paper and put them on black paper.
6. Measure the fluorescence intensity of tissues with the IVIS system.

3.11 Histology

1. Fix fresh liver or kidney tissues in the Bouin's solution for 48 h.
2. Wash the tissues with running tap water.
3. Immerse the tissues in xylene for 30 min, followed by absolute, 95%, 85%, and 70% ethanol for 30 min each time.
4. Immerse the tissues in thawed paraffin at 65 °C for 5–6 h.
5. Embed the tissues in paraffin blocks.
6. Cut 6-μm-thick paraffin sections.
7. Incubate the slides at 65 °C for 1 h.
8. Immerse the slides in xylene for 40 min, followed by absolute, 95%, 85%, and 70% ethanol and water for 5 min each time.
9. Stain the slides in hematoxylin for 10–15 min.
10. Wash the back of the slides in running tap water.
11. Stain the slides in eosin for 30–60 s.
12. Wash the back of the slides in running tap water.

13. Immerse the slides in 70% ethanol for 30 s, 85% ethanol for 30 s, 95% ethanol for 2 min, and absolute ethanol for 2 min.
14. Immerse the slides in xylene for 10 min.
15. Mount the slides with neutral balsam.

3.12 Isolation of Primary Myoblasts

1. Dissect the hindlimb muscles including gastrocnemius, quadriceps, and TA muscles.
2. Wash the tissues with pre-cooled D-Hanks buffer 5 times.
3. Mince all muscles and digest it with the digestion buffer for 2 h at 37°C.
4. Vortex the digested fiber pellets gently for 30 s.
5. Filter the mixture through 70-mm and 40-mm cell strainers, respectively, followed by centrifugation at 750 *g* for 5 min.
6. Discard the supernatant and resuspend the pellets with the primary myoblast culture medium.
7. Seed the cells in Matrigel-coated (1/100) Petri dishes and change the medium with a fresh cell culture medium after 24 h.

3.13 Myoblast Transplantation in mdx Mice

1. Prepare green fluorescence protein (GFP)-positive primary myoblasts.
2. 5% glycine is intravenously injected into *mdx* mice every other day for 1 week prior to cell transplantation, and maintain the same dosing regimen during the experimental period.
3. *Mdx* mice are fed with 7.5 mg/L FK506 in the drinking water every day during the experimental period.
4. Inject 2 μg of CTX solution into TA muscles of *mdx* mice.
5. Transplant 1 x 10^6 GFP-positive primary myoblasts into the same injured TA muscle 1 day later.
6. Treated *mdx* mice are monitored at weeks 1, 2, and 3 after transplantation with the IVIS imaging system.
7. Three weeks after cell administration, the mice may be sacrificed by cervical dislocation and TA muscles are snap-frozen in liquid nitrogen and stored at -80 °C.

3.14 Generation of RagB Knockout HEK293T Cells

1. Prepare HEK293T cells.
2. Transfect HEK293T cells with CRISPR/Cas9-expressing lentivirus for 24 h.
3. Screen Cas9-expressing cell clones with 2 mg/mL blasticidin in cell culture medium for 2 weeks.
4. Cas9-expressing HEK293T cells are transfected with single-guide RNA (sgRNA)-expressing lentivirus for 24 h.
5. Screen RagB knock-out cell clones with 1 mg/mL puromycin for 2 weeks.

4 Notes

1. Labeled oligonucleotides should be wrapped in foil as it is light sensitive.
2. This substance is highly flammable and therefore should be stored in a cool, dry, and well-ventilated area.
3. It should only be handled in a fume hood, and safety goggles and impervious gloves should be worn at all times. Contaminated waste should be disposed of in special waste appropriately.
4. Once the DAPI has been added to the DAKO, shake the bottle well and wrap in foil as DAPI is light sensitive.
5. Thaw reagents directly before use; however, the enzyme mix or *Taq* should be kept on ice at all times.
6. Keep the container tightly closed and in the designated EB contamination area. Heavily contaminated waste should be disposed of in special waste; however, gels and tips may be discarded in clinical bins.
7. TEMED my decline in quality over time; therefore, it is best to buy small bottles.
8. Transfer buffer can be used for up to 2–3 transfers before it is discarded. In order to keep the transfer buffer cooler than room temperature while running, the experiment and transfer apparatus may be kept on ice or stored in a cold room.
9. The antibody is stable in a freeze-dried conjugate form. It therefore requires reconstitution in 2.5 mL of ddH2O prior to use. Aliquot into tubes and store at −20 °C. Tubes may be removed from the freezer one at a time and stored at 2–8 °C.
10. If the quantification is required, the film may be scanned and programs such as ImageJ can be utilized to compare the scanning densitometry. Alternatively, an instrument such as Fuji-Film LAS-100 plus may be used.
11. It is important to ensure that the samples are kept on dry ice up until the TRIZOL step is finished.
12. Depending on the viscosity of the lysis buffer, equal amounts of PBS may be added to the well prior to fluorescence measurement.
13. The volume of TRIZOL can be adjusted depending on the amount of tissue sections.
14. Some tissues such as muscle may require an additional isolation step due to the high content of proteins, fat, polysaccharides, or extracellular material. Following homogenization, insoluble material is removed by centrifugation at 12,000× *g* for 10 min

at 2–8 °C. The resulting pellet contains extracellular membranes, polysaccharides, and high-molecular-weight DNA, while the supernatant contains RNA.

15. For RT-PCR reactions, nested PCR is not necessary for highly efficient exon skipping, and one round of RT-PCR would be sufficient for the detection of exon skipping bands.
16. The volume of protein lysis buffer can be adjusted depending on the amount of tissue sections.
17. When the stacking gel has set, gels may be removed from the green plastic holders, wrapped in a damp paper towel, and then wrapped in cling film and stored at 4 °C.
18. Soaking the PVDF membrane in methanol prevents proteins from passing straight through the membrane.
19. Ensure the use of sufficient wash solution as the salt assists in the reduction of background.
20. Ensure the film is sealed away before turning on the white light or opening the door as this will expose the film and render it useless.
21. For nNOS staining, the whole procedure is the same as for dystrophin, except that the incubation with the primary antibody is overnight at 4 °C.
22. Avoid the slides dry.
23. A minimal of 30 mL of pre-chilled PBS should be used to flush mice for perfusion.

References

1. Kaspar RW, Allen HD, Ray WC et al (2009) Analysis of dystrophin deletion mutations predicts age of cardiomyopathy onset in Becker muscular dystrophy. Circ Cardiovasc Genet 2(6):544–551. https://doi.org/10.1161/CIRCGENETICS.109.867242
2. Sheikh O, Yokota T (2021) Restoring protein expression in neuromuscular conditions: a review assessing the current state of exon skipping/inclusion and gene therapies for Duchenne muscular dystrophy and spinal muscular atrophy. BioDrugs. https://doi.org/10.1007/s40259-021-00486-7
3. Mendell JR, Rodino-Klapac LR, Sahenk Z et al (2013) Eteplirsen for the treatment of Duchenne muscular dystrophy. Ann Neurol 74(5):637–647. https://doi.org/10.1002/ana.23982
4. Goemans NM, Tulinius M, van den Akker JT et al (2011) Systemic administration of PRO051 in Duchenne's muscular dystrophy. N Engl J Med 364(16):1513–1522. https://doi.org/10.1056/NEJMoa1011367
5. Biressi S, Filareto A, Rando TA (2020) Stem cell therapy for muscular dystrophies. J Clin Invest 130(11):5652–5664. https://doi.org/10.1172/JCI142031
6. Yin H, Lu Q, Wood M (2008) Effective exon skipping and restoration of dystrophin expression by peptide nucleic acid antisense oligonucleotides in mdx mice. Mol Ther 16(1):38–45. https://doi.org/10.1038/sj.mt.6300329
7. Yin H, Moulton HM, Betts C et al (2009) A fusion peptide directs enhanced systemic dystrophin exon skipping and functional restoration in dystrophin-deficient mdx mice. Hum Mol Genet 18(22):4405–4414. https://doi.org/10.1093/hmg/ddp395
8. Yin H, Moulton HM, Seow Y et al (2008) Cell-penetrating peptide-conjugated antisense oligonucleotides restore systemic muscle and cardiac dystrophin expression and function. Hum

Mol Genet 17(24):3909–3918. https://doi.org/10.1093/hmg/ddn293

9. Yin H, Saleh AF, Betts C et al (2011) Pip5 transduction peptides direct high efficiency oligonucleotide-mediated dystrophin exon skipping in heart and phenotypic correction in mdx mice. Mol Ther 19(7):1295–1303. https://doi.org/10.1038/mt.2011.79
10. Gao X, Zhao J, Han G et al (2014) Effective dystrophin restoration by a novel muscle-homing peptide-morpholino conjugate in dystrophin-deficient mdx mice. Mol Ther 22(7):1333–1341. https://doi.org/10.1038/mt.2014.63
11. Hammond SM, Hazell G, Shabanpoor F et al (2016) Systemic peptide-mediated oligonucleotide therapy improves long-term survival in spinal muscular atrophy. Proc Natl Acad Sci U S A 113(39):10962–10967. https://doi.org/10.1073/pnas.1605731113
12. Said Hassane F, Saleh AF, Abes R et al (2010) Cell penetrating peptides: overview and applications to the delivery of oligonucleotides. Cell Mol Life Sci 67(5):715–726. https://doi.org/10.1007/s00018-009-0186-0
13. Biswas S (2021) Polymeric micelles as drug delivery systems in cancer: challenges and opportunities. Nanomedicine (London). https://doi.org/10.2217/nnm-2021-0081
14. Yang H, Liu R, Xu Y et al (2021) Photosensitizer nanoparticles boost photodynamic therapy for pancreatic cancer treatment. Nano-Micro Lett 13(1):35. https://doi.org/10.1007/s40820-020-00561-8
15. Kim Y, Tewari M, Pajerowski JD et al (2009) Polymersome delivery of siRNA and antisense oligonucleotides. J Control Release 134(2):132–140. https://doi.org/10.1016/j.jconrel.2008.10.020
16. Rimessi P, Sabatelli P, Fabris M et al (2009) Cationic PMMA nanoparticles bind and deliver antisense oligoribonucleotides allowing restoration of dystrophin expression in the mdx mouse. Mol Ther 17(5):820–827. https://doi.org/10.1038/mt.2009.8
17. Sirsi SR, Schray RC, Wheatley MA et al (2009) Formulation of polylactide-co-glycolic acid nanospheres for encapsulation and sustained release of poly(ethylene imine)-poly(ethylene glycol) copolymers complexed to oligonucleotides. J Nanobiotechnol 7:1. https://doi.org/10.1186/1477-3155-7-1
18. Williams JH, Schray RC, Sirsi SR et al (2008) Nanopolymers improve delivery of exon skipping oligonucleotides and concomitant dystrophin expression in skeletal muscle of mdx mice. BMC Biotechnol 8:35. https://doi.org/10.1186/1472-6750-8-35
19. Gao X, Ran N, Dong X et al (2018) Anchor peptide captures, targets, and loads exosomes of diverse origins for diagnostics and therapy. Sci Transl Med 10(444). https://doi.org/10.1126/scitranslmed.aat0195
20. Han G, Gu B, Lin C et al (2019) Hexose potentiates peptide-conjugated Morpholino oligomer efficacy in cardiac muscles of dystrophic mice in an age-dependent manner. Mol Ther Nucl Acids 18:341–350. https://doi.org/10.1016/j.omtn.2019.09.012
21. Han G, Gu B, Cao L et al (2016) Hexose enhances oligonucleotide delivery and exon skipping in dystrophin-deficient mdx mice. Nat Commun 7:10981. https://doi.org/10.1038/ncomms10981
22. Lin C, Han G, Ning H et al (2020) Glycine enhances satellite cell proliferation, cell transplantation, and oligonucleotide efficacy in dystrophic muscle. Mol Ther 28(5):1339–1358. https://doi.org/10.1016/j.ymthe.2020.03.003

Chapter 11

Morpholino-Mediated Exons 28–29 Skipping of Dysferlin and Characterization of Multiexon-skipped Dysferlin using RT-PCR, Immunoblotting, and Membrane Wounding Assay

Saeed Anwar and Toshifumi Yokota

Abstract

Dysferlinopathies are a group of disabling muscular dystrophies that includes limb girdle muscular dystrophy type 2B (LGMD2B), Miyoshi myopathy, and distal myopathy with anterior tibial onset (DMAT) as the main phenotypes. They are associated with molecular defects in *DYSF*, which encodes dysferlin, a key player in sarcolemmal homeostasis. Previous investigations have suggested that exon skipping may be a promising therapy for many patients with dysferlinopathies. It was reported that exons 28–29 of *DYSF* are dispensable for dysferlin functions. Here, we present a method for multiexon skipping of *DYSF* exons 28–29 using a cocktail of two phosphorodiamidate morpholino oligomers (PMOs) on cells derived from a dystrophinopathy patient. Also, we describe assays to characterize the multiexon skipped dysferlin at several levels by using one-step RT-PCR, immunoblotting, and a membrane wounding assay.

Key words Morpholino, Dysferlinopathy, LGMD2B, Miyoshi myopathy, Dysferlin, Exon skipping, Multiexon skipping, Antisense therapy, Fibroblasts, Phosphorodiamidate morpholino oligomers (PMO), Immunoblotting, Membrane wounding

1 Introduction

Dysferlinopathy is a group of adult-onset progressive muscular dystrophies that results in progressive weakness in the voluntary muscles [1]. Depending on the initial pattern of the disease, it is classified into three major sub-phenotypes: limb-girdle muscular dystrophy type 2B (LGMD2B, OMIM # 253601), Miyoshi myopathy (OMIM # 254130), and distal myopathy with anterior tibial onset (OMIM # 606768) [2–4]. The clinical representation of the disease is heterogeneous, and disease severity in dysferlinopathy ranges from asymptomatic carriers with elevated serum CK levels and exercise intolerance to severe functional disability

Rika Maruyama and Toshifumi Yokota (eds.), *Muscular Dystrophy Therapeutics: Methods and Protocols*, Methods in Molecular Biology, vol. 2587, https://doi.org/10.1007/978-1-0716-2772-3_11,

[5–8]. Unfortunately, there is yet a cure for dysferlinopathy, and appropriate genetic counseling and management for these patients are not available.

Dysferlinopathy is caused by deficiency or insufficient activity of dysferlin due to mutations in the *DYSF* gene that results in impaired membrane repair, eventually leading to necrotic changes and severe fibrosis in muscles [9, 10]. The gene *DYSF* maps to chromosome 2p12–p14, contains 55 exons, and spans a region of over 230 kb (GenBank NM_003494.3) [11]. The enormous size of the gene makes it highly susceptible to mutations. About half of the people with dysferlinopathy have a homozygous mutation in *DYSF*, while the other half carry compound heterozygote mutations [12]. Mutations are distributed over the transcript with no apparent mutation hotspots; however, a recent study suggested that nearly two-thirds of all patients in Japan have mutations in the inner DysF domain (exons 28–29) [6, 11, 13, 14]. Most of the reported mutations are point mutations, small deletion/insertions, and intronic mutations [13–15]. The knowledge regarding genotype–phenotype correlation is still limited.

Exon skipping therapy has been explored as a novel therapeutic approach for dysferlinopathies. Dysferlin contains 7 Ca^{2+}-dependent phospholipid-binding C2 domains [9]. Previous reports suggest that patients carrying dysferlin lacking one or more internal C2 domains, e.g., the fourth C2 domain, may experience very mild muscle weakness with only elevated serum creatine kinase [14, 16]. This evidence inspires the use of antisense-mediated therapeutic exon skipping as a potential treatment strategy for patients with mutations on *DYSF*. Antisense-mediated exon skipping using synthetic DNA-like molecules, called antisense oligonucleotides (AOs), can be effectively used to remove mutation-carrying exons and flanking exons to maintain the reading frame [17]. This strategy is well-known for treating Duchenne muscular dystrophy (DMD) as the US Food and Drug Administration (FDA) has recently approved four exon skipping drugs for DMD patients with skip-amenable mutations [18–21]. These drugs facilitate in-frame skipping of dystrophin exon 51 (eteplirsen), exon 53 (golodirsen and viltolarsen), and exon 45 (casimersen) and utilize the phosphorodiamidate morpholino oligomer (PMO) antisense chemistry [18–21].

To date, only a few reports, mostly from in vitro studies, are available on *DYSF* exon skipping [22–25]. A major interest of these studies is to evaluate the restoration of dysferlin at the protein level and the ability of the cells to respond to membrane stresses. In a recent study, our lab identified two multiexon combinations, including exons 28–29, which are dispensable for membrane resealing in patient-derived fibroblasts [25]. We designed PMOs that mediated the skipping of exons 28–29 and showed that the multi-exon skipping led to the restoration of dysferlin at the protein level and membrane resealing ability of the cells [25]. In this chapter, we present the step-by-step process to induce multiexon skipping of *DYSF* exons 28–29 skipping using a cocktail of two PMOs and the evaluation of the efficacy using one-step RT-PCR. Also, the restoration of the dysferlin at the protein level and membrane resealing ability of the PMO-treated patient cells using immunoblotting and two-photon laser microscope, respectively, have been described.

2 Materials

2.1 Cell Cultures and Morpholino Transfections

1. Human dysferlinopathy patient fibroblasts (MM-Pt2 cells harboring trans-heterozygous c.1958delG/c.2997G > T mutations induced pluripotent stem cell bank, Kyoto University).
2. Healthy human fibroblasts (GM23815, Coriell Institute).
3. Cell growth media: Dulbecco's modified Eagle's medium/nutrient mixture F-12 (DMEM/F-12, Gibco) growth media containing 10% fetal bovine serum (FBS, Sigma-Aldrich) and 0.5% penicillin and streptomycin.
4. Trypsin–EDTA (0.25%), phenol red (Gibco).
5. T75 EasYFlask™ Cell Culture Flasks (Thermo Scientific).
6. CO_2 incubator.
7. Phosphorodiamidate morpholino oligomers (PMOs) targeting *DYSF* exons 28 and 29 (PMO targeting *DYSF* exon 28, Ac11: 5′ – GGG CAC TCA ATG TCA TCC TTG GGA AGC ACC – 3′; PMO targeting *DYSF* exon 29, Ac32: 5′ – CTC AGC AGG GAC CCA GTG CTT CGG CTT CCG – 3′; mock PMO with a random nonspecific sequence) (Gene Tools).
8. Endo-porter transfection reagent (Gene Tools).
9. Ultrapure water.
10. Transfection media: Dulbecco's modified Eagle's medium/nutrient mixture F-12 (DMEM/F-12, Gibco) growth media supplemented with 10% fetal bovine serum (FBS) (*see* **Note 1**).
11. 12- or 24-well cell culture plates.
12. 35-mm collagen-coated glass-bottom dishes (MatTek).

2.2 One Step RT-PCR and cDNA Sequencing

1. TRIzol™ Reagent (Invitrogen).
2. Chloroform.
3. Ethanol.
4. Isopropanol.
5. Glycogen, RNA grade (Thermo Fisher).
6. DNase/RNase-free distilled water.
7. NanoDrop™ Lite Spectrophotometer (Thermo Fisher).

8. SuperScript™ III One-Step RT-PCR System with Platinum Taq High Fidelity DNA polymerase (Invitrogen).
9. Primers (Forward: 5′ – CGA GAC TAA GTT GGC CCT TG – 3′, reverse: 5′ – CGT TTC CCT GAA CCA TGA CT – 3′).
10. 2% agarose gel with 0.01% GelRed (Biotium).
11. Scalpel blade and scalpel handle.
12. Open UV box.
13. QIAquick gel extraction kit (QIAGEN).

2.3 Protein Extraction and Dysferlin Immunoblotting Analysis

1. Heat block/water bath set.
2. Pierce® RIPA (radioimmunoprecipitation assay) buffer (Thermo Scientific).
3. Complete™, Mini, EDTA-free Protease Inhibitor Cocktail tablet (Roche).
4. Sterile cell scraper.
5. 21G needles
6. Syringes, preferably of 1 mL size.
7. Centrifuge.
8. Pierce™ BCA Protein Assay Kit (Thermo Scientific).
9. 96-well plates.
10. XCell4 SureLock™ Midi-Cell electrophoresis system (Invitrogen).
11. Novex™ Semi-Dry Blotter (Life Technologies).
12. Sample loading buffer: 70 mM Tris–HCl at pH 6.8, 5 mM EDTA, 10% SDS, 20% glycerol, 0.004% w/v bromophenol blue, and 5% β-mercaptoethanol.
13. NuPAGE™ Novex™ 3–8% Tris–Acetate Midi gel (Invitrogen).
14. NuPAGE™ Tris–Acetate SDS running buffer, 20× stock (Invitrogen).
15. NuPAGE™ Antioxidant (Invitrogen).
16. Appropriate protein ladders, e.g., HiMark™ pre-stained protein standard (Thermo Fisher).
17. Extra thick (2.5 mm) filter paper sheets, 8.0 cm × 13.5 cm (Thermo Fisher).
18. Methanol.
19. Concentrated anode buffer: 0.3 M Tris–HCl, 20% methanol.
20. Anode buffer: 0.03 M Tris–HCl, 20% methanol.
21. Cathode buffer: 25 mM Tris–HCl, 40 mM 6-amino-n-hexanoic acid, 20% methanol, 0.01% SDS.

22. Immobilon polyvinylidene difluoride (PVDF) membrane, 0.45 μm pore size (Millipore).
23. Blocking solution: phosphate-buffered saline with 0.05% Tween 20 (PBST) and 5% skimmed milk.
24. Amersham™ ECL Prime Blocking Agent (GE Healthcare).
25. Anti-dysferlin antibodies: NCL Hamlet monoclonal antibody (Leica Biosystems) or alpha-tubulin ab7291 antibody (Abcam).
26. Secondary antibodies: goat anti-mouse immunoglobulin G (IgG) (H+L) secondary antibody, horseradish peroxidase (HRP) (Thermo Fisher).
27. Amersham ECL Select™ Western Blotting Detection Reagent (Cytiva).

2.4 Two-Photon Laser Wounding Assay

1. Dulbecco's phosphate-buffered saline (Sigma-Aldrich).
2. FM 4-64 dye (Invitrogen).
3. Tyrode's salt solution (Sigma-Aldrich).
4. Confocal laser scanning microscope (Carl Zeiss).
5. Chameleon Vision two-photon laser (Coherent).

3 Methods

3.1 Cell Cultures and Morpholino Transfections

1. In a sterile T75 EasYFlask™ Cell Culture Flask, grow and maintain dysferlinopathy patient-derived fibroblasts (e.g., MM-Pt2) in the cell growth medium. When the confluency reaches 80–100%, and passage the cells using 0.25% trypsin–EDTA. Note that healthy fibroblasts (e.g., GM23815) can be cultured as well to serve as controls, particularly for the dysferlin quantification experiments (*see* **Note 2**).
2. Incubate the cells at 37 °C with 5% CO_2.
3. For RT-PCR and immunoblotting analysis, seed cells at a density of $4 \times 10^4/\text{cm}^2$ for each well in a 12- or 24-well plate. For two-photon membrane wounding assays, seed 2 mL of patient and healthy fibroblasts at 1×10^5 and 1×10^4 cells/mL concentration, respectively, into 35-mm collagen-coated glass-bottom dishes.
4. Grow the cells to 50–70% confluence (*see* **Note 3**).
5. Dilute the PMOs to 1.0 mM in ultrapure water. Preheat PMO stocks at 65 °C for 10–20 min prior to transfection to ensure their complete disaggregation (*see* **Note 4**), and vortex well at high speed and spin down.
6. Prepare the PMO transfection media. Aspirate the old media out and transfect cells with the PMOs at the desired dose using

6 μM/mL of the Endo-porter transfection reagent. For 10 μM transfection in a 12-well plate, add 5 μL of each PMO solutions and 3 μL of Endo-porter to 487 mL transfection media. Transfection with similar doses in a 35 mm collagen-coated glass-bottom dish would require 10 μL of each PMO solutions and 6 μL of Endo-porter to 974 mL of transfection media.

7. Incubate for 2 days in a CO_2 incubator and harvest cells for downstream applications, e.g., RT-PCR, cDNA sequencing, immunoblotting, and two-photon laser membrane wounding assay.

3.2 One Step RT-PCR and cDNA Sequencing

1. Aspirate out as much medium as possible from the wells. Add 1.0 mL of ice-cold TRIzol to the cells, and pipette up and down to lyse cells thoroughly.
2. Transfer cell suspensions to 1.5 mL sterile tubes. Vortex for 15 s to further homogenize the cells (*see* **Note 5**).
3. Thaw the samples at room temperature and add 200 μL of chloroform and vortex for 30 s (*see* **Note 6**). Leave the samples at room temperature for 5 min.
4. Centrifuge the sample at 12,000 × *g* for 15 min at 4 °C.
5. Transfer as much of the upper aqueous phase as possible to a fresh 1.5 mL tube. Carefully ensure not to pick up the middle and bottom layers into a new tube.
6. Add 500 μL of isopropanol and 1.0 μL of RNA-grade glycogen and vortex well. Incubate for 10 min at room temperature (*see* **Note 7**).
7. Centrifuge at 12,000 × g for 10 min at 4 °C. A small whitish pellet should be seen at the bottom of the tube at this point.
8. Discard as much supernatant as possible while carefully not disturbing the pellet. Add 1.0 mL of cold 75% ethanol, vortex just enough to see the pellet float.
9. Centrifuge at 7500 × *g* for 5 min at 4 °C.
10. Remove as much ethanol as possible and air-dry the samples for at least 5 min (*see* **Note 8**).
11. Once dry, resuspend the RNA pellet in 20–30 μL of RNase-free water.
12. Incubate the RNA solutions at 65 °C for 10 min and keep them on ice.
13. Measure concentrations of the RNA solutions using NanoDrop Lite Spectrophotometer at 260 nM. The RNA samples can be diluted with RNase-free water to a desired concentration and then kept on ice until use or stored at −80 °C for future use.

14. Prepare one-step RT-PCR reaction. For 25 μL reaction, mix 12.5 μL of 2× reaction mix, 5 μL of RNA (200 ng), 0.5 μL of 10 μM forward primer, 0.5 μL of 10 μM reverse primer, 1 μL of SuperScript III RT/Platinum Taq Mix, and 5.5 μL of RNase-free water.
15. Set up and run the following PCR program: 1 cycle of 50 °C for 20 min, 1 cycle of 94 °C for 2 min, then 34 cycles of 94 °C for 15 s, 60 °C for 30 s, 68 °C for 40 s, 1 cycle of 68 °C for 5 min, and hold at 4 °C. The PCR products can be stored at 4 °C or −20 °C (*see* **Note 9**).
16. Run 5 μL of the PCR products onto a 2% agarose gel pre-stained with 0.01% GelRed and image the gel. The band intensities can be quantified using a suitable gel image analyzer, e.g., ImageJ (*see* **Note 10**).
17. Calculate the exon skipping efficiencies for the tested PMOs using the following formula:

$$\%\text{multiexon skipping} = \frac{\text{intensity of skipped band}}{\text{intensity of nonskipped band} + \text{intensity of skipped band}} \times 100\,(\%)$$

Compare the % multiexon skipping in different treatment groups to nontreated or healthy using one-way ANOVA and Tukey–Kramer multiple comparison tests.

18. Cut the bands carefully with a sharp scalpel blade on an open UV box and collect them into 1.5 mL tubes (*see* **Note 11**).
19. Purify the DNA from the gels using the QIAquick gel extraction kit following the manufacturer's instructions.
20. Sequence the purified PCR product using the same primer as for the one-step RT-PCR.
21. Compare the obtained DNA sequences to *DYSF* reference sequence using the Nucleotide BLAST tool and confirm the correct skipping of the administered PMOs.

3.3 Protein Extraction and Dysferlin Immunoblotting Analysis

1. Aspirate out as much medium as possible from the wells. Wash the cells twice with PBS for 1 min.
2. Add one cOmplete™, Mini, EDTA-free Protease Inhibitor Cocktail table to 7 mL of Pierce® RIPA Buffer to prepare protein extraction solution.
3. Add 170 μL of the protein extraction solution to each well.
4. Detach cells from the well using a sterile cell scraper and collect the sample in a 1.5 mL tube.
5. Incubate the samples in ice for 30 min.
6. Further homogenize the samples by passing them carefully through a 21G needle 10–15 times (*see* **Note 12**).
7. Spin down using a centrifuge at 14,000 × *g* for 15 min at 4 °C.

8. Carefully transfer supernatants (contains protein) to 1.5-mL fresh tubes without taking any of the pelleted debris (*see* **Note 13**) and keep them in ice until use.
9. Quantify protein concentration in the obtained extracts using Pierce™ BCA Protein Assay Kit.
10. Prepare 10 μL of protein extracts for SDS-PAGE by adding 2× sodium dodecyl sulfate (SDS) sample loading buffer.
11. Prepare a master mix by adding 5× NuPAGE LDS Sample Buffer to 2× NuPAGE Sample Reducing Agent. For example, 50 μL of NuPAGE LDS Sample Buffer mixed with 20 μL of NuPAGE Sample Reducing Agent is sufficient for 10 samples.
12. Add 7 μL of master mix and 3 μL of distilled water to each sample. Heat mixture at 70 °C for 10 min in a hot water bath (*see* **Note 14**).
13. While the samples are heating, prepare the XCell4 SureLock™ Midi-Cell electrophoresis tank by setting a precast 3–8% Tris–Acetate Midi gel in the apparatus. Fill the outer buffer chamber with 1× Tris–Acetate SDS running buffer to the indicated level. Fill the inner buffer chamber with the same buffer but supplemented with the antioxidant until the gel's wells are fully submerged.
14. Load at least 1.5 μg of protein into the designated wells (*see* **Notes 15** and **16**). Also, load protein samples from healthy cells at different dilutions to serve as standards for constructing a calibration curve for dysferlin quantification later.
15. Similarly, load an appropriate protein ladder as well to allow for protein size determination—we normally use the HiMark™ protein ladder.
16. Run gel electrophoresis for 70 min at 150 V.
17. Transfer the protein from the gel to a PVDF membrane using a semidry blotting system.
    (i) Soak 3 extra thick blotting filter papers in buffer solutions in each buffer: Concentrated anode buffer: 0.3 M Tris–HCl, 20% methanol; anode buffer: 0.03 M Tris–HCl, 20% methanol; cathode buffer: 25 mM Tris, 20% methanol, 40 mM 6-amino-n-hexanoic acid, 0.01% SDS.
    (ii) Prewet the PVDF membrane in methanol for 10 s, and then incubate in anode buffer for at least 10 min.
    (iii) Soak the gel in the cathode buffer (*see* **Note 17**).
    (iv) Stack transfer components in the following order on top of the anode plate: concentrated anode buffer paper–anode buffer paper–PVDF membrane–protein gel–cathode buffer paper (*see* **Notes 18** and **19**).

(v) Cover the stack firmly with the cathode plate. Ensure the cover is placed evenly, perhaps using a bubble-liquid level indicator (*see* **Note 20**).

(vi) Run for 45 min at 20 V.

18. Carefully obtain the PVDF membrane and wash it with PBST for 5 min at room temperature with shaking.
19. Immerse the PVDF membrane in the blocking solution (PBS containing 0.05% Tween 20—PBST and 5% skimmed milk) and incubate with either the NCL Hamlet monoclonal antibody or alpha-tubulin ab7291 antibody overnight at 4 °C with shaking. Use a dilution of 1:4000 for NCL Hamlet monoclonal antibody and 1:8000 for alpha-tubulin ab7291 antibody, with PBST as diluent.
20. Wash the PVDF membrane thrice with PBST for 10 min each at room temperature with shaking.
21. Incubate the membrane in the goat anti-mouse immunoglobulin G (IgG) (H+L) secondary antibody, horseradish peroxidase (HRP) solution for an hour at room temperature with shaking. Use a dilution of 1:10,000, with PBST as the diluent.
22. Wash the PVDF membrane thrice with PBST for 10 min each at room temperature with shaking.
23. Prepare sufficient detection solution by mixing solutions A (luminol) and B (peroxide) of the Amersham ECL Select™ Western Blotting Detection Reagent kit in a 1:1 ratio. For example, a total volume of 3 mL is sufficient for a midi-sized gel.
24. Incubate the membrane in the newly prepared detection solution for 5 min at room temperature and then visualize the blot using a chemiluminescent imaging system.
25. Using ImageJ, quantify the intensities of the observed dystrophin bands for all the samples. Construct a calibration curve using the obtained intensities for the healthy control protein standards.
26. Use the equation obtained from the calibration curve to quantify the amount of dysferlin rescued in PMO-treated samples compared to healthy levels.
27. Compare the amount of dysferlin in different treatment groups to nontreated using one-way ANOVA and Tukey–Kramer multiple comparison tests.

3.4 Two-Photon Laser Wounding Assay

1. Aspirate out as much medium as possible from the 35 mm collagen-coated glass-bottom dishes.
2. Wash the cells with 1 mL of Tyrode's solution for 1 min.
3. Prepare 2.5 mM working solution of FM 4-64 dye (*see* **Note 21**).

4. Add 1 mL of fresh Tyrode's solution containing 1 μL of 2.5 mM FM 4-64 dye, i.e., the final concentration of FM 4-64 dye is 2.5 μM (*see* **Notes 21** and **22**).
5. Using an inverted confocal scanning microscope and associated program software, create a 0.1 μm × 1 μm target and place it close to the edge of the cell membrane. The target line must overlap and remain perpendicular to the direction of the cell membrane.
6. Use a 543 nm HeNe laser to excite FM 4-64 dye signal and set the detection range to 600–760 nm. Create a 5-min time series of sequential image scans, imaging cells every 5 s. Set the Chameleon Vision two-photon laser set to 820 nm and cell bleach to 25 s after the beginning of the time series. Set the membrane wounding experiment using 15% laser power with ten iterations. Set the cell bleach 25 s after the beginning of the time series.
7. To quantify FM 4-64 dye fluorescence (S), use the Zeiss Zen software to draw a 6 μm × 6 μm region of interest (ROI) and place it close to the wounding location and within the cell (*see* **Note 23**).
8. Run the experiment and wait until it completes (*see* **Note 24**). Export raw fluorescence values into a spreadsheet for analysis.
9. Calculate relative fluorescence (S_r) values for each time point using the following formula:

$$S_r = \frac{\Delta S}{S_0}$$

Here, $\Delta S = S$ at a given time point – background S, $S_0 = S$ at $t = 0$.

10. Repeat the procedure for some 15 cells per treatment group and scatter plot the mean S_r values against time.
11. Analyze the S_r values at the final time point and time to steady-state statistically in the treatment groups to nontreated using one-way ANOVA and Tukey–Kramer multiple comparison test.

4 Notes

1. Many antibiotics, e.g., penicillin–streptomycin, tend to cause a toxic effect. Use the growth media without them for transfection experiments.
2. For the dysferlin quantification experiment, do not subject the healthy fibroblasts to PMO transfection.
3. This can take ~2 days, based on our experience.

4. This sometimes takes longer and/or requires higher temperature.
5. Samples can be stored at −80 °C at this point, and extraction can be finished at a later time.
6. This step is essential for the proper formation of aqueous and organic phases after centrifugation later in the protocol.
7. The samples can be stored at −20 °C at this point for up to a week. Do not incubate the samples if the samples are to be stored.
8. Make sure that no ethanol is left in the tubes before resuspending the RNA pellet. Residual ethanol can affect downstream applications, e.g., RT-PCR.
9. This may have to be further optimized depending on specific system/equipment requirements.
10. The size of the nonskipped PCR product is 613 bp. The exon 28–29 skipped PCR-product is 364 bp.
11. Gel-purification is most efficient with lower % agarose gel. Hence, try to stay in the 0.8–1.0% range if possible. Also, using a wider gel comb and running the gel at a lower voltage will give crisp bands. Try to remove as much excess gel around the band as possible. To make the band cutting easier and less prone to contamination, leave out a blank lane between each sample. In addition, use long-wavelength UV for as short as possible to minimize the risk of DNA damage. Do not forget to use personal protective equipment when using the open UV box.
12. Try to keep the top of the needle as close to the bottom of the tube as possible to avoid the formation of bubbles as much as possible. Bubbles often lead to protein degradation.
13. The samples can be stored in a −80 °C freezer at this point, and protein quantification can be done later. However, we recommend quantifying the protein as soon as possible after extraction.
14. When heating is done, keep the prepared samples on ice until loading.
15. Flush the wells with the running buffer prior to loading.
16. The protein amount can be changed as desired—we usually get satisfactory results with 1.5 μg of proteins. But note that, a maximum of 20 μL of the prepared sample can be loaded onto each well, and all the wells should be loaded with an equal amount of prepared sample.
17. It is recommended to remove the stacking gel portion and then incubate it for 5 min in the cathode buffer first before the transfer.

18. Make sure there are no bubbles between any of the layers while stacking, as this may interfere with the transfer in parts where these bubbles occur.
19. All equipment, e.g., forceps, transfer plate, and gel knife used for stacking, that touches the gel, should be kept in cathode buffer as this creates a nonstick surface and help prevent possible tears that can occur from the gel being stretched.
20. Placing the cover evenly is critical to prevent inconsistent transfer across the membrane.
21. Keep the dye and Tyrode's salt-dye solution on ice. Save the dye, Tyrode's salt-dye solution, and glass-bottom dish after adding the Tyrode's salt-dye solution from light.
22. The cells will start to dye soon after the addition of Tyrode's salt-dye solution. Try to complete the following steps within the next 30–40 min.
23. Regions of pixel oversaturation can be sidestepped by using the "Range Indicator" function in the software—in this application, red pixels represent oversaturated pixels.
24. Fluorescence signal from FM 4-64 dye should be observed as it infiltrates the cell through membrane rupture and binds to negatively charged phospholipids within the cytoplasm. Sometimes membrane does not rupture due to issues with the target placement, and FM 4-64 dye fails to infiltrate the cell and fluoresce. In our experience, fluorescent signals from ruptured membranes are detectable within 15 s in most cases. If no detectable fluorescent signals can be observed in 20–30 s, the experiment can be stopped. If the cells are still in good condition, a new experiment can be started with a new cell.

Acknowledgments

This work was supported by the University of Alberta Faculty of Medicine and Dentistry, the Friends of Garrett Cumming Research Chair Fund, HM Toupin Neurological Science Research Chair Fund, Muscular Dystrophy Canada, Canada Foundation for Innovation (CFI), Alberta Advanced Education and Technology (AET), Canadian Institutes of Health Research (CIHR), Jesse's Journey—the Foundation for Gene and Cell Therapy, the Women and Children's Health Research Institute (WCHRI), and Alberta Innovates Health Solutions (AIHS). S.A. is supported by scholarships from the Maternal and Child Health (MatCH) Program and the Alberta Innovates Graduate Student Scholarship (AIGSS).

References

1. Bushby KMD (1999) Making sense of the limb-girdle muscular dystrophies. Brain 122: 1403–1420. https://doi.org/10.1093/brain/122.8.1403
2. Liu J, Aoki M, Illa I, Wu C, Fardeau M, Angelini C, Serrano C, Andoni Urtizberea J, Hentati F, Ben HM, Bohlega S, Culper EJ, Amato AA, Bossie K, Oeltjen J, Bejaoui K, McKenna-Yasek D, Hosler BA, Schurr E, Arahata K, De Jong PJ, Brown RH (1998) Dysferlin, a novel skeletal muscle gene, is mutated in Miyoshi myopathy and limb girdle muscular dystrophy. Nat Genet 20:31–36. https://doi.org/10.1038/1682
3. Bashir R, Britton S, Strachan T, Keers S, Vafiadaki E, Lako M, Richard I, Marchand S, Bourg N, Argov Z, Sadeh M, Mahjneh I, Marconi G, Passos-Bueno MR, Moreira E d S, Zatz M, Beckmann JS, Bushby K (1998) A gene related to Caenorhabditis elegans spermatogenesis factor fer-1 is mutated in limb-girdle muscular dystrophy type 2B. Nat Genet 20:37–42
4. Illa I, Serrano-Munuera C, Gallardo E, Lasa A, Rojas-García R, Palmer J, Gallano P, Baiget M, Matsuda C, Brown RH (2001) Distal anterior compartment myopathy: a dysferlin mutation causing a new muscular dystrophy phenotype. Ann Neurol 49:130–134. https://doi.org/10.1002/1531-8249(200101)49:1<130::aid-ana22>3.3.co;2-s
5. Harris E, Bladen CL, Mayhew A, James M, Bettinson K, Moore U, Smith FE, Rufibach L, Cnaan A, Bharucha-Goebel DX, Blamire AM, Bravver E, Carlier PG, Day JW, Díaz-Manera J, Eagle M, Grieben U, Harms M, Jones KJ, Lochmüller H, Mendell JR, Mori-Yoshimura M, Paradas C, Pegoraro E, Pestronk A, Salort-Campana E, Schreiber-Katz O, Semplicini C, Spuler S, Stojkovic T, Straub V, Takeda S, Rocha CT, Walter MC, Bushby K (2016) The clinical outcome study for dysferlinopathy. Neurol Genet 2:e89. https://doi.org/10.1212/NXG.0000000000000089
6. Nguyen K, Bassez G, Krahn M, Bernard R, Laforêt P, Labelle V, Urtizberea JA, Figarella-Branger D, Romero N, Attarian S, Leturcq F, Pouget J, Lévy N, Eymard B (2007) Phenotypic study in 40 patients with dysferlin gene mutations: high frequency of atypical phenotypes. Arch Neurol 64:1176–1182. https://doi.org/10.1001/archneur.64.8.1176
7. Xi J, Blandin G, Lu J, Luo S, Zhu W, Béroud C, Pécheux C, Labelle V, Lévy N, Urtizberea JA, Zhao C, Krahn M (2014) Clinical heterogeneity and a high proportion of novel mutations in a chinese cohort of patients with dysferlinopathy. Neurol India 62:635–639. https://doi.org/10.4103/0028-3886.149386
8. Fanin M, Nascimbeni AC, Aurino S, Tasca E, Pegoraro E, Nigro V, Angelini C (2009) Frequency of LGMD gene mutations in Italian patients with distinct clinical phenotypes. Neurology 72:1432–1435. https://doi.org/10.1212/WNL.0b013e3181a1885e
9. Therrien C, Di FS, Pickles S, Sinnreich M (2009) Characterization of lipid binding specificities of dysferlin C2 domains reveals novel interactions with phosphoinositides. Biochemistry 48:2377–2384. https://doi.org/10.1021/bi802242r
10. Anderson LVB, Davison K, Moss JA, Young C, Cullen MJ, Walsh J, Johnson MA, Bashir R, Britton S, Keers S, Argov Z, Mahjneh I, Fougerousse F, Beckmann JS, Bushby KMD (1999) Dysferlin is a plasma membrane protein and is expressed early in human development. Hum Mol Genet 8:855–861. https://doi.org/10.1093/hmg/8.5.855
11. Liu J, Aoki M, Illa I, Wu C, Fardeau M, Angelini C, Serrano C, Andoni Urtizberea J, Hentati F, Hamida MB, Bohlega S, Culper EJ, Amato AA, Bossie K, Oeltjen J, Bejaoui K, McKenna-Yasek D, Hosler BA, Schurr E, Arahata K, de Jong PJ, Brown RH (1998) Dysferlin, a novel skeletal muscle gene, is mutated in Miyoshi myopathy and limb girdle muscular dystrophy. Nat Genet 20:31–36. https://doi.org/10.1038/1682
12. Bansal D, Miyake K, Vogel SS, Groh S, Chen CC, Williamson R, McNeil PL, Campbell KP (2003) Defective membrane repair in dysferlin-deficient muscular dystrophy. Nature 423: 168–172. https://doi.org/10.1038/nature01573
13. Izumi R, Takahashi T, Suzuki N, Niihori T, Ono H, Nakamura N, Katada S, Kato M, Warita H, Tateyama M, Aoki Y, Aoki M (2020) The genetic profile of dysferlinopathy in a cohort of 209 cases: genotype–phenotype relationship and a hotspot on the inner DysF domain. Hum Mutat 41:1540–1554. https://doi.org/10.1002/humu.24036
14. Krahn M, Béroud C, Labelle V, Nguyen K, Bernard R, Bassez G, Figarella-Branger D, Fernandez C, Bouvenot J, Richard I, Ollagnon-Roman E, Bevilacqua JA, Salvo E, Attarian S, Chapon F, Pellissier JF, Pouget J, Hammouda EH, Laforêt P, Urtizberea JA, Eymard B, Leturcq F, Lévy N (2009) Analysis

of the DYSF mutational spectrum in a large cohort of patients. Hum Mutat 30. https://doi.org/10.1002/humu.20910

15. Blandin G, Beroud C, Labelle V, Nguyen K, Wein N, Hamroun D, Williams B, Monnier N, Rufibach LE, Urtizberea JA, Cau P, Bartoli M, Lévy N, Krahn M (2012) UMD-DYSF, a novel locus specific database for the compilation and interactive analysis of mutations in the dysferlin gene. Hum Mutat 33. https://doi.org/10.1002/humu.22015
16. Sinnreich M, Therrien C, Karpati G (2006) Lariat branch point mutation in the dysferlin gene with mild limb-girdle muscular dystrophy. Neurology 66:1114–1116. https://doi.org/10.1212/01.wnl.0000204358.89303.81
17. Niks EH, Aartsma-Rus A (2017) Exon skipping: a first in class strategy for Duchenne muscular dystrophy. Expert Opin Biol Ther 17: 225–236. https://doi.org/10.1080/14712598.2017.1271872
18. Lim KR, Maruyama R, Yokota T (2017) Eteplirsen in the treatment of Duchenne muscular dystrophy. Drug Des Dev Ther 11:533–545. https://doi.org/10.2147/DDDT.S97635
19. Anwar S, Yokota T (2020) Golodirsen for Duchenne muscular dystrophy. Drugs Today (Barcelona, Spain: 1998) 56:491–504. https://doi.org/10.1358/dot.2020.56.8.3159186
20. Dhillon S (2020) Viltolarsen: first approval. Drugs. https://doi.org/10.1007/s40265-020-01339-3
21. Shirley M (2021) Casimersen: first approval. Drugs Epub ahead. https://doi.org/10.1007/s40265-021-01512-2
22. Wein N, Avril A, Bartoli M, Beley C, Chaouch S, Laforêt P, Behin A, Butler-Browne G, Mouly V, Krahn M, Garcia L, Lévy N (2010) Efficient bypass of mutations in dysferlin deficient patient cells by antisense-induced exon skipping. Hum Mutat 31: 136–142. https://doi.org/10.1002/humu.21160
23. Barthélémy F, Blouin C, Wein N, Mouly V, Courrier S, Dionnet E, Kergourlay V, Mathieu Y, Garcia L, Butler-Browne G, Lamaze C, Lévy N, Krahn M, Bartoli M (2015) Exon 32 skipping of dysferlin rescues membrane repair in patients' cells. J Neuromuscul Dis 2:281–290. https://doi.org/10.3233/JND-150109
24. Dominov JA, Uyan Ö, Sapp PC, McKenna-Yasek D, Nallamilli BRR, Hegde M, Brown RH (2014) A novel dysferlin mutant pseudoexon bypassed with antisense oligonucleotides. Ann Clin Transl Neurol 1:703–720. https://doi.org/10.1002/acn3.96
25. Lee JJA, Maruyama R, Duddy W, Sakurai H, Yokota T (2018) Identification of novel antisense-mediated exon skipping targets in DYSF for therapeutic treatment of dysferlinopathy. Mol Ther Nucleic Acids 13:596–604. https://doi.org/10.1016/j.omtn.2018.10.004

Chapter 12

Knocking Down *DUX4* in Immortalized Facioscapulohumeral Muscular Dystrophy Patient-Derived Muscle Cells

Kenji Rowel Q. Lim and Toshifumi Yokota

Abstract

The third most common muscular dystrophy in the world, facioscapulohumeral muscular dystrophy (FSHD), is an inherited disorder characterized by distinct asymmetric, progressive skeletal muscle weakness that begins in the upper body and spreads to other regions with age. It is caused by mutations that induce aberrant expression of the *DUX4* gene in skeletal muscle. DUX4 is highly cytotoxic in skeletal muscle, dysregulating numerous signaling pathways as a result of its transcription factor activity. A promising set of approaches being developed to treat FSHD uses antisense oligonucleotides (AOs) to inhibit *DUX4* transcript expression. Both steric-blocking and gapmer AOs have been shown to induce efficient *DUX4* transcript knockdown in vitro and in vivo. Here, we describe a protocol that allows reliable screening of *DUX4*-targeting AOs through the evaluation of *DUX4* transcript expression by quantitative real-time polymerase chain reaction. We also describe methods to assess the efficacy of these AOs by looking at their effect on the expression of DUX4 downstream target and potential off-target genes, as well as on the amelioration of in vitro muscle cell phenotypes.

Key words Facioscapulohumeral muscular dystrophy, Antisense oligonucleotides, Gapmers, DUX4, Gene knockdown, Immortalized patient-derived muscle cells, Transfection, qPCR, Muscle cell fusion, Muscle cell diameter

1 Introduction

Facioscapulohumeral muscular dystrophy (FSHD) is an autosomal dominant disorder characterized by a distinctively asymmetric and progressive weakening of skeletal muscles that begins in the upper body and spreads to other regions with age [1]. It is considered the third most common muscular dystrophy in the world, with an estimated prevalence of 1:8,000–22,000 [1, 2]. Disease severity and age of onset are both highly variable in FSHD. Extramuscular symptoms may occur, particularly for patients with the more severe, early-onset form of the disease [3, 4]. There is consensus in the field that aberrant expression of the *double homeobox protein 4* (*DUX4*) gene in skeletal muscle is the primary genetic cause of FSHD

[5, 6]. *DUX4* is part of the D4Z4 macrosatellite repeat array on chromosome 4q35, which is normally 11–100 units long and hypermethylated in most healthy, differentiated tissues [5, 7, 8]. The pathogenic, full-length *DUX4* isoform involved in FSHD is composed of three exons; each D4Z4 repeat carries the first two exons, whereas the third exon is located immediately after the most downstream D4Z4 repeat in the array [6]. In FSHD patients, *DUX4* expression is activated by mutations that induce de-repression of the D4Z4 array (e.g., array contraction, mutations in genes coding for epigenetic regulators), coupled with the presence of a stabilizing polyadenylation signal in *DUX4* exon 3 in the permissive 4qA haplotype [5, 9–14]. Owing to its activities as a transcription factor, the presence of DUX4 protein in differentiated skeletal muscle leads to the dysregulation of numerous signaling pathways, most of which ultimately converge in muscle cell death [15].

While there is currently no cure for FSHD, research is actively being done to search for potential therapeutic candidates [16]. One promising strategy is to use short, single-stranded antisense oligonucleotides (AOs) to knock down *DUX4* transcript expression. Steric-blocking AOs, which are usually composed of a single base chemistry throughout the molecule (e.g., phosphorodiamidate morpholino oligomers and phosphorothioated 2′-*O*-methyl RNAs), have previously been designed to block access to splice sites, the *DUX4* polyadenylation signal, and other downstream elements and have resulted in approximately 25–90% *DUX4* transcript knockdown in vitro [17–20]. *DUX4* knockdown was also observed in vivo using xenograft and *DUX4*-transduced mouse models [19, 20]. More recently, our group demonstrated that *DUX4* knockdown in vitro could likewise be achieved with gapmer AOs. Gapmer AOs consist of a central DNA segment flanked by chemically modified bases [21]. Upon hybridization of gapmers to their targets, they form an RNA/DNA complex that is recognized by RNase H, which then degrades the target mRNAs. We have shown that gapmers of both the locked nucleic acid (LNA) and 2′-*O*-methoxyethyl (2′-MOE) chemistries could reduce *DUX4* transcript expression by nearly 100% in vitro, as well as knock down *DUX4* upon local treatment in a *DUX4*-transgenic mouse model [22, 23].

Regardless of the approach used, a reliable means to evaluate the effects of AOs on *DUX4* transcript expression is needed to accurately select candidates for further preclinical testing. This is made difficult by the rare expression of *DUX4* in muscle cells, with one study showing that only 1/1000 myoblasts and 1/200 myotubes actually have detectable DUX4 protein in their nuclei [24, 25]. In our previous work with gapmers, we developed a screening pipeline for the very purpose of robustly quantifying *DUX4* transcript expression in patient-derived myotubes

[22]. This pipeline is composed of two main parts. The first is an optimized cell culture protocol that uses KnockOut™ Serum Replacement (KOSR, Gibco), a reagent that has been previously shown to increase endogenous *DUX4* expression in patient-derived myotubes [26, 27]. In our pipeline, we specifically use patient-derived myotubes immortalized via *CDK4/hTERT* integration since they overcome issues associated with replicative senescence, providing more reproducible results [28]. The second is a quantitative real-time polymerase chain reaction (qPCR) protocol that has been optimized to detect *DUX4* expression from KOSR-supplemented patient-derived myotubes. We have shown that this qPCR protocol was sufficiently quantitative to distinguish small differences in *DUX4* knockdown efficacy between tested AOs [22, 23]. Even though our screening pipeline as described in this chapter was designed specifically for gapmers, we believe it could be easily adapted for *DUX4*-targeting AOs of other configurations and chemistries. Aside from screening AOs based on their ability to knock down *DUX4*, they can also be evaluated based on how well they restore the expression of DUX4 downstream target genes (e.g., *ZSCAN4*, *TRIM43*, *MBD3L2*) back to normal levels, avoid the induction of sequence-dependent off-target effects, and ameliorate pathological cellular phenotypes associated with FSHD in immortalized patient-derived myotubes. We detail our AO screening pipeline and these other experimental procedures in this chapter, which we summarize in Fig. 1.

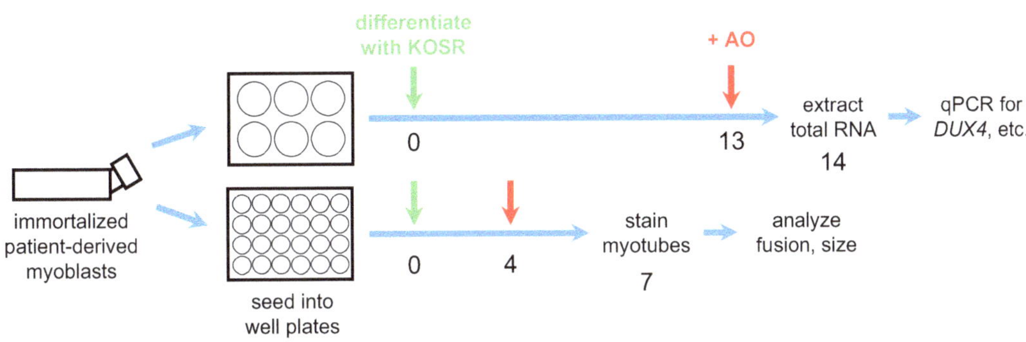

Fig. 1 Screening procedure for *DUX4*-targeting antisense oligonucleotides. An overview of the methods described in this chapter is shown. Immortalized myoblasts derived from facioscapulohumeral muscular dystrophy patients are seeded onto gelatin-coated 6-well or 24-well plates for quantitative real-time polymerase chain reaction (qPCR) or immunocytochemistry applications. These myoblasts are then differentiated using a medium containing KnockOut™ Serum Replacement (KOSR) for different durations (indicated by the numbers, in days post-differentiation), transfected with antisense oligonucleotides (AOs) and then either harvested for total RNA or immunostained for evaluation of AO therapeutic efficacy

2 Materials

2.1 Cell Culture and Transfection of Gapmers

1. Immortalized FSHD patient-derived and healthy control myoblasts (can be obtained from the Wellstone Program, University of Massachusetts Medical School).
2. Basal medium: 20% Medium 199 (Gibco, #11150059), 0.03 μg/mL $ZnSO_4$, 1.4 μg/mL vitamin B12, 2.5% penicillin–streptomycin antibiotics in high-glucose Dulbecco's modified Eagle's medium with 4-(2-hydroxyethyl)-1-piperazineethanesulfonic acid (HEPES) (Gibco, #12430054).
3. Growth medium: 15% fetal bovine serum (Sigma), 2.5 ng/mL recombinant human hepatocyte growth factor (EMD Millipore), 10 ng/mL recombinant human fibroblast growth factor (BioPioneer), 0.055 μg/mL dexamethasone in basal medium.
4. Differentiation medium: 15% KOSR (Gibco), 10 μg/mL insulin (Sigma), 100 μg/mL human apo-transferrin (R&D Systems) in basal medium.
5. 0.05% trypsin–EDTA with phenol red (Gibco).
6. 1× phosphate-buffered saline (PBS), sterile.
7. 0.1% gelatin in MilliQ water.
8. 0.4% trypan blue stain (Invitrogen).
9. *DUX4*-specific and mock gapmers (Exiqon or Eurogentec).
10. Lipofectamine™ RNAiMAX transfection reagent (Invitrogen).
11. Opti-MEM™ reduced serum medium (Gibco).
12. Equipment: cell culture flasks and tubes, 6- or 24-well cell culture plates, microscope, Countess™ Cell Counting Chamber slides (Invitrogen), Countess™ II Automated Cell Counter (Applied Biosystems).

2.2 RNA Extraction, cDNA Synthesis, and qPCR Analysis of Transcript Knockdown

1. RNeasy Mini Kit (Qiagen).
2. QIAshredder spin-column homogenizer (Qiagen).
3. RNase-free DNase set (Qiagen).
4. 0.5 μg/μL Oligo$(dT)_{12–18}$ primer (Invitrogen).
5. SuperScript™ IV Reverse Transcriptase (Invitrogen).
6. RNaseOUT™ recombinant ribonuclease inhibitor (Invitrogen).
7. *DUX4*-specific primers, forward: 5′-CCCAGGTACCAGCAGACC-3′, reverse: 5′-TCCAGGAGATGTAACTCTAATCCA-3′ (IDT).

8. *GAPDH*-specific primers, forward: 5′-GCAAATTCCATGG CACCGT-3′, reverse: 5′-AGGGATCTCGCTCCTGGAA-3′ (IDT).
9. Pre-designed TaqMan™ Gene Expression assays (Thermo Fisher) for DUX4 downstream target genes or off-target genes of choice (*see* **Note 1**).
10. SsoAdvanced™ Universal SYBR® Green Supermix (Bio-rad).
11. TaqMan™ Fast Advanced Master Mix (Applied Biosystems).
12. UltraPure™ DNase/RNase-free distilled water (Invitrogen).
13. Equipment: plastic cell scrapers, 1.5 mL nuclease-free and sterile microcentrifuge tubes, centrifuge, vortex, NanoDrop™ Spectrophotometer (Thermo Scientific), QuantStudio 3 Real-Time PCR System (Applied Biosystems), qPCR-applicable strip tubes or plates.

2.3 Immunocytochemistry and Analysis of Muscle Cell Phenotypes

1. 4% paraformaldehyde.
2. 1× PBS.
3. 1× PBS with 0.5% Triton X-100.
4. Blocking agent: 20% fetal bovine serum (Sigma) in 1× PBS.
5. Primary antibody: 1:200 rabbit polyclonal anti-desmin (Abcam) in blocking agent.
6. Secondary antibody: 1:100 Alexa594-conjugated goat anti-rabbit IgG (H+L) (Invitrogen) in 1× PBS.
7. SlowFade™ Gold Antifade Mountant with DAPI (Invitrogen).
8. ImageJ software (NIH).
9. Equipment: fluorescence microscope.

3 Methods

3.1 Cell Culture and Transfection of Gapmers

1. Thaw immortalized patient-derived and healthy control myoblasts into separate cell culture flasks with growth medium and grow at 37°C, 5% CO_2. Passage and multiply as needed to obtain desired quantity of cells for seeding into well plates (*see* **Note 2**).
2. While the cells are growing, coat 6-well (for RNA extraction) or 24-well (for immunocytochemistry) cell culture plates with gelatin by adding enough 0.1% gelatin solution to cover the surface of each well, incubating at 37 °C for 30 min, and then removing as much solution as possible from each well. Allow the plates to dry overnight at room temperature, and then store at 4 °C until use (*see* **Note 3**).

3. Once enough cells are available, remove the growth medium, wash the cells twice with a sufficient volume of 1× PBS, and trypsinize the myoblasts by adding a volume of trypsin-EDTA that is one-fifth the volume of the growth medium used in the flask. Incubate at 37 °C, 5% CO_2 for 5 min. Inspect the cells after 5 min to see if they have successfully detached; if not, incubate for a longer time in trypsin.
4. Resuspend the trypsinized myoblasts in growth medium in a 50 mL tube and centrifuge for 5 min at 300 rcf to pellet down the cells. Decant the supernatant.
5. Resuspend the cells in 5 mL of growth medium (*see* **Note 4**). Take a 15 μL aliquot and mix well with 15 μl of trypan blue in a 1.5 mL tube. Pipette 10 μL of this solution into each side of a Countess™ chamber slide, and then count the cells in each side using the Countess™ II Automated Cell Counter.
6. For 6-well gelatin-coated plates, dilute the cells to 2×10^5 cells/ml with growth medium and seed 2 mL per well (*see* **Note 5**). For 24-well gelatin-coated plates, dilute the cells to 1×10^5 cells/ml with growth medium, seed 250 μL per well, and top up to 500 μL with growth medium.
7. Incubate at 37 °C, 5% CO_2 overnight, and then replace the growth medium with an equal amount of differentiation medium to initiate differentiation of myoblasts into myotubes (*see* **Note 6**).
8. Once the cells have reached 13 days post-differentiation (for RNA extraction) or 4 days post-differentiation (for immunocytochemistry), transfect them with the gapmer AOs at the desired concentrations using Lipofectamine™ RNAiMAX, as below. The protocol described is for 2 wells of a 6-well plate and can be scaled up (or down, for a 24-well plate) as required.
   (a) In one tube, dilute the selected *DUX4*-specific or mock gapmer in Opti-MEM™ at the desired concentration, with a final volume of 300 μL. For nontreated controls, use 300 μL of Opti-MEM™ instead.
   (b) In another tube, dilute 6.6 μL Lipofectamine™ RNAiMAX reagent in 323.4 μL Opti-MEM™ (*see* **Note 7**).
   (c) Take 300 μL of the solution in (b) and combine it with the solution in (a). Mix well by pipetting up and down.
   (d) Incubate the solution for 5 min at room temperature (*see* **Note 8**).
   (e) Add 2400 μL of differentiation medium to the solution.
   (f) Add 1.4 mL of the final solution each to 2 wells of a 6-well plate (*see* **Note 9**).

9. Incubate in the transfection solution overnight. For RNA extraction, the cells can be harvested right after this overnight incubation. For immunocytochemistry, replace the transfection solution with a differentiation medium and stain the cells three days post-transfection (*see* **Note 10**).

3.2 RNA Extraction, cDNA Synthesis and qPCR Analysis of Transcript Knockdown

1. For one sample, total RNA will be extracted from two wells' worth of cells to obtain sufficient yield. First, remove or aspirate out as much culture medium as possible from the pair of wells to be extracted from. Then, add 500 μL of Buffer RLT from the RNeasy Mini Kit to the cells in one well, thoroughly scrape off the cells, and transfer this lysate to the other well. Scrape off the cells, and finally transfer the suspension to a sterile 1.5 mL tube (*see* **Note 11**).
2. Homogenize the cells further by transferring the lysate into a QIAshredder spin column and centrifuging for 2 min at maximum speed.
3. Follow the rest of the RNeasy Mini Kit manufacturer's instructions for the rest of the extraction, making sure to perform the on-column DNase digestion step using the RNase-free DNase set. Use 20–25 μL of the provided RNase-free water for elution (*see* **Note 12**).
4. Quantify the obtained total RNA using NanoDrop™, and dilute as necessary. Total RNA extracts can be stored at −80 °C until use.
5. For cDNA synthesis, use the SuperScript™ IV Reverse Transcriptase system following the manufacturer's instructions. For the primer-annealing step, use 1 μL of the oligo(dT) primer and use UltraPure™ distilled water to bring the solution to volume (*see* **Note 13**). cDNA can be stored at −20 °C until use.
6. To detect *DUX4* and *GAPDH* expression by qPCR, prepare the following (for 1 reaction): 10 μL SYBR® Green Supermix, 0.8 μL each of 10 μM forward and reverse primer, 5.4 μL of UltraPure™ distilled water, and 3 μL of cDNA template (*see* **Note 14**). Make sure to prepare enough for three technical replicates per sample. Carefully pipette prepared solutions onto qPCR-applicable strip tubes or plates (*see* **Note 15**).
7. Spin down the strip tubes or plates, and then load them onto the qPCR machine. Select the appropriate settings and run using the default "Fast" cycling program: 1 cycle of 95 °C, 20 s, 40 cycles of a) 95 °C, 1 s and b) 60 °C, 20 s, and an additional step for constructing the melt curve.
8. Analyze the data and calculate relative *DUX4* expression using the ΔΔCt method, using *GAPDH* expression for

normalization and the nontreated controls as reference samples. When screening gapmers, perform statistical testing against mock gapmer-treated cells.

9. To detect the expression of DUX4 downstream target genes or potential off-target genes by qPCR using pre-designed TaqMan™ assays, prepare the following (for 1 reaction): 10 μL of TaqMan™ Fast Advanced Master Mix, 1 μL of the pre-designed assay, 7 μL of UltraPure™ distilled water, and 2 μL of cDNA template; if a SYBR®-based assay is used, follow the reaction recipe as indicated in **step 6**. Follow the rest of the procedure in **steps 6–8**, but calculate the relative expression of the gene/s of choice as normalized to *GAPDH* expression.

3.3 Immunocytochemistry and Analysis of Muscle Cell Phenotypes

1. Remove or aspirate out as much culture medium as possible from each well, and incubate cells with 4% paraformaldehyde for 5 min at room temperature to fix them. All steps below will also be done at room temperature.
2. Wash the cells once with 1× PBS, and then incubate them with PBS containing 0.5% Triton X-100 for 5 min.
3. Replace the PBS/Triton X-100 solution with the blocking agent, and incubate for 1 h.
4. Replace the blocking agent with diluted anti-desmin primary antibody and incubate for 1 h (*see* **Note 16**).
5. Wash cells three times with 1× PBS, 5 min each.
6. Incubate the cells in diluted Alexa594-conjugated anti-rabbit IgG secondary antibody and incubate for 1 h.
7. Repeat **step 5**.
8. Mount cells with DAPI-containing SlowFade Gold Antifade Mountant, allow to sit at room temperature for 15 min, and then keep at 4 °C until visualization.
9. Visualize stained cells using a fluorescence microscope and take at least three images for each well (*see* **Note 17**).
10. To calculate the myogenic fusion index, count the number of nuclei in myotubes, divide it by the total number of nuclei in desmin-positive cells in the field of view, and then multiply by 100 (*see* **Note 18**). Nuclei can be counted using ImageJ. See Fig. 2a–b for a sample result.
11. To measure myotube diameters, take the average of three diameter measurements along the widest parts of a myotube, excluding areas near branch points or that are blocked by other myotubes. Myotube diameters can be measured using ImageJ. See Fig. 2a, c, d for a sample result.

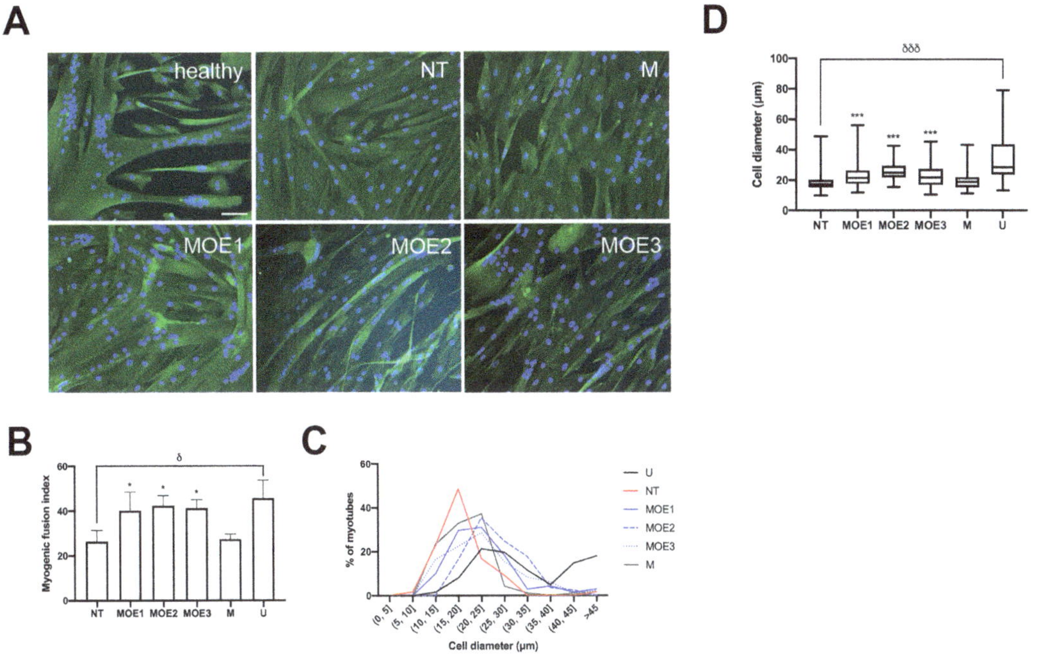

Fig. 2 Sample result for quantitative evaluation of in vitro muscle cell phenotypes after treatment with *DUX4*-targeting antisense oligonucleotides. (**a**) Representative immunocytochemistry images of healthy immortalized control myotubes (U) and nontreated (NT), mock 2′-MOE gapmer-treated (M), and *DUX4*-specific 2′-MOE gapmer-treated (MOE1, MOE2, MOE3) immortalized FSHD patient-derived myotubes stained for nuclei (blue) and desmin (green). In this case, patient-derived myotubes were transfected with 10 nM of the various 2′-MOE gapmers at 4 days post-differentiation and then stained 3 days later. Scale bar: 100 μm. (**b**) Myogenic fusion index quantification for the various treatment groups. (**c**) Frequency distribution of myotube diameters across the different treatment groups. $n = 3$ independent experiments; 382 nuclei and 26 myotubes on average for each replicate, per condition, were counted for quantification of myogenic fusion indices and muscle cell diameters, respectively. (**d**) Individual myotube diameters from (**c**) were plotted. (**b**) Error bars: SD; $n = 3$ independent experiments. (**d**) The box represents the P_{25}–P_{75} with the central line marking the median, and the whiskers represent the range; $n = 3$. $^{*}p < 0.05$, $^{***}p < 0.0005$ versus M, one-way ANOVA with Dunnett's test. $^{\delta}p < 0.05$, $^{\delta\delta\delta}p < 0.0005$, unpaired, two-tailed t test. (Figure (modified, with the removal of a panel) and legend reproduced from Lim et al. 2020 [23] with permission from Molecular Therapy)

4 Notes

1. Alternatively, primers can be designed in-house for SYBR® or TaqMan™ applications.
2. The size of the cell culture flask used will depend on the behavior and number of cells in the stock vial, as well as how many cells will be needed for seeding. We usually use a T175 flask for growing the cells prior to seeding. If the stock vial has a low number of cells or if the cells appear to be growing slowly, start with smaller cell culture flasks and then scale up as needed. Also, never let the myoblasts reach more than 90% confluence, as overcrowding will induce premature differentiation.

3. We recommend doing this at least a day before seeding to ensure that the gelatin has had sufficient time to dry and coat each well.
4. To resuspend, we usually flick the bottom of the tube first to physically dissociate the pellet. We then add 1 mL of the growth medium, mix, and then add the remaining 4 mL of the growth medium followed by a final mix. Cells are mixed by pipetting the suspension up and down a number of times. It is critical that this resuspension step be done well in order to get as accurate a cell count as possible.
5. Total RNA will be extracted from two wells' worth of cells for each condition later on to obtain sufficient yield for downstream analysis, so take this into consideration when deciding how many wells to seed.
6. In our hands, an overnight incubation is usually sufficient for the cells to reach the desired confluence of 70–80% for differentiation. The length of incubation may have to be optimized if this is not achieved.
7. We have optimized this amount of RNAiMAX to minimize toxicity to the cells.
8. This step is very critical to allow the proper formation of gapmer–lipid complexes.
9. Less than the total volume of prepared transfection solution is added to ensure that the same amount is received by all the wells.
10. Transfection times can be varied as needed.
11. Total RNA extraction should be performed in a sterile, RNase-free environment to maximize yield and purity.
12. For higher yield, we usually perform two-step elution, where the eluate from the first round is put back onto the surface of the column and eluted once more.
13. The amount of total RNA used for cDNA synthesis can vary, but we usually use >1000 ng of the extract.
14. An initial qPCR experiment using a dilution series of the cDNA is recommended to determine the optimal dilution to use for each gene to be assayed.
15. Excellent pipetting skill is critical for the success of qPCR. Make sure to vortex all solutions well before pipetting from them, especially before pipetting into the strip tubes or plate to be used for running the qPCR. As much as possible, do not touch the bottom or top parts of the strip tubes or plates so that no smudges will interfere with fluorescence detection. Also, a reference sample (usually the nontreated control) will have to be included in every qPCR run to allow the calculation of relative expression.

16. Any antibody that stains the cytoplasm of myoblasts and myotubes would be suitable.
17. This should be done blinded.
18. We define a myotube as a muscle cell where at least two nuclei are shared by the same cytoplasm. If a batch effect is observed, i.e., myogenic fusion varying greatly across different batches even for the same condition, the myogenic fusion index can be expressed relative to a reference sample that must be present in all batches tested.

Acknowledgments

This work was supported by the Friends of Garrett Cumming Research & Muscular Dystrophy Canada HM Toupin Neurological Science Research Chair Fund, Muscular Dystrophy Canada (853669), Canadian Institutes of Health Research (PS 175261), University Hospital Foundation, Women and Children's Health Research Institute (3279), and the Alberta Innovates Graduate Studentship.

References

1. Wang LH, Tawil R (2016) Facioscapulohumeral dystrophy. Curr Neurol Neurosci Rep 16: 66. https://doi.org/10.1007/s11910-016-0667-0
2. Mah JK, Korngut L, Fiest KM et al (2016) A systematic review and meta-analysis on the epidemiology of the muscular dystrophies. Can J Neurol Sci/J Can Des Sci Neurol 43:163–177. https://doi.org/10.1017/cjn.2015.311
3. Klinge L, Eagle M, Haggerty ID et al (2006) Severe phenotype in infantile facioscapulohumeral muscular dystrophy. Neuromuscul Disord 16:553–558. https://doi.org/10.1016/j.nmd.2006.06.008
4. Goselink RJM, Voermans NC, Okkersen K et al (2017) Early onset facioscapulohumeral dystrophy – a systematic review using individual patient data. Neuromuscul Disord 27: 1077–1083. https://doi.org/10.1016/j.nmd.2017.09.007
5. Lemmers RJLF, van der Vliet PJ, Klooster R et al (2010) A unifying genetic model for facioscapulohumeral muscular dystrophy. Science 329:1650–1653. https://doi.org/10.1126/science.1189044
6. Himeda CL, Jones PL (2019) The genetics and epigenetics of facioscapulohumeral muscular dystrophy. Annu Rev Genomics Hum Genet 20:265–291. https://doi.org/10.1146/annurev-genom-083118-014933
7. Gabriëls J, Beckers MC, Ding H et al (1999) Nucleotide sequence of the partially deleted D4Z4 locus in a patient with FSHD identifies a putative gene within each 3.3 kb element. Gene 236:25–32
8. Hewitt JE, Lyle R, Clark LN et al (1994) Analysis of the tandem repeat locus D4Z4 associated with facioscapulohumeral muscular dystropothhy. Hum Mol Genet 3:1287–1295. https://doi.org/10.1093/hmg/3.8.1287
9. Lemmers RJLF, de Kievit P, Sandkuijl L et al (2002) Facioscapulohumeral muscular dystrophy is uniquely associated with one of the two variants of the 4q subtelomere. Nat Genet 32: 235–236. https://doi.org/10.1038/ng999
10. Hewitt JE (2015) Loss of epigenetic silencing of the DUX4 transcription factor gene in facioscapulohumeral muscular dystrophy: figure 1. Hum Mol Genet 24:R17–23. https://doi.org/10.1093/hmg/ddv237
11. van Overveld PGM, Lemmers RJFL, Sandkuijl LA et al (2003) Hypomethylation of D4Z4 in 4q-linked and non-4q-linked facioscapulohumeral muscular dystrophy. Nat Genet 35: 315–317. https://doi.org/10.1038/ng1262

12. Lemmers RJLF, Tawil R, Petek LM et al (2012) Digenic inheritance of an SMCHD1 mutation and an FSHD-permissive D4Z4 allele causes facioscapulohumeral muscular dystrophy type 2. Nat Genet 44:1370–1374. https://doi.org/10.1038/ng.2454
13. van den Boogaard ML, Lemmers RJLF, Balog J et al (2016) Mutations in DNMT3B modify epigenetic repression of the D4Z4 repeat and the penetrance of facioscapulohumeral dystrophy. Am J Hum Genet 98:1020–1029. https://doi.org/10.1016/j.ajhg.2016.03.013
14. Hamanaka K, Šikrová D, Mitsuhashi S et al (2020) Homozygous nonsense variant in LRIF1 associated with facioscapulohumeral muscular dystrophy. Neurology 94: e2441–e2447. https://doi.org/10.1212/WNL.0000000000009617
15. Lim KRQKRQ, Nguyen Q, Yokota T (2020) DUX4 signalling in the pathogenesis of facioscapulohumeral muscular dystrophy. Int J Mol Sci 21:729. https://doi.org/10.3390/ijms21030729
16. Lim KRQ, Yokota T (2021) Genetic approaches for the treatment of facioscapulohumeral muscular dystrophy. Front Pharmacol 12:642858. https://doi.org/10.3389/fphar.2021.642858
17. Vanderplanck C, Ansseau E, Charron S et al (2011) The FSHD atrophic myotube phenotype is caused by DUX4 expression. PLoS One 6:e26820. https://doi.org/10.1371/journal.pone.0026820
18. Marsollier A-C, Ciszewski L, Mariot V et al (2016) Antisense targeting of 3′ end elements involved in DUX4 mRNA processing is an efficient therapeutic strategy for facioscapulohumeral dystrophy: a new gene-silencing approach. Hum Mol Genet 25:1468–1478. https://doi.org/10.1093/hmg/ddw015
19. Chen JC, King OD, Zhang Y et al (2016) Morpholino-mediated knockdown of DUX4 toward facioscapulohumeral muscular dystrophy therapeutics. Mol Ther 24:1405–1411. https://doi.org/10.1038/mt.2016.111
20. Ansseau E, Vanderplanck C, Wauters A et al (2017) Antisense oligonucleotides used to target the DUX4 mRNA as therapeutic approaches in faciosscapulohumeral muscular dystrophy (FSHD). Genes (Basel) 8:93. https://doi.org/10.3390/genes8030093
21. Lim KRQ, Yokota T (2020) Invention and early history of gapmers. In: Yokota T, Maruyama R (eds) Gapmers methods protoc. Springer, pp 3–19
22. Lim KRQ, Maruyama R, Echigoya Y et al (2020) Inhibition of DUX4 expression with antisense LNA gapmers as a therapy for facioscapulohumeral muscular dystrophy. Proc Natl Acad Sci 117:16509–16515. https://doi.org/10.1073/pnas.1909649117
23. Lim KRQ, Bittel A, Maruyama R et al (2020) DUX4 transcript knockdown with antisense 2′-o-methoxyethyl gapmers for the treatment of facioscapulohumeral muscular dystrophy. Mol Ther. https://doi.org/10.1016/j.ymthe.2020.10.010
24. Snider L, Geng LN, Lemmers RJLF et al (2010) Facioscapulohumeral dystrophy: incomplete suppression of a retrotransposed gene. PLoS Genet 6:e1001181. https://doi.org/10.1371/journal.pgen.1001181
25. Tassin A, Laoudj-Chenivesse D, Vanderplanck C et al (2013) DUX4 expression in FSHD muscle cells: how could such a rare protein cause a myopathy? J Cell Mol Med 17:76–89. https://doi.org/10.1111/j.1582-4934.2012.01647.x
26. Pandey SN, Khawaja H, Chen Y-W (2015) Culture conditions affect expression of DUX4 in FSHD myoblasts. Molecules 20: 8304–8315. https://doi.org/10.3390/molecules20058304
27. Block GJ, Narayanan D, Amell AM et al (2013) Wnt/β-catenin signaling suppresses DUX4 expression and prevents apoptosis of FSHD muscle cells. Hum Mol Genet 22:4661–4672. https://doi.org/10.1093/hmg/ddt314
28. Stadler G, Chen JCJ, Wagner K et al (2011) Establishment of clonal myogenic cell lines from severely affected dystrophic muscles – CDK4 maintains the myogenic population. Skelet Muscle 1:12. https://doi.org/10.1186/2044-5040-1-12

Chapter 13

Peptide-Conjugated PMOs for the Treatment of Myotonic Dystrophy

Jessica Stoodley, David Seoane Miraz, Yahya Jad, Mathieu Fischer, Matthew J. A. Wood, and Miguel A. Varela

Abstract

Antisense oligonucleotides (ASOs) have shown great therapeutic potential in the treatment of many neuromuscular diseases including myotonic dystrophy 1 (DM1). However, systemically delivered ASOs display poor biodistribution and display limited penetration into skeletal muscle. The conjugation of cell-penetrating peptides (CPPs) to phosphorodiamidate morpholino oligonucleotides (PMOs), a class of ASOs with a modified backbone, can be used to enhance ASO skeletal muscle penetration. Peptide–PMOs (P-PMOs) have been shown to be highly effective in correcting the DM1 skeletal muscle phenotype in both murine and cellular models of DM1 and at a molecular and functional level. Here we describe the synthesis and conjugation of P-PMOs and methods for analyzing their biodistribution and toxicity in the HSA-LR DM1 mouse model and their efficacy both in vitro and in vivo using FISH and RT-PCR splicing analysis.

Key words Myotonic dystrophy, Antisense oligonucleotides, Cell-penetrating peptide, DM1, MBNL1, PMO, Muscle, Splicing

1 Introduction

Myotonic dystrophy types 1 and 2 (DM1 and DM2, respectively) are autosomal dominant genetic disorders that affect multiple body systems. DM1 is the most common form of muscular dystrophy with a prevalence of 0.5–18.1 per 100,000 people worldwide [1]. The severity and nature of symptoms of the most common type of myotonic dystrophy, DM1, are highly variable but typically include distal muscle weakness and hypotonia, cardiac defects, cataracts, insulin resistance, and cognitive impairment. The genetic cause of DM1 is the repeat expansion of a trinucleotide CTG repeat in the 3′ UTR of the dystrophia myotonica protein kinase (DMPK) gene on chromosome 19q 13.3 [2–4]. Transcription of the expanded CTG DNA region gives rise to expanded CUG mRNA,

Rika Maruyama and Toshifumi Yokota (eds.), *Muscular Dystrophy Therapeutics: Methods and Protocols*, Methods in Molecular Biology, vol. 2587, https://doi.org/10.1007/978-1-0716-2772-3_13,

which form nuclear foci [5]. CUG expanded nuclear foci are able to sequester muscleblind-like family proteins (MBNL) and interfere with other splicing regulators, including altering levels of CUGBP/Elav-like (CELF) proteins [6–8]. The overall result is global splicing misregulation, with numerous alternative splicing events directly correlating to the diverse clinical symptoms of DM1 [9–14].

There is currently no approved treatment to cure or slow the disease progression of DM1. However, there are over 40 known drugs in the development pipeline for DM1, with six therapies in clinical trials [15]. Over one-third of the drugs in development are small molecules that aim to inhibit the toxic, gain-of-function CUG RNA, inhibit subsequent mis-spliced proteins, or upregulate MBNL levels [15–17]. However, with advancements in RNA therapeutics, a more specific, alternative therapeutic approach is to directly target toxic CUG expanded RNA foci using modified antisense oligonucleotides.

Antisense oligonucleotides (ASOs) are short synthetic sequences of nucleic acid analogs that bind to target RNA via Watson–Crick base pairing. This elicits either altered RNA splicing events or reduced gene expression via inhibition of RNA translation. As nucleotides are rapidly degraded in vivo, chemical modifications are used in order to increase their resistance to nucleases and improve serum half-life [18]. Phosphorodiamidate morpholino oligomers (PMOs) are a class of modified ASO with a nucleotide backbone comprised of morpholino rings [19]. Morpholino rings are neutrally charged, which decreases the interactions of ASOs with other proteins and increases their serum stability [19, 20]. To date, 11 modified ASOs have been approved by the FDA, including eteplirsen, a PMO approved in 2016 for the treatment of the neuromuscular disease Duchenne muscular dystrophy (DMD) [21, 22]. ASOs in development for the treatment of DM1 target the expanded CUG RNA repeat and act by either facilitating the RNAseH degradation of toxic RNA foci [23–26] or RNAse H independent mechanisms that allow the release of sequestered splicing regulators, such as MBNL proteins [27–30]. It has been shown that utilizing ASOs in this manner can reverse splicing defects and alleviate DM1-associated phenotypes in muscles [23–26]. Despite this potential, a major limitation of ASO therapy for DM1 is achieving therapeutically relevant concentrations in severely affected target tissues, including skeletal muscle where over 40 individual mis-splicing events have been reported [31]. A phase I/II clinical trial (ClinicalTrials.gov identifier: NCT02312011), using a naked modified ASO that had shown efficacy in both mouse and non-human-primate studies [32], failed due to low skeletal muscle concentration [33].

One method to improve ASO biodistribution to target tissues is to use cell-penetrating peptides (CPPs) as a delivery vehicle. CPPs are short peptides that facilitate the delivery of molecules across biological membranes and have been used to successfully deliver PMOs into relevant tissues for several neuromuscular diseases [34–38]. A study published in 2019 demonstrated that the conjugation of a CPP (pip6a) to a DM1 PMO targeting CUG expanded RNA is able to enhance PMO delivery into the skeletal muscles >20-fold in comparison with naked (unconjugated) PMOs [27]. The peptide-conjugated PMO (P-PMO), acting via an RNAse independent mechanism, demonstrated enhanced activity within the skeletal muscles of systemically treated HSA-LR mice. Long-term correction of mis-splicing events was observed within the mice and skeletal muscle myotonia was also reduced. Likewise, Sarepta has recently found in a phase II clinical trial (MOMENTUM, ClinicalTrials.gov identifier: NCT04004065) that their P-PMO (SRP-5051) administered monthly at 30 mg/kg resulted in 18 times the exon skipping and eight times the dystrophin production as eteplirsen (unconjugated PMO) administered weekly for 24 weeks.

The HSA-LR mouse is a well-characterized, transgenic mouse model for DM1 [39]. The mouse skeletal muscles selectively express a human skeletal muscle actin transgene (HSA) that contains a 250 base pair CTG repeat in the 3′UTR. Transcription of the HSA transgene gives rise to untranslated, CUG RNA repeats that form nuclear foci. Consequently, DM1-associated muscular phenotypes such as MBNL protein sequestration, splicing defects, and myotonia can be observed in the skeletal muscles, making it a good model for assaying the efficacy of P-PMOs. Furthermore, while P-PMO administration to HSA-LR mice will only demonstrate desired activity in the skeletal muscles, systemic delivery of P-PMOs still allows for the analysis of biodistribution across tissues and toxicity markers (e.g. urinary KIM-1 [40]), both of which are crucial to the preclinical evaluation of potential therapies.

The development of a robust, immortalized human DM1 muscle cell line also allows for mid-high-throughput in vitro screening of ASOs prior to in vivo validation in HSA-LR mice [41]. The immortalized DM1 patient cell line displays disease-associated molecular phenotypes, including RNA foci, that sequester mbnl1 and alternative mis-splicing [41]. Analysis of these phenotypic readouts can be used to evaluate P-PMO efficacy and allow the selection of promising compounds for testing in animal models. Furthermore, the use of an immortalized cell line overcomes the issues of availability and limited proliferative capability associated with DM1 patient primary muscle cells.

Using the combination of the cellular and animal DM1 models described above, we provide methodologies in this chapter for designing P-PMOs targeting CUG expanded RNA and measuring their efficacy in skeletal muscle both in vitro and in vivo. We further describe methods for evaluating in vivo biodistribution and toxicity to give an overall profile of P-PMOs designed for the treatment of DM1.

2 Materials

2.1 Preparation of Peptide–PMO Conjugates

1. Fmoc amino acids.
2. Fmoc-β-Ala-Wang polystyrene resin 0.4 mmol/g loading.
3. (Benzotriazol-1-yloxy)tripyrroldinophosphonium hexafluorophosphate (PyBOP).
4. Diisopropylethylamine (DIPEA).
5. N,N-Dimethylformamide (DMF), peptide synthesis grade.
6. Acetic anhydride.
7. Trifluoroacetic acid (TFA).
8. Triisopropylsilane (TIS).
9. Diethyl ether.
10. Reversed-phase HPLC equipped with Zorbax 300SB-C18 column (4.6 50 mm, 3.5 μm) for analytical analysis and Luna C18 [2] (21.2 × 250 mm, 5 μm) for preparative purification. Buffer A is 0.1% TFA in Milli-Q water and buffer B is 0.1% TFA in acetonitrile (HPLC grade).
11. Lyophilized morpholino oligonucleotide purchased from Gene Tools.
12. Dimethyl sulfoxide (DMSO) (*see* **Note 1**).
13. N-Methyl-2-pyrrolidone (NMP) (*see* **Note 1**).
14. Ammonia solution.
15. Hydrazine hydrate.
16. 0.1 M HCl in aqueous solution.
17. Cation-exchange purification using an FPLC or similar system equipped with Resource S column (GE Healthcare). Buffer A: 25 mM sodium phosphate buffer (pH = 7.0) containing 20% acetonitrile, Buffer B: 25 mM sodium phosphate buffer (pH = 7.0) containing 20% acetonitrile, 1 M sodium chloride.
18. Amicon® ultra-15 3K centrifugal filter/desalting units.
19. 2 mL Spin-X 0.22 μm cellulose acetate filtration units.
20. Freeze dryer.
21. Spectrophotometer.

2.2 Measuring P-PMO Efficacy In Vitro: Fluorescent In Situ Hybridization (FISH) of RNA Foci in DM1 Myotubes

2.2.1 Sterile Culture of Control and DM1 Myotubes

1. DM1 and control myoblasts [41].
2. Antibiotic-Antimycotic (Gibco).
3. Skeletal muscle cell growth medium (PromoCell): Transfer the entire content of growth medium supplement to skeletal muscle cell growth medium and add 5 mL of antibiotic/antimycotic.
4. Skeletal muscle differentiation medium (PromoCell): Transfer the entire content of differentiation medium supplement to skeletal muscle cell differentiation medium and add 5 mL of antibiotic/antimycotic.
5. PBS buffer (Gibco).
6. TripLE (Gibco).
7. Pipette buoy and 5, 10, and 25 mL stripettes in individual casings.
8. T25, T75, and T175 flasks for tissue culture.
9. 50-mL Corning tubes.
10. 70% ethanol in a spray bottle.
11. Benchtop centrifuge.
12. Cell culture incubator.
13. Laminar flow cabinet.
14. Micropipettes and 10 μL sterile, filtered tips.
15. Neubauer cell counting chamber.

2.2.2 Myotube Fixation

1. Micropipettes and 1000, 200, 10, and 2 μL sterile, filtered tips.
2. 4-well Nunc™ Lab-Tek™ II Chamber Slide™ System (Thermo Scientific).
3. 1X phosphate-buffered saline (PBS) (Gibco).
4. 4% formaldehyde in PBS (ChemCruz).

2.2.3 Fluorescent In Situ Hybridization

1. Micropipettes and 1000, 200, 10, and 2 μL sterile, filtered tips.
2. 15 mL and 50 mL sterile falcon tubes.
3. Pipette buoy and 5, 10, and 25 mL stripettes.
4. 0.5% Triton X-100 solution: 0.5% Triton X-100 (Sigma) in 1X PBS (Gibco).
5. 50% 20X SSC/formamide solution: 0.3 mL of SSC Buffer 20X (Sigma), 1.2 mL of RNase-free water, 1.5 mL of formamide (Sigma). Store at 4 °C.
6. 2% yeast extract tRNA: 2% yeast extract tRNA (Invitrogen) dissolved in RNase-free water, aliquoted and stored at −20 °C.
7. 2% BSA: 2% bovine serum albumin (Sigma) in RNase-free water, aliquoted and stored at −20 °C.

8. 40% dextran sulfate: 40% dextran sulfate (Fischer Scientific) in RNase-free water, mixed well, aliquoted and stored at −20 °C.
9. Probe (IDT): 2′OMe Cy3-CAG7 (5′-mCmAmGmCmAmGm CmAmGmCmAmGmCmAmGmCmAmGmCmAmG/ 3Cy3Sp/-3′), diluted at 2 μM in RNase-free water, according to the manufacturer's instructions.
10. Hybridization solution: 2 μL of 2% yeast extract tRNA, 5 μL of 40% dextran sulfate, 4.5 μL of RNase-free water, 0.4 μL of 2% BSA, 4 μL of SSC Buffer 20X (Sigma), 20 μL of formamide (Sigma), 0.1 μL of RNaseOUT™ Recombinant Ribonuclease Inhibitor (3–4U) (Invitrogen), 4 μL of 2′OMe Cy3-CAG7 probe at 2 μM. Mix well and use immediately.
11. Parafilm.
12. Tweezers.
13. Aluminum foil.
14. 1X PBS (Gibco).
15. PBS-T: 0.1% Tween-20 (Sigma) in 1X PBS (Gibco).
16. Blocking solution: 0.1% Tween-20 (Sigma), 3% bovine serum albumin (Sigma) in 1X PBS (Gibco).
17. Primary antibody (Developmental Studies Hybridoma Bank, Antibodies at the University of Iowa for use in research).
18. Secondary antibody AlexaFluor® 488—goat anti-mouse IgG, IgM (H+L).
19. Wet chamber.
20. Coverslip.
21. DAPI (5 mg/mL): Dissolve 5 mg of DAPI (Sigma) in 1 mL of RNase-free water and aliquot and store at −20 °C. Keep protected from light.
22. DAPI solution: 0.75 μL of DAPI 5 mg/mL in 10 mL of 1X PBS (Gibco). Keep protected from light.

2.2.4 Slide Preparation

1. Micropipettes and 10 μL sterile, filtered tips.
2. ProLong™Gold antifade mountant (Invitrogen).
3. Coverslip.
4. Clear nail polish.

2.3 Preparation and Intravenous Administration of P-PMO to Adult Mice

1. Benchtop centrifuge.
2. Spectrophotometer (NanoDrop 2000, Thermo Scientific).
3. Heated ultrasonic water bath.
4. Mouse restrainer.
5. U-100 insulin syringe with 29-30G needle (Omnican® 50, 0.5 mL, 30G).

6. Warming cabinet (Mini-thermacage, Datesand).
7. Autoclaved paper tissue with 70% ethanol.
8. P-PMO.
9. 9% (w:v) saline.
10. Sterile distilled water.
11. Micropipettes and 1000, 200, 10, and 2 μL sterile, filtered tips.

2.4 Measuring P-PMO Toxicity in HSA-LR Mice

2.4.1 Evaluating Toxicity Markers in HSA-LR Mice Urine

Collection of Urine Using Metabolic Cages

1. Metabolic cages (Tecniplast).
2. Collection chiller (Tecniplast).
3. Refrigerated microcentrifuge.
4. Micropipettes and 1000 and 200 μL tips.

Urinary KIM-1 Quantification

1. Mouse TIM-1/KIM-1/HAVCR Immunoassay (Quantikine®ELISA, MKM100, R&D Systems®, Inc).
2. Microplate reader capable of measuring absorbance at 450 nm, with the correction wavelength set at 540 nm or 570 nm.
3. Deionized or distilled water.
4. 100 mL and 500 mL graduated cylinders.
5. Horizontal orbital microplate shaker (0.12" orbit), capable of maintaining a speed of 500 ± 50 rpm.

2.4.2 Evaluating Toxicity Markers in HSA-LR Mice Serum

1. Blood collection tube with clotting activator (Microvette® CB 300 Serum, Sarstedt).
2. CO_2 euthanasia chamber.
3. Dissection tools.
4. 70% ethanol.
5. 1 mL syringes with 23G needles.

2.4.3 Collection and Preparation of Liver and Kidney Tissues for Histopathological Analysis

1. CO_2 euthanasia chamber.
2. Dissection tools.
3. 70% ethanol.
4. 1X PBS.
5. Autoclaved paper tissue.
6. Neutral buffered formalin solution 10% (v/v).
7. Large collection tubes or clean air-tight containers (>15 mL) for the formalin fixation.

8. Small air-tight collection tubes (1.5–5 mL) for long-term storage in ethanol 70%.
9. Cryotubes.
10. Liquid nitrogen in an insulated container.

2.5 Measuring PMO Tissue Concentration Using an Enzyme-Linked Immunosorbent Assay (ELISA)

2.5.1 Tissue Homogenization

1. Micropipettes and 1000, 200, 10, and 2 μL sterile, filtered tips.
2. 0.2, 0.5, and 1.7 mL sterile, microcentrifuge tubes.
3. Ice buckets, fresh ice and dry ice.
4. Absorbent paper.
5. Tissue homogenizer and 1.5–2.0 mL homogenization tubes with metal beads (Precellys, Bertin Instruments).
6. Tissues.
7. PBS buffer (Gibco).
8. Forceps and scissors.
9. High precision scale.
10. Weighing boats.
11. RIPA buffer (Thermo Scientific) supplemented with protease and phosphatase inhibitor tablets (Thermo Scientific): Dilute one tablet of protease inhibitors and one tablet of phosphatase inhibitors in 10 mL of RIPA buffer. Shake until completely dissolved and store at 4 °C.
12. Benchtop centrifuge.
13. Dry bath.

2.5.2 Probes and PMO Preparation

1. Micropipettes and 1000, 200, 10, and 2 μL sterile, filtered tips.
2. 0.2, 0.5, and 1.7 mL sterile, microcentrifuge tubes.
3. Specific probe with digoxigenin and biotin ends modifications (IDT).
4. Probe diluting buffer: 0.1 mM Tris–EDTA.
5. Peptide-conjugated phosphorodiamidate morpholino oligonucleotides (PMOs) (GeneTools).
6. Distilled water.

2.5.3 Performing the ELISA

1. Micropipettes and 1000, 200, 10, and 2 μL sterile, filtered tips.
2. 0.2, 0.5, and 1.7 mL sterile microcentrifuge tubes.
3. 15 mL and 50 mL sterile falcon tubes.
4. Pipette buoy and 5, 10, and 25 mL stripettes.
5. Ice bucket and fresh ice.
6. Polymerase chain reaction film.
7. Aluminum foil.

8. Dry bath.
9. Vortex.
10. Incubator.
11. 1X Tris–EDTA buffer: 10 mM Tris–HCl, 1 mM EDTA, pH 8.
12. PMO diluting buffer: 0.1% v/v Triton X-100 in 1X Tris–EDTA buffer.
13. Hybridization buffer: 1M NaCl, 0.1% v/v Triton X-100 in 1X Tris–EDTA buffer.
14. 1% tissue background solution: Supplement 4545 μL of PMO diluting buffer with 55 μL of a saline tissue homogenate prepared as sample tissues.
15. Specific probe with digoxigenin and biotin ends modifications (IDT).
16. PMO.
17. 96-well plate.
18. 96-well plate NeutrAvidin coated (Thermo Scientific).
19. Washing buffer: 50 mM Tris–HCl, 150 mM NaCl, 0.1% v/v Tween-20, pH 7.6.
20. Portable aspiration system (Vacusip, Integra).
21. Micrococcal nuclease buffer: 50 mM Tris–HCl, 200 mM NaCl, 5 mM $CaCl_2$, 0.1 mg/mL bovine serum albumin, pH 8.2.
22. Micrococcal nuclease solution: 0.2 U/μL of micrococcal nuclease (New England Biolabs) in micrococcal nuclease buffer.
23. Antibody solution: 1/5000 dilution of anti-digoxigenin antibody conjugated with alkaline phosphatase (Roche), 0.25% v/v Tween-20 in SuperBlock (TBS) Blocking Buffer (Thermo Scientific).
24. AttoPhos® solution: 60 mL of AttoPhos® buffer (Promega) transferred into one entire bottle of AttoPhos® substrate (36 mg) (Promega).
25. Fluorescence plate reader (BMG Labtech).

2.5.4 Analyzing PMO Concentration

1. Plate reader data.
2. Spreadsheet software (Excel, Microsoft).

1. Tissues harvested from P-PMO-treated mice or P-PMO-transfected cells.
2. Micropipettes and 1000, 200, 10, and 2 μL sterile, filtered tips.
3. Tissue homogenizer and 1.5–2.0 mL homogenization tubes with metal beads (Precellys, Bertin Instruments).

2.6 Measuring P-PMO Activity In Vitro And In Vivo: Analyzing Splicing Correction

2.6.1 RT-PCR of P-PMO-Treated Human Cells or Mouse Tissue

4. Maxwell® RSC Instrument.
5. Maxwell® RSC simplyRNA Cells Kit (Promega).
6. Maxwell® RSC simplyRNA Tissue Kit (Promega).
7. Benchtop vortex mixer.
8. Benchtop centrifuge.
9. Spectrophotometer (NanoDrop 2000, Thermo Scientific).
10. High-Capacity cDNA Reverse Transcription Kit (Applied Biosystems).
11. PCR reagents: DreamTaq DNA Polymerase, 10X DreamTaq Buffer, dNTP Mix 2 mM each (Thermo Scientific).
12. Primers (Table 1).
13. Nuclease-free water.
14. Thermocycler.
15. PCR tubes.

2.6.2 Gel Electrophoresis and Splicing Analysis

1. Gel Loading Dye: Purple (6X), no SDS (New England Biolabs).
2. Agarose.
3. Ethidium bromide.
4. Electrophoresis set including generator, tank, tray, and comb.
5. 1X Tris–acetate–EDTA (TAE) buffer.
6. UV gel imaging system.
7. ImageJ software (NIH) [42].
8. Spreadsheet software (Excel, Microsoft).

3 Methods

3.1 Preparation of Peptide–PMO Conjugates

3.1.1 Synthesis of Peptides

1. Synthesize peptides by following the standard Fmoc-solid-phase peptide synthesis protocol [43] with an automated microwave-assisted peptide synthesizer (CEM Liberty Blue synthesizer) [44].
2. After the last Fmoc removal, acetylate the N-terminus by adding acetic anhydride/DIPEA (10 equiv each) in DMF for 15 min.
3. Wash the resin with DMF (5 x 30 s) and DCM (5 x 30 s).
4. After drying the resin, add the TFA cleavage cocktail (TFA/H_2O/TIS 95:2.5:2.5) and allow for shaking for 2–4 h at room temperature.
5. Precipitate the peptide in cold diethyl ether.

6. Centrifuge the peptide, remove the ether layer, and repeat the ether wash twice.
7. Purify the peptide using RP-HPLC and collect the pure fractions together.
8. Lyophilize to remove the water and remaining acetonitrile.

3.1.2 Synthesis of Peptide–PMO Conjugates on a 1.5-μmol Scale (See Note 2)

1. Dissolve the lyophilized PMO in DMSO to a concentration of 10 mM.
2. Prepare 50-μL solutions of the peptide (100 mM), PyBOP (300 mM), and HOAt (300 mM) in NMP.
3. Mix the peptide (37.5 μL, 100 mM, 2.5 equiv to PMO), PyBOP (28.8 μL, 300 mM, 5.75 equiv to PMO), and HOAt (25 μL, 300 mM, 5 equiv to PMO) in a 2 mL Eppendorf tube.
4. Add DIPEA (1.5 μL, 5.75 equiv to PMO) to the mixture and allow it to activate for 1 min (*see* **Note 3**).
5. Add 150 μL of the PMO solution to the activated peptide and allow the conjugation to incubate at 40 °C for 1–2 h, monitoring to completion via HPLC (*see* **Note 1**, Fig. 1a, b).
6. Quench the reaction by adding hydrazine hydrate and 25% ammonia solution for 15 min at 40 °C (*see* **Note 4**).
7. Neutralize the ammonia with 10% AcOH and then dilute the reaction to 2 mL by adding deionized water.
8. Purify the conjugate on an (HR)-16 cation exchange column using a gradient elution with a sodium phosphate buffer system (buffer A: sodium phosphate buffer (25 mM, pH 7.0), 20% acetonitrile; buffer B: sodium phosphate (25 mM, pH = 7.0), 20% acetonitrile, 1 M sodium chloride). Elute the conjugate at a flow rate of 6 mL min^{-1}. Set the detector to record at 220 nm and 260 nm for the detection of peaks (*see* **Note 5**).

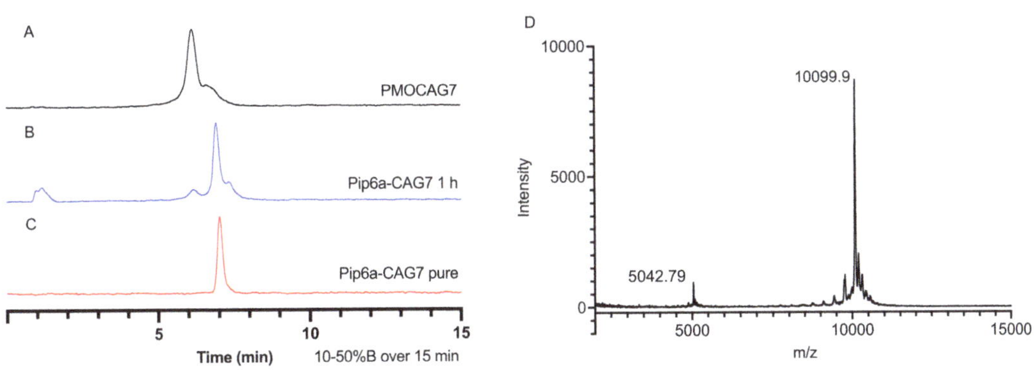

Fig. 1 HPLC and MALDI-TOF profiles for the synthesis and preparation of the peptide–PMO Pip6a-CAG7. The HPLC analysis of (**a**) PMO antisense oligonucleotide PMOCAG7 (starting material), (**b**) 90% conversion into the desired conjugated product (peptide–PMO) after 1 h, and (**c**) peptide–PMO after ion-exchange purification. (**d**) MALDI-TOF spectra for Pip6a-CAG7. Calculated $[M+H]^+$: 10092.59, found: 10099.9

9. Collect and pool the purified peaks together and desalt using Amicon® Ultra-15 3 K centrifugal units as per the manufacturer's instructions.
10. Lyophilize the desalted product.
11. For sterilization, dissolve the peptide–PMO conjugate in sterile, RNAse-free water and filter through a sterile Spin-X 0.22 μm cellulose acetate filtration unit.
12. Analyze the product by reverse-phase HPLC using a gradient elution at a flow rate of 1 mL min^{-1}, monitoring at 220 and 260 nm and by MALDI-TOF mass spectrometry using sinapinic acid as the matrix (*see* Fig. 1c, d, respectively).
13. Quantify the peptide–PMO conjugate by measuring the UV absorbance at 265 nm in 0.1 M HCl as per the manufacturer's instructions (*see* **Note 6**).

3.2 Measuring P-PMO Efficacy In Vitro: Fluorescent In Situ Hybridization (FISH) of DM1 Myotubes RNA Foci

3.2.1 Sterile Cell Culture of Control and DM1 Myotubes (See Note 7)

1. Wash all surfaces of a laminar flow cabinet with 70% ethanol. Work in this area for all steps in Subheading 3.2.1 (*see* **Note 8**).
2. Culture myoblasts in a T25 flask with 10 mL of skeletal muscle growth medium. Change the medium every 2 days.
3. Once myoblasts reach the 90–100% confluence, remove the skeletal muscle growth medium and wash with 10 mL of sterile 1X PBS.
4. Remove the 1X PBS and add 3 mL of TripLE. Incubate at 37 °C for 5 min in a cell culture incubator.
5. Neutralize the TripLe by adding 9 mL of skeletal muscle growth medium.
6. Transfer the final 12 mL to a 50 mL corning tube and centrifuge at 500 x *g* for 5 min.
7. Discard the supernatant and resuspend the myoblasts in 5 mL of skeletal muscle growth medium.
8. Plate the 5 mL of myoblasts in a new T75 flask and add 10 mL of skeletal muscle growth medium.
9. Softly shake the flasks for homogenous distribution of the myoblasts.
10. Once myoblasts reach 90–100% confluence again, repeat **steps 3–9** using 5 mL of TripLE in **step 4** and 15 mL of skeletal muscle growth medium in **step 5** for a final volume of 20 mL in **step 6**. In **step 8**, add 20 mL of skeletal muscle growth medium instead of 10 mL and use a T175 flask.
11. Once myoblasts reach 90–100% confluence in a T175 flask, remove the skeletal muscle growth medium and wash with 10 mL of 1X PBS.

12. Remove 1X PBS, add 5 mL of TripLE, and incubate at 37 °C for 5 min.
13. Neutralize TripLE with 15 mL of skeletal muscle growth medium and transfer the 20 mL into a 50 mL corning tube. Centrifuge myoblasts at 500 x *g* for 5 min.
14. Discard the supernatant and resuspend the myoblasts in skeletal muscle differentiation medium.
15. Determine the myoblasts concentration in a 50 mL Corning tube by taking a sample of 10 μL of resuspended myoblasts and counting in a Neubauer cell counting chamber.
16. Plate between 400,000 and 500,000 cells in each well of a Nunc™ Lab-Tek™ II Chamber Slide.
17. Shake gently the chamber slide to distribute the myoblasts evenly.
18. Leave the chamber slide inside a cell culture incubator at 37 °C for 5 to 8 days to differentiate myoblasts into myotubes, changing the skeletal muscle differentiation media every 2–3 days.
19. After 5–8 days, differentiation is complete. The myotubes should show multinucleated fibers and are ready to be fixed.

3.2.2 Cell Fixation

1. Gently wash cells with 1 mL of 1X PBS (*see* **Note 9**).
2. Remove the 1X PBS from the wells.
3. Add 1 mL of 4% formaldehyde/PBS.
4. Incubate 10 min at room temperature.
5. Wash twice with 1 mL of 1X PBS and leave in the second wash. Fixed cells can be kept at 4 °C for at least a week if the slide chambers are sealed with Parafilm.

3.2.3 Fluorescent In Situ Hybridization (FISH)

1. Prepare fresh 0.5% Triton/1X PBS and 2X SSC/50% formamide solution and keep at 4 °C.
2. Cut one piece of Parafilm per well to be processed, fitted to the size of a Lab-Tek chamber slide well.
3. Remove the 1X PBS.
4. Add 500 μL of 0.5% Triton X-100 solution.
5. Incubate at room temperature for 10 min.
6. Aspirate 0.5% Triton X-100 solution.
7. Wash with 1 mL of 2X SSC/50% formamide for 5 min.
8. Aspirate the 2X SSC/50% formamide.
9. Add 40 μL of hybridization solution directly onto cells and place a cut piece of Parafilm over it so that the liquid covers the entire surface. Probe not denatured.

10. Protect the samples from light from this point on.
11. Seal the plate with Parafilm and incubate at 37 °C for a minimum of 2 h to overnight.
12. Wash 2X with 1 mL of 2X SSC/50% formamide at 37 °C for 30 min and remove the Parafilm on the first wash. Use tweezers to remove the Parafilm if required.
13. Wash with 1 mL of 1X PBS.
14. Rinse cells twice with PBS-T.
15. Incubate cells in blocking solution for 1 h at room temperature.
16. Dilute the primary antibody in blocking solution (2.5 μg/mL).
17. Pipette 30 μL of antibody into each well and replace the cover.
18. Incubate for 1 h at 37 °C, in wet chamber if necessary.
19. Wash 3 times with PBS-T for 10 min.
20. Dilute the secondary antibody in blocking solution (1:250).
21. Add 30 μL per well of antibody and replace the cover.
22. Incubate for 1 h at room temperature, in wet chamber if necessary.
23. Wash once with PBS-T.
24. Wash once with 1 mL of 1X PBS.
25. Remove the 1X PBS.
26. Stain the nucleus with 1 mL of DAPI solution for 10 min.
27. Remove the DAPI solution from the wells.
28. Wash twice with 1 mL of 1X PBS and leave the last wash in the wells.

3.2.4 Slide Preparation

1. Remove the chambers from the slide.
2. Add one drop of ProLong™Gold antifade mountant to each well.
3. Place a coverslip on top of the slide.
4. Leave to cure in the dark, on a flat, dry surface, for 24 h at room temperature.
5. Once the curing is complete, seal the coverslip perimeter with clear nail polish.
6. Leave to dry for 30 min.
7. Observe the slide using a confocal microscope or keep the slide at −20 °C for long-term storage.

3.3 Preparation and Intravenous Administration of P-PMO to Adult Mice

1. Thaw the P-PMO and vortex briefly.
2. Dilute the P-PMO to the desired concentration with distilled water.
3. Incubate the P-PMO for 30 min in a 37 °C water bath (or equivalent).
4. Sonicate the P-PMO aliquot for 5 min with an ultrasonic water bath set at 37 °C.
5. Briefly vortex and spin the P-PMO.
6. Immediately before injection, add 9% saline to the P-PMO to give a final concentration of 0.9% saline (P-PMO final concentration: 2–9 μg/μL).
7. Briefly vortex and pulse spin the P-PMO, and then keep the PPMO solution at room temperature until the injection.
8. Weigh the mouse.
9. Transfer the mouse to a warming cabinet set at 38 °C.
10. Calculate the injection volume, ensuring that the volume and dosage to be injected are within the parameters of any relevant regulations or guidelines.
11. Using a U-100 insulin syringe that has a low plunger resistance, remove any bubbles and aspirate the injection volume (*see* **Note 10**).
14. After 5 min at 38 °C (maximum 10 min), transfer the mouse to the restrainer. Check that the nose of the mouse is not obstructed by the moving part of the restrainer.
15. Swab the tail of the mouse with sterile paper tissue dampened with 70% ethanol.
16. Rotate the restrainer by 90° in order to have the most visible lateral vein facing upward toward you.
17. Hold the tail over your index finger with the thumb of your nondominant hand.
18. Hold the syringe in your dominant hand with the bevel of the needle facing upward and insert the needle into the vein with an angle as close to 0° as possible. The needle should be almost parallel to the vein.
19. Push the plunger of the syringe slowly until the full volume is injected (*see* **Note 11**).
20. Remove the needle from the tail and apply pressure to the injection site with a sterile and dry paper tissue for at least 20 s and until the bleeding has ceased.
21. Transfer the mouse to its cage and clean the restrainer.
22. Observe the mouse until it has recovered by evaluating its condition and behavior (*see* **Note 12**). It may be necessary to cull the mouse if it shows side effects from the injection that fail to improve (*see* **Note 12**).

3.4 Measuring P-PMO Toxicity in HSA-LR Mice

3.4.1 Evaluating Toxicity Markers in HSA-LR Mice Urine

Collection of Urine Using Metabolic Cages

The urine is collected for 16 h within 24–48 h and 6–7 days post-injection.

1. Remove the collection chiller from the freezer at least 3 h before transferring the mouse to the metabolic cage.
2. Assemble the metabolic cages according to manufacturer's instructions. Take care to not touch the surface that will be in contact with the urine without gloves.
3. Fill the feeder with small pieces of the mouse's regular food (*see* **Note 13**).
4. Add the drinking bottle filled with water and ensure that there will be no leakage.
5. Encase the collection chiller in its insulating shell.
6. Place the urine collection tube in the deepest eccentric hole of the collection chiller and place everything under the metabolic cage. Ensure that the urine collection tube is aligned with the spout of the urine ring.
7. Attach the hatch to the lower chamber.
8. Weigh the mouse and transfer it gently to the support grid of the metabolic cage.
9. Replace the cage cover and leave the mouse in the cage overnight for a typical period of 16 h.
10. Weigh the mouse and transfer it back to its home cage (*see* **Note 14**).
11. Transfer the urine from the collection tube to a 1.5 mL microtube pre-chilled on ice. A minimum of 200 μL is expected.
12. Wash all parts of the metabolic cage and autoclave if required.
13. Centrifuge the urine at 4 °C at 10,000 x *g* for 10 min.
14. Transfer the supernatant to a new tube.
15. Keep the urine on ice or at 4 °C if it is going to be analyzed on the day of the collection or store the urine at –20 °C for further analysis.
16. When ready, proceed to Subheading 3.6.1 for urinary KIM-1 quantification and/or carry out clinical chemistry analysis of urinary metabolites (*see* **Note 15**).

Urinary KIM-1 Quantification

1. Prepare all reagents, working standards, and control as described in the Quantikine®ELISA manual.
2. Dilute the urine samples with an appropriate volume of Calibrator Diluent RD5-26 from the Quantikine®ELISA kit (diluted 1:4), according to the toxicity expected (*see* **Note 16**). Prepare a minimum of 150 μL to ensure that there is sufficient volume to analyze the samples in duplicates.

3. Follow the assay steps as described in the Quantikine®ELISA manual.
4. Calculate the KIM-1 concentration using the standard curve from the assay and compare concentrations across all samples.
5. If clinical chemistry analysis of urinary metabolite levels has also been performed on the samples, normalize sample KIM-1 concentration to creatinine concentration for further sample comparison (*see* **Note 15**).

3.4.2 Evaluating Toxicity Markers in HSA-LR Mouse Serum

1. Cull mouse 2 weeks after the final injection with a rising concentration of CO_2 (*see* **Notes 17** and **18**) and proceed immediately to the next steps to start blood collection.
2. Spray 70% ethanol over the entire mouse.
3. Place the mouse on its back on a dissection mat.
4. Dissect the skin all along the ventral surface.
5. Cut along the inferior margin of the rib cage and then along the lateral walls of the rib cage. **Be careful not to puncture the heart.**
6. Fold the rib cage over the head to expose the heart.
7. Collect blood by inserting a 23G needle attached to a 1-mL syringe into the right ventricle of the heart (*see* **Note 19**).
8. Remove the needle and transfer the blood into collection tubes that are opened at both ends (maximum 300 μL per tubes) and keep the tubes tilted to avoid the leakage of blood from the bottom of the tubes.
9. Close the bottom followed by the top of the collection tubes.
10. Briefly shake the collection tubes by hand.
11. Allow the blood samples to clot at room temperature for 30 min. During this time period, any tissue collection from the mouse for histopathological analysis (*see* Subheadings 2.4.3 and 3.4.3) or for other downstream applications (*see* Subheadings 2.5, 2.6, 3.5, and 3.6) can be performed (*see* **Note 20**).
12. After 30 min, centrifuge the blood samples at 10,000 x *g* for 5 min at room temperature.
13. Carefully transfer the serum (supernatant) to a clean microcentrifuge tube (*see* **Note 21**). Keep serum on dry ice or transfer to 80 °C for long-term storage.
14. Analyze serum for metabolites or enzymes of interest using a clinical chemistry analyzer or equivalent. This can be done in-house or at a clinical chemistry facility (*see* **Note 15**).

3.4.3 Collection and Preparation of Liver and Kidney Tissues for Histopathological Analysis

1. Collect one kidney and one lobe of the liver from a mouse culled by rising CO_2 concentration (*see* **Notes 17** and **18**). You may wish to additionally collect and snap-freeze in liquid nitrogen any tissues for further experiments (*see* **Note 20**).
2. Wash tissues with 1X PBS and remove the excess liquid with a sterile paper tissue.
3. Transfer the kidney and piece of the liver lobe to large collection tubes or containers pre-filled with cold 10% formalin (*see* **Note 22**).
4. Leave the kidney and liver lobe in formalin for 24 h at 4 °C under a constant, gentle agitation (*see* **Note 23**).
5. After 24 h, quickly wash the fixated tissues by replacing the formalin with cold 1X PBS.
6. Gently invert the tubes twice and discard the 1X PBS.
7. Proceed to wash the tissues twice for 10 min with cold 1X PBS and constant agitation.
8. Transfer the samples to a small collection tube filled with ethanol 70%.
9. Store tissue samples at 4 °C, until ready for in-house or external histopathological analysis (*see* **Note 24**).

3.5 Measuring PMO Tissue Concentration Using ELISA

3.5.1 Tissue Homogenization

1. Wash tissues (*see* **Note 20**) in 1X PBS buffer and dry them in absorbent paper before cutting them.
2. Use a precision scale to weigh 20 mg of non-CNS tissues and 30 mg of CNS tissues.
3. Transfer cut tissue pieces into tissue grinding tubes with metal beads.
4. Add 100 μL of RIPA buffer, supplemented with protease and phosphatase inhibitors, per 10 mg of the cut tissue.
5. Using a tissue homogenizer, homogenize tissues by shaking samples 4 times at 200 rpm for 20 s.
6. Centrifuge samples at 12,000 x *g* for 3 min. Transfer supernatants into new clean microcentrifuge tubes of an appropriate volume.
7. Incubate the tubes at 55 °C overnight in a dry bath.
8. Centrifuge the samples at 13,000 x *g* for 15 min. Transfer supernatants into new clean Eppendorf tubes and proceed to the ELISA. Alternatively, store samples at −20 °C for long-term analysis.

3.5.2 Preparation of Probes and PMOs

1. Suspend the lyophilized probes in TE buffer at 0.5 nM.
2. Suspend PMOs in distilled water at 1 μM.
3. Keep all probes and PMOs at 4 °C prior to use. Aliquots of probes and PMOs can be stored at −20 °C.

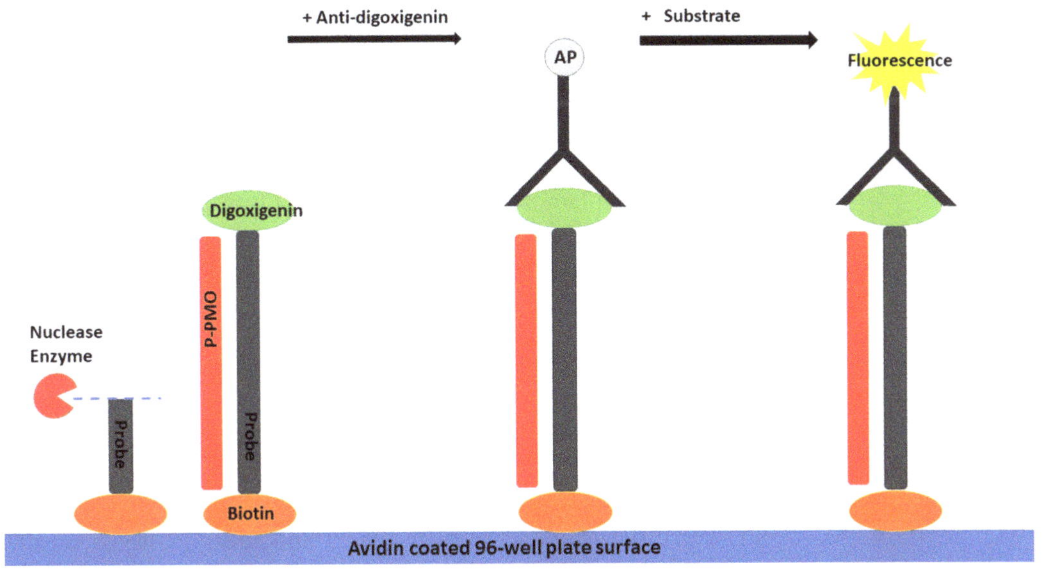

Fig. 2 Overview of ELISA method to detect oligonucleotides. The probe is conjugated to biotin at the 3′-end and digoxigenin at the 5′-end. Probes are hybridized to the antisense oligonucleotide (PMO) and bind the avidin surface of the 96-well plates via biotin. Addition of the nuclease enzyme cleaves the digoxigenin end of any nonhybridized probes to prevent their signal on the addition of subsequent reagents. An anti-digoxigenin mAb linked to an alkaline phosphate enzyme (AP) binds to the probe digoxigenin. On addition of the enzyme substrate, a fluorescent product can be detected on a plate reader

3.5.3 Performing the ELISA (See Note 25 and Fig. 2)

1. Dilute samples and controls in PMO diluting buffer and vortex for 10 s. Kidney samples should be diluted at 1:100–1:500, other non-CNS tissues at 1:10–1:50, and CNS tissues should not be diluted.
2. Incubate PMO at 65 °C for 15 min, vortex for 1 min, and prepare standards by serial dilutions of a known PMO concentration in 1% tissue background solution to give standard curve points ranging from 25 pM to 2000 pM.
3. Incubate the probe at 65 °C for 15 min, vortex for 1 min, and then dilute in hybridization buffer to give a 0.5 nM solution.
4. Hybridize samples, controls, and PMO standards with a probe in a 96-well plate. Add 100 μL of each per well in duplicate (triplicate for standards) followed by 100 μL of the probe diluted to 0.5 nM in hybridization buffer.
5. Seal the plate with polymerase chain reaction film and incubate at 37 °C for 30 min to allow the probe to hybridize with the PMO.
6. Transfer 100 μL/well of the hybridized solution to a 96-well NeutrAvidin-coated plate.
7. Seal the plate with polymerase chain reaction film and incubate at 37 °C for 30 min to allow the biotin probe to bind to the NeutrAvidin-coated plate.

8. Wash the plate 3 times with 400 μL of washing buffer (see **Note 26**) and add 150 μL/well of micrococcal nuclease solution per well.
9. Seal the plate with polymerase chain reaction film and incubate at 37 °C for 1 h.
10. Wash the plate 3 times with 400 μL of washing buffer and add 150 μL/well of antibody solution.
11. Seal the plate with polymerase chain reaction film and incubate at 37 °C for 30 min.
9. Wash the plate 3 times with 400 μL of washing buffer and add 150 μL/well of AttoPhos® solution.
10. Seal the plate with polymerase chain reaction film, wrap in aluminum foil to protect from light, and incubate at 37 °C for 30 min.
11. Place the plate in the plate reader and shake gently at double orbital shaking at 600 rpm for 10 s before measuring the fluorescent intensity.
12. Determine the endpoint of fluorescence intensity in a fluorescence microplate reader at 444 nm excitation and 555 nm emission.

3.5.4 Analyzing PMO Concentration

1. Generate a standard curve by plotting fluorescence intensity data on the y-axis and PMO or P-PMO concentration on the x-axis in an Excel sheet (*see* **Note 27**).
2. Use the standard curve to determine the PMO or P-PMO concentration within the samples.

3.6 Measuring P-PMO Activity In Vitro and In Vivo: Analyzing Splicing Correction

3.6.1 RT-PCR of P-PMO-Treated Human Cells or Mouse Tissue

1. Prepare all reagents as described in the Maxwell® RSC simplyRNA Cells and simplyRNA Tissue Kits protocols (*see* **Note 28**).
2. Add prepared chilled 1-thioglyceriol/homogenization solution from the kits to samples.

   For cells: Remove the medium from the wells of transfected human cells, add the 1-thioglyceriol/homogenization and vortex until the pellet is dispersed and cells appear lysed.

   For tissues: Cut 20 mg of frozen tissues (*see* **Note 20**) from treated mice, transfer to homogenization tubes, and homogenize in 200 μL of prepared chilled 1-thioglycerol/homogenization solution.
3. Proceed to the RNA purification by following all further steps according to the manufacturer's kit protocols, with the amendment of adding to the lysis buffer 15 μL of proteinase K per sample.
4. Measure the RNA concentration of each sample with a spectrophotometer.

5. Dilute all RNA samples in nuclease-free water to a standard concentration and store at –80 °C.
6. Carry out the reverse transcription with the High-Capacity cDNA Reverse Transcription Kit (*see* **Note 29**) according to the manufacturer's protocol and using 400 ng of RNA in 20 μL reaction volume.
7. Dilute the cDNA samples 1:2 in nuclease-free water to give a final concentration of 10 ng/μL.
8. Prepare PCR mastermixes for each primer set (Table 1) according to Thermo's Scientific DreamTaq DNA Polymerase protocol (*see* **Note 30**). Final sample reaction volume: 25 μL, final primer concentration: 0.5 μM.
9. Pipette 17 μL per sample of each mastermix into separate PCR tubes and add 8 μL of diluted sample cDNA.
10. Vortex and centrifuge samples.
11. Run the PCR reactions on a thermocycler with the following parameters for all primers sets:

    95 °C for 2 min, followed by 30 cycles of 30 s 95 °C, 30 s at annealing temperature specified in Table 1 and 1 min at 72 °C. End reaction with 5 min at 72 °C.
12. Keep amplicon DNA at 4 °C or store long-term at −20 °C.

3.6.2 Gel Electrophoresis and Splicing Analysis

1. Prepare a solution 1X TAE, 2% agarose gel solution by microwaving agarose in the 1X TAE until boiling and all the agarose has melted (*see* **Note 31**).
2. Cool the agarose solution under a stream of cold tap water by holding the container with a heat-resistant glove or holder until it reaches a temperature close to 50 °C (*see* **Note 32**).
3. Add 0.5 μg/mL ethidium bromide (*see* **Note 33**) directly to the gel solution and mix well.
4. Wait for all bubbles to disappear.
5. Pour the gel into a gel tray with a well comb of a size appropriate for the number of samples.
6. Remove immediately all bubbles on the poured gel with a pipette tip.
7. Allow the gel to solidify for at least 30 min.
8. Place gel into gel tank and remove well combs.
9. Remove the gel from its tray and put it in an electrophoresis tank with fresh 1X TAE. Remove well combs and add additional 1X TAE to ensure wells are filled with buffer and that the whole gel is just covered.
10. Dilute the amplicon DNA samples with 5 μL of gel loading dye.

Table 1
Primers, annealing temperatures, and expected bands for RT-PCRRT-PCR splicingSplicing analysis

Transcript targeted	Primer type	Primer sequence (5′–3′)	Bands expected	DM1 splicing event	Annealing temperature (°C)
Mouse					
Mbnl1	Forward	GCTGCCCAATACCAGG TCAAC	270; 216	ex5 inclusion	61.7
	Reverse	TGGTGGGAGAAATGCTG TATGC			
Clcn1	Forward	TTCACATCGCCAGCATCTG TGC	325 ; 246	ex7a inclusion	64.1
	Reverse	CACGGAACACAAAGGCAC TGAATGT			
Atp2a1	Forward	GCTCATGGTCCTCAAGA TCTCAC	218 ; 176	ex22 exclusion	60.6
	Reverse	GGGTCAGTGCCTCAGC TTTG			
Human					
MBNL1	Forward	GCTGCCCAATACCAGG TCAAC	270; 216	ex5 inclusion	61.7
	Reverse	TGGTGGGAGAAATGCTG TATGC			
INSR	Forward	CCAAAGACAGACTC TCAGAT	165; 129	ex11 exclusion	53.2
	Reverse	AACATCGCCAAGGGACC TGC			
SOS1	Forward	CAGTACCACAGATG TTTGCAGTG	192; 147	ex21 exclusion	60.3
	Reverse	TCTGGTCGTCTTCG TGGAGGAA			
DMD	Forward	TTAGAGGAGGTGA TGGAGCA	148; 116	ex78 exclusion	54
	Reverse	GATACTAAGGACTCCA TCGC			

11. Load an appropriate volume of samples and DNA ladder for the gel system used.
12. Close the tank and run the electrophoresis at 6 V/cm for 50 min.
13. Use UV gel imaging systems to take a picture of the gel, adjusting the exposure to obtain the maximum signal for your bands without reaching saturation.

14. Using ImageJ [42], convert all images into 8 bits and perform a sliding paraboloid background subtraction (*see* **Note 34**).
15. Use the ImageJ gel analysis features to quantify the density of the bands obtained (Analyze>Gels) following the steps in the ImageJ online manual [45].
16. Using spreadsheet software, e.g., Microsoft Excel, for each sample, calculate the percent-splice-in value (PSI value) by calculating the percentage of the heaviest band density (indicating an inclusion) over the density of all bands.

4 Notes

1. DMSO and NMP are anhydrous solvent grade and must be stored in inert conditions. We have noticed that water decreases the overall yield of the conjugation.
2. This is a very efficient method to produce the desired P-PMO conjugate as HPLC analysis showed approximately 90% conversion in 1 h.
3. The solution turns pale yellow once DIPEA is added due to the formation of HOAt anion. This indicates that the reaction has started, 0.5 equiv of DIPEA can be added if the color does not develop.
4. MALDI-TOF analysis showed that the addition of hydrazine hydrate and/or ammonia solution contributes to improving the overall purity of the final P-PMO product by removing any unwanted alkylation side reactions on the PMO.
5. Our method of choice is 0–75% B gradient over 30 min. The PMO has an early retention time on the cation-exchange column (1–2 min). The P-PMO usually appears at the range 12–18 min depending on the number of Arg. A higher Arg content reflects a longer retention time. Lastly, the unreacted peptide comes at around 20–25 min.
6. The overall yield of P-PMO varies from 30% to 50% based on the PMO starting material.
7. The protocol is adapted from Jauvin et al. (2017) [25].
8. All steps should be carried out inside a laminar flow cabinet except for incubations performed in a cell culture incubator, as stated. The cell culture incubated should be at 37 °C and 5.0% CO_2.
9. Working under the biological hood is unnecessary at this stage, and there is no waiting time during washes unless mentioned. For the remainder of Subheading 3.2, all volumes are adapted for 4-well Lab-Tek chamber slides.

10. At the appropriate syringe plunger resistance, you should be able to easily push the plunger with the side of your little finger when holding the syringe between the index finger and thumb. Any bubbles in the syringe can be removed by aspirating and subsequently ejecting a small volume of P-PMO twice.
11. If the injection is successful, you will be able to observe the injected solution spreading in the veins downstream from the injection site. If there is high resistance, it is likely that the vein has been missed. To re-attempt the injection, inject again closer to the base of the tail; this can be done for a maximum of three punctures. The other lateral vein can be used if the collection of tail vein blood is not required on the same day. Every puncture to the tail (maximum of three) makes the bevel of the needle less sharp, which will make subsequent injections more challenging and also cause more pain to the mouse. Therefore, we advise using a new needle/syringe for each injection attempt.
12. The mouse is considered to have recovered when it shows spontaneous exploration or grooming behavior, no cyanosis, normal breathing, and no or mild pain signs. We recommend evaluating the recovery of a mouse by observing the five key features in Table 2 and looking for any signs of the listed side effects. It may be necessary to humanely cull the mouse if it has a prolonged flattened position, prolonged lethargy, cyanotic skin, or difficulty to breathing and shows no signs of improvement.
13. As food consumption is not measured, powdered food is not required. Moreover, as mice used for this study are not habituated to the metabolic cages before the first urine collection, they will be unused to powdered food and may not initially consume it. Furthermore, as food and water intake are not measured, it is not necessary to weigh the feed drawer and the drinking bottle prior to or after the experiment.

Table 2
Mouse features to evaluate for signs of adverse drug effects during post-injection recovery

Mouse feature	Signs of adverse drug effects
Movement	Lethargy, reluctance to move, lack of exploration
Skin color	Cyanosis
Breathing	Dyspnea and/or tachypnea
Behavior	Flattened or hunched posture, allogrooming, self-grooming
Pain	Orbital tightening, nose bulging, unusual ear position

Table 3
List of clinical chemistry tests performed in urine and serum of P-PMO-treated HSA-LR mice

Urine	Serum	
Sodium	Sodium	Glucose
Potassium chloride	Potassium chloride	Triglycerides Glycerol
Urea	Urea	Total bilirubin
Creatinine	Creatinine	Lactate dehydrogenase
Calcium	Total calcium	Iron
Inorganic phosphorus	Inorganic phosphorus	α-Amylase
Glucose	Alkaline phosphatase	Creatine kinase
Uric acid	Alanine aminotransferase	Uric acid
Magnesium	Aspartate aminotransferase	Magnesium
Urinary protein	Total protein	Lipemia
	Hemolysis	Icterus
	Albumin	
	Total cholesterol	

14. A bodyweight loss is expected, but this should not exceed 15% of the mouse's weight before the experiment. As a result of stress, mice may eat or drink less than normal; however, this loss is usually rapidly recovered.
15. Clinical chemistry of urine and serum samples is performed with a Beckman Coulter AU680 clinical chemistry analyzer and carried out by the Clinical Chemistry service of the Mary Lyon Centre (MRC Harwell, UK). For a list of tests carried out on urine and serum samples, see Table 3.
16. To remain within the range of detection, day 2 urine samples of mice expected to suffer from severe kidney injury should be diluted at 1:100. This can be assumed for all mice that had a long post-injection recovery time. All the other urine samples should be diluted at 1:20.
17. Once the mouse stops showing life signs, leave the mouse for 1 min more in the CO_2-filled chamber. Confirm the cessation of respiration, the faded eye color (eyes become transparent in albinos), and the cessation of the heartbeat (palpation). Extend the time in the chamber if required but no more than 3 min before starting with the dissection.

    Avoid using cervical dislocation as a culling method as it usually leads to significant internal bleeding and insufficient blood collection for later analyses.
18. The death of the mouse in this protocol is confirmed by the observation of a permanent cessation of the circulation, followed by exsanguination during the collection of blood for serum analysis. If blood collection is not being carried out, ensure that death is confirmed in accordance with the regulations and standards of the company or institutions' respective national legislation.

19. Pull the syringe plunger slowly to avoid the ventricle collapsing. The post-mortem low venous blood pressure still allows the collection of up to 600 μl of blood from a 25 g mouse. An alternative or complement to ventricular collection is to cut into the right atrium and collect the blood flowing directly into the rib cage with the tip of the collection tube. The mouse, and especially the rib cage filled with blood, must be held above the tip of the collection tube in order to allow its filling by gravity and capillary action.
20. If there are multiple serum samples from the same mouse, we advise pooling them after centrifugation. Avoid pooling if the sample is visibly hemolyzed (red in color). It is possible to collect a serum volume equivalent to at least one-third of the blood volume.
21. We recommend the collection of the following tissues for the investigation of in vivo P-PMO efficacy (*see* Subheadings 2.6 and 3.6) and biodistribution (*see* Subheadings 2.5 and 3.5): heart, lung, spleen, liver, kidney, diaphragm, brain, triceps, quadriceps, gastrocnemius, soleus, and tibialis anterior. Upon collection, tissues should be snap-frozen in liquid nitrogen and kept on dry ice until long-term storage at −80 °C is possible.
22. When transferring harvested tissues for histopathological analysis into 10% formalin, we recommend selecting containers of a sufficient size for at least 15 volume equivalents of formalin per volume of tissue.
23. The 24 h fixation and subsequent steps can be carried out at room temperature if the tissues are for histopathological analysis only.
24. Formalin-fixed livers and kidneys (stored in ethanol 70%) are trimmed to standardized trimming planes, placed in tissue cassettes. The tissues are embedded in paraffin wax according to standardized blocking patterns. 5 μm sections from each block are collected on SnowCoat microscope slides, dried at 37 °C, and stained with hematoxylin and eosin. On completion of staining, all slides are digitally scanned at x20 using a Hamamatsu NanoZoomer 2.0 HT. The slides are assessed by a pathologist for liver and kidney damage.
25. The ELISA protocol is adapted from Burki et al. (2015) [46].
26. After each final washing of the ELISA plate, before the addition of the next reagent, it is highly advisable to cover the plate with absorbent paper and invert and shake it with the paper. Any remaining drops of washing buffer will be absorbed by the paper, which will improve the working of the reagent next added.

27. For optimum visualization of the standard curve, we recommend performing log–log transformations of both axis and adding a power trend line.
28. Other purification methods can be used, e.g., guanidinium thiocyanate–phenol–chloroform extraction or Spin column-based nucleic acid purification.
29. Alternative cDNA synthesis kit can be used, e.g., the SuperScript™ First-Strand Synthesis System for RT-PCR (Invitrogen) with random hexamers.
30. It is important to avoid adding additional $MgCl_2$ to the mastermix as nonspecific bands are expected with a $MgCl_2$ concentration exceeding 2 mM.
31. When fully melted, the agarose solution should be fully transparent.
32. We consider the temperature of 50 °C reached once the agarose container can be touched with a rubber-gloved hand without a burning sensation.
33. The ethidium bromide in the gel will migrate to the negative pole during the electrophoresis. Avoid the use of the bottom of the gel because of the ethidium bromide depletion.
34. We use a radius of 50 px for the background subtraction. This value needs to be adapted the first time depending on the size of your bands in pixel. Keep the same parameters for all further analyses.

Acknowledgments

This work was supported by Vinnova, NNF Center for Biosustainability, WCPR Wallenberg Center for Protein Research, and the Knut and Alice Wallenberg foundation.

References

1. Theadom A, Rodrigues M, Roxburgh R et al (2014) Prevalence of muscular dystrophies: a systematic literature review. Neuroepidemiology 43:259–268
2. Brook JD, McCurrach ME, Harley HG et al (1992) Molecular basis of myotonic dystrophy: expansion of a trinucleotide (CTG) repeat at the 3′ end of a transcript encoding a protein kinase family member. Cell 68:799–808
3. Mahadevan M, Tsilfidis C, Sabourin L et al (1992) Myotonic dystrophy mutation: an unstable ctg repeat in the 3′ untranslated region of the gene. Science 255(5049): 1253–1255
4. Fu Y-H, Pizzuni A, Fenwick RG et al (1992) An unstable triplet repeat in a gene related to myotonic muscular dystrophy. Science (New York, N.Y.) 255(5049):1256–1258
5. Taneja KL, McCurrach M, Schalling M et al (1995) Foci of trinucleotide repeat transcripts in nuclei of myotonic dystrophy cells and tissues. J Cell Biol 128:995–1002
6. Davis BM, Mccurrach ME, Taneja KL et al (1997) Expansion of a CUG trinucleotide repeat in the 3′ untranslated region of myotonic dystrophy protein kinase transcripts results in nuclear retention of transcripts. Proc Natl Acad Sci 94:7388–7393

7. Lee K, Li M, Manchanda M et al (2013) Compound loss of muscleblind-like function in myotonic dystrophy. EMBO Mol Med 5: 1887–1900
8. Miller JW, Urbinati CR, Teng-Umnuay P et al (2000) Recruitment of human muscleblind proteins to (CUG)(n) expansions associated with myotonic dystrophy. EMBO J 19: 4439–4448
9. Wang ET, Treacy D, Eichinger K et al (2019) Transcriptome alterations in myotonic dystrophy skeletal muscle and heart. Hum Mol Genet 28:1312–1321
10. Mankodi A, Takahashi MP, Jiang H et al (2002) Expanded CUG repeats trigger aberrant splicing of ClC-1 chloride channel pre-mRNA and hyperexcitability of skeletal muscle in myotonic dystrophy. Mol Cell 10: 35–44
11. Charizanis K, Lee KY, Batra R et al (2012) Muscleblind-like 2-mediated alternative splicing in the developing brain and dysregulation in myotonic dystrophy. Neuron 75:437–450
12. Fugier C, Klein AF, Hammer C et al (2011) Misregulated alternative splicing of BIN1 is associated with T tubule alterations and muscle weakness in myotonic dystrophy. Nat Med 17: 720–725
13. Freyermuth F, Rau F, Kokunai Y et al (2016) Splicing misregulation of SCN5A contributes to cardiac-conduction delay and heart arrhythmia in myotonic dystrophy. Nat Commun 7:25
14. López-Martínez A, Soblechero-Martín P, De-La-puente-ovejero L et al (2020) An overview of alternative splicing defects implicated in myotonic dystrophy type I. Genes 11:1109
15. Porter B, Drug development pipeline for myotonic dystrophy type 1 (DM1) and myotonic dystrophy type 2 (DM2), https://www.myotonic.org/sites/default/files/pages/files/Myotonic-Dystrophy-Drug-Development-Pipeline-as-of-Feb-01-2021.pdf
16. Reddy K, Jenquin JR, Cleary JD, et al (2019) Mitigating RNA toxicity in myotonic dystrophy using small molecules. www.mdpi.com/journal/ijms
17. López-Morató M, Brook JD, Wojciechowska M (2018) Small molecules which improve pathogenesis of myotonic dystrophy type:1. www.frontiersin.org
18. Roberts TC, Langer R, Wood MJA (2020) Advances in oligonucleotide drug delivery. Nat Rev Drug Discov 19:673–694.
19. Summerton J, Weller D (1997) Morpholino antisense oligomers: design, preparation, and properties. Antisense Nucleic Acid Drug Dev 7:187–195
20. Hudziak RM, Barofsky E, Barofsky DF et al (1996) Resistance of Morpholino phosphorodiamidate oligomers to enzymatic degradation. Antisense Nucleic Acid Drug Dev 6:267–272
21. FDA grants accelerated approval to first drug for Duchenne muscular dystrophy | FDA. https://www.fda.gov/news-events/press-announcements/fda-grants-accelerated-approval-first-drug-duchenne-muscular-dystrophy
22. Aartsma-Rus A, Krieg AM (2017) FDA approves eteplirsen for duchenne muscular dystrophy: the next chapter in the eteplirsen saga. Nucleic Acid Ther 27:1–3. /pmc/articles/PMC5312460/
23. Wheeler TM, Leger AJ, Pandey SK et al (2012) Targeting nuclear RNA for in vivo correction of myotonic dystrophy. Nature 488:111–117
24. Lee JE, Bennett CF, Cooper TA (2012) RNase H-mediated degradation of toxic RNA in myotonic dystrophy type 1. Proc Natl Acad Sci U S A 109:4221–4226
25. Jauvin D, Chrétien J, Pandey SK et al (2017) Targeting DMPK with antisense oligonucleotide improves muscle strength in myotonic dystrophy type 1. Mice 7:465–474
26. Yadava RS, Yu Q, Mandal M et al (2020) Systemic therapy in an RNA toxicity mouse model with an antisense oligonucleotide therapy targeting a non-CUG sequence within the DMPK 3′UTR RNA. Hum Mol Genet 29:1440–1453
27. Klein AF, Varela MA, Arandel L et al (2019) Peptide-conjugated oligonucleotides evoke long-lasting myotonic dystrophy correction in patient-derived cells and mice. J Clin Investig 129:4739–4744
28. Mulders SAM, van den Broek WJAA, Wheeler TM et al (2009) Triplet-repeat oligonucleotide-mediated reversal of RNA toxicity in myotonic dystrophy. Proc Natl Acad Sci U S A 106: 13915–13920
29. Wheeler TM, Sobczak K, Lueck JD et al (2009) Reversal of RNA dominance by displacement of protein sequestered on triplet repeat RNA. Science 325:336–339
30. Wojtkowiak-Szlachcic A, Taylor K, Stepniak-Konieczna E et al (2015) Short antisense-locked nucleic acids (all-LNAs) correct alternative splicing abnormalities in myotonic dystrophy. Nucleic Acids Res 43:3318–3331
31. Nakamori M, Sobczak K, Puwanant A et al (2013) Splicing biomarkers of disease severity in myotonic dystrophy. Ann Neurol 74: 862–872
32. Pandey SK, Wheeler TM, Justice SL et al (2015) Identification and characterization of modified antisense oligonucleotides targeting

DMPK in mice and nonhuman primates for the treatment of myotonic dystrophy type 1s. J Pharmacol Exp Ther 355:329–340

33. Ionis Pharmaceuticals Reports DMPKRx Phase 1/2 Trial Results. https://us8.campaign-archive.com/?u=8f5969cac3271759ce78c8354&id=8cc67ae9b8&e=cd1f4d18fe
34. Betts C, Saleh AF, Arzumanov AA et al (2012) Pip6-PMO, a new generation of peptide-oligonucleotide conjugates with improved cardiac exon skipping activity for DMD treatment. Mol Ther Nucl Acids 1:e38
35. Yin H, Moulton HM, Seow Y et al (2008) Cell-penetrating peptide-conjugated antisense oligonucleotides restore systemic muscle and cardiac dystrophin expression and function. Hum Mol Genet 17:3909–3918
36. Gao X, Zhao J, Han G et al (2014) Effective dystrophin restoration by a novel muscle-homing peptide-morpholino conjugate in dystrophin-deficient mdx mice. In: Molecular therapy. Nature Publishing Group, pp 1333–1341
37. Leger AJ, Mosquea LM, Clayton NP et al (2013) Systemic delivery of a peptide-linked morpholino oligonucleotide neutralizes mutant RNA toxicity in a mouse model of myotonic dystrophy. Nucleic Acid Ther 23:109–117
38. Hammond SM, Hazell G, Shabanpoor F et al (2016) Systemic peptide-mediated oligonucleotide therapy improves long-term survival in spinal muscular atrophy. Proc Natl Acad Sci USA 113:10962
39. Mankodi A, Logigian E, Callahan L et al (2000) Myotonic dystrophy in transgenic mice expressing an expanded CUG repeat. Science 289:1769–1772
40. Vaidya VS, Ozer JS, Dieterle F et al (2010) Kidney injury molecule-1 outperforms traditional biomarkers of kidney injury in preclinical biomarker qualification studies. Nat Biotechnol 28:478–485
41. Arandel L, Espinoza MP, Matloka M et al (2017) Immortalized human myotonic dystrophy muscle cell lines to assess therapeutic compounds. Dis Model Mech 10:487–497
42. Schneider CA, Rasband WS, Eliceiri KW (2012) NIH Image to ImageJ: 25 years of image analysis. https://www.nature.com/articles/nmeth.2089
43. Behrendt R, White P, Offer J (2016) Advances in Fmoc solid-phase peptide synthesis. J Pep Sci 22:4–27
44. Collins JM, Porter KA, Singh SK et al (2014) High-efficiency solid phase peptide synthesis (he -Spps). Org Lett 16:940
45. Ferreira T, Rasband W (2012) ImageJ User Guide ImageJ User Guide IJ 1.46r
46. Burki U, Keane J, Blain A et al (2015) Development and application of an ultrasensitive hybridization-based ELISA method for the determination of peptide-conjugated phosphorodiamidate morpholino oligonucleotides. Nucleic Acid Ther 25:275

Chapter 14

Developing Therapeutic Splice-Correcting Antisense Oligomers for Adult-Onset Pompe Disease with c.-32-13T>G Mutation

Kristin A. Ham, Russell D. Johnsen, Michel Tchan, Steve D. Wilton, and May T. Aung-Htut

Abstract

The mutation c.-32-13T>G in the *GAA* gene impacts normal exon 2 splicing and is found in two-thirds of late-onset Pompe disease cases. We have explored a therapeutic strategy using splice modulating phosphorodiamidate morpholino oligomers to enhance *GAA* exon 2 inclusion in the mature mRNA of patients carrying this common mutation. We performed in silico analysis of the *GAA* gene transcript for potential splicing silencers and designed oligomers targeting motifs predicted to enhance exon 2 retention in the mature mRNA. Two patient-derived fibroblasts were obtained from Coriell Institute for Medical Research, and seven fibroblast strains from unrelated patients were supplied by Westmead Hospital in Sydney, Australia. Both fibroblasts and forced-myogenic cells were treated with optimized phosphorodiamidate morpholino oligomers supplied by Sarepta Therapeutics. Total RNA and protein were extracted from the cells after incubation with phosphorodiamidate morpholino oligomers, and RT-PCR and RT-qPCR were performed to confirm exon 2 inclusion is enhanced. Acid α-glucosidase activity and expression levels were also assessed to confirm therapeutic potential.

Key words Antisense oligomers, Pompe disease, Exon inclusion, c.-32-13T>G, Acid α-glucosidase, Phosphorodiamidate morpholino oligomers

1 Introduction

Pompe disease is an autosomal recessive disorder caused by mutations in the *GAA* gene that lead to insufficient levels of acid α-glucosidase (GAA) activity. The primary function of GAA is to degrade stored lysosomal glycogen into its simpler form, glucose, so it can be readily used by cells for energy. Without sufficient GAA, glycogen accumulates in the lysosomes of cells, primarily muscle cells, which leads to cellular malfunction and damage. Ultimately, this damage triggers muscle weakness and wasting. Pompe disease severity is inversely related to the amount of functioning GAA

Rika Maruyama and Toshifumi Yokota (eds.), *Muscular Dystrophy Therapeutics: Methods and Protocols*, Methods in Molecular Biology, vol. 2587, https://doi.org/10.1007/978-1-0716-2772-3_14,

enzyme produced [1–3]. Generally, the lower the amount of GAA present, the earlier symptoms arise, the faster the disease progresses, and the greater the disease severity.

The progressive nature of adult-onset Pompe disease phenotype and the correlation of the disease severity with the level of residual GAA activity indicate that even a small increase in GAA activity may delay disease progression and be beneficial for adult Pompe patients. The c.-32-13T>G mutation is found in more than 70% of adult patients [4] and also a substantial number of juvenile-onset patients and affects exon 2 splicing during *GAA* pre-mRNA processing [5, 6]. This splice motif mutation is not absolute, and although the majority of the *GAA* transcripts are missing whole or partial exon 2, approximately 10% of the *GAA* transcripts are correctly spliced and able to produce functional GAA enzyme. We propose that increasing the levels of these correctly spliced *GAA* transcripts is achievable using splice-manipulating antisense oligomers (AOs), and this will provide some therapeutic relief for adult-onset Pompe disease patients. This strategy of silencing a splice silencer element has been successful for the treatment of spinal muscular atrophy disease, where exon 7 inclusion into the *SMN2* transcript was enhanced via antisense oligomer-mediated splice manipulation [7].

Historically, proof-of-concept studies on splice correction were performed using minigene assays, where cell lines are transfected with plasmids carrying select regions of the target gene, sometimes inserted within another gene. Obvious limitations of this system include inappropriate cell/tissue-specific splicing and the common inability to include flanking exons and the entire introns flanking the exon of interest. Derived from over two decades of experience in developing splice manipulating AOs for the treatment of several diseases, we have seen that in vitro splice manipulation can differ between individual cell strains as the genetic background can influence the splicing process. Therefore, we strongly recommend testing splice-modulating AOs in individual patient cell strains, preferably cells reflecting the disease tissue/organ if possible. One alternative, although costly, is to reprogram skin or blood cells to iPSC and subsequently differentiate them into relevant cell types.

In the method we describe here, patient-derived fibroblasts were sourced from the Coriell Institute for Medical Research or collected at Westmead Hospital in Sydney, Australia. Phosphorodiamidate morpholino oligomers (PMOs) were delivered into the patient-derived fibroblasts using the Nucleofector transfection system. Initially, PMO screening was performed in the fibroblast lineage, but since Pompe disease is generally regarded as a progressive muscle disease, lead PMO candidates were tested in myogenic cells where the patient-derived fibroblasts were transduced using a MyoD expression viral construct [8].

The efficacy of lead candidate PMOs in correcting exon 2 splicing in the *GAA* transcript was measured using RT-PCR and RT-qPCR analysis in each patient-derived cell type. Cell lysates were collected to assess the changes in mRNA transcripts, amount of GAA produced, and the GAA activity after treatment. Here we describe the step-by-step process to assess the efficacy of PMOs promoting exon 2 inclusion in the *GAA* transcript in both fibroblasts and forced-myogenic cells derived from Pompe patients.

2 Materials

2.1 Cell Lines, Culture Reagents, and PMO Delivery Reagents

We examined seven patient-derived fibroblasts collected at Westmead Hospital in Sydney, Australia and two fibroblast cell strains obtained from Coriell Institute for Medical Research. All cells are reported to carry the common c.-32-13T>G allele in the *GAA* gene.

(a) GM00443.

(b) GM11661.

1. Culture medium: Dulbecco's modified Eagle's medium (DMEM) high glucose supplemented with 10% fetal bovine serum (FBS) and 2 mM GlutaMAX™-1.
2. Differentiation medium: DMEM low glucose supplemented with 5% horse serum (HS).
3. Fibroblast transfection incubation medium: DMEM supplemented with 1% FBS.
4. Phosphate-buffered saline (PBS): 137 mM NaCl, 2.7 mM KCl, 10 mM Na_2HPO_4, 1.8 mM KH_2PO_4.
5. Trypsin: Trypsin–EDTA (0.5%) diluted in PBS.
6. Trypsin inactivation medium: DMEM high glucose supplemented with 10% HS.
7. Poly-D-lysine: Tissue culture grade Poly-D-lysine comes as a 5 mg lyophilized powder. Resuspend in 100 mL of sterile ultrapure water to prepare 50 μg/mL stock.
8. Matrigel: Matrigel is supplied at 8–12 mg/mL concentration and is used at 100 μg/mL diluted in DMEM (*see* **Note 1**).
9. MyoD expressing adenoviral vector: Ad5.f50.AdApt.MyoD (The Native Antigen Company).
10. Sucrose storage solution: 10 mM Tris–HCl, pH 8, 2 mM $MgCl_2$, 4% sucrose.
11. Phosphorodiamidate morpholino oligomers (PMOs): Resuspend to 5 mM in sterile ultrapure water (water was filter sterilized using a 0.2 μm filter) and store at 4 °C.

(a) Negative control oligomer: Standard control morpholino oligomer (5′-CCTCTTACCTCAGTTACAATTTATA-3′) from Gene Tools, LLC that targets a human beta-globin intronic mutation.

(b) Positive control oligomer: Targeting integrin alpha 4 (*ITGA4*) (ITGA4 H3A(+30+49); 5′-TCTCTCTCTT CCAAACAAGT-3′) [9] or survival motor neuron (*SMN*) (SMN H7A(+13+32); 5′- CACCTTCCTTCTTTTT GATT-3′) [10].

12. 4D-Nucleofector™ X Unit (Lonza).
13. P3 Primary Cell 4D-Nucleofector™ X Kit S (Lonza).
14. T75 cm^2 flasks and 24-well plates.
15. Hemocytometer.

2.2 GAA Activity and Expression

1. 1.5 mL Eppendorf tubes.
2. Dry ice.
3. GAA lysis buffer: 10 mM HEPES, 70 mM sucrose, 220 mM mannitol supplemented with 1× protease inhibitors, pH 7.4 [11].
4. Reaction buffer 0.2 M acetate buffer, pH 3.9: Dissolve 3.28 g sodium acetate anhydrous in 150 mL of water, adjust pH to 3.9 with glacial acetic acid dropwise, and make to a final volume of 200 mL.
5. Reaction buffer 0.2 M acetate buffer, pH 6.5: Prepare as mentioned in **step 4,** except adjusting pH to 6.5 with glacial acetic acid.
6. 70 mM artificial substrate 4-methylumbelliferyl-β-D-glucopyranoside (4-MUG) dissolved in dimethyl sulfoxide (DMSO) (*see* **Note 2**): Prepare the 70 mM stock by dissolving 4.732 mg of 4-MUG in 200 μL of DMSO. Dilute the 70 mM stock at a ratio of 1:50 in the appropriate reaction buffer (pH 3.9 or pH 6.5) to give a working concentration of 1.4 mM.
7. 96-well plate, black polystyrene flat bottom.
8. Stop buffer 0.5 M sodium carbonate buffer, pH 10.7: Adjust to pH 10.7 by titration of approximately 50 mL of 0.5 M sodium bicarbonate against 450 mL of 0.5 M sodium carbonate buffer.
9. Pierce BCA protein assay kit (ThermoFisher Scientific).
10. Western loading dye: 0.2 M Tris–HCl, pH 6.8, 0.4 M DTT, 8.0% (w/v) SDS, 6 mM bromophenol blue, 4.3 M glycerol.
11. NuPAGE Novex 4–12% Bis-Tris gels 1 mm Midi Protein gels, 20 well (ThermoFisher Scientific).
12. PVDF transfer membrane.

13. 1× TBST: 20 mM Tris–HCl, pH 8, 150 mM NaCl, 0.1% (w/v) Tween® 20.
14. Blocking solution: 5% skim milk in 1× TBST.
15. Anti-GAA antibody (Abcam, cat. no.137068).
16. HRP-labeled anti-rabbit antibody (Dako, cat. no P0448).
17. HRP-labeled anti-mouse antibody (Dako, cat. no D0447).
18. Mouse monoclonal anti-β-tubulin antibody (DSHB, cat. no. E7).
19. Rabbit anti-desmin antibody (ThermoFisher Scientific, cat. no. PA5-16705).
20. Luminata Crescendo Western HRP substrate (Merk Millipore).
21. Fusion FX system using the Fusion-Capt software (version 17.03) (Vilber Lourmat, Marne-la-Vallée, France).
22. FLUOstar Omega (BMG LABTECH, Mornington, Australia) with 355 nm excitation and 460 nm emission filters.

2.3 RNA Isolation, RT-PCR, and RT-qPCR

1. MagMax™-96 Total RNA Isolation Kit (AM1830) including a DNase treatment step (ThermoFisher Scientific).
2. KingFisher 96 Magnetic Particle Processor.
3. NanoDrop™ 1000 Spectrophotometer.
4. SuperScript™ IV Reverse Transcriptase Kit containing reaction buffer (5X), DTT 100 mM, and reverse transcriptase (200 U/μL) (ThermoFisher Scientific).
5. dNTP set, 100 mM solutions diluted to 5 mM total (1.25 mM each dATP, dCTP, dGTP, and dTTP) (ThermoFisher Scientific).
6. Random Primers (3 μg/μL) diluted to 100 ng/μL in ultrapure water (ThermoFisher Scientific).
7. RNaseOUT™ Recombinant Ribonuclease Inhibitor (40 U/μ L) (ThermoFisher Scientific).
8. Primers (Integrated DNA Technologies): Resuspend in water to 100 μM.
   (a) RT-PCR: GAA_1F 5′-GGAAACTGAGGCACGGAGCG-3′; GAA_5R 5′- GGACCACATCCATGGCATTGC -3′; CCND_1F 5′- TGTTCGTGGCCTCTAAGATGAA -3′; CCND_1R 5′-GTCCGGGTCACACTTGATCAC-3′.
   (b) RT-qPCR: GAA_1F(q) 5′-TGGGAAAGCTGAGGTTG TCG-3′; GAA_1-2R(q) 5′-TCCTACAGGCCCGCTCC-3′; TBP_1-2F(q) 5′-TCTTTGCAGTGACCCAGCAT-CAC-3′; TBP_2R(q) 5′-CCTAGAGCATCTCCAG CACACTCT-3′.

9. TaKaRa LA Taq® DNA Polymerase with GC Buffer II containing reaction buffer (2X), dNTPs (10 mM), and *LA Taq* (5 U/μL).
10. Standard agarose gel equipment.
11. RedSafe™ Nucleic Acid Staining Solution (20,000X) (Scientifix).
12. Fast SYBR™ Green Master Mix (2X) (ThermoFisher Scientific).
13. CFX384 Touch Real-Time PCR detection system using the CFX Manager 3.1 software (BioRad).

3 Methods

3.1 Cell Culture

1. Propagate approximately 4×10^5 primary human fibroblasts in 75-cm^2 non-coated flasks using an appropriate culture medium.
2. To subculture, aspirate media from flasks and wash cells once with PBS using the same volume of culture medium. Add enough trypsin solution to cover the bottom of the flask and place it in the incubator until the cells have lifted off the bottom. Stop trypsinization by adding approximately double the volume of trypsin inactivation medium. Centrifuge cells at 1× RCF for 2.5 min, remove the supernatant, and resuspend the cell pellet in 1 mL of culture medium. Count the number of cells per 1 mL using a hemocytometer. Cells can be used for propagation or for nucleofection experiments outlined below.

3.2 Nucleofection of PMOs into Fibroblasts

Cells should be passaged 2–4 days prior to nucleofection after reaching approximately 90% confluency.

1. Pre-warm PMOs for approximately 30 min at 37 °C. Vortex and centrifuge before using. Use an appropriate volume of PMO solutions (not more than 2 μL) to achieve the desired PMO concentration in 20 μL total volume. For example, addition of 2 μL of PMO (1 mM stock) into a Nucleocuvette with 18 μL of cell suspension would amount to 50 μM final PMO concentration in 20 μL total volume.
2. Prepare P3 Primary Cell Nucleofector Solution and Supplement 1 at a ratio of 82% and 18%, respectively (*see* **Note 3**). Prepare an 18 μL volume per treatment, multiplied by the number of proposed treatments, and then include an excess of 10%.
3. Harvest fibroblasts as above and transfer the equivalent of 3.0×10^5 cells per treatment into a 1.5 mL Eppendorf tube. Centrifuge at 0.9× RCF for 10 min and remove all the supernatant.

4. Resuspend the pellet in the required Lonza P3 solution from **step 2**, ensuring to remove clumps and avoid generating any bubbles.
5. Add 18 μL of the cell suspension into individual wells of a 16-well Nucleocuvette strip using a reverse pipetting technique to avoid air bubbles. Add 2 μL of each PMO stock solution. As a negative control, use a scrambled PMO or Gene Tools Control PMO as well as the untreated ZAP control where no PMO is added. As a positive control to monitor transfection efficiencies, one can use an *ITGA4* exon 3 skipping PMO to induce exon 3 from the mature *ITGA4* transcript [9] or a *SMN* exon 7 skipping PMO to induce the removal of exon 7 from the mature *SMN* transcript [10].
6. Gently tap the Nucleocuvette strip to ensure that there are no air bubbles, and the liquid covers the bottom of the wells.
7. Place the Nucleocuvette strip with cells into the 4D-Nucleofector X Unit of the 4D-Nucleofector System. For primary human fibroblasts, use the CA137 program. Programs for other cell types can be found in the manufacturer's guidelines or from [9].
8. After the program has been run, incubate the Nucleocuvette strip at room temperature for 10 min, followed by the addition of 80 μL of pre-warmed fibroblast transfection incubation medium.
9. Transfer the entire volume of each Nucleocuvette well into 5 mL of pre-warmed fibroblast transfection incubation medium in a 15 mL falcon tube, mix, and place 500 μL into 10 wells of a 24-well plate. Collect six wells for protein studies and four wells for RNA on day 4 (*see* **Note 4**).

3.3 Nucleofection with PMOs: Myogenic Lineages

Pre-coat 24-well plates with 250 μL of 50 μg/mL poly-d-lysine for 1 h at room temperature. Remove excess solution, and then add 250 μL of 100 μg/mL Matrigel and incubate for 1 h at 37 °C (*see* **Note 5**). Follow **steps 1** to 7 in Subheading 3.2.

1. After the program has been run, incubate the Nucleocuvette strip at room temperature for 10 min, followed by the addition of 80 μL of pre-warmed differentiation medium for propagation into a myogenic cell lineage.
2. Dilute (Ad5.f50.AdApt.MyoD) 10^{11} pfu/μL in sucrose storage solution to 10^{10} pfu/μL. Prepare 5 mL of pre-warmed differentiation medium with MyoD expressing adenoviral vector at an MOI of 200 for each treatment in 15-mL falcon tubes. Move the entire volume of each Nucleocuvette well into the tubes, mix, and place 500 μL into 10 wells of a 24-well plate.

3. On day 4, collect and combine cells from four wells for RNA analysis (continue to Subheading 3.8). Add 500 μL of differentiation medium without MyoD expressing adenoviral vector to the remaining 6 wells, which are to be harvested for protein studies on day 6.

3.4 Cell Collection for GAA Activity and Expression

1. Remove the medium and wash once with PBS.
2. Add 100 μL of 1× Trypsin in PBS into each well of the 24-well plate and incubate for 2 min at 37 °C.
3. Add 400 μL of trypsin inactivation medium and combine all wells (6 wells × 500 μL/well = 3 mL) into a 15 mL falcon tube.
4. Centrifuge at 1× RCF for 2.5 min, remove the supernatant, and resuspend in 2 mL of fresh PBS.
5. Repeat **step 4**, remove supernatant, and snap-freeze the cell pellet in dry ice. Store cell lysate at −80 °C.

3.5 Cell Lysis for GAA Activity Assay and Expression

It is preferable to perform the GAA activity assay within a few days and no more than five days after the collection. Cell lysates should always be kept on ice during handling and analysis.

1. Freeze-thaw the cell pellet on dry ice three times before resuspending in 50 μL of the GAA lysis buffer.
2. Sonicate at a frequency of 30 kHz six times for 1 s.
3. Centrifuge the cell lysate at 1.1× RCF for 10 min at 4 °C.
4. Transfer the supernatant into a fresh tube and measure the total protein concentration using Pierce BCA protein assay kit.

3.6 GAA Activity Assay

1. Acid GAA enzyme activity: Add approximately 3–5 μg of the total protein lysate from each sample into separate wells of a 96-well plate to a final volume of 10 μL (made up with GAA lysis buffer) to undertake acid-α-glucosidase activity assays. Include a blank sample with GAA lysis buffer only (no protein added). Initiate GAA enzyme activity by adding 20 μL of reaction buffer 0.2 M acetate, pH 3.9, containing 1.4 mM 4-MUG substrate. Reactions should be performed in triplicate.
2. Determine basic GAA enzyme activity in the same as **step 1** except using the pH 6.5 reaction buffer and only one-tenth of the protein used for acid GAA enzyme activity.
3. Incubate both reactions at 37 °C for 1 h.
4. Add 200 μL of stop buffer (0.5 M sodium carbonate buffer, pH 10.7).
5. Measure fluorescent signals using FLUOstar Omega with 355 nm excitation and 460 nm emission filters.

6. Calculate the ratio of signals generated at pH 3.9 to those at pH 6.5 after subtracting the corresponding blanks from both measurements.

3.7 GAA Expression by Western Blotting

1. Load approximately 4 μg of total protein as measured by BCA, diluted 1:1 in western loading dye, onto NuPAGE Novex 4–12% Bis-Tris gels (*see* **Note 6**).
2. Perform electrophoresis at 200 volts until the bromophenol blue marker is at the bottom of the gel (approximately 55 min).
3. Transfer the fractionated proteins from the gel onto PVDF transfer membranes at 350 mA for 1 h at 4 °C.
4. Immerse the membrane in 10–15 mL of blocking solution for 1 h at room temperature with gentle shaking.
5. Remove the blocking solution and then incubate the membrane with rabbit polyclonal anti-GAA antibody at 1:1000 dilution in fresh blocking solution overnight at 4 °C.
6. Wash the membrane for 1 h in 1× TBST at room temperature with gentle shaking.
7. Incubate with HRP-labeled anti-rabbit antibody at 1:10,000 dilution in blocking solution for 1 h at room temperature.
8. Repeat **step 6**.
9. Incubate with Luminata Crescendo Western HRP substrate for 5 min at room temperature.
10. Acquire images with the Fusion FX system using 30-second serial scans until the desired image intensities are acquired.
11. For housekeeping and confirming myogenicity, incubate with mouse monoclonal anti-β-tubulin antibody at 1:5,000 dilution and rabbit anti-desmin antibody at 1:1,000 dilution, respectively, both in blocking solution overnight at 4°C.
12. Repeat **steps 6–10** using HRP-labeled anti-mouse/rabbit antibody at a dilution of 1:10,000 in blocking solution.

3.8 RNA Isolation and Analysis

MagMax™-96 Total RNA Isolation Kit (AM1830), including a DNase step, can be most convenient for RNA extraction of a large number of samples. This kit was used according to the manufacturer's guidelines on a KingFisher 96 Magnetic Particle Processor using the program "AM1830" with some modifications to the protocol (*see* **Note 7**).

1. Remove the medium and wash the cells once with PBS.
2. Add 100 μL of 1× Trypsin in PBS into each well of the 24-well plate and incubate for 2 min at 37 °C.
3. Add 250 μL of trypsin inactivation medium to each well and combine all wells (4 wells × 350 μL/well = 1.4 mL) into a 1.5 mL Eppendorf tube.

4. Centrifuge at 1× RCF for 2.5 min, remove the supernatant, and add 140 μL of the supplied lysis buffer from the MagMax^{TM}-96 Total RNA Isolation Kit (AM1830). Follow the remaining steps according to the manufacturer's guidelines.
5. Elute RNA with 42 μL of elution buffer and use 2 μL to determine RNA concentration using the NanoDrop™ 1000 Spectrophotometer. Samples are to be kept on ice at all times.
6. Generate cDNA in a 20 μL reverse transcription (RT) reaction using SuperScript™ IV Reverse Transcriptase.
   (a) For each sample, prepare the following mix: Random primer (100 ng/μL) 1 μL; dNTPs (5 mM total) 2 μL; RNA template 125 ng; sterile ultrapure water up to a final volume of 13 μL.
   (b) Thermocycler annealing program: 5 min at 65 °C, then on ice for 5 min.
   (c) After the annealing step, add the following to each sample: SSIV reaction buffer (5X) 4 μL; DTT 1 μL; RNaseOUT Recombinant RNase Inhibitor 1 μL; SuperScript IV RTase 1 μL.
   (d) Thermocycler cDNA synthesis program: 10 min at 23 °C, 10 min at 50 °C, 10 min at 80 °C.
   (e) Dilute the cDNA 1:1 in sterile ultrapure water.
   (f) Use samples immediately or store them at −20 °C.
7. End-point PCR amplification from the cDNA above.
   (a) For each sample, prepare the following mix for a 12.5 μL reaction: GC Buffer II (2X) 6.25 μL; forward primer (10 μM) 0.25 μL; reverse primer (10 μM) 0.25 μL; dNTPs (10 mM) 1 μL; LA *Taq* 0.125 μL; template cDNA 1 μL; sterile ultrapure water up to a final volume of 12.5 μL.
      (i) Forward primer GAA_1F and reverse primer GAA_5R are used to amplify *GAA* transcript from exons 1 to 5. *See* Fig. 1 for expected amplicon sizes, as determined by the splice variants.
      (ii) Forward primer CCND_1F and reverse primer CCND_1R are used to amplify *CCND1* transcript from exons 2 to 4/5 and are used as a reference transcript.
   (b) Thermocycler PCR conditions: initial denaturation 95 °C 5 min; X cycles of 95 °C 30 s, 60 °C 30 s and 72 °C 2 min; final extension of 72 °C 5 min.
      (i) X = 34 cycles to amplify the *GAA* transcript.
      (ii) X = 23 cycles to amplify the *CCND1* transcript.

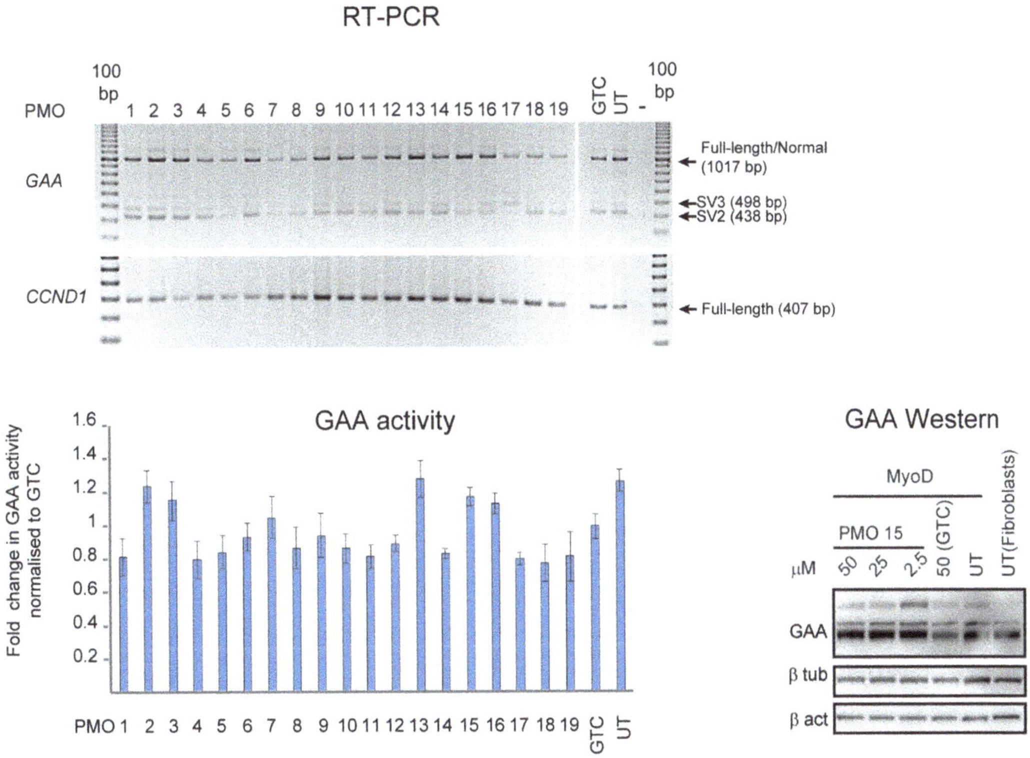

Fig. 1 Analysis of *GAA* transcripts and GAA activities and expression during the development of splice-correction PMOs for adult-onset Pompe patients with c.-32-13T>G mutation. *GTC* Gene Tools control, *UT* untreated, *SV* splice variant, *β tub* beta tubulin, *β act* beta actin [8] (creative commons)

8. Fractionate PCR products on 2% agarose gels in Tris–acetate–EDTA buffer 100V 1 h until the bromophenol blue dye front is near the bottom of the gel. Incubate the gel with RedSafe stain for 10 min at room temperature and visualize using the Fusion FX system.
   (a) Prepare RedSafe stain by diluting 10 μL in 250 mL TAE buffer. Do not discard the stain as it can be used several times (*see* **Note 8**).
9. Perform quantitative PCR (qPCR) amplifications (in triplicate) using the cDNA.
   (a) *GAA* transcript amplification: For each sample, prepare the following mix for a 10-μL reaction: Fast SYBR® Green (2X) 5 μL; forward primer GAA_1F(q) (5 μM) 0.2 μL; reverse primer GAA_1-2R(q) 0.2 μL; sterile ultrapure water 1.6 μL; template cDNA 2 μL (*see* **Note 9**).
   (b) *TBP* transcript amplification: For each sample, prepare the following mix for a 10 μL reaction: Fast SYBR® Green (2X) 5 μL; forward primer TBP_1-2F(q) (5 μM) 1 μL; reverse primer TBP_2R(q) 1 μL; template cDNA 2 μL (*see* **Note 10**).

(c) Perform reactions in a CFX384 Touch Real-Time PCR detection system for 95 °C 1 min, 39 cycles of 95 °C 3 s, 60 °C 15 s, and 72 °C 30 s.

(d) The relative expression of *GAA* to *TBP* mRNA was determined using the 2-ΔΔCT method and presented as a fold change compared to the healthy or Gene Tools control-treated patients-derived fibroblasts.

4 Notes

1. Matrigel® basement membrane mix is a preparation derived from the Engelbreth–Holm–Swarm mouse sarcoma. The major components are laminin, collagen IV, and heparan sulfate proteoglycans, entactin/nidogen, and several growth factors.
2. Initial trials encountered difficulty in dissolving high concentrations of MUG into aqueous solutions. This can be resolved using a stock containing 70 mM MUG in DMSO and then dispensing this into aqueous buffers using a 1:50 v/v ratio with vortexing to provide a 1.4 mM working solution [12].
3. The manufacturer's protocol states that the entire supplement is added to the Nucleofector solution. However, after they are combined, the mixture has a 3-month shelf life at 4 °C. Therefore, only add the required volume of Supplement 1 to the Nucleofector solution in an Eppendorf tube for each experiment to prolong the shelf life of the kit. Make enough for 20 μL for each treatment plus a 10% excess for pipetting errors.
4. Preparing a total of 10 wells per treatment, six wells for protein analysis and four wells for RNA analysis will allow for enough sample to perform the necessary experiments twice.
5. Plates can be coated ahead of time and dried in a sterile hood before sealing with clingwrap and storing at 4 °C for no longer than a week.
6. Generally, the volumes required for all samples are similar (~10 μL) if a number of cells are kept constant and no cell death is observed.
7. Modifications to the AM1830 KingFisher program include an additional 5 min of mixing during the lysis binding step and an extra 90 s for all washing steps. The extra lysing and washing steps increase the quality and quantity of the RNA extracted from cells plated onto a matrix.
8. Diluted RedSafe stain can be used to stain five individual agarose gels when used as a post-stain. Post-stain the agarose gel for 10 min on a shaker and then de-stain in water for 10–30 min. Store diluted RedSafe stain in the dark at room temperature.

9. For the *GAA* transcript amplification standard curve, a twofold serial dilution starting at 20 ng is an optimal range of cDNA concentrations, as calculated from the RNA concentrations.
10. For the *TBP* transcript amplification standard curve, a twofold serial dilution starting at 40 ng is an optimal range of cDNA concentrations, as calculated from the RNA concentrations.

References

1. Mehler M, DiMauro S (1977) Residual acid maltase activity in late-onset acid maltase deficiency. Neurology 27(2):178–184
2. Umapathysivam K, Hopwood JJ, Meikle PJ (2005) Correlation of acid alpha-glucosidase and glycogen content in skin fibroblasts with age of onset in Pompe disease. Clin Chim Acta 361(1–2):191–198. https://doi.org/10.1016/j.cccn.2005.05.025
3. van der Ploeg AT, Reuser AJ (2008) Pompe's disease. Lancet 372(9646):1342–1353. https://doi.org/10.1016/S0140-6736(08)61555-X
4. Huie ML, Chen AS, Tsujino S et al (1994) Aberrant splicing in adult onset glycogen storage disease type II (GSDII): molecular identification of an IVS1 (-13T-->G) mutation in a majority of patients and a novel IVS10 (+1GT-->CT) mutation. Hum Mol Genet 3(12): 2231–2236
5. Boerkoel CF, Exelbert R, Nicastri C et al (1995) Leaky splicing mutation in the acid maltase gene is associated with delayed onset of glycogenosis type II. Am J Hum Genet 56(4):887–897
6. Raben N, Nichols RC, Martiniuk F et al (1996) A model of mRNA splicing in adult lysosomal storage disease (glycogenosis type II). Hum Mol Genet 5(7):995–1000
7. Ottesen EW (2017) ISS-N1 makes the first FDA-approved drug for spinal muscular atrophy. Transl Neurosci 8:1–6. https://doi.org/10.1515/tnsci-2017-0001
8. Aung-Htut MT, Ham KA, Tchan M et al (2020) Splice modulating antisense oligonucleotides restore some acid-alpha-glucosidase activity in cells derived from patients with late-onset Pompe disease. Sci Rep 10(1): 6702. https://doi.org/10.1038/s41598-020-63461-2
9. Aung-Htut MT, McIntosh CS, Ham KA et al (2019) Systematic approach to developing splice modulating antisense oligonucleotides. Int J Mol Sci 20(20):doi:10.3390/ijms20205030
10. Flynn LL, Mitrpant C, Adams A et al (2021) Targeted SMN exon skipping: a useful control to assess in vitro and in vivo splice-switching studies. Biomedicines 9(5):doi:10.3390/biomedicines9050552
11. Nilsson MI, Samjoo IA, Hettinga BP et al (2012) Aerobic training as an adjunctive therapy to enzyme replacement in Pompe disease. Mol Genet Metab 107(3):469–479. https://doi.org/10.1016/j.ymgme.2012.09.010
12. Kallwass H, Carr C, Gerrein J et al (2007) Rapid diagnosis of late-onset Pompe disease by fluorometric assay of alpha-glucosidase activities in dried blood spots. Mol Genet Metab 90(4):449–452. https://doi.org/10.1016/j.ymgme.2006.12.006

Part IV

Gene Replacement Therapies

Chapter 15

Molecular and Biochemical Assessment of Gene Therapy in the Canine Model of Duchenne Muscular Dystrophy

Chady H. Hakim, Dennis Pérez-López, Matthew J. Burke, James Teixeira, and Dongsheng Duan

Abstract

Mutations in the dystrophin gene result in Duchenne muscular dystrophy (DMD), a progressive muscle-wasting disease. Adeno-associated virus (AAV) mediated gene replacement, and CRISPR/Cas9-mediated genome editing hold the potential to treat DMD. Molecular and biochemical analyses are essential to determine gene transfer efficiency and therapeutic efficacy. In this chapter, we present a series of methods routinely used in our laboratory to extract and quantify DNA, RNA, and protein in gene therapy studies performed in the canine DMD model.

Key words Duchenne muscular dystrophy, DMD, Canine model, DNA, Viral DNA, RNA, cDNA, Western blot, RT-qPCR, ddPCR

1 Introduction

Duchenne muscular dystrophy (DMD) is a fatal X-linked muscle disease affecting 1 in every 5000 male births. DMD is caused by mutations in the *DMD* gene, one of the largest known human genes. The dystrophin protein is a 427 kilodalton (kDa) subsarcolemmal cytoskeletal protein. Dystrophin stabilizes the sarcolemma during muscle contraction. Lack of dystrophin renders the sarcolemma susceptible to contraction-induced injury.

Gene therapy was proposed as a promising strategy to treat DMD at its genetic root soon after the discovery of the *DMD* gene [1]. After years of studies in murine DMD models, especially the dystrophin-deficient mdx mouse, it becomes clear that adeno-associated virus (AAV)-mediated systemic delivery of micro-dystrophin may significantly ameliorate muscle and heart disease [2, 3]. To bridge the translation between the mouse model and human patients, we and others evaluated AAV micro-dystrophin gene therapy in the canine DMD model. These studies provide

Rika Maruyama and Toshifumi Yokota (eds.), *Muscular Dystrophy Therapeutics: Methods and Protocols*, Methods in Molecular Biology, vol. 2587, https://doi.org/10.1007/978-1-0716-2772-3_15,

critical proof of the principle that a single injection of AAV micro-dystrophin vector can improve muscle function and reduce muscle pathology [4–9]. Built on the success of animal studies, three clinical trials sponsored by three companies were initiated simultaneously in December 2017 and January 2018 [10]. These studies have revealed encouraging signs of moderate disease amelioration [11–13]. Clinical findings suggest that there is still room for improvement. The canine model will undoubtedly be crucial for developing the optimized next-generation systemic AAV micro-dystrophin gene therapy. Besides gene replacement therapy, recently emerged CRISPR editing technology has opened the door to correct the mutations in the *DMD* gene and generate larger, functionally superior dystrophin proteins to treat DMD. Two AAV CRISPR studies have been conducted in the canine DMD model [14, 15]. These studies have revealed the promises and challenges of the CRISPR therapy for DMD. They also call for further development of the CRISPR therapy in the canine model before a human trial is initiated.

Accurate quantification of gene transfer efficiency and dystrophin restoration at the DNA, RNA, and protein levels is important to fully assess gene therapy performed in the canine model. We have developed and validated a series of molecular and biochemical assays suitable for this purpose. In this chapter, we present protocols to extract and quantify DNA, RNA, and protein from canine tissues. We also present protocols to quantify the vector genome and dystrophin transcript. Finally, we present the protocol to quantify dystrophin protein expression in the canine muscle lysate.

2 Materials

2.1 Nucleic Acid and Protein Extraction

2.1.1 DNA Extraction

1. Personal protective equipment (PPE), including a laboratory coat or disposable gown, safety glasses or splash goggles, and gloves.
2. Tris(hydroxymethyl)aminomethane (Tris) (Sigma-Aldrich, St. Louis, MO, USA).
3. Ethylenediaminetetraacetic acid (EDTA) (Sigma-Aldrich, St. Louis, MO, USA).
4. Sodium chloride (NaCl) (Sigma-Aldrich, St. Louis, MO, USA).
5. Sodium dodecyl sulfate (SDS) (Sigma-Aldrich, St. Louis, MO, USA).
6. Proteinase K (Research Products International, Mount Prospect, IL, USA).
7. Screw-top tubes (2 mL) with O-ring cap (ThermoFisher Scientific, Hampton, NH, USA).

8. Rotary hybridization oven (ThermoFisher Scientific, Waltham, MA, USA).
9. Benchtop centrifuge (ThermoFisher Scientific, Waltham, MA, USA).
10. Sample mixer (ThermoFisher Scientific, Waltham, MA, USA).
11. Microcentrifuge tube (ThermoFisher Scientific, Hampton, NH, USA).
12. Isopropyl alcohol (Isopropanol) (Sigma-Aldrich, St. Louis, MO, USA).
13. Ethyl alcohol (ethanol) (Sigma-Aldrich, St. Louis, MO, USA).
14. Incubator (ThermoFisher Scientific, Waltham, MA, USA).

2.1.2 RNA Extraction (Fig. 1)

1. Personal protective equipment (PPE), including a laboratory coat or disposable gown, safety glasses or splash goggles, and gloves.
2. RNeasy fibrous tissue mini kit. The kit includes RNeasy Mini spin column, collection tube, RNase-Free microcentrifuge tube, proteinase K, RLT buffer, RW1 buffer, RPE buffer, RNase-free water, RNase-Free DNase I (lyophilized), and RDD buffer (Qiagen, Hilden, Germany).

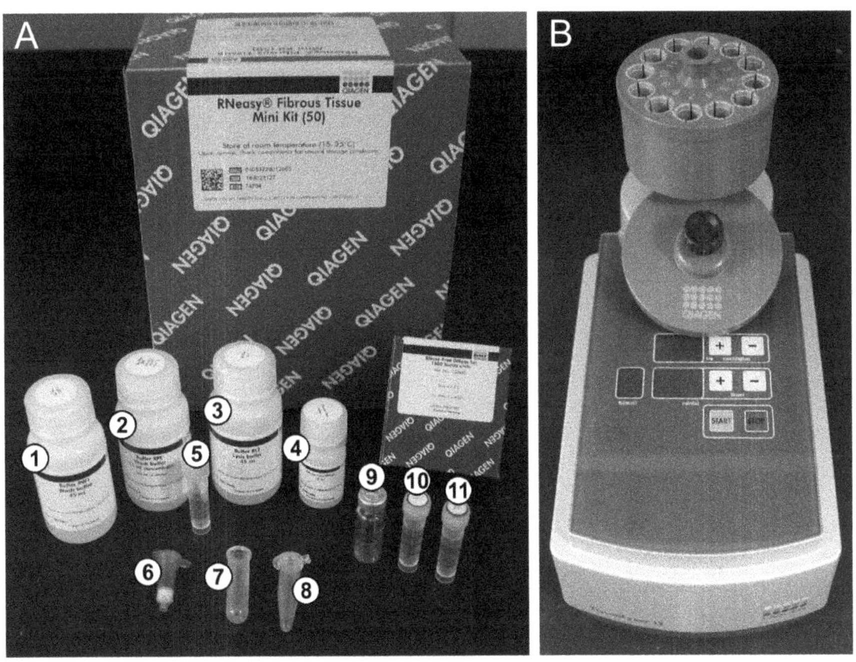

Fig. 1 Materials required for RNA extraction. (**a**) RNeasy fibrous tissue mini kit. The kit consists of (1) RW1 buffer, (2) RPE buffer, (3) RLT buffer, (4) RNase-free water, (5) proteinase K, (6) RNeasy Mini Spin column, (7) collection tube, (8) RNase-Free microcentrifuge tube, (9) RNase-Free DNase I (lyophilized), (10) RDD buffer, and RNase-free water. (**b**) TissueLyser LT

3. Ethyl alcohol (ethanol) (Sigma-Aldrich, St. Louis, MO, USA).
4. Screw-top tubes (2 mL) with O-ring cap (ThermoFisher Scientific, Hampton, NH, USA).
5. TissueLyser LT (Qiagen, Hilden, Germany).
6. Stainless steel beads (5 mm) (Qiagen, Hilden, Germany).
7. Microcentrifuge tubes (ThermoFisher Scientific, Hampton, NH, USA).
8. Water bath (ThermoFisher Scientific, Hampton, NH, USA).

2.1.3 Protein Extraction (Fig. 2)

Bullet Blender Materials

1. Personal protective equipment (PPE), including a laboratory coat or disposable gown, safety glasses or splash goggles, and gloves.
2. Sodium dodecyl sulfate (SDS) (Sigma-Aldrich, St. Louis, MO, USA).
3. Ethylenediaminetetraacetic acid (EDTA) (Sigma-Aldrich, St. Louis, MO, USA).

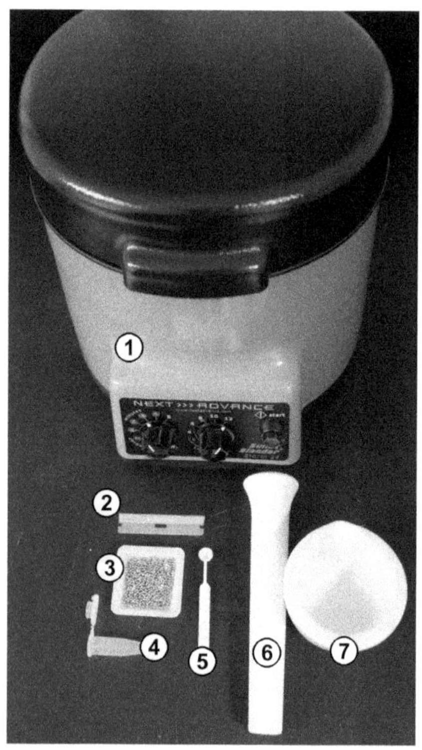

Fig. 2 Materials required for protein extraction. (1) Bullet Blender tissue homogenizer, (2) carbon steel single edge razor blade, (3) homogenization beads, (4) microcentrifuge tubes with safe-lock, (5) microspoon, (6) pestle, and (7) mortar

4. Tris(hydroxymethyl)aminomethane (Tris) (Sigma-Aldrich, St. Louis, MO, USA).
5. Halt™ Protease and Phosphatase inhibitor cocktail (100X) (ThermoFisher Scientific, Hampton, NH, USA).
6. Carbon steel single edge razor blade (Razor Blade Co., Van Nuys, CA, USA).
7. Microcentrifuge tubes with safe-lock (ThermoFisher Scientific, Hampton, NH, USA).
8. Homogenization beads (Next Advance Inc., Troy, NY, USA).
9. Microspoon (Next Advance Inc., Troy, NY, USA).
10. Bullet Blender tissue homogenizer (Next Advance Inc., Troy, NY, USA).
11. Microcentrifuge tubes (ThermoFisher Scientific, Hampton, NH, USA).

Mortar and Pestle Materials

1. Personal protective equipment (PPE), including a laboratory coat or disposable gown, safety glasses or splash goggles, and gloves.
2. Sodium dodecyl sulfate (SDS) (Sigma-Aldrich, St. Louis, MO, USA).
3. Ethylenediaminetetraacetic acid (EDTA) (Sigma-Aldrich, St. Louis, MO, USA).
4. Tris(hydroxymethyl)aminomethane (Tris) (Sigma-Aldrich, St. Louis, MO, USA).
5. Halt™ Protease and Phosphatase inhibitor cocktail (100X) (ThermoFisher Scientific, Hampton, NH, USA).
6. Carbon steel single edge razor blade (Razor Blade Co., Van Nuys, CA, USA).
7. Mortar and pestle (ThermoFisher Scientific, Hampton, NH, USA).
8. Microcentrifuge tubes (ThermoFisher Scientific, Hampton, NH, USA).

2.2 Complementary DNA (cDNA) Generation

1. Personal protective equipment (PPE), including a laboratory coat or disposable gown, safety glasses or splash goggles, and gloves.
2. SuperScript™ IV VILO reverse transcriptase kit with ezDNase™ enzyme. The kit includes SuperScript IV VILO master mix, SuperScript IV VILO master mix "No RT" control, ezDNase enzyme, ezDNase buffer, and nuclease-free water (ThermoFisher Scientific, Hampton, NH, USA).
3. Thermocycler (ThermoFisher Scientific, Hampton, NH, USA).

2.3 Nucleic Acid and Protein Quantification

2.3.1 DNA Quantification

Qubit Fluorometer

1. Personal protective equipment (PPE), including a laboratory coat or disposable gown, safety glasses or splash goggles, and gloves.
2. Qubit™ 1× dsDNA high-sensitivity (HS) Assay Kit. The kit includes Qubit™ 1× dsDNA HS working solution, Qubit™ 1× dsDNA HS standard #1, and Qubit™ 1× dsDNA HS standard #2 (ThermoFisher Scientific, Hampton, NH, USA).
3. Qubit™ assay tubes (ThermoFisher Scientific, Hampton, NH, USA).
4. Vortex mixer (ThermoFisher Scientific, Hampton, NH, USA).
5. Qubit™ Fluorometer (ThermoFisher Scientific, Hampton, NH, USA) (Fig. 3a).

Nanodrop Spectrophotometer

1. NanoDrop™ One/One^{C} Microvolume UV-Vis Spectrophotometer (ThermoFisher Scientific, Hampton, NH, USA) (Fig. 3b).
2. Kimwipes (ThermoFisher Scientific, Hampton, NH, USA).

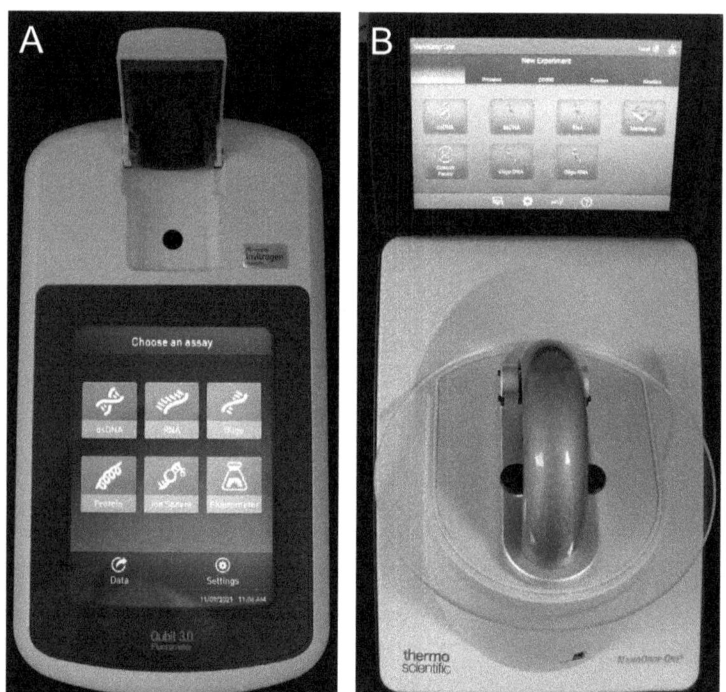

Fig. 3 Equipment used for nucleic acid and protein quantification. (**a**) Qubit™ Fluorometer. (**b**) NanoDrop™ One/One^{C} Microvolume UV-Vis Spectrophotometer

2.3.2 *RNA Quantification*

Qubit Fluorometer

1. Personal protective equipment (PPE), including a laboratory coat or disposable gown, safety glasses or splash goggles, and gloves.
2. Qubit™ RNA high-sensitivity (HS) Assay Kit. The kit includes Qubit™ RNA HS dye, Qubit™ RNA HS buffer, Qubit™ 1× RNA HS standard #1, and Qubit™ 1× RNA HS standard #2 (ThermoFisher Scientific, Hampton, NH, USA).
3. Qubit™ assay tubes (ThermoFisher Scientific, Hampton, NH, USA).
4. Vortex mixer (ThermoFisher Scientific, Hampton, NH, USA).
5. Qubit™ Fluorometer (ThermoFisher Scientific, Hampton, NH, USA) (Fig. 3a).

Nanodrop Spectrophotometer

1. NanoDrop™ One/One^{C} Microvolume UV-Vis Spectrophotometer (ThermoFisher Scientific, Hampton, NH, USA) (Fig. 3b).
2. Kimwipes (ThermoFisher Scientific, Hampton, NH, USA).

2.3.3 *cDNA Quantification*

Qubit Fluorometer

1. Personal protective equipment (PPE), including a laboratory coat or disposable gown, safety glasses or splash goggles, and gloves.
2. Qubit™ ssDNA Assay Kit. The kit includes Qubit™ ssDNA dye, Qubit™ ssDNA dye buffer, Qubit™ 1× ssDNA standard #1, and Qubit™ ssDNA standard #2 (ThermoFisher Scientific, Hampton, NH, USA).
3. Qubit™ assay tubes (ThermoFisher Scientific, Hampton, NH, USA).
4. Vortex mixer (ThermoFisher Scientific, Hampton, NH, USA).
5. Qubit™ Fluorometer (ThermoFisher Scientific, Hampton, NH, USA) (Fig. 3a).

Nanodrop Spectrophotometer

1. NanoDrop™ One/One^{C} Microvolume UV-Vis Spectrophotometer (ThermoFisher Scientific, Hampton, NH, USA) (Fig. 3b).
2. Kimwipes (ThermoFisher Scientific, Hampton, NH, USA).

2.3.4 Protein Quantification

Qubit Fluorometer

1. Personal protective equipment (PPE), including a laboratory coat or disposable gown, safety glasses or splash goggles, and gloves.
2. Qubit™ Fluorometer (ThermoFisher Scientific, Hampton, NH, USA) (Fig. 3a).
3. Qubit™ Protein broad range (BR) assay kit. The kit includes Qubit™ Protein BR dye, Qubit™ Protein BR buffer, Qubit™ Protein BR standard #1, Qubit™ Protein BR standard #2, and Qubit™ Protein BR standard #3 (ThermoFisher Scientific, Hampton, NH, USA).
4. Qubit™ assay tubes (ThermoFisher Scientific, Hampton, NH, USA).
5. Vortex mixer (ThermoFisher Scientific, Hampton, NH, USA).

DC Protein Assay

1. Personal protective equipment (PPE), including a laboratory coat or disposable gown, safety glasses or splash goggles, and gloves.
2. Bovine Serum Albumin (BSA) (purified and lyophilized) (Millipore Sigma, Burlington, MA, USA).
3. Bio-Rad DC Protein Assay Kit™. The kits include an alkaline copper tartrate solution (Reagent A), a dilute Folin reagent (Reagent B), and surfactant solution (Reagent S) (Bio-Rad Laboratories, Inc., Hercules, CA, USA).
4. Microcentrifuge tubes (ThermoFisher Scientific, Hampton, NH, USA).

2.4 Quantitative Real-Time Polymerase Chain Reaction (qPCR) (Fig. 4)

1. QG buffer (Qiagen, Hilden, Germany).
2. QIAquick PCR Purification Kit: The kit includes a QIAquick spin column, PE buffer, EB buffer, collection tubes, and microcentrifuge tubes (Qiagen, Hilden, Germany).
3. Vacuum manifold (ThermoFisher Scientific, Hampton, NH, USA).
4. Benchtop centrifuge (ThermoFisher Scientific, Hampton, NH, USA).
5. Qubit fluorometer (ThermoFisher Scientific, Hampton, NH, USA) (Fig. 3a).
6. NanoDrop™ One/One^{C} Microvolume UV-Vis Spectrophotometer (ThermoFisher Scientific, Hampton, NH, USA) (Fig. 3b).
7. TaqMan™ Universal Master Mix (ThermoFisher Scientific, Hampton, NH, USA).

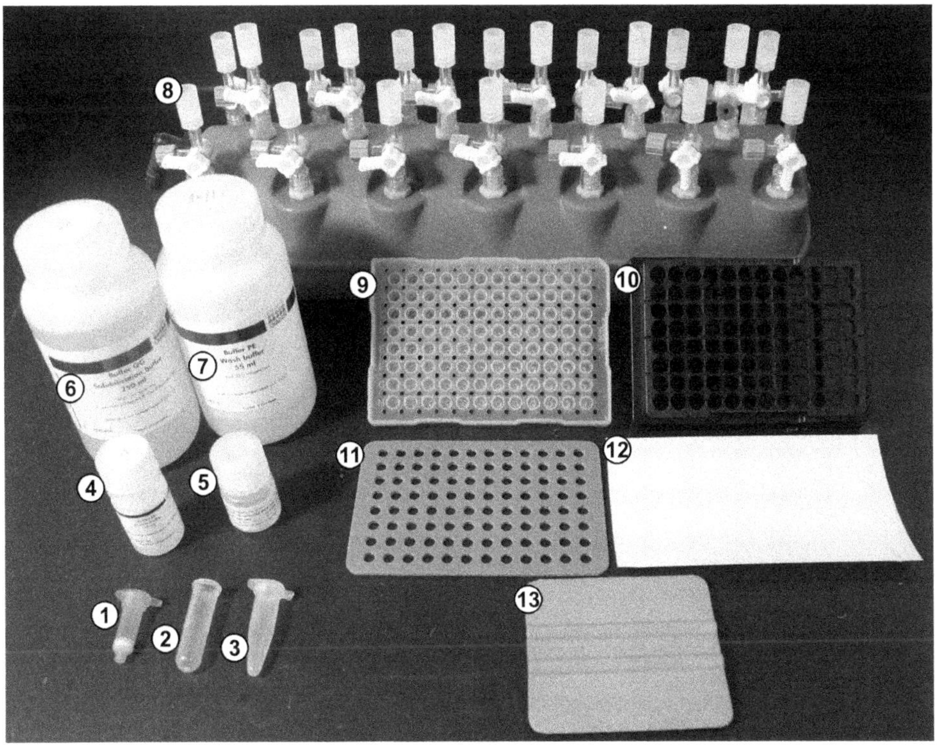

Fig. 4 Materials required for the quantitative real-time polymerase chain reaction. (1) QIAquick column, (2) collection tube, (3) microcentrifuge tube, (4) EB buffer, (5) TaqMan™ Universal Master Mix, (6) QG buffer, (7) PE buffer, (8) vacuum manifold, (9) MicroAmp™ optical 96-well reaction plate, (10) precision plate holder for the MicroAmp™ optical plate, (11) MicroAmp™ optical cover compression pad, (12) MicroAmp™ optical adhesive film, and (13) applicator

8. MicroAmp™ optical 96-well reaction plate (ThermoFisher Scientific, Hampton, NH, USA).
9. Precision plate holder for the MicroAmp™ optical plate (ThermoFisher Scientific, Hampton, NH, USA).
10. MicroAmp™ optical adhesive film kit. The kit includes MicroAmp™ optical adhesive films, applicator, and MicroAmp™ optical cover compression pad (ThermoFisher Scientific, Hampton, NH, USA).
11. SYBR™ green master mix (ThermoFisher Scientific, Hampton, NH, USA).

2.5 Droplet Digital PCR (ddPCR) (Fig. 5)

1. ddPCR Supermix for probe (no dUTP) (Bio-Rad Laboratories, Inc., Hercules, CA, USA).
2. 96-well PCR plate (ThermoFisher Scientific, Hampton, NH, USA).
3. Adhesive film for a 96-well PCR plate (ThermoFisher Scientific, Hampton, NH, USA).

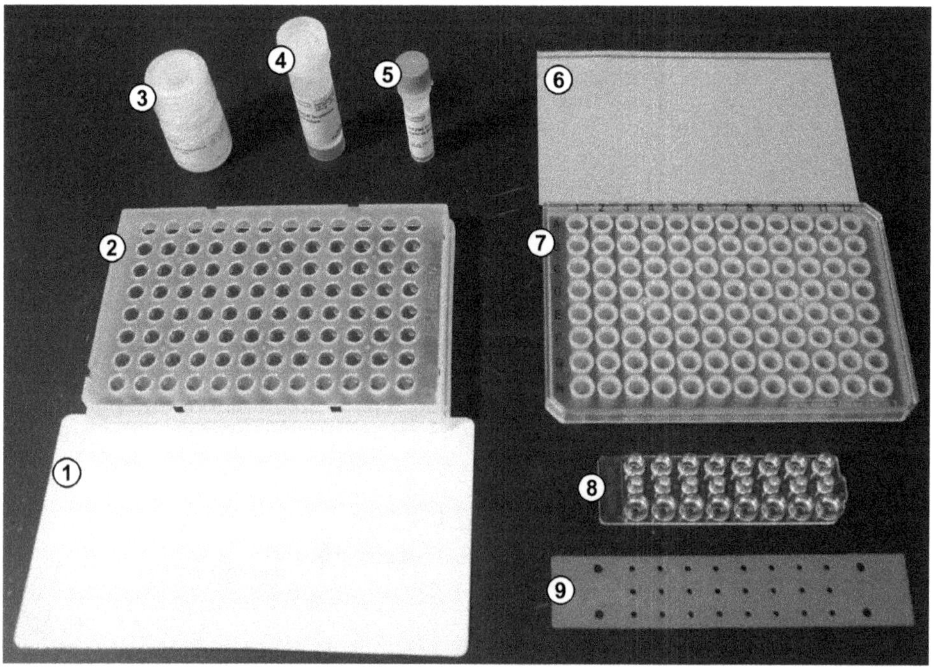

Fig. 5 Materials required for the droplet digital PCR. (1) Adhesive film for a 96-well PCR plate, (2) 96-well PCR plate, (3) droplet generation oil, (4) ddPCR supermix for probe, (5) ddPCR EvaGreen supermix, (6) PCR plate heat seal foil, (7) ddPCR 96-well plate, (8) DG8 cartridge, and (9) DG8 cartridge gasket

4. Cartridge holder (Bio-Rad Laboratories, Inc., Hercules, CA, USA).
5. DG8 cartridges and gaskets (Bio-Rad Laboratories, Inc., Hercules, CA, USA).
6. Droplet Generation Oil (Bio-Rad Laboratories, Inc., Hercules, CA, USA).
7. QX200 Droplet Generator (Bio-Rad Laboratories, Inc., Hercules, CA, USA).
8. ddPCR 96-Well Plates (Bio-Rad Laboratories, Inc., Hercules, CA, USA).
9. PCR plate heat seal foil (Bio-Rad Laboratories, Inc., Hercules, CA, USA).
10. PX1 PCR plate sealer (Bio-Rad Laboratories, Inc., Hercules, CA, USA).
11. C1000 Touch™ thermal cycler (Bio-Rad Laboratories, Inc., Hercules, CA, USA).
12. QX200 Droplet reader (Bio-Rad Laboratories, Inc., Hercules, CA, USA).

13. QuantaSoft™ Analysis Pro software version 1.0 (Bio-Rad Laboratories, Inc., Hercules, CA, USA).
14. ddPCR EvaGreen supermix (Bio-Rad Laboratories, Inc., Hercules, CA, USA).

2.6 Western Blot

1. 2X Laemmli sample buffer (Bio-Rad Laboratories, Inc., Hercules, CA, USA).
2. β-mercaptoethanol (Sigma-Aldrich, St. Louis, MO, USA).
3. Dry block heater (ThermoFisher Scientific, Hampton, NH, USA).
4. Benchtop centrifuge (ThermoFisher Scientific, Hampton, NH, USA).
5. Tris(hydroxymethyl)aminomethane (Tris–base) (Sigma-Aldrich, St. Louis, MO, USA).
6. Glycine (Sigma-Aldrich, St. Louis, MO, USA).
7. Sodium dodecyl sulfate (SDS) (Sigma-Aldrich, St. Louis, MO, USA).
8. Methyl alcohol (methanol) (Sigma-Aldrich, St. Louis, MO, USA).
9. Sodium chloride (NaCl) (Sigma-Aldrich, St. Louis, MO, USA).
10. Hydrochloric acid (HCl) (Sigma-Aldrich, St. Louis, MO, USA).
11. TWEEN® 20 (Sigma-Aldrich, St. Louis, MO, USA).
12. Mini-PROTEAN tetra cell gel electrophoresis system. The system includes a spacer plate, short plate, casting frame, casting stand, comb, buffer dam, electrophoresis module, tank, and lid (Bio-Rad Laboratories, Inc., Hercules, CA, USA) (Figs. 6a and 7a).
13. Ethyl alcohol (Ethanol) (Sigma-Aldrich, St. Louis, MO, USA).
14. Kimwipes™ (ThermoFisher Scientific, Hampton, NH, USA).
15. Glass flask (ThermoFisher Scientific, Hampton, NH, USA).
16. 40% acrylamide and bis-acrylamide solution, 29:1 (Bio-Rad Laboratories, Inc., Hercules, CA, USA).
17. Ammonium persulfate (APS) (Bio-Rad Laboratories, Inc., Hercules, CA, USA).
18. *N*, *N*, *N*, *N*′-tetramethyl-ethylenediamine (TEMED) (Bio-Rad Laboratories, Inc., Hercules, CA, USA).
19. Isopropyl alcohol (isopropanol) (Bio-Rad Laboratories, Inc., Hercules, CA, USA).

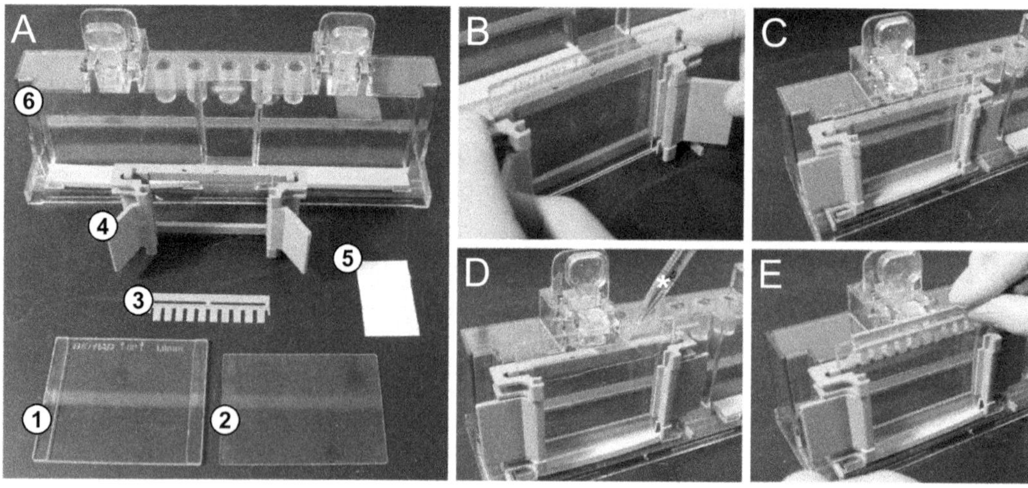

Fig. 6 Gel preparation for western blot. (**a**) Materials required for preparing the western blot gel. (1) Space plate, (2) short plate, (3) comb, (4) casting frame, (5) filter paper, and (6) casting stand. (**b**) A photograph showing the assembly of the glass cassette sandwich into the casting frame. (**c**) A photograph showing the attachment of the casting frame into the casting stand. (**d**) A photograph showing the filling the glass cassette sandwich with gel solution using a transfer pipette. * transfer pipette. (**e**) A photograph showing the seating of the comb into the stacking gel solution

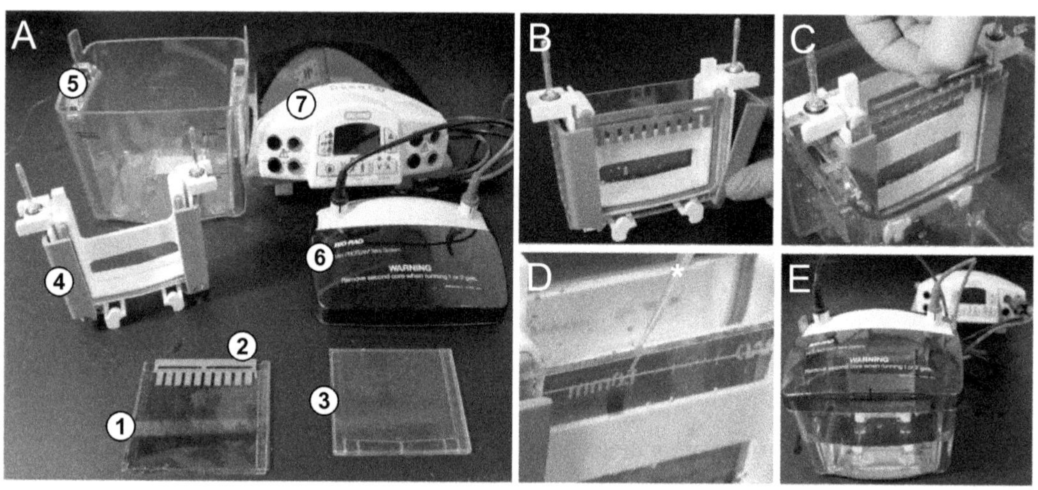

Fig. 7 Gel electrophoresis for western blot. (**a**) Materials required for gel electrophoresis. (1) Gel cassette sandwich, (2) comb, (3) buffer dam, (4) electrophoresis module, (5) tank, (6) lid, and (7) PowerPac universal power supply. (**b**) A photograph showing the assembly of the electrophoresis module. (**c**) A photograph showing the removal of the comb from the gel cassette sandwich. (**d**) A photograph showing the loading of a sample in one of the gel lanes using an elongated gel loading pipette tip. * elongated pipette tip. (**e**) A phototgraph showing the final assembly of the Mini-PROTEAN Tetra cell gel electrophoresis unit

20. Precision Plus Protein™ Kaleidoscope™ Prestained Protein (Bio-Rad Laboratories, Inc., Hercules, CA, USA).
21. PowerPac universal power supply (Bio-Rad Laboratories, Inc., Hercules, CA, USA) (Fig. 7a).
22. Trans-Blot® Turbo™ system (Bio-Rad Laboratories, Inc., Hercules, CA, USA) (Fig. 8B.a).
23. Trans-Blot® Turbo™ transfer pack. The pack includes ion reservoir stacks and a blotting membrane (Bio-Rad Laboratories, Inc., Hercules, CA, USA) (Fig. 8B.a).
24. Mini Trans-Blot cell blotting system. The system includes gel holder cassettes, fiber pads, electrode module, tank, and lid (Bio-Rad Laboratories, Inc., Hercules, CA, USA) (Fig. 8C.a).
25. Filter paper (ThermoFisher Scientific, Hampton, NH, USA) (Fig. 8C.a).
26. Blotting membrane (Bio-Rad Laboratories, Inc., Hercules, CA, USA) (Fig. 8C.a).
27. Intercept® Blocking Buffer (LI-COR Biosciences, Lincoln, NE, USA).
28. Orbital shaker (ThermoFisher Scientific, Hampton, NH, USA).
29. HulaMixer™ Sample Mixer (ThermoFisher Scientific, Hampton, NH, USA).
30. Pierce™ enhanced chemiluminescence (ECL) western blotting substrate reagent kit. The kit contains detection reagent 1 (Peroxide solution) and detection reagent 2 (Luminol enhancer solution) (ThermoFisher Scientific, Hampton, NH, USA).

3 Methods

3.1 Nucleic Acid and Protein Extraction

The following protocol describes the genomic DNA extraction from canine tissues using a salt precipitation approach (*see* **Note 1**).

3.1.1 DNA Extraction

1. Prepare 50 mL of the lysis stock buffer by mixing the following chemicals: 2.5 mL of 1 M Tris (pH 8.0), 10 mL of 0.5 M EDTA, 1 mL of 5 M NaCl, 5 mL of 10% SDS, and 29 mL of sterile double-distilled water (ddH_2O) (*see* **Note 2**).
2. Prepare the working lysis buffer by adding 0.5 mg of the Proteinase K for each mL of stock lysis buffer needed according to the total number of processed samples (*see* **Note 3**).
3. Obtain 50–100 mg of a tissue sample and place it in a clean 2 mL screw-top tube (*see* **Note 4**).
4. Cover the tissue with 750 μL of working lysis buffer and screw the tube cap firmly. Incubate the sample overnight in a 55 °C rotary hybridization oven.

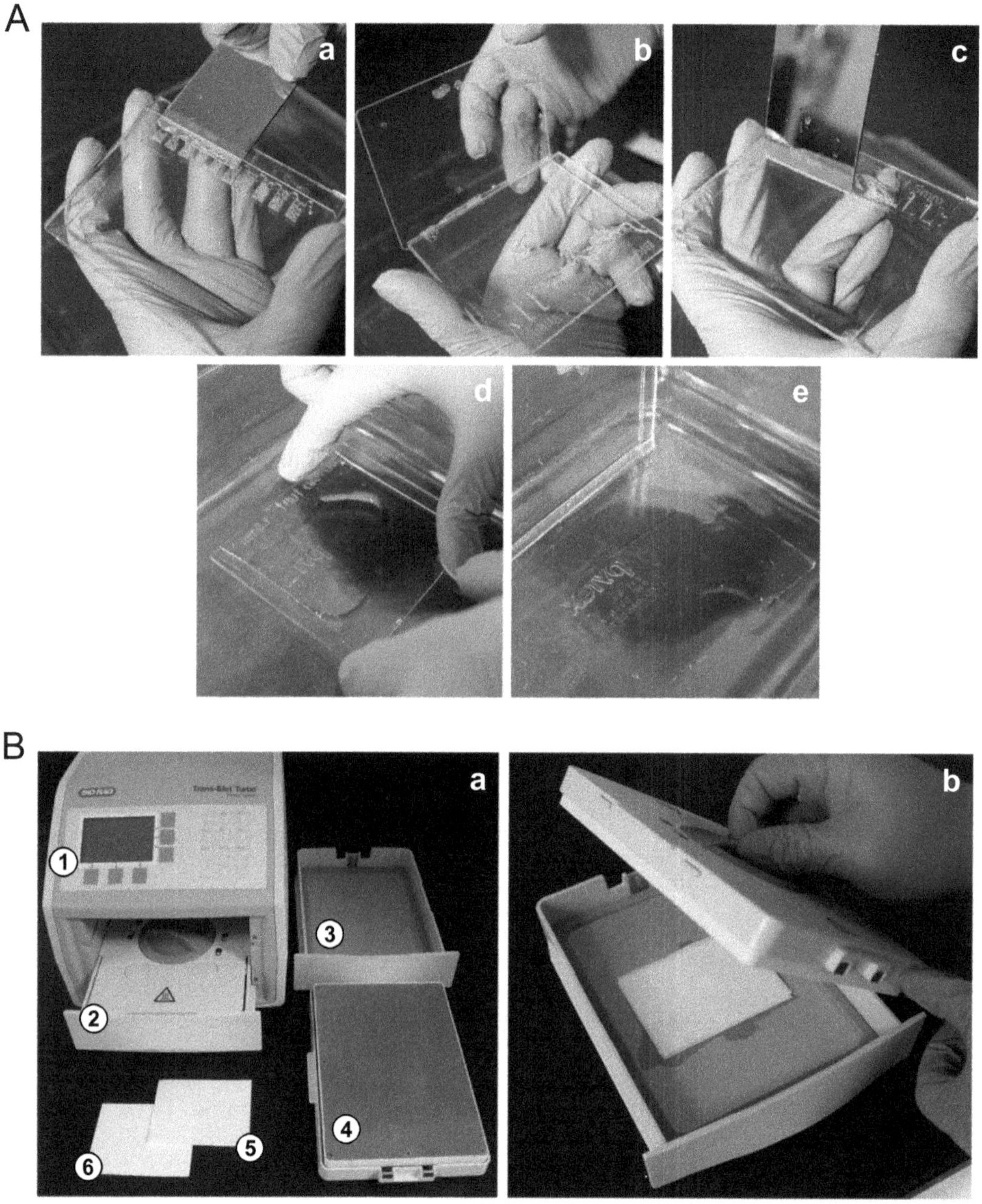

Fig. 8 Protein transfer/blotting for western blot. (**A**) Step-by-step to separate the gel from the gel cassette. (a) Separate the short plate from the cassette using a spatula. (b) Remove the short plate completely from the cassette. (c) Trim and discard the gel lanes from the gel. (d) Submerge the spacer plate with the gel in 1X transfer buffer. (e) Allow the gel to float off the plate into the transfer buffer. (**B**) Protein transfer using the Trans-Blot® Turbo™ system. (a) Materials required for Trans-Blot® Turbo™ system. (1) Trans-Blot® TurboTM system, (2) assembled cassette, (3) bottom portion of the cassette (anode(+)), (4) inside view of the top portion of the cassette (cathode (−)), (5) blotting membrane, and (6) ion reservoir stack. (b) Assembled Trans-Blot® Turbo™ transfer pack in the cassette. (**C**) Protein transfer using the Mini Trans-Blot cell blotting system. (a) Materials required for Mini Trans-Blot cell blotting system. (1) Tank, (2) PowerPac universal power supply, (3) electrode module, (4) lid, (5) gel holder cassette, (6) fiber pad, (7) filter paper, and (8) blotting membrane. (b) A photograph showing the assembly of the gel sandwich in the gel holder cassette. (c) A photograph showing the assembly of the gel holder cassette. (d) A photograph showing the placement of the gel holder cassette in the electrode module. (e) A photograph showing the final assembly of the Mini Trans-Blot cell blotting unit. * stir bar

C

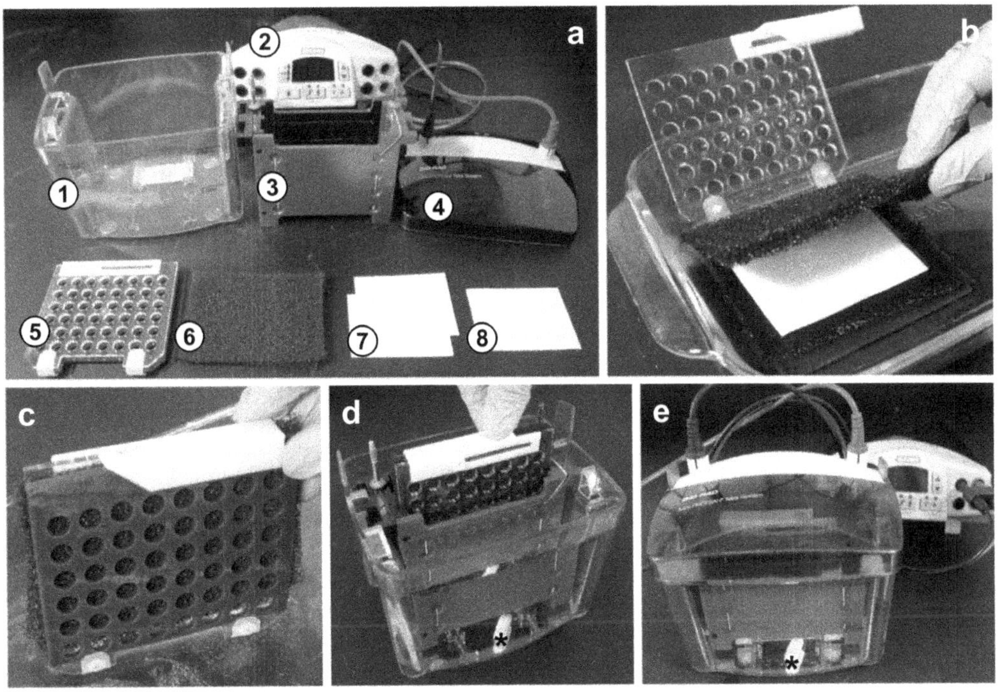

Fig. 8 (continued)

5. Spin the tube briefly to collect the solution at the bottom. Add 250 μL of 6 M NaCl and incubate the sample on the sample mixer for 1 h at room temperature.
6. Centrifuge the tube at max speed (~14,000 RPM) for 30 min at room temperature.
7. Transfer 850 μL of supernatant to a sterile screw-top tube and add 600 μL of isopropanol. Mix for 5 min by inverting the tube.
8. Centrifuge the tube at max speed (~14,000 RPM) for 5 min at room temperature.
9. Remove supernatant without disturbing the DNA pellet and add 1 mL of 70% ethanol to wash the pellet. Vortex gently to suspend the DNA pellet in ethanol. Centrifuge the tube at max speed (~14,000 RPM) for 2 min at room temperature.
10. Repeat **step 9** to wash the DNA pellet one more time.
11. Remove the supernatant without disturbing the DNA pellet. Allow the DNA pellet to air dry in a 37 °C incubator for 30 min.
12. Resuspend the pellet in the desired amount of DNAse-free water. Store DNA in a screw cap tube at –80 °C for long-term storage.

3.1.2 RNA Extraction

The following protocol describes the RNA extraction from canine tissues using the RNeasy fibrous tissue mini kit from Qiagen (*see* **Note 5**).

1. Prepare the RNeasy fibrous tissue mini kit reagents by adding (1) 10 μL of β-mercaptoethanol for each mL of RTL buffer, (2) 44 mL of 100% ethanol to the RPE buffer, and (3) 550 μL of RNase-free water to the lyophilized DNase I (*see* **Note 6**).
2. Obtain 20–30 mg of a tissue sample and place it in a clean 2 mL screw-top tube (*see* **Note 7**).
3. Place one 5 mm sterilized stainless steel bead into the tube and then add 300 μL of Buffer RLT. Screw the tube cap firmly and homogenize the tissue using the TissueLyser LT for 2 min at 50 Hz (*see* **Note 8**) (Fig. 1b).
4. Transfer the tissue lysate into a clean microcentrifuge tube. Add 590 μL of RNase-free H_2O and 10 μL of Proteinase K to the lysate. Mix thoroughly by pipetting and tube inversion (do not vortex). Incubate the tissue lysate in a 55 °C water bath for 10 min.
5. Centrifuge the tube at 10,000 rpm for 3 min at room temperature. Transfer the supernatant to a clean microcentrifuge tube. Add 450 μL of ethanol to the supernatant and mix by pipetting.
6. Transfer 700 μL of the sample to the RNeasy Mini spin column placed in a 2 mL collection tube. Centrifuge at 10,000 rpm for 15 s at room temperature. Discard the flow-through.
7. Repeat **step 6** to pass the remaining sample through the RNeasy Mini spin column.
8. Add 300 μL of RW1 buffer to the RNeasy Mini spin column. Centrifuge at 10,000 rpm for 15 s at room temperature and discard the flow-through.
9. Add 80 μL of the DNAse I working solution to the RNeasy Mini spin column and incubate for 15 min at room temperature (*see* **Note 9**).
10. Add 350 μL of RW1 buffer to the RNeasy Mini spin column. Centrifuge at 10,000 rpm for 15 s at room temperature and discard the flow-through.
11. Add 500 μL of RPE buffer to the RNeasy Mini spin column. Centrifuge at 10,000 rpm for 15 s at room temperature and discard the flow-through.
12. Add 500 μL of RPE buffer to the RNeasy Mini spin column. Centrifuge at 10,000 rpm for 2 min at room temperature and replace the collection tube with a new one.

13. Centrifuge at maximum speed for 1 min at room temperature and discard the collection tube. Place the RNeasy Mini spin column into an RNase-Free microcentrifuge tube.
14. Add 40 μL of RNase-free water directly to the center of the RNeasy Mini spin column. Centrifuge at 10,000 RPM for 1 min to elute RNA. Store RNA at −80 °C for long-term storage.

3.1.3 Protein Extraction

The following protocol describes the protein extraction from canine tissue using two methods: the bullet blender and the mortar and pestle (*see* **Note 10**).

Bullet Blender Method

1. Prepare the homogenizing buffer stock by mixing 10 g of SDS, 1 mL of 0.5 M EDTA, 12.5 mL of 0.5 M Tris together and bringing the final volume to 100 mL using distilled water. Store at room temperature for up to 3 months.
2. Add 10 μL of 100X Halt™ Protease and Phosphatase inhibitor cocktail per 990 μL of the homogenizing buffer stock to make the working homogenizing buffer.
3. Obtain 30–40 mg of tissue and cut it into small cubic pieces using a sterilized razor blade. Place the tissue pieces in a clean microcentrifuge tube with a safe lock feature (*see* **Note 11**).
4. Add 600–800 μL (20× the sample weight) of working homogenizing buffer to the sample. Add a scoop of the homogenization beads to the sample and lock the tube firmly.
5. Place the tube in the Bullet Blender tissue homogenizer and homogenize the sample for 5 min at max speed at 4 °C.
6. Centrifuge the tube at max speed for 1 min to collect the lysate at the bottom of the tube.
7. Transfer the supernatant to a clean microcentrifuge tube and centrifuge at maximum speed for 4 min at room temperature.
8. Transfer the supernatant to a clean microcentrifuge tube and store it at −80 °C for long-term storage (*see* **Note 12**).

Mortar and Pestle Method

1. Prepare the homogenizing buffer stock by mixing 10 g of SDS, 1 mL of 0.5 M EDTA, and 12.5 mL of 0.5 M Tris together and bringing the final volume to 100 mL using deionized water. Store at room temperature for up to 3 months.
2. Add 10 μL of 100X Halt™ Protease and Phosphatase inhibitor cocktail per 990 μL of the homogenizing buffer stock to make the working homogenizing buffer.

3. Obtain 40–50 mg of tissue and cut it into small cubic pieces using a sterilized razor blade (*see* **Note 11**).
4. Cool the mortar and the pestle with liquid nitrogen and place the tissue pieces in the cooled mortar. Grind the tissue using the cooled pestle until the tissue becomes a fine powder. Wait for 5 min to allow the mortar and pestle to reach room temperature. Add 800–1000 μL (20× the sample weight) of the working homogenizing buffer to the ground sample. Grind the sample in the working homogenizing buffer using the pestle.
5. Transfer the homogenized tissue to a clean microcentrifuge tube and vortex the tube for 30 s. Centrifuge the tube at max speed for 10 min at room temperature.
6. Transfer the supernatant to a clean microcentrifuge tube and store it at −80 °C for long-term storage (*see* **Note 12**).

3.2 Complementary DNA (cDNA) Generation

The following protocol describes the generation of complementary DNA (cDNA) from RNA extracted from canine tissue using the SuperScript™ IV VILO reverse transcriptase kit (*see* **Note 13**).

1. Mix 1 μL of ezDNase enzyme and 1 μL of 10X ezDNAse buffer with 1–8 μL of the RNA sample. Bring the final volume to 10 μL using RNase-free water (*see* **Note 14**).
2. Gently mix the content and incubate for 2 min at 37 °C in a thermocycler.
3. Place the sample on ice for 2 min and spin briefly.
4. Add 4 μL of the SuperScript™ IV VILO master mix and 6 μL of nuclease-free water to the DNase-treated sample. Gently mix the content and incubate in a thermocycler using the following thermal cycling parameters: 10 min at 25 °C, 10 min at 50 °C, and 5 min at 85 °C (*see* **Note 15**).
5. Store the cDNA at −80 °C for long-term storage.

3.3 Nucleic Acid and Protein Quantification

3.3.1 DNA Quantification

The following protocol describes the measurement of double-stranded DNA (dsDNA) concentration using the Qubit™ fluorometer or the NanoDrop™ Spectrophotometer (*see* **Note 16**).

Qubit Fluorometer

1. Add 10 μL of each Qubit™ standard into a separate Qubit™ assay tube (*see* **Note 17**).
2. Add 1–20 μL of the DNA sample into a Qubit™ assay tube (*see* **Note 18**).
3. Add the Qubit™ 1× dsDNA HS working solution to the standards and DNA sample to a final volume of 200 μL (*see* **Note 19**).

4. Vortex each tube for 3–5 s and incubate for 2 min at room temperature while protecting it from direct light.
5. Determine the DNA quantity using the Qubit fluorometer according to the manufacturer's instructions (*see* **Note 20**) (Fig. 3a).

Nanodrop Spectrophotometer

1. Lift the instrument arm and wipe the upper and lower pedestals with a clean Kimwipe wet with deionized water (*see* **Note 21**).
2. From the home screen, select "dsDNA" (Fig. 3b).
3. Add 1–3 μL of blanking solution onto the lower pedestal and carefully lower the instrument arm. Select "blank" on the home screen and wait for the machine to complete the blank measurement (*see* **Note 22**).
4. Clean both pedestals with a clean Kimwipe wet with deionized water, and add 1–3 μL of the dsDNA sample to the lower pedestal and lower the instrument arm. Select "Measure" on the home screen and wait for the machine to complete the measurement (*see* **Note 23**).
5. Repeat **step 4** to measure more samples. Otherwise, select "End experiment" and clean both pedestals with a clean Kimwipe wet with deionized water.

3.3.2 RNA Quantification

The following protocol describes the measurement of the RNA concentration using the Qubit™ fluorometer or the NanoDrop™ Spectrophotometer (*see* **Note 24**).

Qubit Fluorometer

1. Prepare the Qubit™ working solution by diluting the Qubit™ RNA HS dye 1:200 in Qubit™ RNA HS buffer (*see* **Note 25**).
2. Add 10 μL of each Qubit™ standard into a separate Qubit™ assay tube (*see* **Note 26**).
3. Add 1–20 μL of the RNA sample into a Qubit™ assay tube (*see* **Note 27**).
4. Add the Qubit™ working solution to the standards and RNA sample to a final volume of 200 μL (*see* **Note 28**).
5. Vortex each tube for 3–5 s and incubate for 2 min at room temperature while protecting it from direct light.
6. Determine the RNA quantity using the Qubit fluorometer according to the manufacturer's instructions (*see* **Note 29**) (Fig. 3a).

Nanodrop Spectrophotometer

1. Lift the instrument arm and wipe the upper and lower pedestals with a clean Kimwipe wet with deionized water (*see* **Note 30**).
2. From the home screen, select "RNA" (Fig. 3b).
3. Add 1–3 μL of blanking solution onto the lower pedestal and carefully lower the instrument arm. Select "blank" on the home screen and wait for the machine to complete the blank measurement (*see* **Note 31**).
4. Clean both pedestals with clean Kimwipe wet with deionized water and add 1–3 μL of the RNA sample to the lower pedestal and lower the instrument arm. Select "Measure" on the home screen and wait for the machine to complete the blank measurement (*see* **Note 32**).
5. Repeat **step 4** to measure more samples. Otherwise, select "End experiment" and clean both pedestals with a clean Kimwipe wet with deionized water.

3.3.3 cDNA Quantification

The following protocol describes the measurement of the cDNA concentration using the Qubit™ fluorometer or the NanoDrop™ Spectrophotometer (*see* **Note 33**).

Qubit Fluorometer

1. Prepare the Qubit™ working solution by diluting the Qubit™ ssDNA dye 1:200 in Qubit™ ssDNA buffer (*see* **Note 34**).
2. Add 10 μL of each Qubit™ standard into a separate Qubit™ assay tube (*see* **Note 35**).
3. Add 1–20 μL of the cDNA sample into a Qubit™ assay tube (*see* **Note 36**).
4. Add the Qubit™ working solution to the standards and cDNA sample to a final volume of 200 μL (*see* **Note 37**).
5. Vortex each tube for 3–5 s and incubate for 2 min at room temperature while protecting it from direct light.
6. Determine the cDNA concentration using the Qubit fluorometer according to the manufacturer's instructions (*see* **Note 38**) (Fig. 3a).

Nanodrop Spectrophotometer

1. Lift the instrument arm and wipe the upper and lower pedestals with a clean Kimwipe wet with deionized water (*see* **Note 39**).
2. From the home screen, select "ssDNA" (Fig. 3b).
3. Add 1–3 μL of blanking solution onto the lower pedestal and carefully lower the instrument arm. Select "blank" on the home screen and wait for the machine to complete the blank measurement (*see* **Note 40**).

4. Clean both pedestals with clean Kimwipe wet with deionized water, and add 1–3 μL of the cDNA sample to the lower pedestal and lower the instrument arm. Select "Measure" on the home screen and wait for the machine to complete the blank measurement (*see* **Note 41**).
5. Repeat **step 4** to measure more samples. Otherwise, select "End experiment" and clean both pedestals with a clean Kimwipe wet with deionized water.

3.3.4 Protein Quantification

The following protocol describes the measurement of protein concentration using the Qubit™ fluorometer, or the colorimetric method using the Bio-Rad DC Protein Assay Kit™.

Qubit Fluorometer (See Note 42)

1. Prepare the Qubit™ protein working solution by diluting the Qubit™ Protein BR dye 1:200 in Qubit™ Protein BR buffer (*see* **Note 43**).
2. Add 10 μL of each Qubit™ Protein standard (#1–3) into a separate Qubit™ assay tube (*see* **Note 44**).
3. Add 1–20 μL of the sample into a Qubit™ assay tube (*see* **Note 45**).
4. Add the Qubit™ working solution to the standards and samples to make a final volume of 200 μL (*see* **Note 46**).
5. Vortex each tube for 3–5 s and incubate for 15 min at room temperature while protecting it from direct light.
6. Determine the protein concentration in the sample using the Qubit fluorometer according to the manufacturer's instructions (*see* **Note 47**) (Fig. 3a).

DC Protein Assay (See Note 48)

1. Standard preparation (*see* **Note 49**):
   (a) Prepare a 10 mg/ml standard stock by dissolving 10.0 mg of BSA in 1.0 mL of protein homogenization buffer used in protein extraction (*see* **Note 50**).
   (b) Dilute the BSA stock standard to achieve the following dilutions: 10, 5, 2.5, 1.25, 1, 0.6, 0.4, and 0.2 mg/ml (*see* **Note 50**).
2. Prepare the working reagent by adding 20 μL of DC Protein Assay™ Reagent S for each 1 mL of DC Protein Assay™ Reagent A needed for the assay (*see* **Note 51**).
3. Add 20 μL of each standard, sample, and homogenization buffer (blank) in a clean microcentrifuge tube (*see* **Note 52**).
4. Add 100 μL of the working reagent to each standard, sample, and blank. Vortex each tube for 1–2 s at room temperature.

5. Add 800 μL of reagent B to each standard, sample, and blank. Vortex each tube for 1–2 s at room temperature. Incubate the tubes for 15 min at room temperature while protected from light (*see* **Note 53**).
6. Set the absorbance on a spectrophotometer or microplate reader at 650–750 nm absorbance. Use the blank to calibrate the measuring instrument (*see* **Note 54**).
7. Measure each standard and sample, and then record the absorbance value.
8. Plot a standard curve using a quadratic regression curve based on the absorbance and concentrations of the standards used in the assay (*see* **Note 55**).
9. To determine unknown sample concentrations, solve for the y-value of the curve's quadratic function using the sample absorbances as the x-value (*see* **Note 56**).

3.4 Quantitative Real-Time Polymerase Chain Reaction (qPCR)

The real-time polymerase chain reaction (PCR) protocol described here quantifies the absolute copy number of a known target sequence from a genomic or complementary DNA (cDNA) sample. The quantification is determined based on a standard curve method. The standard curve is made from a plasmid containing the target sequence and serially diluted to a specific copy number. The amplification of the target sequence in a sample or standard is achieved using either TaqMan or SYBR Green real-time PCR approach. In the TaqMan approach, a target-specific probe and primers are used to amplify the target sequence, whereas target-specific primers are only used in the SYBR Green approach (*see* **Note 57**).

1. *Standard Preparation.*
   (a) Plasmid linearization
      (i) Mix 100 μg of the circular plasmid with the restriction enzyme of interest and its buffer overnight at the designated digestion temperature (*see* **Note 58**).
      (ii) Add 500 μL of QG buffer to the digestion mix and vortex for 2–3 s (*see* **Note 59**).
      (iii) Attach a QIAquick column to a vacuum manifold, add 500 μL of QG buffer to the column, and apply vacuum (*see* **Note 60**) (Fig. 4).
      (iv) Add the digested sample to the QIAquick column and apply vacuum (*see* **Note 61**).
      (v) Add 750 μL of PE buffer to the QIAquick column and wait for 5 min and then apply vacuum.
      (vi) Add 750 μL of PE buffer to the QIAquick column and apply vacuum.

(vii) Transfer the QIAquick column into a collection tube and centrifuge for 1 min at maximum speed (~14,000 RPM).

(viii) Transfer the QIAquick column into a clean microcentrifuge tube. Add 40 μL of DNAse free water or EB buffer to the center of the QIAquick column membrane and wait for 2 min. Centrifuge for 1 min at maximum speed (~14,000 RPM) (*see* **Note 62**).

(ix) Confirm the plasmid linearization through gel electrophoresis (*see* **Note 63**).

(b) Plasmid dilution

(i) Determine the linearized plasmid concentration using the Qubit fluorometer or the Nanodrop spectrophotometer method sections.

(ii) Calculate the plasmid molecular weight (MW) using the following equation: [(number of base pairs (bp) in the plasmid $\times$ 607.4) + 157.9] (*see* **Note 64**).

(iii) Calculate the number of plasmid copies per μL using the following equation: [((Plasmid concentration determined in **step i** (ng/μl) $\times$ 10^{-9})/MW determined in **step ii**} $\times$ 6.022 $\times$ 10^{23}) $\times$ 2] (*see* **Note 65**).

(iv) Make a standard stock by diluting the linearized plasmid to 1 $\times$ 10^9 copy/μl according to the number of copies/μl determined in **step iii** (*see* **Note 66**).

(v) Perform tenfold serial dilutions of the standard stock to make a qPCR standard ranging from 1 $\times$ 10^9 to 1 $\times$ 10^2 copy/μl in an 8-well PCR tube strip.

(vi) Store the standard at 4 °C for 1 week or −20 °C for 1 month.

2. *Sample Preparation*

(a) Determine the concentration of the target sample using the Qubit fluorometer or the Nanodrop spectrophotometer method sections.

(b) Dilute the sample to the preferred concentration for the real-time PCR reaction (*see* **Note 67**).

3. *Real-Time PCR Setup*

TaqMan Approach

(a) Prepare the master mix by mixing the following: 10 μL of the TaqMan™ Universal Master Mix, 1 μL of the 20X TaqMan primer and probe mix, and 6 μL of DNAse-free water for each reaction (*see* **Note 68**).

(b) Place the MicroAmp™ optical 96-well reaction plate in a plate holder and place them on ice (*see* **Note 69**).

(c) Add 17 μL of the master mix to the designated well for each standard, sample(s), and NTC.

(d) Add 3 μL of standard, sample, or water (for NTC) to the designated well in the plate (*see* **Note 70**).

(e) Cover all wells in the plate with MicroAmp™ optical adhesive film and press it firmly using the applicator to seal the wells. Place the MicroAmp™ optical cover compression pad on top of the film and align it with the wells.

(f) Set up the real-time PCR machine to run the following thermal cycling parameters: 10 min at 95 °C, followed by 40 cycles: 15 s at 95 °C and 1 min at 60 °C.

(g) Load the MicroAmp™ optical 96-well reaction plate into the real-time PCR machine and start the PCR run.

SYBR Green Approach

(a) Prepare the master mix by mixing the following: 10 μL of the 2X SYBR™ green master mix, 1 μL of each primer, and 5 μL of DNAse-free water for each reaction (*see* **Note 71**).

(b) Place the MicroAmp™ optical 96-well reaction plate in a plate holder and place them on ice.

(c) Add 17 μL of the master mix to each standard and sample well.

(d) Add 3 μL of standard or sample to the designated well.

(e) Cover all wells in the plate with MicroAmp™ optical adhesive film and press firmly using the applicator to seal the wells. Place the MicroAmp™ optical cover compression pad on top of the film and align it with the wells.

(f) Set up the real-time PCR machine to run the following thermal cycling parameters: 10 min at 95 °C, then 40 cycles of 15 s at 95 °C and 1 min at 60 °C, followed by a melt curve cycle of 15 s at 95 °C, 1 min at 60 °C, and 15 s at 95 °C (*see* **Note 72**).

(g) Load the MicroAmp™ optical 96-well reaction plate into the real-time PCR machine and start the run.

4. *Real-Time PCR Data Analysis*

(a) Analyze the data according to the manufacturer's guideline to determine the absolute target copy number in each sample (*see* **Note 73**).

(b) Convert each sample's threshold cycle (Ct) value to copy number using the known amount of copies in the plasmid standard.

(c) To calculate the copy number/ng, divide the total sample copy number by the total amount (ng) of the sample added to the reaction.

(d) To calculate the copy number/diploid cell, first, determine the amount of diploid genome in 1 ng of genomic DNA using the following equation: {1 ng/((C value of the species × 2)/1000). Then, divide the total copy number of the sample target sequence by the total amount of the sample (ng) added to the reaction (*see* **Note 74**).

3.5 Droplet Digital PCR (ddPCR)

The Droplet Digital Polymerase Chain Reaction (ddPCR) protocol described here quantifies the absolute copy number of a known target sequence from a genomic or complementary DNA (cDNA) sample. The amplification of the target sequence is achieved using target-specific TaqMan probe and primers (TaqMan approach) or only target-specific primers (EvaGreen approach) (*see* **Note 75**).

1. *Sample Preparation*

   (a) Determine the concentration of the target sample using the Qubit fluorometer or the Nanodrop spectrophotometer method sections.

   (b) Dilute the sample to the preferred concentration for the ddPCR reaction (*see* **Note 76**).

2. *ddPCR Setup*

TaqMan Approach

   (a) Prepare the master mix by mixing the following: 11 μL of the ddPCR supermix for probes, 1.1 μL of the 20X TaqMan primer and probe mix, and 6.9 μL of DNAse-free water for each reaction (*see* **Note 77**).

   (b) Place a 96-well PCR plate in a plate holder and add 19 μL of the master mix in the designated well for the sample(s), NTC, and positive control (*see* **Note 78**).

   (c) Add 3 μL of each sample, positive control, or water (for NTC) to the designated well in the plate (*see* **Note 79**).

   (d) Cover all wells in the plate with adhesive film and press it firmly using the applicator to seal the wells. Vortex the plate to mix and spin it at max speed for 30 s.

   (e) Generate the ddPCR droplets using the QX200 droplet generator. First, place the DG8 cartridge into the cartridge holder and add 20 μL of each sample, positive control, or NTC, and 70 μL of the droplet generation oil into the designated wells in the cartridge. Then, cover the cartridge with the gasket, place it in the droplet generator, and run the machine to generate droplets (*see* **Note 80**).

(f) Carefully transfer 40 μL of each sample, positive control, or NTC droplets into the designated well in the ddPCR 96-well plate (*see* **Note 81**).

(g) Seal all the wells in the plate with a heat seal foil using the PX1 ddPCR plate sealer.

(h) Set up the C1000 TouchTM thermal cycler to run the following thermal cycling parameters: 10 min at 95 °C, followed by 40 cycles: 15 s at 95 °C and 1 min at 60 °C (*see* **Note 82**).

(i) Load the ddPCR 96-well plate in the C1000 TouchTM thermal and start the PCR run.

(j) Transfer the ddPCR 96-well plate into the QX200 droplet reader machine to analyze the droplets according to the manufacturer's instructions.

(k) Analyze the data using the QuantaSoft™ Analysis Pro software according to the manufacturer's guidelines to determine the target copy number in each sample (*see* **Note 83**).

EvaGreen Approach

(a) Prepare the master mix by mixing the following: 11 μL of the ddPCR EvaGreen supermix, 2.2 μL of each primer, and 3.6 μL of DNAse-free water for each reaction (*see* **Note 84**).

(b) Place a 96-well PCR plate in a plate holder and add 19 μL of the master mix in the designated well for the sample(s), NTC, and positive control.

(c) Add 3 μL of each sample, positive control, or water (for NTC) to the designated well in the plate (*see* **Note 79**).

(d) Cover all wells in the plate with adhesive film and press it firmly using the applicator to seal the wells. Vortex the plate to mix and spin it at max speed for 30 s.

(e) Generate the ddPCR droplets using the QX200 droplet generator. First, place the DG8 cartridge into the cartridge holder and add 20 μL of each sample, positive control, or NTC, and 70 μL of the droplet generation oil into the designated wells in the cartridge. Then, cover the cartridge with the gasket, place it in the droplet generator, and run the machine to generate droplets (*see* **Note 80**).

(f) Carefully, transfer 40 μL of each sample, positive control, or NTC droplets into the designated well in the ddPCR 96-well plate (*see* **Note 81**).

(g) Seal all the wells in the plate with a heat seal foil using the PX1 ddPCR plate sealer.

(h) Set up the C1000 TouchTM thermal cycler to run the following thermal cycling parameters: 5 min at 95 °C, followed by 40 cycles: 30 s at 95 °C, 1 min at 59 °C, and 1 min at 72 °C (*see* **Note 82**).

(i) Load the ddPCR 96-well plate in the C1000 TouchTM thermal and start the PCR run.

(j) Transfer the ddPCR 96-well plate into the QX200 droplet reader machine to analyze the droplets according to the manufacturer's instructions.

(k) Analyze the data using the QuantaSoft™ Analysis Pro software to determine the absolute target copy number in each sample (*see* **Note 83**).

3.6 Western Blot

The following Western blot (immunoblotting) protocol describes the detection and characterization of a specific protein molecule among a mixture of proteins. The protein mixture is separated according to size by gel electrophoresis and transferred/blotted to a membrane. Then, the protein of interest is detected using a labeled antibody. The protocol contains five steps: sample preparation, reagent preparation, gel electrophoresis, proteins transfer/blot, and protein detection and visualization (*see* **Note 85**).

1. *Sample Preparation*

(a) Determine the protein concentration in a sample using the Qubit fluorometer or DC protein assay method sections.

(b) Dilute the sample to double the amount of the desired protein concentration using the homogenization buffer. Then, re-dilute the sample in 2X Laemmli sample buffer containing β-mercaptoethanol (*see* **Note 86**).

(c) Heat the diluted sample for 5 min at 95 °C in a dry block heater and place it on ice for 2 min. Centrifuge at 2000× *g* for 30 s and place it back on ice (*see* **Note 87**).

2. *Reagent Preparation*

(a) Prepare 10X running buffer by dissolving 300 g of Tris–base, 1441.6 g of glycine, and 100 g of SDS in 10 liters of distilled water. Mix 100 mL of the 10X stock buffer with 900 mL distilled water to make a fresh 1X running buffer before starting the electrophoresis step.

(b) Prepare 10X transfer buffer by dissolving 606 g of Tris–base, 2883.2 g of glycine, and 100 g of SDS in 10 liters of distilled water. Mix 100 mL of the 10X stock buffer and 200 mL of methanol with 700 mL of distilled water to make a fresh 1X transfer buffer before starting the transfer step.

(c) Prepare 10X Tris-buffered saline (TBS) buffer by dissolving 24 g of Tris–base and 88 g of NaCl in 900 mL of distilled water. Adjust the pH to 7.6 with 12 M HCl and bring the final volume to 1 liter. Mix 400 mL of 10X TBS, 4 mL of Tween-20, and 3600 mL of distilled water to make fresh 0.1X TBST before starting the protein detection step.

3. *Gel Electrophoresis*

Native and sodium dodecyl sulfate–polyacrylamide gel electrophoresis (SDS-PAGE) gels are the primary gel types for western blot. They are used to separate proteins based on their size and charge. Native PAGE separates proteins in their native state without denaturation during the electrophoresis process. However, in SDS-PAGE, the sodium dodecyl sulfate (SDS) detergent denatures the protein during the electrophoresis process. There are two types of SDS-PAGE gels: gradient and fixed concentration gels. Gradient concentration gels are made from one layer with a higher acrylamide concentration at the bottom that gradually decreases toward the top of the gel. Therefore, the size of the pores is larger at the top and gradually decreases in size toward the bottom. This makes gradient gels ideal for separating proteins with a wide range of molecular weights. Fixed concentration gels have two layers: stacking and resolving. The stacking layer forms the top 1/3 portion of the gel, which contains a lower concentration of acrylamide, making the pore larger. The resolving layer comprises 2/3 of the lower portion of the gel and includes a higher acrylamide concentration, making the pore size smaller. Since the size of the pores is uniform in each layer, fixed gels are ideal for separating proteins with similar molecular weights. Here, we present a method to prepare a fixed SDS-PAGE gel with a 3% acrylamide concentration in the stacking layer and a 6% acrylamide concentration in the resolving layer.

(a) Clean the comb, spacer plate, short plate, casting frame, and casting stand thoroughly with a Kimwipe soaked in 70% ethanol. Allow all components to air dry before assembly (Fig. 6a).

(b) Place the casting frame upright with the pressure cams in the open position and facing forward on a flat surface. Place the short plate on top of the spacer plate to form a glass cassette sandwich. Slide the cassette inside the casting frame and ensure that both glass plates are flush at the bottom with the flat surface. Engage the pressure cams to secure the glass cassette sandwich in the casting frame (Fig. 6b). Place the casting frame into the casting stand (with the locked pressure cams facing out) and secure it by engaging the spring-loaded lever of the casting stand onto the glass cassette sandwich (*see* **Note 88**) (Fig. 6c).

(c) Place the comb into the glass cassette sandwich and mark the short plate 1 cm below the comb teeth. This is the max level to which the resolving gel is poured. Remove the comb (*see* **Note 89**).

(d) Prepare the resolving gel solution in a glass flask by mixing 5.85 mL of water, 2.50 mL of 1.5 M Tris (pH 8.8), and 1.5 mL of 40% acrylamide and bis-acrylamide solution (29:1). Swirl the flask gently between each addition to mix. Incubate the mix at room temperature for 15 min.

(e) Add 125 μL of 10% SDS, 125 μL of 10% APS, and 7.5 μL of TEMED to the resolving gel solution mix. Swirl the flask gently between each addition to mix (*see* **Note 90**).

(f) Pour the resolving gel solution gently into the glass cassette sandwich and avoid air bubbles (Fig. 6d). Cover the solution with 500 μL of isopropanol. Wait for 15–20 min to allow the gel to polymerize.

(g) Prepare the stacking gel solution in a glass flask by mixing 1.67 mL of water, 625 μL of 0.5 M Tris (pH 6.8), and 187.5 μL of 40% acrylamide and bis-acrylamide solution (29:1). Incubate the mix at room temperature for 15 min.

(h) Discard the isopropanol. Rinse the resolving gel surface with distilled water and dry it with filter paper.

(i) Add 25 μL of 10% SDS, 12.5 μL of 10% APS, and 5.0 μL of TEMED to the stacking gel solution mix (*see* **Note 90**).

(j) Pour the stacking gel solution gently into the glass cassette sandwich to the top of the short plate. Avoid air bubbles. Seat the desired comb slowly into the stacking gel solution (Fig. 6e). Wait for 15–20 min to allow the gel to polymerize (*see* **Note 91**).

(k) Position the electrophoresis module on the flat surface and open the frame clamp. Place the gel cassette sandwich with the short plate facing inward into one side of the module and a second gel cassette sandwich or a buffer dam on the other side of the module. Gently close the frame clamp to secure the gel cassettes in the module (Fig. 7b). Place the assembled module into the Mini-PROTEAN tank at the designated orientation. Place the tank on a flat surface.

(l) Fill the inside of the assembled module with 1X running buffer to just under the edge of the outer spacer plate. Gently remove the comb from the gel cassette sandwich and add more 1X running buffer inside the electrophoresis module if needed (*see* **Note 82**) (Fig. 7c).

(m) Load the protein standard and sample to the designated lanes in the gel cassette sandwich using elongated gel loading pipette tips (*see* **Note 93**) (Fig. 7d). Fill the tank with 1X running buffer to the indicated level.

(n) Place the lid on the tank while matching the color-coded electrodes of both the lid and module. Apply power to the Mini-PROTEAN Tetra cell to begin electrophoresis (*see* **Note 94**) (Fig. 7e).

(o) Following the completion of the electrophoresis, turn off the power supply and retrieve the electrophoresis module. Discard the buffer and remove the gel cassette. Separate the short plate from the cassette, submerge the spacer plate with the gel in 1X transfer buffer, and allow the gel to float off the plate into the buffer (Fig. 8Aa–e). Rinse the gel with fresh 1X transfer buffer (*see* **Note 94**).

4. *Protein Transfer/Blotting*

The following method describes two systems for blotting: the Trans-Blot® Turbo™ system and the Mini Trans-Blot cell blotting system.

(a) Trans-Blot® Turbo™ system.

(i) Submerge the blotting membrane in 100% ethanol for 20 s and then place it in 1X transfer buffer (*see* **Notes 96** and **97**).

(ii) Soak ion reservoir stacks in 1X transfer buffer for 3 min at room temperature.

(iii) Assemble the Trans-Blot® Turbo™ transfer pack in the system cassette by stacking the following from bottom to top: bottom (+) cassette electrode (anode), bottom ion reservoir stack, blotting membrane, gel, top ion reservoir stack, and top (−) cassette electrode (cathode) (*see* **Note 98**) (Fig. 8B.b).

(iv) Lock the top (−) cassette electrode and insert the cassette into the machine. Run the transfer protocol (*see* **Note 99**).

(v) Following the completion of the transfer, disassemble the cassette and retrieve the blotting membrane.

(b) Mini Trans-Blot cell blotting system (*see* **Note 100**).

(i) Cut the filter paper and blotting membrane to the dimensions of the gel.

(ii) Submerge the blotting membrane in methanol for 1 min and then in distilled water for 10 s. Place the membrane in 1X transfer buffer (*see* **Note 96**).

(iii) Soak the fiber pad and filter paper in transfer buffer for 2–3 min.

(iv) Assemble the gel sandwich in the gel holder cassette on a flat surface in the following order from bottom to top: black side of the cassette, soaked fiber pad, 3 soaked filter papers, gel, soaked blotting membrane, 3 soaked filter papers, soaked fiber pad, and clear side of the cassette (Fig. 8C.b). Lock the cassette and place it in the electrode module (Fig. 8C.c).

(v) Place the module in the buffer tank and fill the tank to the blotting mark with transfer buffer (*see* **Note 101**) (Fig. 8C.d).

(vi) Cover the tank with the lid while matching the electrodes of both the lid and module (Fig. 8C.e). Apply power to begin the transfer (*see* **Note 102**).

(vii) Following the completion of the transfer, disassemble the cassette and retrieve the blotting membrane (*see* **Notes 103** and **104**).

5. *Protein Detection and Visualization*

(a) Cover the blotting membrane with Intercept® blocking buffer and incubate on an orbital shaker for 1 h at room temperature.

(b) Place the blotting membrane in a pouch and add the primary antibody to the membrane. Seal the pouch and incubate the membrane on a plate mixer at 4 °C (*see* **Note 105**).

(c) Recover the primary antibody and wash the blotting membrane with 0.1% TBST three times for 15 min each at room temperature.

(d) Discard the 0.1% TBST and cover the blotting membrane with the secondary antibody. Incubate on orbital shaker according to hybridization conditions of the antibody and the target protein (*see* **Note 105**).

(e) Discard the secondary antibody and wash the blotting membrane with 0.1% TBST three times for 10 min each at room temperature.

(f) Prepare the Pierce™ ECL western blotting substrate working solution by mixing an equal volume of detection reagents 1 and 2 to yield a sufficient volume to cover the entire blotting membrane. Add the chemiluminescent development working solution to the membrane and incubate for 5 min at room temperature while protecting from light (*see* **Note 106**).

(g) Detect the chemiluminescence signal using an imaging system. Perform the necessary quantification to determine the expression level of the target protein.

4 Notes

1. The presented DNA extraction protocol was adapted from previously published studies [16]. We have modified it to extract genomic DNA from different canine tissues. This protocol has many advantages over commercially available DNA extraction kits. It does not require expensive, environmentally hazardous reagents or expensive equipment. Also, it is simple and fast, produces high-quality DNA, and can be easily adapted to extract DNA from different types of tissues. We have used this protocol to extract genomic DNA from canine skeletal muscle, heart, internal organs, smooth muscle, and nervous system. The protocol must be performed in a DNAse-free environment.
2. We recommend using molecular-grade reagents in this protocol. We also recommend filtering the stock lysis buffer using a Stericup® vacuum filter system (ThermoFisher Scientific, Hampton, NH, USA). Store the buffer at room temperature for a period no longer than 1 month.
3. The working buffer should be prepared fresh before each use.
4. Sample can be obtained from snap freezing tissue in Optimal Cutting Temperature (OCT) media or snap freezing tissue in liquid nitrogen. When using an OCT sample, tissue sections must be rinsed in sterile 1× phosphate-buffered saline (PBS) to remove any residual OCT from the tissue.
5. The protocol described here is a summarized version of the detailed manufacturer's protocol. We highly recommend that the user read and follow the manufacturer's guidelines and instructions in performing the assay. The RNA extraction must be performed in an RNase-free environment by spraying all benchtops, tube racks, pipets, and gloves with RNase Zap spray (ThermoFisher Scientific, Waltham, MA, USA).
6. Once the lyophilized DNAse is resonstituted, it can be aliquoted and stored at −20 °C for up to 9 months. It is not recommended to thaw and re-freeze aliquots.
7. Sample can be obtained from snap freezing tissue in Optimal Cutting Temperature (OCT) media, RNA later stored tissue, or snap freezing tissue in liquid nitrogen. When using an OCT

sample, tissue sections must be rinsed in sterile 1× phosphate-buffered saline (PBS) to remove any residual OCT from the tissue. We have used this protocol to extract RNA from canine skeletal muscle, heart, internal organ, smooth muscle, and nervous system.

8. If the tissue is not thoroughly homogenized, repeat the homogenization step.
9. To prepare the DNase working solution, mix 10 μL of the DNase I solution with 70 μL of the RDD buffer for each sample in a clean microcentrifuge. Flick the tube gently to mix (do not vortex).
10. We have used both protocols described here to extract protein extraction from canine skeletal muscle, heart, internal organs, smooth muscle, and nervous system.
11. Sample can be obtained from snap freezing tissue in Optimal Cutting Temperature (OCT) media or snap freezing tissue in liquid nitrogen. When using an OCT sample, tissue sections must be rinsed in sterile 1× phosphate-buffered saline (PBS) to remove any residual OCT from the tissue.
12. The protein concentration of the tissue lysate can be determined following the extraction, and lysate can be aliquoted accordingly for future experiments.
13. The protocol described here is a summarized version of the detailed manufactured protocol. Therefore, we highly recommend that the user read and follow the manufacturer's guidelines before performing the assay. The cDNA synthesis must be performed in an RNase-free environment by spraying all benchtops, tube racks, pipets, and gloves with RNase Zap spray.
14. The amount of RNase-free water added depends on the RNA amount added in the reaction. The manufacturer allows up to 2.5 μg of RNA per reaction. To run a negative control, add 4 μL of SuperScript IV VILO "No RT" control mix instead of RNA in this step.
15. A thermocycler can be set up to perform all three incubation steps continuously.
16. The protocol described here is a summarized version of the detailed manufacturer protocol. We highly recommend that the user read and follow the manufacturer's guidelines and instructions in performing the assay. The protocol must be performed in a DNase-free environment.
17. The manufacturer recommends preparing new standards to calibrate the machine each time it is used.

18. The volume of the DNA sample in the reaction is determined based on the sample availability and the expected total amount of DNA in the volume added. The quantification range of the Qubit™ 1× dsDNA HS kit is 0.2–120 ng. If the total DNA amount in the DNA sample volume added exceeds 100 ng, the DNA sample must be diluted. Alternatively, the company offers the Qubit™ 1X dsDNA broad range (BR) assay kit with a 4–4000 ng quantification range.
19. For each standard, add 190 μL of the Qubit™ 1× dsDNA HS working solution. For the DNA sample, add 199–180 μL depending on the DNA volume added to the reaction.
20. Do not label the side of the tube or introduce air bubbles in the reaction solution as it will interfere with the sample reading.
21. Never use alcohol, bleach, or any other detergents to clean the pedestals.
22. The blanking solution should be the same buffer or water in which dsDNA was extracted, purified, or diluted. Ionic solutions may interfere with concentration and purity measurements. The blank value should not read more than 0.04 absorbance different from the baseline of the 260 nm. If the value is higher, repeat the cleaning process of the pedestal and re-measure the blanking solution.
23. Review the spectral image generated to determine sample and measurement quality. The device automatically generates purity ratios at 260/280 and 260/230 absorbance to evaluate purity. The 260/280 ratio is a primary measure for dsDNA purity. A value of ~1.8 is considered pure. The 260/230 ratio value is used as a secondary measure of dsDNA purity. A value between 1.8 and 2.2 is considered pure.
24. The protocol described here is a summarized version of the detailed manufacturer protocol. We highly recommend that the user read and follow the manufacturer's guidelines and instructions in performing the assay. The protocol must be performed in an RNAse-free environment.
25. To measure 10 samples (8 samples and 2 standards), prepare 2 mL (10 samples × 200 μL final volume of each sample or standard) of Qubit™ working solution. Add 10 μL of the Qubit™ RNA HS to 1990 μL of Qubit® buffer to have a total of 2 mL working solution. Mix by inverting multiple times and avoid vortexing. Store in the dark until use.
26. The manufacturer recommends preparing new standards to calibrate the machine each time it is used.
27. The volume of the RNA sample in the reaction is determined based on the sample availability and the expected total amount of RNA in the volume added. The quantification range of the

Qubit™ RNA HS kit is 5–100 ng. If the total RNA amount in the sample volume added exceeds 100 ng, the RNA sample must be diluted. Alternatively, the company offers the Qubit™ RNA broad range assay kit with a 20–1000 ng quantification range or the Qubit™ RNA extended range assay kit with a 200 ng to 10 μg quantification range.

28. For each standard, add 190 μL of the Qubit™ RNA HS working solution. For the RNA sample, add 199–180 μL depending on the RNA volume added to the reaction.
29. Do not label the side of the tube or introduce air bubbles in the reaction solution as it will interfere with the sample reading.
30. Never use alcohol, bleach, or any other detergents to clean the pedestals.
31. The blanking solution should be the same buffer or water in which dsDNA was extracted, purified, or diluted. Ionic solutions may interfere with concentration and purity measurements. The blank value should not read more than 0.04 absorbance different from the baseline of 260 nm. If the value is higher, repeat the cleaning process of the pedestal and re-measure the blanking solution.
32. Review the spectral image generated to determine sample and measurement quality. The device automatically generates purity ratios at 260/280 and 260/230 absorbance to evaluate purity. The 260/280 ratio is a primary measure of RNA purity. A value of ~2.0 is considered pure. The 260/230 ratio value is used as a secondary measure of RNA purity. A value between 2.0 and 2.2 is considered pure.
33. The protocol described here is a summarized version of the detailed manufacturer protocol. We highly recommend that the user read and follow the manufacturer's guidelines and instructions in performing the assay.
34. To measure 10 samples (8 samples and 2 standards), prepare 2 mL (10 samples × 200 μL final volume of each sample or standard) of Qubit™ working solution. Add 10 μL of the Qubit™ ssDNA dye to 1990 μL of Qubit® buffer to have a total of 2 mL working solution. Mix by inverting multiple times and avoid vortexing. Store in the dark until use.
35. The manufacturer recommends preparing new standards to calibrate the machine each time it is used.
36. The volume of the cDNA sample in the reaction is determined based on the sample availability and the expected total amount of ssDNA in the volume added. The quantification range of the Qubit™ ssDNA kit is 5–100 ng. If the total cDNA amount in the sample volume added exceeds 100 ng, the sample must be diluted.

37. For each standard, add 190 μL of the Qubit™ ssDNA working solution. For the cDNA sample, add 199–180 μL depending on the sample volume added to the reaction.
38. Do not label the side of the tube or introduce air bubbles in the reaction solution as it will interfere with the sample reading.
39. Never use alcohol, bleach, or any other detergents to clean the pedestals.
40. The blanking solution should be the same buffer or water in which cDNA was prepared or diluted. Ionic solutions may interfere with concentration and purity measurements. The blank value should not read more than 0.04 absorbance different from the baseline of 260 nm. If the value is higher, repeat the cleaning process of the pedestal and re-measure the blanking solution.
41. Review the spectral image generated to determine sample and measurement quality. The device automatically generates purity ratios at 260/280 and 260/230 absorbance to evaluate purity. The 260/280 ratio is a primary measure of cDNA purity. A value of ~1.8 is considered pure. The 260/230 ratio value is used as a secondary measure of cDNA purity. A value between 1.8 and 2.2 is considered pure.
42. The protocol described here is a summarized version of the detailed manufacturer protocols. We highly recommend that the user read and follow the manufacturer's instructions in performing the Qubit assay.
43. To measure 10 samples (7 samples and 3 standards), prepare 2 mL (10 samples × 200 μL final volume of each sample or standard) of Qubit™ working solution. Add 10 μL of the Qubit™ Protein BR dye to 1990 μL of Qubit® Protein BR buffer to have a total of 2 mL working solution at 1:200 dilution. Mix by inverting multiple times and avoid vortexing. Store in the dark until use.
44. The Qubit™ Protein BR Assay requires the use of three standards to generate the internal standard curve. The manufacturer recommends preparing new standards to calibrate the machine each time it is used.
45. The volume of the protein sample in the reaction is determined based on the sample availability and the expected concentration of the sample. The quantification range of the Qubit™ Protein BR Assay is 12.5 μg/ml to 5 mg/ml. If the protein concentration exceeds 5 mg/ml, the sample must be diluted in the same homogenization buffer used in protein extraction.
46. For each standard, add 190 μL of the Qubit™ working solution. For each protein sample, add 199–180 μL depending on the sample volume added to the reaction.

47. Do not label the side of the tube or introduce air bubbles in the reaction solution as it will interfere with the sample reading.
48. The DC Protein Assay™ can be used for protein samples with a minimum concentration of 0.2 mg/ml. The protocol described here is an optimized version of the manufacture protocol to quantify protein from whole canine tissue lysate. We highly recommend that the user read and follow the manufacturer's safety instructions in performing this assay. The DC Protein Assay™ should be compatible with most protein homogenization buffer formulations, including those containing detergents. We also highly recommend adding a protease inhibitor to the homogenization buffer.
49. We recommend using pure BSA diluted in the homogenization buffer to make standards. Users may also choose another purified protein to create standards or use commercially prepared protein standards. Standard concentrations should span the entire expected concentration range of the samples. The minimum number of standards recommended by the manufacturer is one blank and three standards. However, we highly recommend one blank and up to nine standards ranging from 10.0 to 0.2 mg/ml. This will provide a more accurate standard curve and encompass a greater range of sample concentrations.
50. Unused BSA standards can be stored at 4 °C for up to 1 week.
51. Working reagent is stable for 1 week at room temperature. If a precipitate forms, vortex and warm gently to dissolve.
52. If the sample protein concentration is expected to be greater than 10 mg/ml, dilute the sample accordingly using the same homogenization buffer used in the extraction.
53. The color development is completed after 15 min, and the color is stable for up to 1 h at room temperature. We highly recommend measuring samples shortly after the 15 min.
54. Any spectrophotometer or microplate reader capable of measuring absorbance at 650–750 nm will be suitable for use with the DC protein assay. In addition, the Nanodrop One/One^c is capable of measuring DC assay samples, using its "Protein Lowry" function. A specific protocol for the use of the Nanodrop One/One^c for measuring DC Assay (Lowry) samples is available from the manufacturer.
55. To generate a standard curve, create an XY scatter plot in Microsoft Excel, with standard absorbance on the x-axis and standard concentration on the y-axis. Generate a second-order polynomial (quadratic) regression line and display the equation and R^2 value. The R^2 value determines how well the regression line fits the standard concentrations. An R^2 value should be as

close to 1.0 as possible and at least greater than 0.9 for valid results. The equation of the regression line is a quadratic function in standard format: ($y = ax^2 + bx + c$).

56. To determine the concentration of unknown samples, solve the quadratic function for y, using [x = (sample absorbance)]. If the sample was diluted before or during the procedure, multiply the sample concentration by the sample's dilution factor.

57. The real-time PCR protocol must be performed in a DNase-free environment. The TaqMan approach uses a fluorogenic probe and requires a highly specific hybridization between the probe and target sequence to generate a fluorescent signal for quantification. However, the SYBR Green approach uses a double-stranded DNA binding dye that binds to specific or nonspecific PCR products. Therefore, it is important to confirm the primer efficiency and optimize the amplification condition in the SYBR green approach. We recommend using the TaqMan approach as it is highly specific to the target sequence.

58. We recommend cutting the plasmid in the backbone, or 300–600 bp away from the target sequence to preserve the target site.

59. The purification protocol of the linearized plasmid described here is a summarized version of the detailed manufactured protocol. Therefore, we highly recommend that the user read and follow the manufacturer's guidelines in performing the protocol.

60. Alternative to using a vacuum manifold, the column can be placed in a collection tube and centrifuged at max speed to pass the solution throughout the column membrane. Discard the flow-through following the centrifugation or replace the collection tube with a new one.

61. Since the maximum loading volume of the column is 800 μL, the digested sample volume (~1200 μL) will be divided into two loads.

62. The amount of DNAse-free water used to elute the DNA from the column membrane can vary based on the amount of circular plasmid used in the digestion and the desired plasmid concentration following the elution.

63. Ideally, one band should be present on the gel at the expected size of the plasmid. However, if two bands are present, this will suggest that the plasmid has not been completely linearized.

64. In the presented equation, 607.4 is the average molecular weight (g/mol) of 1 dsDNA in polynucleotide form. This average is determined from an individual MW for A, T, G, and C in polymer form. The addition of 157.9 to the ddDNA MW takes into account the 5′ monophosphate left by most restriction enzymes.

65. The plasmid concentration (ng/μl) is converted to g/μl by multiplying with 10^{-9}.6.022 × 10^{23} is Avogadro's number, defined as molecules per mole.

66. We recommend storing the stock plasmid (1 × 10^9 copy/μl) into 100 μL aliquots and storing it at −80 °C. Once an aliquot is thawed and diluted, store at 4 °C for up to 1 week.

67. Refer to the master mix manufacturer's guideline for the amount of template needed in the reaction. For example, the user guide for TaqMan® Universal PCR master mix recommends using 1–100 ng of genomic DNA or cDNA in the real-time PCR reaction (https://www.thermofisher.com/document-connect/document-connect.html?url=https://assets.thermofisher.com/TFS-Assets%2FLSG%2Fmanuals%2F4304449_TaqManPCRMM_UG.pdf)

68. The 20X TaqMan primer and probe mix contains 18 μM of each primer and 5 μM of the probe. We found these concentrations to be optimal for our needs. However, we recommend that the user determine the optimal probe and primer concentration that meets their needs. Also, we recommend mixing the primers and the probe and aliquoting it into a small volume (ex. 100 μL). The aliquots can be stored at −20 °C for long-term storage. Avoid the repeated freeze and thaw of a single aliquot.

69. To avoid contaminating the thermocycler block of the real-time PCR machine, it is highly recommended to place the PCR plate in a plate holder to prevent the contact of the plate base with any surface.

70. We highly recommend adding at least 3 μL of standard or the sample in the PCR reaction to avoid any pipetting error.

71. We found that 900 nM of each primer is ideal for the SYBR green PCR reaction. However, we recommend that the user determine the optimal primer concentration that meets their needs.

72. Since the SYBR Green approach uses a double-stranded DNA binding dye that binds to specific or nonspecific PCR products, it is highly recommended to perform a melting curve cycle at the end of the PCR to confirm the specificity of the primer and PCR condition and the absence of any unspecific amplicons.

73. Since 3 μL standard was added per reaction, the copy number should be multiplied by three when setting the data acquisition software. For example, the well that received 3 μL of the 1 × 10^9 copy/μl should be marked in the software as 3 × 10^9 copies of the standard. Also, 3 μL of the sample was added per reaction. Therefore, the total copy number of a target sequence in this sample should be divided by three to achieve a copy/μl value.

74. In the presented equation, the C value is the amount of DNA contained within a haploid nucleus. For canine (*Canis familiaris*), the C value range from 2.88 to 3.54 pg with an average of 3.12 pg (http://www.genomesize.com/). The C value is multiplied by 2 to reflect the amount of DNA within a diploid nucleus. Then, it is divided by 1000 to change from pg to ng. Therefore, the number of canine diploid genomes in 1 ng of DNA is 160.26.
75. The ddPCR protocol must be performed in a DNAse-free environment. The TaqMan approach uses a fluorogenic probe and requires a highly specific hybridization between the probe and target sequence to generate a fluorescent signal for quantification. However, the EvaGreen approach uses a double-stranded DNA binding dye that binds to specific or nonspecific PCR products. Therefore, it is important to confirm the primer efficiency and optimize the amplification condition in the EvaGreen approach. We recommend using the TaqMan approach as it is highly specific to the target sequence.
76. The user guide for the ddPCR Supermix for Probes (no dUTP) recommends using up to 330 ng of DNA or cDNA template per reaction (https://www.bio-rad.com/sites/default/files/webroot/web/pdf/lsr/literature/10026868.pdf). We found that 100 ng of DNA or cDNA is enough for our needs. However, we highly recommend that the user determine the optimal amount of template that meets their needs.
77. The 20X TaqMan primer and probe mix contains 18 μM of each primer and 5 μM of the probe. We found these concentrations to be optimal for our needs. However, we recommend that the user determine the optimal probe and primer concentration that meets their needs. Also, we recommend mixing the primers and the probe and aliquoting it into a small volume (ex. 100 μL). The aliquots can be stored at −20 °C for long-term storage. Avoid the repeated freeze and thaw of a single aliquot.
78. The mixing of the sample, positive control, or the NTC with the master mix can be performed at room temperature.
79. We highly recommend adding at least 3 μL of standard or the sample in the PCR reaction to avoid any pipetting error.
80. Fill any unused well in the DG8 cartridge with droplet generation oil as an empty well may cause uneven vacuum pressure to the cartridge and thus may affect the droplet generation in the other wells. The gasket must be secured to the cartridge according to the manufacturer's guidelines to avoid introducing air into the wells during the droplet generation.

81. We highly recommend using low retention pipette tips when transferring the droplets into the ddPCR 96-well plate to avoid losing droplets. The aspiration and ejection of the droplet using the pipette must be performed at a very slow and constant rate to prevent mechanical shearing of the droplets.
82. The thermal cycling parameters should be optimized to meet the user's needs.
83. Since 3 μL of the sample was added per reaction. Therefore, the total copy number of a target sequence in this sample should be divided by three to achieve a copy/μl value.
84. The user guide for the ddPCR EvaGreen Supermix recommends using up to 100 nM of each primer per 20 μL reaction. We found that 10 nM of each primer is ideal for our needs. However, we recommend that the user determine the optimal primer concentration that meets their needs.
85. The western blot protocol requires using reagents and equipment from different manufacturers. Here, we present a summarized version of each manufacturer's protocols. We highly recommend that the user read and follow the manufacturers' guidelines and instructions when using any reagents or equipment used in this protocol. Also, the five steps of the protocol may not be completed in 1 day. Therefore, it is essential to plan accordingly before starting the protocol.
86. The dilution of the sample is determined based on the total amount of protein extracted and the amount needed to detect the target protein. We recommend diluting the protein to 5 ng/μl, aliquot it into 30 μL aliquots, and store at −80 °C for long-term storage. One aliquot would provide up to 150 ng (30 μL × 5 ng/μl) protein per lane. Also, it is important to consider the max volume that can be added in each reservoir (well) of the gel lane when aliquoting the diluted protein.
87. Following this step, the diluted protein sample can be stored at −80 °C for long-term storage. The sample can be later thawed on ice for electrophoresis, and there will be no need to re-heat it at 95 °C to denature the protein. Avoid thawing and heating the sample more than once.
88. A leak may occur if the base of both plates is not flushed, misaligned, or oriented incorrectly.
89. Bio-Rad offers a wide selection of combs with different thicknesses and a number of wells.
90. The gel solution mix must be prepared fresh before use. Also, the 10% APS must be made fresh before adding to the mix. The APS and TEMED must be added to the mixture just before pouring it into the cassette. Otherwise, it will polymerize prematurely.

Table 1
Electrophoresis conditions

Target protein	Protein size (kDa)	Total protein loaded (μg)	Gel type	Stacking gel percentage	Running gel percentage	Voltage	Run time
Full-length dystrophin	427	100	Fixed Conc. SDS-PAGE	3%	6%	60* 140	15* 90
Micro-dystrophin	140	100	Fixed Conc. SDS-PAGE	3%	6%	60* 140	15* 90
Vinculin	117	100	Fixed Conc. SDS-PAGE	3%	6%	60* 140	15* 90

kDa kilodalton, *μg* microgram, *Conc.* concentration, * initial voltage and time to allow migration of the protein from the gel lanes into the stacking gel

91. The gel cassette sandwich can be wrapped with wet paper towels, placed in a plastic bag, and stored at 4 °C for up to 10 days.
92. We recommend flushing the gel wells after removing the comb with fresh 1X running buffer to remove any residual chemicals or debris. This will ensure an even migration of the protein into the stacking gel.
93. Bio-Rad offers a sample loading guide device. It is placed between the two gel cassettes in the electrophoresis module and facilitates locating the sample wells when loading samples.
94. The voltage and duration of electrophoresis may vary based on the target protein. Table 1 presents the electrophoresis conditions of some of the target proteins that we have adopted using the described protocol. We recommend surrounding the tank with ice to cool down the running buffer, especially when high voltage and/or longer duration is used during electrophoresis.
95. The gel can be stained with Coomassie blue staining to verify the transfer quality and validate the equal loading of all the samples. To perform the staining, incubate the gel in Coomassie protein staining solution overnight on an orbital shaker at room temperature. On the next day, wash the gel three times with tap water for 15 min each on an orbital shaker at room temperature. Take a picture of the gel at 700 nm wavelength using an imaging system.
96. There are two types of blotting membranes for western blots: polyvinylidene difluoride (PVDF) or nitrocellulose membrane. The PVDF membrane is more commonly used in western blot than the nitrocellulose membrane. It has a higher binding protein capacity, is less fragile, and can sustain stripping and re-probing better than the nitrocellulose membrane. Both

Table 2
Transfer conditions for the Trans-Blot® Turbo™ system

Target protein	Protein size (kDa)	Membrane type	Transfer current and voltage	Transfer time
Full-length dystrophin	427	PDVF	1.2 A 25 V	10 min
Micro-dystrophin	140	PDVF	1.2 A 25 V	10 min
Vinculin	117	PDVF	1.2 A 25 V	10 min

kDa kilodalton, *A* amplifier (amp), *V* voltage

membranes are available in 0.2 and 0.45 μm pore sizes. The 0.45 μm is suitable for transferring proteins with a molecular weight larger than 20 kDa, while the 0.2 μm is used for proteins with less than 20 kDa weights.

97. Bio-Rad offers the Trans-Blot® Turbo™ transfer packs with PVDF or nitrocellulose membrane.
98. Use the blotting roller to remove any air bubbles, to provide consistent contact between the layers.
99. The voltage and duration of the transfer will vary based on the target protein. Table 2 presents the electrophoresis conditions of some of the target proteins that we have adopted using the Trans-Blot® Turbo™ system. Refer to the manufacturer's guidelines on how to set up a transfer protocol on the Trans-Blot® Turbo™ system.
100. Since the Mini Trans-Blot cell blotting and the Mini-PROTEAN Tetra cell electrophoresis module are interchangeable systems, the tank and lid of both systems can be used for electrophoresis and/or blotting.
101. We recommend stirring the buffer during transfer to maintain even buffer temperature and ion distribution. To stir the buffer, add a stir bar to the tank and place the tank on a stir plate (Fig. 8C.d). Set the stirring speed as fast as possible during the transfer.
102. The voltage and duration of the transfer will vary based on the target protein. Table 3 presents the transfer conditions of some of the target proteins that we have optimized using the described protocol.
103. The gel can be stained with Coomassie blue staining to confirm the transfer of all the proteins from the gel to the blotting membrane.

Table 3
Transfer conditions for the Mini Trans-Blot® cell blotting system

Target protein	Protein size (kDa)	Membrane type	Transfer voltage (V)	Transfer time (Hrs.)
Full-length dystrophin	427	PDVF	50	8
Micro-dystrophin	140	PDVF	50	8
Vinculin	117	PDVF	50	8

kDa kilodalton, *Hrs.* hours

104. The blotting membrane can be stained with Ponceau S stain to detect protein bands or confirm equal loading in all samples. To perform the stain, place the blotting membrane in ethanol for 30 s. Transfer the membrane into Ponceau S solution and shake for 30 s at room temperature. Discard the stain and wash the membrane three times with water for 1 min each at room temperature.
105. We recommend diluting the primary and secondary antibodies in the blocking buffer to increase the efficiency of blocking all potential nonspecific protein binding sites. We have used the described protocol to evaluate 65 epitope-specific dystrophin monoclonal antibodies and other antibodies specific to other intracellular and cellular expressed proteins [15, 17–19]. Table 4 presents the working dilution and the source of primary and secondary antibodies for some of these proteins that we have adopted using the described protocol. It also includes the hybridization conditions: incubation time and temperature for each target protein. We highly recommend optimizing the dilution of each primary antibody according to the target protein.
106. Infrared fluorescence dye–conjugated secondary antibodies can be used to detect the target protein instead of the ECL substrate [18].

Acknowledgments

The research on the canine DMD model in the Duan lab is currently supported by the National Institutes of Health (NS-90634 and AR-70517), Jesse Davidson Foundation-Defeat Duchenne Canada, Hope for Javier, Jackson Freel DMD Research Fund, Parent Project Muscular Dystrophy, Jett Foundation, Michael's Cause, Ryan's Quest, Solid Biosciences Inc., and the University of Missouri.

Table 4
Primary and secondary antibodies used in western blot

		Primary antibody hybridization conditions							Secondary antibody hybridization conditions					
Antibody name	Clone	Diluent	Dilution	Clonality	Host species	Isotype	Source	Catalog number	Antibody information*	Dilution	Clonality	Host species	Source	Catalog number
Dystrophin (C-terminal)	N/A	IBB:RB (1:1)	1:250	Polyclonal	Rabbit	IgG	ThermoFisher Scientific	RB-9024-P0	Goat anti-Rabbit IgG	1: 10000	Polyclonal	Goat	Millipore Sigma	AP132P
Dystrophin (exon 59)	4E10	IBB:RB (1:1)	1:250	Monoclonal	Mouse	IgG1	DSHB	MANEX59B (4E10)	Goat anti-Mouse IgG (H+L)	1:10000	Polyclonal	Goat	Millipore Sigma	AP124P
Dystrophin (exon 8)	5D12	IBB:RB (1:1)	1:250	Monoclonal	Mouse	IgG1	DSHB	MANHINGE1C (5D12)	Goat anti-Mouse IgG (H+L)	1:10000	Polyclonal	Goat	Millipore Sigma	AP124P
Vinculin	hVIN-1	IBB:RB (1:1)	1: 1000	Monoclonal	Mouse	IgG1	Millipore Sigma	V9131	Goat anti-Mouse IgG (H+L)	1:10000	Polyclonal	Goat	Millipore Sigma	AP124P

N/A not applicable, *IBB* Intercept® Blocking Buffer (TBS), *RB* 1X running buffer, *DSHB* Developmental Studies Hybridoma Bank, * peroxidase conjugated antibody

Disclosure DD is a member of the scientific advisory board for Solid Biosciences and equity holders of Solid Biosciences. DD is a member of the scientific advisory board for Sardocor Corp. In the last 3 years, the Duan lab has received research supports unrelated to this project from Solid Biosciences and Edgewise.

References

1. Kunkel LM (1989) The Wellcome lecture, 1988. Muscular dystrophy: a time of hope. Proc R Soc Lond B Biol Sci 237(1286):1–9
2. Duan D (2016) Systemic delivery of adeno-associated viral vectors. Curr Opin Virol 21: 16–25. https://doi.org/10.1016/j.coviro.2016.07.006
3. Duan D (2018) Systemic AAV micro-dystrophin gene therapy for Duchenne muscular dystrophy. Mol Ther 26(10):2337–2356. https://doi.org/10.1016/j.ymthe.2018.07.011
4. Shin J-H, Pan X, Hakim CH et al (2013) Microdystrophin ameliorates muscular dystrophy in the canine model of Duchenne muscular dystrophy. Mol Ther 21(4):750–757. https://doi.org/10.1038/mt.2012.283
5. Yue Y, Pan X, Hakim CH et al (2015) Safe and bodywide muscle transduction in young adult Duchenne muscular dystrophy dogs with adeno-associated virus. Hum Mol Genet 24(20):5880–5890
6. Le Guiner C, Servais L, Montus M et al (2017) Long-term microdystrophin gene therapy is effective in a canine model of Duchenne muscular dystrophy. Nat Commun 8:16105. https://doi.org/10.1038/ncomms16105
7. Hakim CH, Pan X, Kodippili K et al (2016) Intravenous delivery of a novel micro-dystrophin vector prevented muscle deterioration in young adult canine Duchenne muscular dystrophy dogs. Mol Ther 24(Suppl 1): S198–S199
8. Hakim CH, Kodippili K, Jenkins G et al (2018) AAV micro-dystrophin therapy ameliorates muscular dystrophy in young adult Duchenne muscular dystrophy dogs for up to 30 months following injection. Mol Ther 26(S1):5
9. Wang Z, Kuhr CS, Allen JM et al (2007) Sustained AAV-mediated dystrophin expression in a canine model of Duchenne muscular dystrophy with a brief course of immunosuppression. Mol Ther 15(6):1160–1166
10. Duan D (2018) Micro-dystrophin gene therapy goes systemic in Duchenne muscular dystrophy patients. Hum Gene Ther 29(7): 733–736. https://doi.org/10.1089/hum.2018.012
11. Mendell JR, Sahenk Z, Lehman K et al (2020) Assessment of systemic delivery of rAAVrh74.MHCK7.micro-dystrophin in children with Duchenne muscular dystrophy: a nonrandomized controlled trial. JAMA Neurol. https://doi.org/10.1001/jamaneurol.2020.1484
12. Solid Biosciences (2021) solid biosciences reports efficacy and safety data from the ongoing IGNITE DMD clinical trial and resumption of patient dosing in the 2E14 vg/kg cohort, March 15. https://investors.solidbio.com/news-releases/news-release-details/solid-biosciences-reports-efficacy-and-safety-data-ongoing
13. Pfizer (2020) Pfizer's new phase 1b results of gene therapy in ambulatory boys with Duchenne muscular dystrophy (DMD) support advancement into pivotal phase 3 study, May 15. https://investors.pfizer.com/investor-news/press-release-details/2020/Pfizers-New-Phase-1b-Results-of-Gene-Therapy-in-Ambulatory-Boys-with-Duchenne-Muscular-Dystrophy-DMD-Support-Advancement-into-Pivotal-Phase-3-Study/default.aspx
14. Amoasii L, Hildyard JCW, Li H et al (2018) Gene editing restores dystrophin expression in a canine model of Duchenne muscular dystrophy. Science 362(6410):86–91. https://doi.org/10.1126/science.aau1549
15. Hakim CH, Kumar SRP, Perez-Lopez DO et al (2021) Cas9-specific immune responses compromise local and systemic AAV CRISPR therapy in multiple dystrophic canine models. Nat Commun 12(1):6769. https://doi.org/10.1038/s41467-021-26830-7
16. Biase FH, Franco MM, Goulart LR et al (2002) Protocol for extraction of genomic DNA from swine solid tissues. Genet Mol Biol 25(3):313–315
17. Kodippili K, Vince L, Shin JH et al (2014) Characterization of 65 epitope-specific dystrophin monoclonal antibodies in canine and

murine models of Duchenne muscular dystrophy by immunostaining and western blot. PLoS One 9(2):e88280. https://doi.org/10.1371/journal.pone.0088280

18. Hakim CH, Wasala NB, Nelson CE et al (2018) AAV CRISPR editing rescues cardiac and muscle function for 18 months in dystrophic mice. JCI Insight 3(23):124297. https://doi.org/10.1172/jci.insight.124297

19. Hakim CH, Clement N, Wasala LP et al (2020) Micro-dystrophin AAV vectors made by transient transfection and herpesvirus system are equally potent in treating mdx mouse muscle disease. Mol Ther Methods Clin. ev 18: 664–678. https://doi.org/10.1016/j.omtm.2020.07.004

Chapter 16

Histological Assessment of Gene Therapy in the Canine DMD Model

Chady H. Hakim, Matthhew J. Burke, James Teixeira, and Dongsheng Duan

Abstract

Assessing histological changes is essential for characterizing muscle disease progression and for studying the response to therapies in Duchenne muscular dystrophy (DMD), an X-linked progressive muscle-wasting disease caused by the loss of the dystrophin protein. Canine models are by far the best-characterized large animal models for DMD. In this chapter, we describe methods for muscle tissue collection and storage, hematoxylin and eosin staining for studying general muscle morphology, and special staining protocols for evaluating fibrosis, calcification, and neuronal nitric oxide synthase (nNOS) activity. We also provide immunofluorescence staining protocols that are often used to characterize the expression and localization of dystrophin and components of the dystrophin-associated glycoprotein complex. Lastly, we presented immunohistochemical staining protocols that we use to assess muscle inflammation and immune responses.

Key words DMD, Canine model, Biopsy, Necropsy, Histology, Tissue staining, Fibrosis, Immunofluorescence, Immunohistochemical, Dystrophin, nNOS, Duchenne muscular dystrophy

1 Introduction

A dog has more than 700 muscles, accounting for more than 45% of the body weight. Unlike other cells in the body, muscle cells are multinucleated and show characteristic striated patterns caused by repeating functional units (sarcomeres). Muscle cells, often called myofibers, are organized in bundles, also known as fascicles. Each myofiber is surrounded by endomysium, and muscle fascicles are surrounded by perimysium. In longitudinal (sagittal plane) sections, myofibers are long and cylindrical. In cross-sections (transverse plane), myofibers appear as uniform polyhedral cells with multiple nuclei.

Rika Maruyama and Toshifumi Yokota (eds.), *Muscular Dystrophy Therapeutics: Methods and Protocols*, Methods in Molecular Biology, vol. 2587, https://doi.org/10.1007/978-1-0716-2772-3_16,

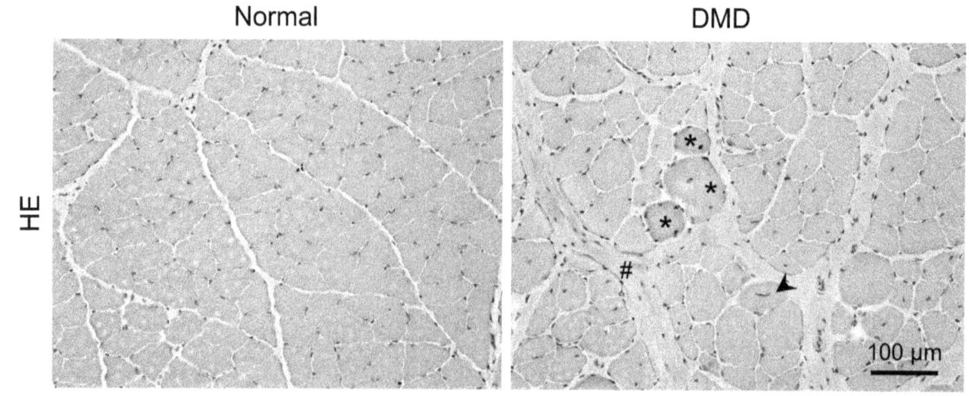

Fig. 1 Representative photomicrograph of H&E stained normal and dystrophic dog muscle. *, heavily stained hyalinated/hypercontracted myofiber; arrow, centrally nucleated myofiber; #, fibrotic tissue deposition

Normal muscle consists of well-organized myofibers and fascicles with peripherally located nuclei, uniform size, minimal fibrosis, and with little-to-no calcification or inflammatory cell infiltration (Fig. 1). The histological appearance of muscle tissues undergoes substantial changes in diseased muscles. Distinctive histological features of different muscle diseases are important biomarkers for differential diagnosis in clinical practice. In preclinical studies, histological evaluation is routinely used to study disease severity and response to experimental therapeutics.

Duchenne muscular dystrophy (DMD) is an X-linked genetic disease [1]. It is the most common form of muscular dystrophy affecting children. DMD is caused by mutations in the *DMD* gene, which encodes thę dystrophin protein, a large subsarcolemmal structural protein. Dystrophin and its associated proteins form the dystrophin-associated glycoprotein complex (DAGC). The DAGC protects muscle from contraction-induced injuries. The absence of dystrophin leads to the disassembly of the DAGC, muscle degeneration, inflammation, calcification, and necrosis. Necrotic fibers are later replaced with fibrotic tissue and fat. Dystrophin-deficient muscles are characterized by the presence of myofibers with centrally located nuclei (a hallmarker of muscle regeneration), hyalinized and calcified myofibers, great variation in the myofiber size, abundant deposition of fibrotic tissue within extracellular spaces and between fascicles, and widespread infiltration of inflammatory cells that are often accumulated at the necrotic foci (Fig. 1).

In this chapter, we illustrate histological and immunological staining techniques that are commonly used to study muscle disease in the canine DMD model [2, 3]. Similar techniques can also be used to study muscle disease in other mammalian models of DMD and DMD patients. Specifically, we provide detailed methods for muscle tissue collection via biopsy and necropsy and muscle tissue storage for histopathological examination. We present a set of protocols to examine (i) the general muscle histology using hematoxylin and eosin (H&E) staining, (ii) muscle fibrosis using Masson's trichrome (MTC) staining, and (iii) calcification using the Alizarin Red S staining. Neuronal nitric oxide synthase (nNOS) is anchored to the sarcolemma by dystrophin [4, 5]. Delocalization of nNOS from the muscle cell membrane contributes to DMD pathogenesis. Hence, we include a special histochemical staining method to study nNOS activity in situ. Finally, we provide a detailed immunofluorescence staining protocol for characterizing dystrophin, DAGC components, muscle fiber type, and a detailed immunohistochemical staining protocol for characterizing inflammation and immune responses.

2 Materials

2.1 Tissue Collection

2.1.1 Needle Biopsy (Fig. 2)

1. Personal protective equipment (PPE), including disposable gowns and/or dedicated scrubs, safety glasses or splash goggles, head cover, shoe cover, surgical mask, and gloves.
2. Sedation agents: Dexdomitor (Zoetis Inc., Kalamazoo, MI, USA); Nalbuphine (Hospira Inc., Lake Forest, IL, USA).
3. Disposable sterile syringe (Becton, Dickinson and Company, Franklin Lakes, NJ, USA).
4. Animal hair clipper with a #50 blade (Oster Professional Products, Sunbeam Products, Inc., Boca Raton, FL, USA).
5. Digital thermometer (ReliOn Inc., Spokane Valley, WA, USA).
6. Stethoscope (3M Littmann, St. Paul, MN, USA).
7. Conductive blankets (Adroit Medical Systems Inc, Loudon, TN, USA) connected to a heated circulating water bath (ThermoFisher Scientific, Hampton, NH, USA).
8. Medical grade oxygen tank (Airgas, Radnor, PA, USA).
9. Canine anesthesia mask (Shopmedvet Inc., Mettawa, IL, USA).
10. Veterinary vital sign monitor (DRE Veterinary, Louisville, KY, USA).
11. Nonwoven gauze sponges (4 inch × 4 inch) (Medline Industries Inc., Northfield, IL, USA).
12. Rugby artificial tears ointment (Walmart Stores Inc., Bentonville, AR, USA).
13. 2% Chlorhexidine gluconate (Vedco Inc., Saint Joseph, MO, USA).
14. 70% Isopropyl alcohol (Walmart Stores Inc., Bentonville, AR).
15. Surgical gloves (Cardinal Health, Dublin, OH, USA).
16. Surgical drape with circular aperture (24″ × 26″) (Medline, Northfield, IL, USA).

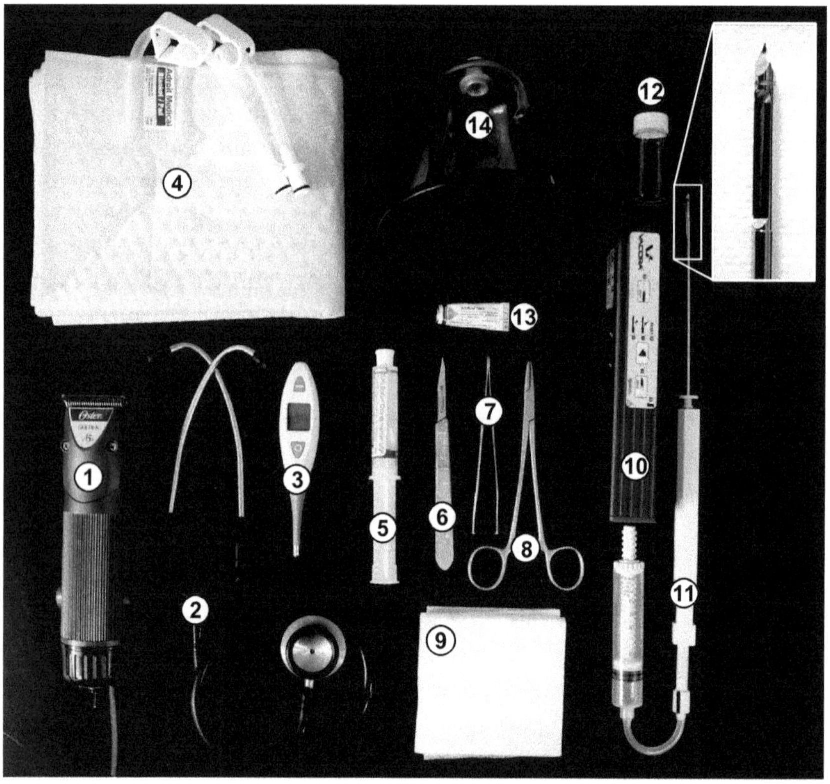

Fig. 2 Materials required for vacuum-assisted biopsy. (1) Animal hair clipper with a #50 blade, (2) stethoscope, (3) digital thermometer, (4) conductive blankets, (5) BD PosiFlush Pre-filed saline (0.9% sodium chloride) syringe for intravenous injection, (6) surgical blade #11 attached to a scalpel handle #3, (7) tissue forceps (1 × 2 teeth), (8) needle holder, (9) nonwoven gauze sponges (4 inch × 4 inch), (10) Vacora® vacuum-assisted biopsy system driver, (11) Vacora® Biopsy probe (inset: an enlarge image of the probe chamber), (12) disposable plastic container with a screw-on cap, (13) eye lubricant, (14) canine anesthesia mask

17. Surgical tools: Needle holder, Scalpel handle #3, and tissue forceps (1 × 2 teeth) (Fine Science Tools, Foster City, CA, USA).
18. Surgical blade #11 (Aspen Surgical, Caledonia, MI, USA).
19. Vacora® vacuum assisted biopsy system driver (Bard Peripheral Vascular Inc., Tempe, Arizona, USA).
20. Vacora® Biopsy probe 10G × 140 mm (Bard Peripheral Vascular Inc., Tempe, Arizona, USA).
21. Vacora® Biopsy probe 14G × 116 mm (Bard Peripheral Vascular Inc., Tempe, Arizona, USA).
22. BD PosiFlush pre-filled saline (0.9% sodium chloride) syringe for intravenous (IV) injection (Becton Dickinson company, Franklin Lakes, NJ, USA).
23. Disposable plastic container with a screw-on cap (Uline, Pleasant Prairie, WI, USA).

24. Pain relief and anti-inflammatory agent: Rimadyl (Zoetis Inc., Kalamazoo, MI, USA).
25. Postanesthetic reversal agent: Antisedan (Zoetis Inc., Kalamazoo, MI, USA).
26. Vetbond tissue adhesive (3M Animal care product, Saint Paul, MN, USA).

2.1.2 Necropsy (Fig. 4)

1. Personal protective equipment (PPE), including disposable gowns and/or dedicated scrubs, safety glasses or splash goggles, head cover, shoe cover, surgical mask, and gloves.
2. Color Atlas of Veterinary Anatomy, The Dog and Cat, book (Elsevier Inc., Amsterdam, Netherlands).
3. Animal hair clipper with a #50 blade (Oster Professional Products, Sunbeam Products, Inc., Boca Raton, FL, USA).
4. Surgical dissection instruments (Fine science tools, Foster City, CA, USA).
5. Weitlaner retractor (Fine Science Tools, Foster City, CA, USA).
6. 1× Phosphate Buffered Saline (PBS) (ThermoFisher Scientific, Hampton, NH, USA).
7. Nonwoven gauze sponges (4 inch × 4 inch) (Medline Industries Inc., Northfield, IL, USA).
8. Petri dish (ThermoFisher Scientific, Hampton, NH, USA).
9. Oscillating saw (Dremel, Racine, WI, USA).

2.2 Tissue Processing

2.2.1 Snap Freezing Tissue in Optimal Cutting Temperature (OCT) Media (Fig. 5a)

1. Personal protective equipment (PPE), including laboratory coat or disposable gown, safety glasses or splash goggles, and gloves.
2. Tissue Plus® Optimal Cutting Temperature O.C.T. media (Scigen Scientific Inc., Gardena, CA, USA).
3. Cork stopper (ThermoFisher Scientific, Hampton, NH, USA).
4. Carbon steel single edge razor blade (Razor Blade Co., Van Nuys, CA, USA).
5. Plastic microscope coverslips (22 mm) (ThermoFisher Scientific, Hampton, NH, USA).
6. Cryo-marker pen (ThermoFisher Scientific, Hampton, NH, USA).
7. Metal beaker (Polar Ware Co., Kiel, WI, USA).
8. Isopentane (2-methylbutane) (ThermoFisher Scientific, Hampton, NH, USA).
9. Styrofoam box (ULINE, Pleasant Prairie, WI, USA).
10. Long-handled forceps (Fine Science Tools, Foster City, CA, USA).

2.2.2 Snap Freezing Tissue in Liquid Nitrogen

1. Personal protective equipment (PPE), including laboratory coat or disposable gown, safety glasses or splash goggles, and gloves.
2. Screw-top tubes (1.8 mL) with O-ring cap (ThermoFisher Scientific, Hampton, NH, USA).
3. Styrofoam box (ULINE, Pleasant Prairie, WI, USA).
4. Dissection tools (Fine science tools, Foster City, CA, USA).
5. Carbon steel single edge razor blade (Razor Blade Co., Van Nuys, CA, USA).
6. Long-handled forceps (Fine Science Tools, Foster City, CA, USA).

2.2.3 Tissue Storage in RNAlater™ Stabilization Solution

1. Personal protective equipment (PPE), including laboratory coat or disposable gown, safety glasses or splash goggles, and gloves.
2. Screw-top tubes (2 mL) with O-ring cap (ThermoFisher Scientific, Hampton, NH, USA).
3. RNA*later*™ stabilization solution (ThermoFisher Scientific, Waltham, MA, USA).
4. Dissection tools (Fine science tools, Foster City, CA, USA).
5. Carbon steel single edge razor blade (Razor Blade Co., Van Nuys, CA, USA).

2.3 Histological Staining

2.3.1 Hematoxylin and Eosin (H&E) Staining

1. Personal protective equipment (PPE), including laboratory coat or disposable gown, safety glasses or splash goggles, and gloves.
2. Cryostat tissue-sectioning machine (Epredia, Kalamazoo, MI, USA).
3. Superfrost™ Plus microscope slides (ThermoFisher Scientific, Waltham, MA, USA).
4. Glacial acetic acid (Sigma-Aldrich, St. Louis, MO, USA).
5. Ammonium hydroxide (Sigma-Aldrich, St. Louis, MO, USA).
6. Eosin Y (Sigma-Aldrich, St. Louis, MO, USA).
7. Phloxine B (Sigma-Aldrich, St. Louis, MO, USA).
8. 95% and 100% Ethanol (Sigma-Aldrich, St. Louis, MO, USA).
9. Tissue-Tek® Manual Slide Staining Set (Tissue-Tek, Torrance, CA, USA).
10. Modified Harris Hematoxylin (ThermoFisher Scientific, Waltham, MA, USA).
11. Xylenes (Sigma-Aldrich, St. Louis, MO, USA).
12. Fisher Chemical™ Permount™ Mounting Medium (ThermoFisher Scientific, Waltham, MA, USA).
13. Glass slide coverslips (ThermoFisher Scientific, Waltham, MA, USA).

2.3.2 Masson's Trichrome (MTC) Staining (Fig. 6)

1. Personal protective equipment (PPE), including laboratory coat or disposable gown, safety glasses or splash goggles, and gloves.
2. Cryostat tissue-sectioning machine (Epredia, Kalamazoo, MI, USA).
3. Superfrost™ Plus microscope slides (ThermoFisher Scientific, Waltham, MA, USA).
4. PAP liquid blocking pen (ThermoFisher Scientific, Waltham, MA, USA).
5. EasyDip slide jar staining system (ThermoFisher Scientific, Waltham, MA, USA).
6. Staining slide tray chamber (ThermoFisher Scientific, Waltham, MA, USA).
7. Epredia™ Richard-Allan Scientific™ Masson Trichrome Kit, containing Bouin's Fluid, Weigerts Iron Hematoxylin, Part A and Part B, Biebrich Scarlet Acid Fuchsin Solution, Phosphotungstic-Phosphomolybdic Acid Solution, and Aniline Blue Stain Solution (ThermoFisher Scientific, Waltham, MA, USA).
8. Ethanol (Sigma-Aldrich, St. Louis, MO, USA).
9. Xylenes (Sigma-Aldrich, St. Louis, MO, USA).
10. Fisher Chemical™ Permount™ Mounting Medium (ThermoFisher Scientific, Waltham, MA, USA).
11. Glass slide coverslips (ThermoFisher Scientific, Waltham, MA, USA).

2.3.3 Alizarin Red Staining

1. Personal protective equipment (PPE), including laboratory coat or disposable gown, safety glasses or splash goggles, and gloves.
2. Cryostat tissue-sectioning machine (Epredia, Kalamazoo, MI, USA).
3. Superfrost™ Plus microscope slides (ThermoFisher Scientific, Waltham, MA, USA).
4. Alizarin Red S (Sigma-Aldrich, St. Louis, MO, USA).
5. Glacial acetic acid (Sigma-Aldrich, St. Louis, MO, USA).
6. Tissue-Tek® Manual Slide Staining Set (Tissue-Tek, Torrance, CA, USA).
7. Ethanol (Sigma-Aldrich, St. Louis, MO, USA).
8. Xylenes (Sigma-Aldrich, St. Louis, MO, USA).
9. Fisher Chemical™ Permount™ Mounting Medium (ThermoFisher Scientific, Waltham, MA, USA).
10. Glass slide coverslips (ThermoFisher Scientific, Waltham, MA, USA).

2.3.4 nNOS Activity Staining

1. Personal protective equipment (PPE), including laboratory coat or disposable gown, safety glasses or splash goggles, and gloves.
2. NADPH (β-Nicotinamide adenine dinucleotide 2′-phosphate reduced tetrasodium salt hydrate, N-1630) (Sigma-Aldrich, St. Louis, MO, USA).
3. 1× Phosphate buffered saline (PBS) (ThermoFisher Scientific, Hampton, NH, USA).
4. NBT (Nitrotetrazolium Blue chloride, N-6876) (Sigma-Aldrich, St. Louis, MO, USA).
5. Methanol (Sigma-Aldrich, St. Louis, MO, USA).
6. Light-sensitive storage tubes (ThermoFisher Scientific, Hampton, NH, USA).
7. Triton X-100 (Sigma-Aldrich, St. Louis, MO, USA).
8. Cryostat tissue-sectioning machine (Epredia, Kalamazoo, MI, USA).
9. Superfrost™ Plus microscope slides (ThermoFisher Scientific, Hampton, NH, USA).
10. PAP liquid blocking pen (ThermoFisher Scientific, Hampton, NH, USA).
11. EasyDip slide jar staining system (ThermoFisher Scientific, Hampton, NH, USA).
12. Staining slide tray chamber (ThermoFisher Scientific, Hampton, NH, USA).
13. Paraformaldehyde (Sigma-Aldrich, St. Louis, MO, USA).
14. Aqua-Mount™ mounting medium (ThermoFisher Scientific, Hampton, NH, USA).
15. Glass slide coverslips (ThermoFisher Scientific, Hampton, NH, USA).
16. DPX (Distyrene plasticizer xylene) mountant (Sigma-Aldrich, St. Louis, MO, USA).

2.3.5 Immunofluorescence Staining

1. Personal protective equipment (PPE), including laboratory coat or disposable gown, safety glasses or splash goggles, and gloves.
2. Cryostat tissue sectioning machine (Epredia, Kalamazoo, MI, USA).
3. Superfrost™ Plus microscope slides (ThermoFisher Scientific, Waltham, MA, USA).
4. PAP liquid blocking pen (ThermoFisher Scientific, Waltham, MA, USA).
5. 1× Phosphate buffered saline (PBS) (ThermoFisher Scientific, Hampton, NH, USA).

6. EasyDip slide jar staining system (ThermoFisher Scientific, Waltham, MA, USA).
7. Staining slide tray chamber (ThermoFisher Scientific, Waltham, MA, USA).
8. CitiFluor AF1 Mounting Media (Electron Microscopy Sciences, Hatfield, PA, USA).
9. DPX (Distyrene plasticizer xylene) mountant (Sigma-Aldrich, St. Louis, MO).
10. Glass slide coverslips (ThermoFisher Scientific, Waltham, MA, USA).

2.3.6 Immunohistochemical Staining

1. Personal protective equipment (PPE), including laboratory coat or disposable gown, safety glasses or splash goggles, and gloves.
2. Cryostat tissue sectioning machine (Epredia, Kalamazoo, MI, USA).
3. Superfrost™ Plus microscope slides (ThermoFisher Scientific, Hampton, NH, USA).
4. PAP liquid blocking pen (ThermoFisher Scientific, Hampton, NH, USA).
5. EasyDip slide jar staining system (ThermoFisher Scientific, Hampton, NH, USA).
6. Staining slide tray chamber (ThermoFisher Scientific, Hampton, NH, USA).
7. Acetone (Sigma-Aldrich, St. Louis, MO, USA).
8. Formaldehyde (Sigma-Aldrich, St. Louis, MO, USA).
9. 1× Phosphate buffered saline (PBS) (ThermoFisher Scientific, Hampton, NH, USA).
10. H_2O_2 (ThermoFisher Scientific, Hampton, NH, USA).
11. ABC peroxidase staining kit (ThermoFisher Scientific, Hampton, NH, USA).
12. Vector substrate kit, peroxidase (HRP) (Vector Laboratories, Burlingame, California, USA).
13. Vector Hemotoxylin QS counterstain (Vector Laboratories, Burlingame, California, USA).
14. Ethanol (ThermoFisher Scientific, Hampton, NH, USA).
15. Xylene (ThermoFisher Scientific, Hampton, NH, USA).
16. Permount™ mounting medium (ThermoFisher Scientific, Hampton, NH, USA).
17. Glass slide coverslips (ThermoFisher Scientific, Hampton, NH, USA).

3 Methods

3.1 Tissue Collection

3.1.1 Needle Biopsy

The following protocol describes acquiring muscle biopsy samples from the right biceps femoris muscle in the upper hindlimb (*see* **Note 1**).

1. Follow the institutional guidelines for animal care and handling. Obtain approval from the institutional animal care and use committee (ACUC). Wear all PPE before working or handling animals. Use aseptic techniques while performing the needle biopsy procedure.
2. Fast the dog the night before the procedure.
3. Prepare the sedation agents cocktail according to the body weight. First, withdraw the calculated dosage of the Dexdomitor (6–15 μg/kg) into a syringe, then withdraw the calculated dosage of the nalbuphine (1 mg/kg) into a separate syringe. Next, add the withdrawn volume of the Dexdomitor into the nalbuphine syringe and gently invert to mix both agents (*see* **Note 2**).
4. Gently restrain the animal and shave the body hair on the right lateral hindlimb, proximal to the stifle, to visualize the musculature of the upper hindlimb (*see* **Note 3**).
5. Perform a physical exam, and record the vital signs (heart rate, respiratory rate and effort, and rectal body temperature), capillary refill time (CRT), and mucous membrane color.
6. Inject the sedation agents cocktail intramuscularly in the epaxial muscle.
7. Place the animal over the preheated thermal blanket in a left lateral recumbency position. Set the oxygen mask over the dog's muzzle to supply the animal with oxygen throughout the procedure.
8. Monitor the heart rate, electrocardiogram, oxygen saturation (SpO_2), blood pressure, and body temperature using the DRE vital sign monitor throughout the entire experiment. Record the vital signs, CRT, mucous membrane color, the palpebral reflex, and the pedal reflex every 15 min. Place eye lubricant in each eye.
9. Disinfect the surgical site using a two-step scrubbing technique. First, place a soaked chlorhexidine gauze at the proposed incision site and then move it in a circular motion. Gradually move your way out from the incision site until the whole shaved area is cleaned with chlorhexidine. Discard the chlorhexidine gauze. Next, repeat the same cleaning process using a soaked isopropanol gauze. Repeat the two-step scrubbing technique three times (*see* **Note 4**).

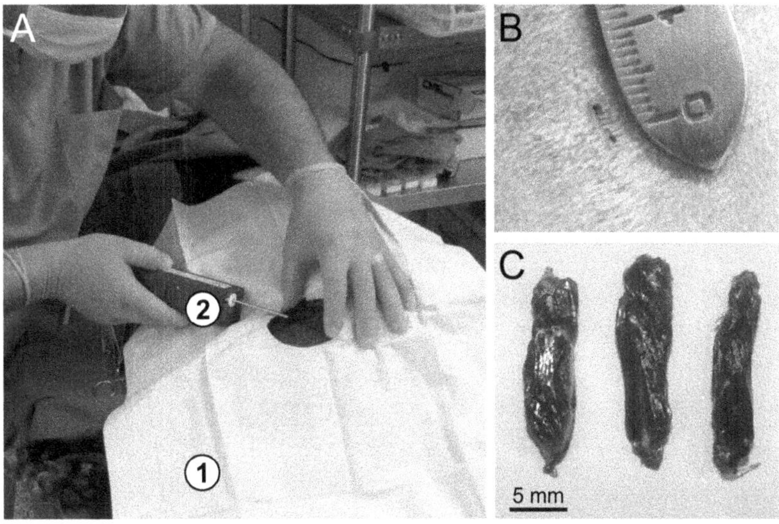

Fig. 3 Dog muscle vacuum-assisted biopsy. (**a**) A photograph showing the insertion of Vacora® Biopsy probe into the biceps femoris muscle to acquire a muscle biopsy. (1) Surgical drape with circular aperture, (2) Vacora® vacuum-assisted biopsy system driver. (**b**) A photograph showing the size of the incision site. (**c**) A photograph showing the muscle biopsy

10. Prepare the surgical site for acquiring the muscle biopsy sample. First, replace gloves with sterile surgical gloves, and place a sterile surgical drape over the disinfected surgical site. Position the center of the circular aperture of the drape over the proposed incision site (Fig. 3a). Next, confirm that the animal has no pedal reflex by squeezing and pulling the distal lower forelimb. Then, using the #11 blade, make a 0.5-cm incision at the proximal end of the biceps femoris muscle (Fig. 3b).
11. Assemble the Vacora® vacuum-assisted biopsy system by inserting the probe into the driver hub. Once the probe is loaded correctly, the drive will automatically prime itself to adjust the position of the cutting sleeve over the biopsy chamber on the probe.
12. Acquire the muscle biopsy sample by first inserting the probe into the biceps femoris muscle through the incision site to be parallel with the muscle fibers (Fig. 3b). Adjust the insertion length and depth of the probe into the muscle as desired. Next, press the multifunction button on the driver to acquire a muscle biopsy sample. Once the driver's motor stops, pull the probe out of the muscle and press the multifunction button again to retract the cutting sleeve away from the biopsy chamber. Using sterile tissue forceps, remove the muscle biopsy sample from the probe chamber and place it over a saline wet gauze in the plastic container (Fig. 3c). Press the multifunction button to reset the probe to acquire additional biopsy samples (*see* **Note 5**).

13. Clean the incision site with chlorhexidine from any bleeding that may have occurred during the procedure and dry it with sterile dry gauze. Seal the incision with 1–2 drops of the 3-M Vetbond.
14. Inject the calculated dosage of Rimadyl (4.4 mg/kg) subcutaneously on the dorsal aspect of the dog's back (*see* **Note 6**).
15. Inject Antisedan intramuscularly on the contralateral side of the epaxial muscle. The administered volume of Antisedan should be equal to the total volume of Dexdomitor given during the procedure.
16. Continue monitoring the dog until it can maintain an unassisted sternal position. Then, perform a postoperative assessment for five days to confirm no adverse effects from the procedure.

3.1.2 Necropsy

1. Follow the institutional guidelines for animal care and handling. Obtain approval from the institutional animal care and use committee (ACUC). Wear all PPE before working or handling animals (*see* **Note 7**).
2. Euthanize the dog according to the *AVMA Guidelines for Euthanasia of Animals* (*see* **Note 8**).
3. *Thoracic and abdominal organs removal*
   (a) Place the animal in dorsal recumbency on the necropsy table and shave the skin hair from the manubrium to the inguinal region.
   (b) Clean the skin of any hair residue, make an incision along the midline from the manubrium to the xyphoid process, and continue the incision through the linea alba to the inguinal ring to expose the thoracic and abdominal contents (*see* **Note 9**).
   (c) Cut along the sternum and use the Weitlaner retractor to separate the rib cage.
   (d) Cut through the pericardial sac to expose the whole heart.
   (e) Clamp the aorta and the vena cava and excise the whole heart (*see* **Note 10**).
   (f) Dissect the individual heart regions and store them in a humidified chamber on a 1× phosphate buffered saline (PBS) wet gauze until processed (*see* **Notes 11** and **12**).
   (g) Collect the desired internal organs and diaphragm muscle. Store the collected tissue in a humidified chamber on a 1× PBS wet gauze until processed (*see* **Note 13**).
4. *Central nervous system dissection*
   (a) Position the animal in ventral recumbency and shave the hair of the dorsal aspect of the head from the frontal bone to the fourth cervical vertebrae.

(b) Clean the skin of any hair residue and make an incision from the nasal bone to the fourth cervical vertebrae. Retract the skin bilaterally to expose the temporalis muscles. Dissect both temporalis muscles and empty the orbital contents.

(c) Cut through the frontal bone and continue through the temporal bone to the occipital bone bilaterally using the oscillating saw (*see* **Note 14**).

(d) Pull the skull bones dorsocadually and cut the cericoauricularis to remove the skull. Then, carefully cut through the meninges to expose the brain. Remove the brain from the olfactory bulb to the cerebellum. Dissect the desired brain parts and store them in a humidified chamber on a 1× PBS wet gauze until processed (*see* **Note 15**).

(e) Dissect the muscles between the vertebral spinous process and the transverse process bilaterally to expose the cervical vertebrae from C_1 to C_4.

(f) Cut along the lateral aspects of the cervical vertebrae to expose the cranial aspect of the spinal cord using the oscillating saw. Then, carefully remove the spinal cord from C_1 to C_4 and store it in a humidified chamber on a 1× PBS wet gauze until processed.

5. *Skeletal muscle collection*

(a) Position the animal in lateral recumbency and shave the hair skin over the target skeletal muscle region (*see* **Note 16**).

(b) Clean the skin of any hair residue and dissect the skin, fascia, and any fat covering the target muscle. Dissect the whole or part of the skeletal muscle and store it in a humidified chamber on a 1× PBS wet gauze until it is processed (Fig. 4).

3.2 Tissue Processing

The following protocol describes the snap freezing of freshly harvested tissue embedded in OCT media (*see* **Note 17**).

3.2.1 Snap Freezing Tissue in Optimal Cutting Temperature (OCT) Media

1. Prepare the cork-coverslip molding by first cutting the cork stoppers into a 50–70 mm disk using a sharp razor blade. Then cut through one of the sides of the disk top to form a shallow channel to insert a plastic coverslip vertically into the cork disk. Label the bottom of the cork disk accordingly using a cryomarker pen (*see* **Note 18**).

2. Fill a metal beaker two-thirds with isopentane and fill a styrofoam container with some liquid nitrogen. Immerse the metal beaker into the liquid nitrogen and add more liquid nitrogen in the styrofoam container until the level of the liquid nitrogen is the same as the isopentane level in the metal beaker. Wait for

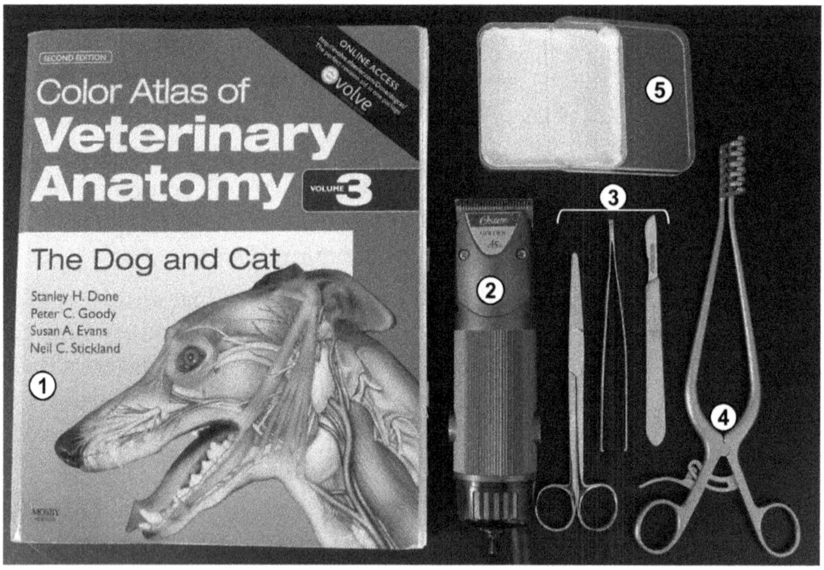

Fig. 4 Materials required for dog necropsy. (1) Color Atlas of Veterinary Anatomy, The Dog and Cat, book, (2) animal hair clipper with a #50 blade, (3) surgical dissection instruments, (4) Weitlaner retractor, (5) humidified chamber to store dissected tissue, the chamber is made of a petri dish containing a 1× PBS wet gauze

5–10 min until a white base is formed at the bottom of the metal beaker to confirm that the isopentane is chilled enough to snap freeze an OCT specimen at −160 °C (*see* **Notes 19** and **20**).

3. Blot any liquid off the harvested tissue and trim it approximately to form a cube of approximately 0.5 cubic inches using a razor blade or dissection tools (*see* **Note 21**).
4. Add a couple of drops of OCT to the coverslip, embed the tissue in the OCT, and orient it accordingly (Fig. 5b). Hold the cork-coverslip molding using long-handled forceps to position the coverslip horizontally and add more OCT to cover the tissue entirely (Fig. 5c, d). Immediately immerse the molding in the chilled isopentane and wait for 30–60 s to snap freeze the OCT embedded tissue (Fig. 5e) (*see* **Note 22**).
5. Retrieve the sample from the chilled isopentane using the long-handled forceps and quickly place it on dry ice. Store the frozen tissue in a −80 °C freezer for long-term storage (*see* **Note 23**).

3.2.2 Snap Freezing Tissue in Liquid Nitrogen

The following protocol describes the snap freezing of freshly harvested tissue in liquid nitrogen (*see* **Note 24**).

1. Label a clean screw-top tube accordingly using a cryo-marker pen.
2. Fill a styrofoam container with enough liquid nitrogen to submerge the tube (*see* **Notes 25** and **26**).

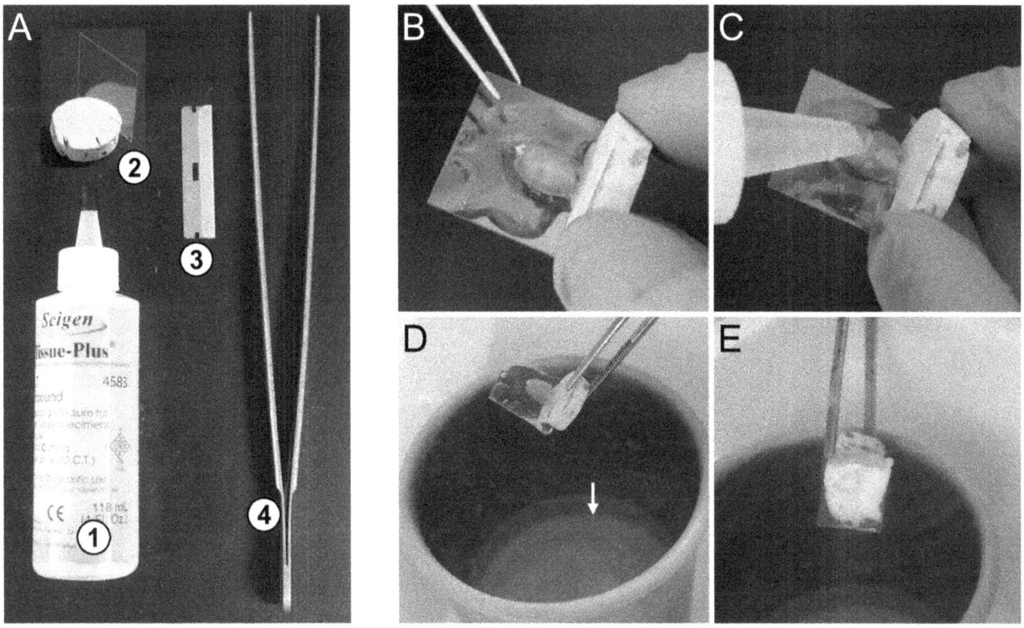

Fig. 5 Snap freezing tissue in optimal cutting temperature (OCT) media. (**a**) Materials required to snap freeze tissue in OCT. (1) Tissue Plus® optimal cutting temperature O.C.T. media, (2) cork-coverslip tissue embedding molding, (3) carbon steel single edge razor blade, (4) long-handled forceps. (**b**) A photograph showing the orientation and embedding of tissue in the cork-coverslip molding. (**c**) A photograph showing the covering of the tissue with OCT before the freezing. (**d**) A photograph showing how to hold the cork-coverslip molding before it is immersed in chilled isopentane. Arrow, frozen isopentane at the base of the metal beaker. (**e**) A photograph showing a frozen OCT-embedded tissue

3. Blot any liquid off the harvested tissue and trim it from any fat or connective tissue using dissection tools. Cut the tissue into small cubicles (0.2 cubic inches) using a sharp razor blade. Fill half the tube with tissue pieces and screw cap it (*see* **Note 27**).
4. Hold the tube with long-handled forceps and immerse it slowly in the liquid nitrogen. Allow the tissue to freeze for 1–3 min (*see* **Note 28**).
5. Retrieve the tube from the liquid nitrogen using the long-handled forceps and quickly place it on dry ice. Store the frozen tissue in a −80 °C freezer for long-term storage.

3.2.3 Tissue Storage in RNAlater™ Stabilization Solution

The following protocol describes storing freshly harvested tissue samples in RNA*later*™ stabilization solution for the RNA extraction approach (*see* **Note 29**).

1. Label a clean screw-top tube accordingly using a fine-point permanent marker.
2. Add 500–750 μL of RNA*later*™ stabilization solution into the tube (*see* **Note 30**).

3. Blot any liquid off the harvested tissue and trim it from any fat or connective tissue using dissection tools. Cut the tissue into small cubicles (0.2 cubic inches) using a sharp razor blade. Submerge the tissue pieces in the RNA*later*™ stabilization solution and screw the cap on the tube (*see* **Notes 31** and **32**).
4. Store sample at room temperature for up to 1 week, 4 °C for up to 1 month, or −20 °C or −80 °C for long-term storage (*see* **Note 33**).

3.3 Histological Staining

3.3.1 Hematoxylin and Eosin (H&E) Staining

The following protocol describes H&E staining using a snap-frozen OCT-embedded canine tissue sample. H&E staining is the gold standard staining approach to investigate general histopathology. Hematoxylin is acidophilic and will stain all nuclei a blue-purple color. Eosin is basophilic and will stain cytoplasm and the extracellular matrix with a pink color. In the context of muscular dystrophy, H&E staining is very useful for observing general dystrophic pathology, including the location of the nucleus in the fiber, muscle fiber size, mononucleated cell infiltration, and fibrosis deposition (*see* **Note 34**).

1. Cut 10 μm thick canine tissue sections using a cryostat machine. Mount the tissue section on a Superfrost™ plus microscope slide. Allow the tissue section to dry completely at room temperature (*see* **Note 35**).
2. Prepare the acetic acid alcohol fixative by mixing 90 mL of 95% ethanol with 10 mL of glacial acetic acid.
3. Prepare the ammonia water solution by mixing 250-mL distilled water with 1.5 drops of ammonium hydroxide.
4. Prepare the eosin staining solution by mixing the following reagents: 25 mL of 1% w/v eosin Y, 2.5 mL 1% w/v phloxine B, 195 mL of 95% ethanol, and 1.25 of glacial acetic acid.
5. Using the Tissue-Tek® Manual Slide Staining Set for H&E staining, immerse the slide in the following reagents order: (*see* **Note 36**).
   (a) Acetic acid alcohol fixative for 5 min.
   (b) Tap water for 15 s.
   (c) Modified Harris hematoxylin for 5 min (*see* **Note 37**). Drain the excess of the hematoxylin by tapping the slide on paper towels.
   (d) Tap water for 30 s and repeat in clean tap water for 30 s.
   (e) Ammonia water solution for 15 s.
   (f) 95% ethanol for 15 s and repeat in clean 95% ethanol for 15 s.
   (g) Eosin stain solution for 15 s. Drain the excess eosin by tapping the slide on paper towels.

(h) 95% ethanol for 15 s and repeat in clean 95% ethanol for 15 s.

(i) 100% ethanol for 15 s and repeat in clean 100% ethanol for 15 s.

(j) Xylenes for 15 s and repeat in clean xylenes for 15 s. Drain the excess xylenes by tapping the slide on paper towels.

6. Add one drop of Permount™ mounting medium to the tissue section and cover with an appropriate-sized coverslip. Drain any excess mounting media using filter paper. Allow the slide to air dry (*see* **Note 38**).
7. Visualize the stained tissue under a light microscope using bright-field illumination.

3.3.2 Masson's Trichrome (MTC) Staining

The following protocol describes MTC staining using a snap-frozen OCT-embedded canine tissue sample. A hallmark of muscular dystrophy disease is the replacement of muscle tissue with fibrotic tissue. Masson trichrome staining is generally used to stain collagenous connective tissue, which is useful to assess fibrosis deposition in muscle tissues. The staining is achieved by immersion the fixed sample into Weigert's iron hematoxylin to stain the nuclei a black color, Biebrich Scarlet-Acid Fuchsin to stain myofibers dark red, and phosphotungstic-phosphomolybdic acid and aniline blue together to stain collagen a dark blue color (*see* **Note 39**).

1. Cut 10-μm thick canine tissue sections using a cryostat machine. Mount the tissue section on a Superfrost™ plus microscope slide. Allow the tissue section to dry completely at room temperature (*see* **Note 40**).
2. Outline around the tissue with a PAP liquid blocking pen (hydrophobic barrier) to hold reagents within the outlined area when applied on the tissue section.
3. Cover the tissue section with ddH_2O for 5 min to hydrate (*see* **Note 41**).
4. Discard the ddH_2O and cover the tissue section with Bouin's fluid for 1 h at 56 °C.
5. Rinse sections with tap water for 5 min until the Bouin's fluid color (yellow color) disappears.
6. Cover the tissue section with Working Weigert's iron hematoxylin stain solution for 10 min at room temperature (*see* **Note 42**).
7. Rinse sections in tap water for 10 min.
8. Cover the tissue section with Biebrich Scarlet-Acid Fuchsin solution for 10 min at room temperature.
9. Rinse sections with ddH_2O for 30 s.

10. Cover the tissue section with phosphotungstic-phosphomolybdic acid solution for 5 min at room temperature.
11. Discard the phosphotungstic-phosphomolybdic acid solution and cover the tissue section with aniline blue solution for 10 min at room temperature.
12. Discard the aniline blue solution and cover the tissue section with 1% acetic acid for 1 min at room temperature.
13. Rinse sections with ddH_2O for 30 s.
14. Immerse the slide in 100% ethanol for 1 min and repeat in clean 100% ethanol for 1 min to dehydrate.
15. Immerse the slide in xylenes for 1 min and repeat two more times in clean xylenes for 1 min each. Drain excess xylenes by tapping the slide on paper towels.
16. Add one drop of Permount™ mounting medium to the tissue section and cover with an appropriate-sized coverslip. Drain any excess mounting media using filter paper. Allow the slide to air dry (*see* **Note 43**).
17. Visualize the stained tissue under a light microscope using bright-field illumination.

3.3.3 Alizarin Red Staining

The following protocol describes Alizarin Red S staining using a snap-frozen OCT-embedded canine tissue sample. Alizarin Red S staining assesses calcium deposition and calcification in muscle tissues and identifies calcified myofibers and lesions in the context of muscular dystrophy (*see* **Note 44**).

1. Cut 10 μm thick canine tissue sections using a cryostat machine. Mount the tissue section on a Superfrost™ plus microscope slide. Allow the tissue section to dry completely at room temperature (*see* **Note 45**).
2. Prepare 2% w/v Alizarin Red S staining solution in deionized water. Sterile filter and adjust the pH to 4.1–4.3 using glacial acetic acid.
3. Using the Tissue-Tek® Manual Slide Staining Set, immerse the slide in the following reagents order:
   (a) 2% Alizarin Red S staining solution for 2 min (*see* **Note 46**).
   (b) Deionized water for 10 s (*see* **Note 47**).
   (c) 70% ethanol for 10 s.
   (d) 90% ethanol for 10 s.
   (e) 100% ethanol for 10 s.
   (f) Xylenes for 1 min and repeat in clean xylenes for 1 min. Drain the excess of the xylenes by tapping the slide on paper towels.

4. Add one drop of Permount™ mounting medium to the tissue section and cover with an appropriate-sized coverslip. Drain any excess mounting media using filter paper. Allow the slide to air dry (*see* **Note 48**).
5. Visualize the stained tissue under a light microscope using bright-field illumination.

3.3.4 nNOS Activity Staining

The following protocol describes a histochemical staining protocol to identify the cellular localization of the neuronal nitric oxide synthase (nNOS) using a snap-frozen OCT-embedded canine tissue sample (*see* **Note 49**).

1. Prepare 10-mM NADPH by dissolving 8.3 mg of NADPH in 1 mL of 1× PBS. Aliquot 100 μL in light-sensitive storage tubes and store at −20 °C.
2. Prepare 16 mg/mL NBT by dissolving 16 mg of NBT (N-6876) in 1-mL methanol. Aliquot 100 μL in light-sensitive storage tubes and store at −20 °C.
3. Prepare 1 mL of the NADPH working solution by adding 100 μL of 2% Triton X-100, 20 μL of 10-mM NADPH, and 10 μL of 16 mg/mL NBT to 870 μL of 1× PBS. The working solution should be freshly prepared before starting the staining.
4. Section a 16 μm thick canine tissue sections using a cryostat machine. Mount the tissue section on a Superfrost™ plus microscope slide and allow it to dry completely at room temperature (*see* **Note 50**).
5. Outline around the tissue with a PAP liquid blocking pen to hold reagents within the outlined area when applied on the tissue section (*see* **Note 51**).
6. Fix the tissue with 4% paraformaldehyde for 2 h at 4 °C.
7. Wash the tissue section three times in 1× PBS for 8 min each at room temperature and drain the slide (*see* **Note 52**).
8. Immerse the slide in a pre-warmed 0.2% Triton X-100 and incubate in a 37 °C water bath for 20 min.
9. Cover the tissue section with a freshly made NADPH working solution and incubate for 3–4 h in a 37 °C incubator.
10. Wash the tissue section twice in ddH_2O for 2 min each at room temperature and drain the slides.
11. Add a couple of drops of Aqua-Mount™ mounting medium to the tissue section and cover with an appropriate-sized coverslip. Drain the excess of the mounting medium using filter paper. Seal the edges of the coverslip to the slide using DPX mountant (*see* **Note 53**).
12. Visualize the stained tissue under a microscope using bright-field illumination.

3.3.5 Immuno-fluorescence Staining

The following protocol provides a general guide on performing immunofluorescent (IF) staining using a snap-frozen OCT-embedded canine tissue sample. It is important to optimize the experimental conditions for each antigen of interest (*see* **Note 54**).

1. Cut 10 μm thick canine tissue sections using a cryostat machine. Mount the tissue section on a Superfrost™ plus microscope slide. Allow the tissue section to dry completely at room temperature (*see* **Note 55**).
2. Outline around the tissue section with a PAP liquid blocking pen to hold reagents within the outlined area when applied on the tissue section (*see* **Note 56**).
3. Cover the tissue section with 1% serum of the secondary antibody host species for 5 min. Discard the serum and repeat this twice for 5 min each (*see* **Note 57**).
4. Discard the 1% serum and cover the tissue section with 20% serum of the secondary antibody host species for 30 min at room temperature to block any unspecific binding sites (*see* **Note 58**).
5. Discard the serum, wash the tissue section three times with 1% serum for 5 min at room temperature, and drain the slide (*see* **Note 59**).
6. Cover the tissue section with the primary antibody and incubate overnight at 4 °C (*see* **Note 60**).
7. Recover the primary antibody and wash the tissue section with 1% serum three times for 5 min each at room temperature and drain the slide (*see* **Note 61**).
8. Cover the tissue section with secondary antibody for 30 min at room temperature and protected from light.
9. Discard the secondary antibody and wash the tissue section two times with 1× phosphate buffered saline (PBS) for 5 min each at room temperature and drain any excess PBS.
10. Add a couple of drops of CitiFluor AF1 media to the tissue section and cover with an appropriate-sized coverslip. Remove any bubbles by gently pressing a pipette tip against the coverslip (*see* **Note 62**).
11. Seal the edges of the coverslip to the glass slide using DPX mountant and allow the DPX to dry for several hours or overnight while protecting from light (*see* **Note 63**).
12. Visualize the stained tissue section using a fluorescence microscope (*see* **Note 64**).

3.3.6 Immunohistochemical Staining

The following protocol provides a general guide on performing immunohistochemistry (IHC) staining using a snap-frozen OCT-embedded canine tissue sample. It is important to optimize the experimental conditions for each antigen of interest (*see* **Note 65**).

1. Cut 10 μm thick canine tissue sections using a cryostat machine. Mount the tissue section on a Superfrost™ plus microscope slide. Allow the tissue section to dry completely at room temperature (*see* **Note 66**).
2. Outline around the tissue section with a PAP liquid blocking pen to hold reagents within the outlined area when applied on the tissue section.
3. Fix the tissue section with −20 °C acetone for 10 min at −20 °C, and then allow the tissue to air dry for 10 min at room temperature (*see* **Note 67**).
4. Fix the tissue with 2% formaldehyde for 10–15 min at room temperature.
5. Wash the tissue section three times in 1× phosphate buffered saline (PBS) for 3 min at room temperature and drain the slide (*see* **Note 68**).
6. Cover the tissue section with a freshly made 0.3% H_2O_2 and incubate for 30 min at room temperature to quench the endogenous peroxidase activity.
7. Discard the H_2O_2, wash the tissue section three times with 1× PBS for 3 min at room temperature, and drain the slide.
8. Block the tissue section with serum matching the secondary antibody species for 30 min at room temperature to block nonspecific hydrophobic interactions between the primary antibody and the tissue (*see* **Note 69**).
9. Discard the blocking solution, wash the tissue section three times with 1× PBS for 3 min at room temperature, and drain the slide.
10. Cover the tissue section with primary antibody and incubate overnight at 4 °C (*see* **Note 70**).
11. Recover the primary antibody, wash the tissue section three times with 1× PBS for 7 min each at room temperature, and drain the slide (*see* **Note 71**).
12. Cover the tissue section with secondary antibody for 30 min at room temperature.
13. Discard the secondary antibody, wash the tissue section five times with 1× PBS for 3 min each at room temperature, and drain the slide.

14. Mix one drop of solution A and one drop of Solution B of the ABC peroxidase staining kit in 2.5 mL of 1× PBS. Cover the tissue with the mix and incubate for 30 min at room temperature to conjugate the biotinylated antibody with a high sensitivity avidin-HRP complex.
15. Discard the mix solution, wash the tissue section five times with 1× PBS for 3 min at room temperature, and drain the slide.
16. Prepare the substrate solution according to the Vector laboratory substrate kit manufacturer's instructions (*see* **Note 72**).
17. Cover the tissue section with the substrate and incubate for 5–10 min at room temperature (*see* **Note 73**).
18. Discard the substrate, wash the tissue section two times with 1× PBS for 5 min at room temperature, and drain the slide.
19. Cover the tissue with Vector Hemotoxylin QS counterstain and incubate for 4 min at room temperature (*see* **Note 74**).
20. Discard the counterstain, rinse the slide in tap water for 1 min at room temperature, and drain the slide.
21. Immerse the slide in 70% ethanol for 1 min at room temperature.
22. Immerse the slide in 95% ethanol for 1 min at room temperature.
23. Immerse the slide in 100% ethanol for 1 min at room temperature.
24. Immerse the slide two times in xylene two times for 1 min at room temperature.
25. Add a couple of drops of Permount^{TM} mounting medium to the tissue section and cover with an appropriate-sized coverslip. Drain excess mounting medium using filter paper. Allow the slide to air dry.
26. Visualize the stained tissue under a microscope using bright-field illumination.

4 Notes

1. Muscle biopsy is a commonly performed procedure for diagnosing neuromuscular diseases and evaluating novel therapeutic approaches in human patients. Bergstrom needle biopsy, open biopsy, and vacuum-assisted biopsy system (VAB) are three commonly used techniques [6–8]. We have previously reported protocols for Bergstrom needle biopsy and open biopsy [9]. Here, we describe the VAB method. The VAB method was first used as a minimally invasive procedure for breast biopsy [10]. It was later adopted for muscle biopsy

[7, 11]. The VAB method has many advantages. It can be performed outside the surgery suite under local anesthesia. It is well tolerated, causes minor discomfort, and results in quick recovery when used in human patients [7, 11]. The VAB method has been used in DMD patients [12]. Here, we describe the Vacora® biopsy system-based VAB method for dog muscle biopsy. The procedure was developed and implemented by the author CHH. The protocol described in this chapter only presents a method for acquiring muscle tissue from the biceps femoris muscle. This procedure can also be used to acquire muscle tissues from other parts of the body. We have successfully performed this procedure in the vastus lateralis, gracilis, cranial tibialis, gastrocnemius, semitendinosus, deltoid, supraspinatus, infraspinatus, and lateral triceps muscles in dogs.

2. We usually start Dexdomitor at 6 μg/kg. Based on the level of sedation determined by the general mentation and pedal reflex, the dosage can be increased gradually at 3 μg/kg up to a maximum of 15 μg/kg during the procedure. We have successfully sedated dogs younger than 16 weeks without adverse events.
3. The person performing the needle biopsy procedure should have received sufficient training on dog anatomy and performed dog muscle surgery before.
4. When scrubbing the skin, the gauze should be folded into quarters by bringing the four corners together and held by the corners. This generates a smaller contact surface area but limits your finger from touching the skin during the scrubbing process. Also, never return to the starting center point after completing the scrubbing center-out technique. Discard the scrubbing gauze immediately once it reaches the fur/shaving line. There is no need to apply excessive pressure on the gauze while scrubbing, as it may compromise the dermal layer of the skin. Any skin abrasions or inflammation may lead to incision healing complications.
5. The person performing the biopsy should have sufficient knowledge of the shape, size, and fiber orientation of the biopsied muscle. We recommend practicing the procedure on carcasses before performing it on a live animal.
6. For dogs that are less than 16 weeks, we recommend intramuscular injection of 1 mL 2% lidocaine at the biopsy site for pain relief. If signs of pain are noted during the 5-day postoperation recovery period, a veterinarian should be consulted.
7. The dog necropsy procedure should be performed by investigators who are well trained in dog anatomy. We have used the following books in dog anatomy training, including Color

Atlas of Veterinary Anatomy, The Dog and Cat (ISBN: 9780-7234-3415-3) (Fig. 4); Dog Anatomy, A Pictorial Approach to Canine Structure (ISBN: 0-85131-636-0); Dog Anatomy, A Coloring Atlas (ISBN: 1-893441-17-2); and Miller's Anatomy of the Dog (ISBN: 978-0-7216-3200-1)

8. Review the AVMA guidelines for the euthanasia of animals (https://www.avma.org/resources-tools/avma-policies/avma-guidelines-euthanasia-animals) periodically for up-to-date information about acceptable euthanasia methods.
9. Exercise caution while cutting through the midline to avoid cutting through the diaphragm muscle, liver, or other internal organs.
10. Rinse the whole heart in 1× PBS to remove clotted blood from the inside of the heart chambers. Place the whole heart on dry paper towels to absorb any excess PBS from the outer walls of the heart.
11. We usually collect samples from the following regions from the heart: the right and left atriums, anterior and posterior papillary muscles, left ventricle anterior and posterior walls, apex, interventricular septum, and the cranial and caudal aspects of the right ventricle.
12. We use a square petri dish lined with 1× PBS wet gauze as a humidified chamber to store dissected tissue (Fig. 4). The collected tissue sample is placed immediately on top of the wet gauze, the plate is covered with the lid, and stored on ice to keep the sample cold and moist until processing.
13. We collect the following internal organs: thymus, lung, esophagus, liver, spleen, kidney, stomach, small and large intestine, pancreas, bladder, sex organs, and lymph nodes. We also collect muscle samples from three regions of the diaphragm: sternal, middle, and lumbar. Following the removal of the heart, diaphragm, and all the internal organs, we fill the thoracic and abdominal cavities with ice packs to keep the body temperature low. This helps preserve skeletal muscles.
14. Avoid cutting too deep into the skull bones to prevent damaging the brain.
15. We collect the following brain regions bilaterally: cerebral cortex, cerebellar cortex, and hippocampus.
16. The skeletal muscle regions are the head, neck, forelimb, thorax, abdomen, pelvic, and hindlimb. The muscles that we collect from each region are presented in Table 1. We collect skeletal muscle using a systematic approach beginning at the head and moving in a caudal direction to the next region. Within each region, we dissect the entire superficial muscles first. This helps maintain the anatomy of deeper muscles.

Table 1
The list of canine skeletal muscle

Skeletal muscle							
Head	**Neck**	**Chest**	**Shoulder**	**Upper arm**	**Back**	**Abdominal**	**Hind limb**
Temporalis	Sternohyoid muscle	Superficial pectoralis	Trapezius	Triceps lateral	Latissimus dorsi	Rectus abdominis	Biceps femoris
Extra-ocular Muscles	Sternocephalicus	Deep pectoralis	Deltoideus	Triceps accessory	Longissimus thoracis	External oblique muscle	Vastus lateralis
Masseter	Laryngeal muscles		Supraspinatus	Triceps long	External intercostal muscle	Internal oblique muscle	Semitendinosus
Digastricus	Cleidocervicalis		Infraspinatus	Triceps medial	Internal intercostal muscle	Transversus abdominis	Semimembranosus
Geniohyoid muscle			Teres major	Biceps brachii			Rectus femoris
Tongue			Teres minor	Brachialis			Vastus intermedius
				Extensor carpi ulnaris			Gluteus
				Extensor digitorum lateralis			Cranial Sartorius
				Extensor digitorum communis			Caudal Sartorius
				Extensor carpi radialis			Paw muscles
				Flexor carpi ulnaris			Gracilis
				Flexor digitorum superficialis			Adductor

(continued)

Table 1 (continued)

Skeletal muscle							
Head	**Neck**	**Chest**	**Shoulder**	**Upper arm**	**Back**	**Abdominal**	**Hind limb**
				Flexor digitorum profundus			Pectineus
				Flexor carpi radialis			Cranial tibialis
							Extensor digitorum longus
							Peroneus longus
							Gastrocnemius medial
							Gastrocnemius lateral
							Flexor digitorum superficialis

17. Different experimental approaches require pre-treatment or fixation of the tissue prior to the OCT snap freezing. We have previously used this protocol to snap freeze tissues fixed with 4% paraformaldehyde or 10% neutral buffered formalin.
18. The insertion of the coverslip into the cork disk forms an L-shaped molding to embed the sample and support it during freezing. The plastic coverslip provides a wide flat platform to orient and supports the tissue sample when it is placed against the cork disk.
19. The isopentane (2-methylbutane) is highly flammable, harmful to human health, an irritant, and dangerous to the environment. It should be handled by trained individuals, in a fume hood, away from any open flames, and disposed of properly.
20. Since liquid nitrogen evaporates, keep adding it to the styrofoam container to maintain it at a similar level to the isopentane while waiting for the isopentane to chill.
21. Collect tissues in accordance with the institutional guidelines for animal care and handling and approval from the Institutional Animal Care and Use Committee (ACUC). Dissect any fat or connective tissue from the sample prior to embedding it in the OCT.
22. When adding the OCT to the coverslip or on top of the tissue, avoid forming air bubbles as it may introduce artifacts when sectioning the sample. The freezing time may vary according to the size and thickness of the tissue. Gently stir the sample in the chilled isopentane to avoid sticking to the bottom of the metal beaker.
23. Prior to cryosectioning, the sample should be retrieved from the −80 °C freezer and placed in a −20 °C freezer for 4–6 h.
24. We do not recommend using snap frozen tissues in liquid nitrogen for histological studies. The snap freezing protocol is ideal for preserving tissue for DNA and protein extraction.
25. Liquid nitrogen must be adequately handled as it is hazardous. It has a boiling point of −320 °F and can quickly cause cryogenic burns and frostbite on the skin. Also, it is odorless and colorless and may result in an oxygen-deficient atmosphere in a confined space.
26. Since the liquid nitrogen evaporates quickly, keep adding it to the styrofoam container to maintain the desired quantity.
27. Collect tissues in accordance with the institutional guidelines for animal care and handling and approval from the Institutional Animal Care and Use Committee (ACUC).
28. Periodically stir the tube gently in the liquid nitrogen using the long-handled forceps to assure that all surface of the tube is immersed in liquid nitrogen.

29. RNA*later*™ Solution is only used with fresh tissue and cannot be used to store fixed tissues.
30. The amount of RNA*later*™ can be adjusted according to the sample amount/volume. It is recommended to fill tubes with RNA*later*™ 5× times the estimated volume of the tissue sample.
31. Collect tissues in accordance with the institutional guidelines for animal care and handling and approval from the Institutional Animal Care and Use Committee (ACUC).
32. Do not over pack the tube with tissue pieces. The tissue sample should move freely in solution, and all surfaces of the tissue should be exposed to the RNA*later*™ solution.
33. For long-term storage, we recommend incubating RNA*later* ™ samples at 4 °C for at least 12 h before transferring it to −20 °C or −80 °C freezer. RNA*later* ™ samples are stable indefinitely at both −20 °C and −80 °C.
34. All steps of this H&E staining procedure must be performed in a properly ventilated chemical fume hood. Many chemicals described in this protocol are skin, eye, and respiratory irritants, harmful to human health, corrosive, flammable, and harmful to the environment.
35. Canine tissue such as skeletal muscle, heart, internal organs, and nervous system can be used with this protocol. The thickness of the tissue section may vary based on the tissue type. We recommend testing 5–15 μm thick tissue sections to identify the optimal tissue thickness for the staining. A tissue section mounted on a slide can be stored long-term in an upright position in a slide box at −20 °C.
36. When using the Manual Slide Staining Set, fill each staining jar slightly below the top with freshly made reagents. Completely immerse the slides in a new staining jar with fresh reagents for each step (do not re-use reagents).
37. Hematoxylin staining time may vary depending on the tissue type.
38. Permount™ residues on the top of the coverslip or bottom of the slide can be removed with a cotton swab dipped in xylenes.
39. All steps of this MTC staining procedure must be performed in a properly ventilated chemical fume hood. Many chemicals described in this protocol are skin, eye, and respiratory irritants, harmful to human health, corrosive, flammable, and harmful to the environment.
40. Canine tissue such as skeletal muscle, heart, internal organs, and nervous system can be used with this protocol. The thickness of the tissue section may vary based on the tissue type. We recommend testing 5–15 μm thick tissue sections to identify

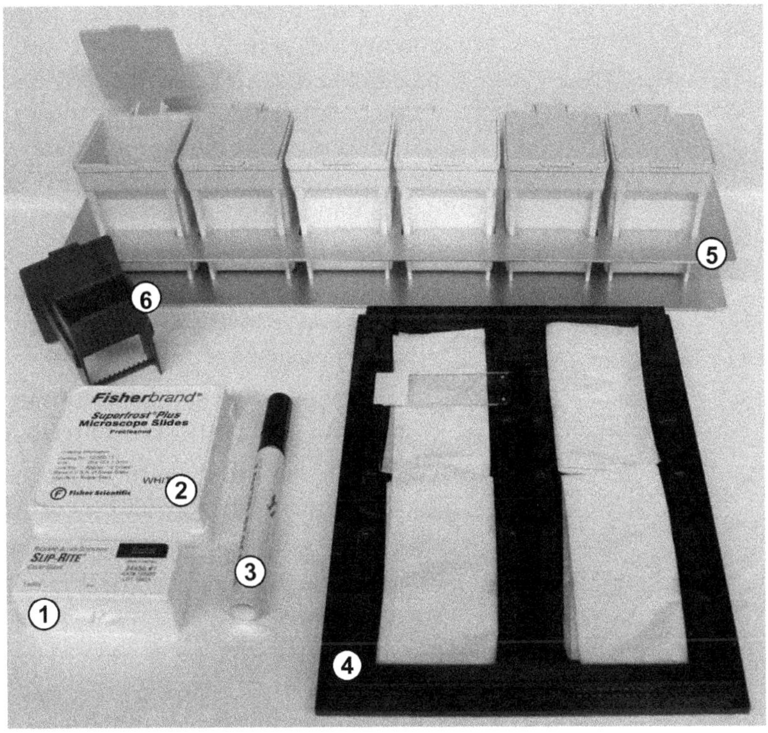

Fig. 6 Materials for histological staining. (1) Glass slide coverslips, (2) Superfrost™ Plus microscope slides, (3) PAP liquid blocking pen, (4) staining slide tray chamber, (5) EasyDip slide jar staining system, (6) slide holder

the optimal tissue thickness for the staining. A tissue section mounted on a slide can be stored long-term in an upright position in a slide box at −20 °C.

41. We recommend using the EasyDip slide jar staining system for the fixation, washing, and mounting steps (Fig. 6). All other steps that require covering the tissue section with a reagent should be carried out in a staining slide tray chamber with water added to the bottom of the tray to provide humidity during incubation (Fig. 6). Typically, 50–70 μL of reagents are sufficient to cover the entire canine tissue section.
42. Working Weigert's Iron Hematoxylin Stain solution is made fresh by mixing Part A and Part B in a 1:1 ratio.
43. Permount™ residues on the top of the coverslip or bottom of the slide can be removed with a cotton swab dipped in xylenes.
44. All steps of this Alizarin Red staining procedure must be performed in a properly ventilated chemical fume hood. Many chemicals described in this protocol are skin, eye, and respiratory irritants, harmful to human health, corrosive, flammable, and harmful to the environment.

45. Canine tissue such as skeletal muscle, heart, internal organs, and nervous system can be used with this protocol. The thickness of the tissue section may vary based on the tissue type. We recommend testing 5–15 μm thick tissue sections to identify the optimal tissue thickness for the staining. A tissue section mounted on a slide can be stored long-term in an upright position in a slide box at −20 °C.
46. This protocol is optimized for staining muscle tissue to determine myofiber calcification. For other tissues, staining time in Alizarin Red S solution may vary. For example, bone requires 30 s, and cartilage requires 1 min of staining time.
47. Immersing the slide in deionized water and ethanol should not exceed 10 s as there is a risk of washing off the stain.
48. Permount™ residues on the top of the coverslip or bottom of the slide can be removed with a cotton swab dipped in xylenes.
49. The nNOS activity staining is based on nicotinamide adenine dinucleotide phosphate (NADPH)-diaphorase activity of NO synthase (NOS).
50. Canine skeletal muscle, heart, and vascular tissues can be used with this protocol. The thickness of the tissue section may vary based on the tissue type and antigen of interest. We recommend testing 10–20-μm thick tissue sections to identify the optimal tissue thickness for the staining. A tissue section mounted on a slide can be stored upright in a slide box at −20 °C until stained.
51. We recommend using the EasyDip slide jar staining system for the fixation and washing steps (Fig. 6). All other steps that require covering the tissue section with a reagent should be carried out in a staining slide tray chamber with water added to the bottom of the tray to provide humidity during incubation (Fig. 6). Typically, 50–70 μL of reagents are sufficient to cover the entire canine tissue section.
52. Use freshly made 1× PBS between washes.
53. DPX is a flammable, toxic skin and eye irritant, and very harmful if ingested or inhaled. Therefore, the coverslip sealing step should be performed in a fume hood.
54. The IF staining protocol presented in this chapter consists of three basic steps: blocking nonspecific hydrophobic interactions between the primary antibodies and the tissue, binding of the primary antibody to the target tissue antigen, and binding a fluorophore-conjugated secondary antibody to the primary antibody. The protocol can also be adopted to paraffin-embedded tissues with minor modifications.

55. Canine tissue such as skeletal muscle, heart, internal organs, and nervous system can be used with this protocol. The thickness of the tissue section may vary based on the tissue type. We recommend testing 5–15 μm thick tissue sections to identify the optimal tissue thickness for the staining. A tissue section mounted on a slide can be stored long-term in an upright position in a slide box at −20 °C.
56. We recommend using the EasyDip slide jar staining system for the washing steps (Fig. 6). All other steps that require covering the tissue section with a reagent should be carried out in a staining slide tray chamber with water added to the bottom of the tray to provide humidity during incubation (Fig. 6). Typically, 50–70 μL of reagents are sufficient to cover the entire canine tissue section.
57. Prepare 1% serum in 1× PBS.
58. Before using an antigen-specific primary antibody, all potential nonspecific protein binding sites in the tissue sample must be blocked. Since serum carries antibodies that bind to reactive sites and prevent nonspecific binding, we highly recommend using serum from the secondary antibody source species. The optimal percentage of the serum should be determined according to the thickness of the tissue section and the antigen of interest. We recommend testing a range of 10–40% serum to determine the optimal serum dilution for blocking. The 20% serum is prepared in 1× PBS.
59. Other staining protocols and/or tissue types may require additional reagents in washing solutions and/or antibody diluent solutions, such as detergents. Other protocols and tissues may also call for various fixation and pre-treatment methods. Additionally, antibody dilutions can vary based on supplier, batch, and intended use. We have determined the described protocol to be optimal for our purposes, but we encourage optimizing the experimental condition for each antigen of interest.
60. We have adopted this IF staining protocol to detect antigens presented in Table 2. We recommend diluting the primary and secondary antibodies in 1% serum of secondary antibody source species to increase the efficiency of blocking all potential non-specific protein binding sites. Table 2 presents the working dilution and the source of some of the primary antibodies for specific protein antigens that we have adopted using the described protocol. We highly recommend optimizing the dilution of each primary antibody according to the staining condition and the type of the tissue.
61. The recovered primary antibody can be stored at 4 °C for one additional use. Before using it, spin at 14,000 RPM for 2 min and only use the supernatant.

Table 2
Primary and secondary antibodies used in immunofluorescence staining

Primary antibody								Secondary antibody				
Antigen target	Antibody name	Clone	Dilution	Type	Host species	Isotype	Supplier	Antibody information	Fluorophore	Dilution	Host species	Supplier
Dystrophin	MANEX44A	5B2	1:300	Monoclonal	Mouse	IgG1	DSHB	F(ab')2-Goat anti-Mouse IgG (H+L)	Alexa Fluor 594	1:100	Goat	ThermoFisher Scientific
Laminin	Anti-Laminin	N/A	1:200	Polyclonal	Rabbit	N/A	Sigma-Aldrich	F(ab')2-Goat anti-Rabbit IgG (H+L)	Alexa Fluor 488	1:100	Goat	ThermoFisher Scientific
Laminin (gamma-1)	2E8	2E8	1:100	Monoclonal	Mouse	IgG2a	DSHB	Goat anti-Mouse IgG2a	Alexa Fluor 647	1:100	Goat	ThermoFisher Scientific
eMyHC	F1.652	F1.652	1:250	Monoclonal	Mouse	IgG1	DSHB	F(ab')2-Goat anti-Mouse IgG (H+L)	Alexa Fluor 594	1:100	Goat	ThermoFisher Scientific
Type I MyHC	BA-D5	BA-D5	1:20	Monoclonal	Mouse	IgG2b	DSHB	Goat anti-Mouse IgG2b	Alexa Fluor 350	1:50	Goat	ThermoFisher Scientific
Type IIa/IIx MyHC	SC-71	SC-71	1:100	Monoclonal	Mouse	IgG1	DSHB	Goat anti-Mouse IgG1	Alexa Fluor 594	1:100	Goat	ThermoFisher Scientific
Type IIb MyHC	BF-F3	BF-F3	1:40	Monoclonal	Mouse	IgM	DSHB	Goat Anti-Mouse IgM, μ chain specific	Fluorescein (FITC)	1:100	Goat	Jackson ImmunoResearch

DSHB Developmental Studies Hybridoma Bank, *N/A* not applicable

62. CitiFluor AF1 can be replaced with other aqueous photobleaching protection media.
63. This step must be performed in a fume hood as DPX (distyrene plasticizer xylene) is a flammable, toxic, skin and eye irritant, and very harmful if ingested or inhaled.
64. Slides can be stored at 4 °C short-term, or −20 °C for long-term storage.
65. The chromogenic IHC staining protocol presented in this chapter consists of seven basic steps: tissue fixation, quenching the endogenous peroxidase activity in the tissue, blocking nonspecific hydrophobic interactions between the primary antibodies and the tissue, binding of the primary antibody to the target tissue antigen, binding the biotinylated secondary antibody to primary antibody, attaching the biotin to high sensitivity avidin-horseradish peroxidase (HRP) conjugate, and visualizing the antibody through the conversion of a chromogenic substrate by HRP enzyme. The protocol can also be adopted to paraffin-embedded tissues with minor modifications.
66. Canine tissue such as skeletal muscle, heart, internal organs, and nervous system can be used with this protocol. The thickness of the tissue section may vary based on the tissue type and antigen of interest. We recommend testing 5–15 μm thick tissue sections to identify the optimal tissue thickness for the staining. A tissue section mounted on a slide can be stored long-term in an upright position in a slide box at −20 °C.
67. We recommend using the EasyDip slide jar staining system for the fixation, washing, and mounting steps (Fig. 6). All other steps that require covering the tissue section with a reagent should be carried out in a staining slide tray chamber with water added to the bottom of the tray to provide humidity during incubation (Fig. 6). Typically, 50–70 μL of reagents are sufficient to cover the entire canine tissue section.
68. Use freshly made 1× PBS between washes.
69. Before using an antigen-specific primary antibody, all potential nonspecific protein binding sites in the tissue sample must be blocked. Since serum carries antibodies that bind to reactive sites and prevent nonspecific binding, we highly recommend using serum from the secondary antibody source species. The optimal percentage of the serum should be determined according to the thickness of the tissue section and the antigen of interest. We recommend testing a range of 10–40% serum to determine the optimal serum dilution for blocking tissue samples.

70. We have adopted this IHC staining protocol for CD4, CD8, FoxP3, granzyme B, neutrophil, macrophage, MHCI, and MHCII staining (Table 3). We recommend diluting the primary and secondary antibodies in 1% serum of secondary antibody source species to increase the efficiency of blocking all potential nonspecific protein binding sites. The dilution information of each antibody is presented in (Table 3).
71. The recovered primary antibody can be stored at 4 °C for one additional use. Before re-using it, spin at 14,000 RPM for 2 min and use the clean supernatant for staining. We recommend re-using the antibody only once following the first use.
72. The Vector laboratory offers multiple HRP chromogenic substrates. We recommend using the Nova Red substrate as it provides the best staining contrast.
73. While incubating the substrate with the tissue sample, a colored precipitate will develop at the antigen expression site as an outcome of converting the chromogenic substrate by the HRP enzyme into an insoluble end product. The incubation period may vary depending on the chromogenic substrate used in the reaction. Therefore, it is critical to monitor the intensity of the substrate color development using a light microscope to determine the optimal incubation time.
74. Prolonged incubation of the hematoxylin counterstain can obscure visualization of targets localized in cell nuclei. Therefore, it is critical to monitor the intensity of hematoxylin staining under a light microscope to determine the optimal incubation time.

Acknowledgments

The research on the canine DMD model in the Duan lab is currently supported by the National Institutes of Health (NS-90634 and AR-70517), Jesse Davidson Foundation-Defeat Duchenne Canada, Hope for Javier, Jackson Freel DMD Research Fund, Parent Project Muscular Dystrophy, Jett Foundation, Michael's Cause, Ryan's Quest, Solid Biosciences Inc., and the University of Missouri.

Disclosure DD is a member of the scientific advisory board for Solid Biosciences and equity holders of Solid Biosciences. DD is a member of the scientific advisory board for Sardocor Corp.In the last 3 years, the Duan lab has received research supports unrelated to this project from Solid Biosciences and Edgewise.

Table 3
Primary and secondary antibodies used in immunohistochemistry staining

Primary antibody								Secondary antibody			
Antigen target	Antibody name	Clone	Dilution	Type	Host species	Isotype	Source	Antibody information[a]	Dilution	Host species	Source
CD4	Rat anti Dog CD4	YKIX302.9	1:1000	Monoclonal	Rat	IgG2a	Bio-Rad	Goat anti-Rat IgG (H+L)	1:1000	Goat	ThermoFisher Scientific
CD8	Rat anti Dog CD8	YCATE55.9	1:200	Monoclonal	Rat	IgG1	Bio-Rad	Goat anti-Rat IgG (H+L)	1:1000	Goat	ThermoFisher Scientific
FoxP3	FOXP3	FJK-16s	1:100	Monoclonal	Rat	IgG2a	ThermoFisher Scientific	Goat anti-Rat IgG (H+L)	1:1000	Goat	ThermoFisher Scientific
Granzyme B	Anti-Granzyme B	N/A	1:50	Polyclonal	Rabbit	N/A	Abcam	Anti-Rabbit IgG (whole molecule)	1:500	Goat	Millipore Sigma
Neutrophil	Canine Neutrophil	CADO48A	1:8000	Monoclonal	Mouse	IgG1	Kingfisher Biotech	Goat anti-Mouse IgG1	1:1000	Goat	ThermoFisher Scientific
Macrophage	Mouse anti Human macrophages	MAC387	1:2000	Monoclonal	Mouse	IgG1	Bio-Rad	Goat anti-Mouse IgG1	1:1000	Goat	ThermoFisher Scientific
MHC I	Canine MHC-I	H58A	1:2000	Monoclonal	Mouse	IgG2a	Kingfisher Biotech	Goat Anti-Mouse IgG (H+L)	1:1000	Goat	ThermoFisher Scientific
MHC II	Mouse anti Dog MHC Class II Monomorphic	CA2.1C12	1:100	Monoclonal	Mouse	IgG1	Bio-Rad	Goat anti-Mouse IgG1	1:1000	Goat	ThermoFisher Scientific

[a] Biotin-conjugated antibody
N/A not applicable

References

1. Duan D, Goemans N, Takeda S et al (2021) Duchenne muscular dystrophy. Nat Rev Dis Primer 7:13
2. McGreevy JW, Hakim CH, McIntosh MA et al (2015) Animal models of Duchenne muscular dystrophy: from basic mechanisms to gene therapy. Dis Model Mech 8(3):195–213. https://doi.org/10.1242/dmm.018424
3. Wasala NB, Chen SJ, Duan D (2020) Duchenne muscular dystrophy animal models for high-throughput drug discovery and precision medicine. Expert Opin Drug Discov 15:1–14. https://doi.org/10.1080/17460441.2020.1718100
4. Lai Y, Thomas GD, Yue Y et al (2009) Dystrophins carrying spectrin-like repeats 16 and 17 anchor nNOS to the sarcolemma and enhance exercise performance in a mouse model of muscular dystrophy. J Clin Invest 119(3):624–635. https://doi.org/10.1172/JCI36612
5. Lai Y, Zhao J, Yue Y et al (2013) alpha2 and alpha3 helices of dystrophin R16 and R17 frame a microdomain in the alpha1 helix of dystrophin R17 for neuronal NOS binding. Proc Natl Acad Sci U S A 110(2):525–530. https://doi.org/10.1073/pnas.1211431109
6. Heckmatt JZ, Moosa A, Hutson C et al (1984) Diagnostic needle muscle biopsy. A practical and reliable alternative to open biopsy. Arch Dis Child 59(6):528–532. https://doi.org/10.1136/adc.59.6.528
7. Akarolo-Anthony SN, Ogundiran TO, Nkwodimmah C et al (2012) Office based muscle biopsy using Vacora vacuum assisted biopsy system. Afr J Med Med Sci 41(3):313–316
8. Mikell CB, Chan AK, Stein GE et al (2013) Muscle and nerve biopsies: techniques for the neurologist and neurosurgeon. Clin Neurol Neurosurg 115(8):1206–1214. https://doi.org/10.1016/j.clineuro.2013.05.008
9. Yue Y, Shin J-H, Duan D (2011) Whole body skeletal muscle transduction in neonatal dogs with AAV-9. Methods Mol Biol 709:313–329. https://doi.org/10.1007/978-1-61737-982-6_21
10. Pistolese CA, Ciarrapico AM, Della Gatta F et al (2009) Cost-effectiveness analysis of two vacuum-assisted breast biopsy systems: Mammotome and Vacora. Radiol Med 114(5):743–756. https://doi.org/10.1007/s11547-009-0404-8
11. Gallo A, Abraham A, Katzberg HD et al (2018) Muscle biopsy technical safety and quality using a self-contained, vacuum-assisted biopsy technique. Neuromuscul Disord 28(5):450–453. https://doi.org/10.1016/j.nmd.2018.02.006
12. Barthelemy F, Woods JD, Nieves-Rodriguez S et al (2020) A well-tolerated core needle muscle biopsy process suitable for children and adults. Muscle Nerve 62(6):688–698. https://doi.org/10.1002/mus.27041

Chapter 17

MRI Evaluation of Gene Therapy in the Canine Model of Duchenne Muscular Dystrophy

Amy R. Zalcman, Chady H. Hakim, Jimmy C. Lattimer, James R. Holland, John R. Dodam, and Dongsheng Duan

Abstract

Magnetic resonance imaging (MRI) is a well-established and widely used technique to characterize and quantify skeletal and cardiac muscle changes in Duchenne muscular dystrophy (DMD). Recently, MRI has been explored to study disease progression and response to gene therapy in the canine DMD model. Using traditional sequences, delayed gadolinium enhancement, novel sequences, and spectroscopy, investigators have begun to (i) establish the baseline MRI characteristics of the muscles in normal and affected dogs and (ii) evaluate gene therapy outcomes in treated dogs. As a noninvasive assay, MRI offers an excellent opportunity to study longitudinal muscle changes in long-term gene therapy studies in the canine model. In this chapter, we outline the MRI method used to study DMD in the canine model.

Key words MRI, DMD, Canine model, Skeletal muscle, Cardiac muscle, Magnetic resonance imaging, Spectroscopy, Heart, Duchenne muscular dystrophy

1 Introduction

Magnetic resonance imaging (MRI) is an imaging technique that uses magnetic fields and radio pulses to generate images of tissues and organs in the body. MRI was developed in the 1970s (Mayor 2003). Today, MRI has become one of the most used techniques in medical diagnosis and biomedical research. MRI has attracted increasing attention in the field of neuromuscular diseases over the last two decades (reviewed in [1–4]). A notable application of MRI is in Duchenne muscular dystrophy (DMD), the most common childhood neuromuscular disease caused by dystrophin deficiency (reviewed in [5–7]).

Dystrophin-deficient canines are currently the best large mammalian model for DMD (reviewed in [8–10]). MRI was first used to study DMD in the canine model in 2007 by Thibaud et al. [11]. The authors performed MRI in 2-month-old normal and

Rika Maruyama and Toshifumi Yokota (eds.), *Muscular Dystrophy Therapeutics: Methods and Protocols*, Methods in Molecular Biology, vol. 2587, https://doi.org/10.1007/978-1-0716-2772-3_17,

golden retriever muscular dystrophy (GRMD) dogs. GRMD dogs carry a point mutation in intron 6 of the dystrophin gene, which disrupts normal splicing and the reading frame (reviewed in [10]). The authors showed a significant increase in the T2-weighted/T1-weighted signal ratio and T2-weighted image heterogeneity (*see* **Note 1**). These changes were attributed to the presence of edema and necrosis in affected dog muscles. The authors also performed gadolinium enhancement and found a higher gadolinium enhancement in affected dog muscle, a finding consistent with the increased permeability of dystrophic myofibers [11]. Subsequent studies from the same group revealed longitudinal MRI changes in GRMD dogs at 2, 4, 6, and 9 months [12]. T1 relaxation abnormality is a good indicator of fatty infiltration, a prominent degenerative change in DMD patients. Interestingly, Thibaud et al. did not find significant fatty infiltration in GRMD skeletal muscle [12]. Recently, Guo et al. applied cardiac MRI (CMR) to study GRMD cardiomyopathy in 17 affected dogs [13]. Compared to dogs that were <30 months of age, dogs that were >45 months of age showed a significant increase in the left ventricular end-diastolic volume, left ventricular end-systolic volume, and myocardial mass, suggesting dilated cardiomyopathy in terminal aged GRMD dogs. The authors also found left ventricular lateral wall lesions and earlier involvement of the anterior septum using late gadolinium enhancement imaging.

In recent years, investigators have begun to use MRI to assess experimental drug and gene therapies in the canine DMD model. Using T2-weighted MRI examination, Yokota et al. showed that antisense oligonucleotide-mediated exon skipping reduced the T2 signal intensity, suggesting less edema and inflammation in treated dogs [14]. MRI has also been used to assess muscle mass following myostatin inhibition gene therapy in GRMD dogs [15] and to study cardiac function following heart-targeted exon-skipping therapy [16, 17]. An intriguing use of MRI is to monitor toxicity in preclinical gene therapy studies in the canine DMD model. In this regard, Kornegay detected an increase in the T2 signal intensity in newborn GRMD puppies treated with systemic AAV micro-dystrophin gene therapy [18]. The increased T2 signal suggests increased inflammation and edema. Subsequent studies revealed that the treated dogs developed marked inflammatory myopathy, likely due to the innate immune response to the particular vector used in the study.

In summary, MRI has become an invaluable tool to study skeletal and cardiac muscle disease and response to therapy in the canine DMD model. In this chapter, we present protocols for (a) anesthesia, (b) MRI imaging of the skeletal musculature, (c) MRI imaging of the cardiac musculature with multiple sequences, and (d) data storage and viewing.

2 Materials

2.1 Sedation and Anesthesia Procedure (Fig. 1)

1. Personal protective equipment (PPE) including disposable gowns and/or dedicated scrubs, head cover, shoe covers, surgical mask, and latex gloves.
2. Stethoscope (3 M Littmann, St. Paul, MN, USA).
3. Digital thermometer (ReliOn Inc., Spokane Valley, WA, USA).
4. MRI-compatible ventilator, Ohmeda Excel 210 (DRE Veterinary, Louisville, KY, USA).
5. Soda lime (carbon dioxide absorption) (Thermo Fisher Scientific, Hampton, NH, USA).
6. Anesthetic gas scavenging filters (Dynasthetics, Salt Lake City, UT, USA).

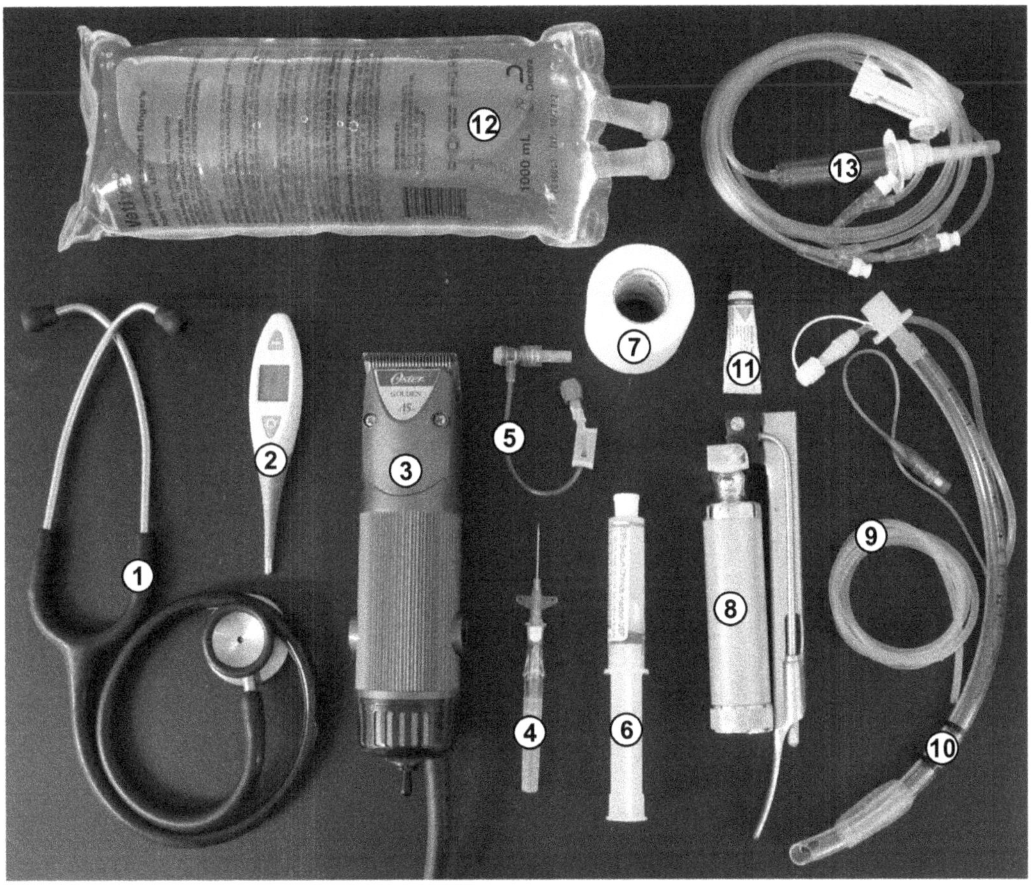

Fig. 1 Materials required for sedation and anesthesia procedures. (1) Stethoscope, (2) digital thermometer, (3) animal hair clipper with a #50 blade, (4) shielded I.V. catheter, (5) T-port extension set with a side clamp, (6) BD PosiFlush pre-filed saline (0.9% sodium chloride) syringe for intravenous injection, (7) medical adhesive tape, (8) laryngoscope, (9) silicone tube to secure the endotracheal tube, (10) sterile endotracheal tube, (11) Rugby artificial tears ointment, (12) lactated Ringer's buffer, and (13) SmartSite gravity set

7. Dexmedetomidine (0.5 mg/ml) (Zoetis Inc., Kalamazoo, MI, USA).
8. Disposable sterile syringes and needles (Becton, Dickinson and Company, Franklin Lakes, NJ, USA).
9. Animal hair clipper with a #40 blade (Oster Professional Products, Sunbeam Products, Inc., Boca Raton, FL, USA).
10. Shielded I.V. Catheter (Becton Dickinson company, Franklin Lakes, NJ, USA).
11. T-port extension set with a side clamp (B.Braun Vet Care, Tuttlingen, Germany).
12. BD PosiFlush Pre-filed saline (0.9% sodium chloride) syringe for intravenous (IV) injection (Becton Dickinson company, Franklin Lakes, NJ, USA).
13. Medical adhesive tape (Thermo Fisher Scientific, Hampton, NH, USA).
14. Diprivan (Propofol) (Fresenius Kabi, Lake Zurich, IL, USA).
15. Laryngoscope (Medline industries, Northfield, IL, USA).
16. Sterile endotracheal tube (Medline industries, Northfield, IL, USA).
17. Surgilube Lubricating Jelly (Allegro Medical supplies, Mesa, AZ, USA).
18. Silicone tube to secure the endotracheal tube (Dow Corning Corp., Midland, MI, USA).
19. MRI-compatible infusion pump (Iradimed Corporation, Winter Springs, FL, USA).
20. MRI-compatible veterinary vital sign monitor, Philips Invivo Expressions system (Koninklijke Philips N.V., Amsterdam, Netherlands).
21. Rugby artificial tears ointment (Walmart Stores Inc., Bentonville, AR, USA).
22. Lactated Ringer's solution (Baxter, Deerfield, IL, USA).
23. SmartSite gravity set (CareFusion, San Diego, CA, USA).
24. Latex gloves (Thermo Fisher Scientific, Hampton, NH, USA).
25. Padded blanket (Standard Textile, Cincinnati, OH, USA).
26. 70% Isopropyl alcohol (Walmart Stores Inc., Bentonville, AR, USA).

2.2 Positioning the Dog for MRI Imaging (Fig. 2)

1. Custom-made trough (University of Missouri, Columbia, MO, USA) (*see* **Note 2**).
2. Positioning wedge (Medline industries, Northfield, IL, USA).
3. Absorbent liner with a non-permeable barrier (Medline industries, Northfield, IL, USA).

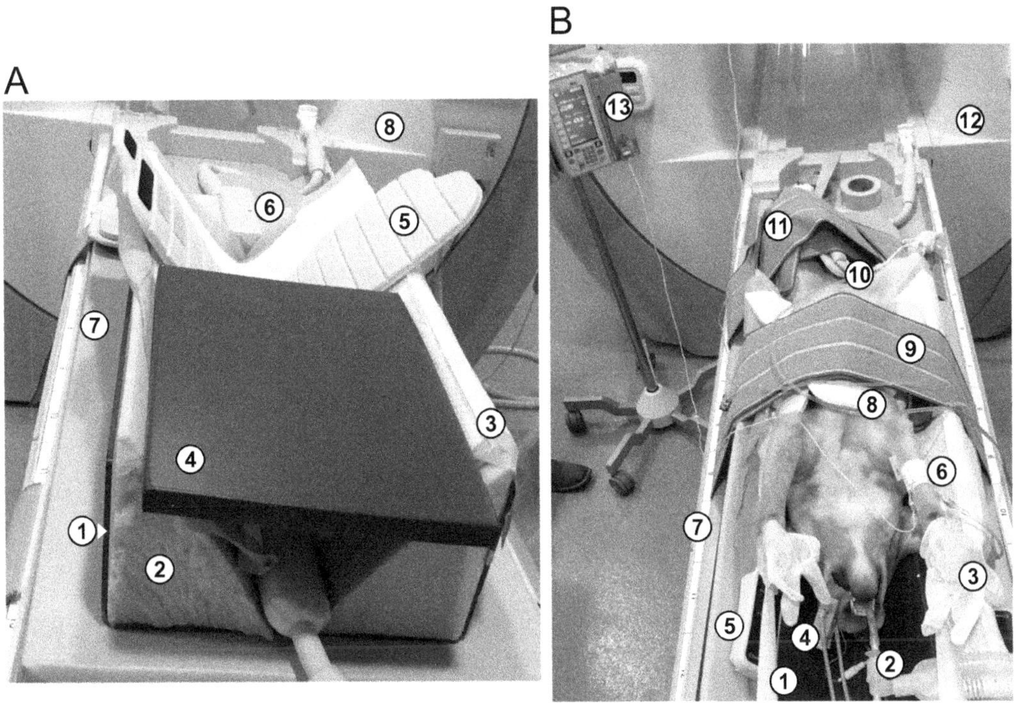

Fig. 2 Materials required to position the dog for MRI imaging. (**a**) The setup of the custom-made trough to position the dog for MRI. (1) Custom-made trough, (2) positioning wedge, (3) absorbent liner with a non-permeable barrier, (4) headrest foam pad, (5) 16 channel transmit/receive coil array for cardiac MRI, (6) four channel transmit/receive coil array for skeletal muscle MRI, (7) patient table, and (8) Toshiba Vantage Titan 3 T MRI machine. (**b**) Positioning the dog for MRI. (1) Medical tape is used to secure the forelimbs away from the cardiac coil array, (2) endotracheal tube is connected to the ventilator circuits, (3) paws are wrapped with the latex glove to help maintain the body temperature, (4) MRI-compatible pulse oximeter is secured to the dog tongue, (5) wireless ECG module transmitter, (6) I.V. catheter connected to the T-port extension set to provides a continuous infusion of propofol and Lactated Ringer's buffer, (7) patient table, (8) 16 channel transmit/receive coil array for cardiac MRI, (9) Velcro straps to secure the cardiac coil array, (10) four channel transmit/receive coil array for skeletal muscle MRI, (11) Velcro straps to secure the skeletal muscle coil array, (12) Toshiba Vantage Titan 3T MRI machine, and (13) MRI-compatible infusion pump

4. Headrest foam pad (Medline industries, Northfield, IL, USA).
5. Magnetic resonance imaging (MRI) machine with spectroscopy capacity, Vantage Titan 3T (Canon Medical Systems USA, Inc., Tustin, CA, USA).
6. Four channel transmit/receive coil array for skeletal muscle MRI (Toshiba Corp., Minato City, Tokyo, Japan).
7. Sixteen channel transmit/receive coil array for cardiac MRI (Toshiba Corp., Minato City, Tokyo, Japan).
8. Nonwoven gauze sponges (4 inch × 4 inch) (Medline Industries Inc., Northfield, IL, USA).

9. 2% Chlorhexidine gluconate (Vedco Inc., Saint Joseph, MO, USA).
10. MRI-compatible ECG cardiac pads (MEDEXsupply, Passaic, NJ, USA).
11. MRI-compatible ECG color-coded leads (MEDEXsupply, Passaic, NJ, USA).
12. Philips Invivo Expression wireless ECG module transmitter (Koninklijke Philips N.V., Amsterdam, the Netherlands).
13. MRI-compatible pulse oximeter (MEDEXsupply, Passaic, NJ, USA).
14. Philips Invivo Expression wireless pulse oximeter transmitter (Koninklijke Philips N.V., Amsterdam, the Netherlands).

2.3 MRI Imaging of the Heart and Skeletal Muscle

1. NovaPlus Omniscan Contrast Media 287 mg/mL (Gadodiamide) (GE Healthcare, Chicago, IL, USA).

2.4 Data Analysis

1. A picture archiving and communication system (PACS, PICOM365) (ScImage, Los Altos, CA, USA).
2. A digital information and communication in medicine (DICOM) viewer (eFilm) (Merge Healthcare, Chicago, IL, USA).

3 Methods

3.1 Sedation and Anesthesia Protocol

1. Follow the institutional guidelines for animal care and handling. Obtain approval from the institutional animal care and use committee (ACUC). Wear all PPE before working or handling animals.
2. Fast the dog the night before the procedure.
3. Perform a physical exam, and record the body weight and the vital signs including heart rate, respiratory rate and effort, rectal body temperature, capillary refill time (CRT), and mucous membrane color.
4. Replace the carbon dioxide absorbent with fresh product. Remove the vaporizer from the machine to prevent accidental administration of inhalant anesthetic. Flush the anesthetic circuit and ventilator with 100% oxygen for 30 min at the flow rate of 4 L/min. Attach activated charcoal filters to the inhalation and exhalation ports of the anesthetic machine (*see* **Note 3**).
5. Sedate the dog with dexmedetomidine (5 μg/kg) injected intramuscularly into a muscle that will not be used in the imaging examination. Monitor the dog for adverse effects while waiting for the sedative effects of the premedication to become apparent (~10–15 min) (*see* **Note 4**).

6. Clip the body hair over the left thoracic wall from the caudal margin of the scapula approximately to the last rib and about 8 in tall centered over the mid-height of the chest (excluding the vertebral column).
7. Clip the hair over the lateral saphenous vein and prepare the skin aseptically for catheter placement. Place an intravenous catheter into the lateral saphenous vein, attach it to a saline-primed T-port, secure the catheter with medical-grade adhesive tape, and assess the catheter for proper placement and patency.
8. Induce anesthesia with propofol given intravenously and to effect (expected dose ~ 6 mg/kg). Place a cuffed endotracheal past the larynx and into the cervical trachea. Secure in place with roll gauze and assess placement. Incrementally inflate the endotracheal tube cuff until a leak around the tube is eliminated at ~ 20 cm H_2O inflation pressure. Lubricate the corneas with a petroleum-based ocular ointment (*see* **Note 5**).
9. Maintain anesthesia with a constant rate intravenous infusion of propofol (0.3 mg/kg/hr) using an MRI-compatible infusion pump. Manual ventilation is initiated to prevent hypoventilation during anesthesia. The animal is monitored with a multi-parameter monitor (electrocardiograph, pulse oximetry, capnography, noninvasive blood pressure, and body temperature). Monitor and record the capillary refill time (CRT), mucous membrane color, palpebral reflex, and pedal reflex every 15 min.
10. Infuse Lactated Ringer's solution intravenously at 5 mL/kg/h.
11. Maintain the core body temperature at 37 °C throughout the experiment (*see* **Note 6**).

3.2 Positioning the Dog for MRI Imaging

1. Position the custom-made trough in the middle of the patient table of the MRI machine. Place two positioning wedges at the bottom of the trough to form a V-shape wedge inside the trough. Overlay the wedges with an absorbent liner. Place the cardiac 16-channel coil array at the center of the trough over the absorbent liner. Place the skeletal muscle 4-channel coil array at the caudal end of the trough (Fig. 2a).
2. Transport the anesthetized dog to the MRI scanner. Position the dog in dorsal recumbency in the prepared trough with the heart at the approximate level of the center point of the cardiac 16-channel flex coil. Position another foam pad underneath the dog head (Fig. 2b). Secure the forelimbs away from the cardiac coil array using medical tape (Fig. 2b).
3. Clean the clipped skin over the left thoracic wall with 2% chlorhexidine and allow the skin to air dry. Attach the MRI-compatible ECG cardiac pads in a diamond orientation over the left thoracic wall and apply moderate but non-compressive pressure to ensure a good contact. Secure

the MRI-compatible ECG color-coded leads to the corresponding pads. Connect the leads to the wireless ECG module transmitter, and place the transmitter on the middle of the bed and in front of the dog (Fig. 2b). Secure the cardiac coil with velcro straps (Fig. 2b). Confirm the quality of the ECG tracing.

4. Secure the MRI-compatible pulse oximeter to the dog tongue and connect it to the wireless pulse oximeter transmitter (Fig. 2b). Place the transmitter in front of the dog on the middle of the bed.
5. Wrap the skeletal muscle 4-channel coil array around the stifle and secure it with Velcro straps (Fig. 2b). Fix the straps with a medical tap to the trough to prevent movement during scanning. Secure the contralateral limb to the trough using medical tape.

3.3 MRI Imaging of the Heart and Skeletal Muscle

The sequences and parameters presented in Tables 1 and 2 were used to acquire images and spectroscopy from skeletal muscle and heart, respectively.

1. Perform the pre-contrast and the spectroscopy scanning sequences of skeletal muscle (Fig. 3, Table 1).
2. Perform the pre-contrast and spectroscopy scanning sequences of the heart (Table 2).
3. Halt the scanning and the propofol infusion. Disconnect the propofol infusion line from the T-Port and flush the port with saline. Connect the gadolinium syringe to the port and inject the gadolinium at a fast rate (average dose of 3.5 mL delivered in less than 3 s). Disconnect the syringe from the port and flush the port with saline. Reconnect the propofol infusion line and restart the infusion (*see* **Note 7**).
4. Perform the post-contrast scanning sequence of the heart 5 min after the administration of the gadolinium (Table 2).
5. Perform the post-contrast scanning sequence of the skeletal muscle 10 min after the administration of the gadolinium (Table 1).

3.4 Animal Recovery Following MRI Imaging

1. After completing the pre- and post-contrast and spectroscopy scanning acquisition, recover the dog from anesthesia and monitor it until it can maintain sternal recumbence. Monitor the vital signs throughout the recovery process.
2. Remove the IV catheter after 15 min from recovery.
3. Continue observing the dog for another 30 min. Return the dog to the kennel and follow with a postoperative assessment for 3 days to confirm no adverse effects from the procedure.

Table 1
MRI scanning sequences for skeletal muscle

Sequence	Weight	Plane	Type	TR (ms)	TE (ms)	FOV (cm)	Matrix	NAQ	Gating	Contrast
#1	T2	SAG	FSE	5372	120	16 × 16	192 × 256	3	No	No
#2	T2	AX	FSE	5372	120	16 × 16	192 × 256	3	No	No
#3	T1	AX	FSE	723	10	16 × 16	192 × 208	1	No	No
#4	Spectroscopy	Vol	FSE	3000	25	1.5 × 1.5 × 1.7		128	No	No
#5	T1	Ax	FSE	723	10	16 × 16	192 × 208	1	No	No

AX axial, *FSE* fast spin echo, *NAQ* number of acquisitions, *SAG* sagittal, *TE* time to echo, *TR* repetition time, *Vol* volume

Table 2
MRI scanning sequences for the heart

Sequence	Weight	Plane	Type	TR (ms)	TE (ms)	FOV (cm)	Matrix	NAQ	Gating	Contrast
#1	2 Chamber Pseudo	SAG	FFE SSFP	3.4	1.7	28.5 × 28.5	192 × 192	1		No
#2	HLA SSFP Cine	AX	FFE SSFP	3.4	1.7	23.8 × 23.8	160 × 160	1	ECG	No
#3	SA Bright Blood	COR	FFE SSFP	3.4	1.7	23.8 × 28.5	160 × 192	1	ECG	No
#4	T1 Black Blood	AX	FASE	1660	10	23.8 × 28.5	192 × 208	1	ECG and BH	No
#5	T2 Black Blood	AX	FASE	1848	70	23.8 × 28.5	192 × 208	1	ECG and BH	No
#6	T2 FS Black Blood	AX	FASE	1984	80	23.8 × 28.5	192 × 208	1	ECG and BH	No
#7	SA 3D delayed FS	Short Axis	FFE3D	5.6	1.9	24.2 × 22.5	128 × 128	1	ECG and BH	No
#8	SA 3D delayed FS	Short Axis	FFE3D	3.7	1.3	24.2 × 22.5	144 × 144	1	ECG and BH	No
#9	SA 3D delayed FS	Short Axis	FFE3D	5.6	1.9	22.1 × 24.1	128 × 128	1	ECG and BH	Yes
#10	SA 3D delayed FS	Short Axis	FFE3D	3.7	1.3	24.1 × 22.5	144 × 144	1	ECG and BH	Yes

AX axial, *BH* breath hold, *COR* coronal, *ECG* electrocardiogram, *FASE* fast advantage spine echo, *FFE* fast field echo, *FFE3D* fast field echo 3 dimension, *FOV* field of view, *FS* fat saturation, *HLA* horizontal long axis, *NAQ* number of acquisitions, *SA* short axis, *SSFP* steady-state free precession, *TE* time to echo, *TR* repetition time

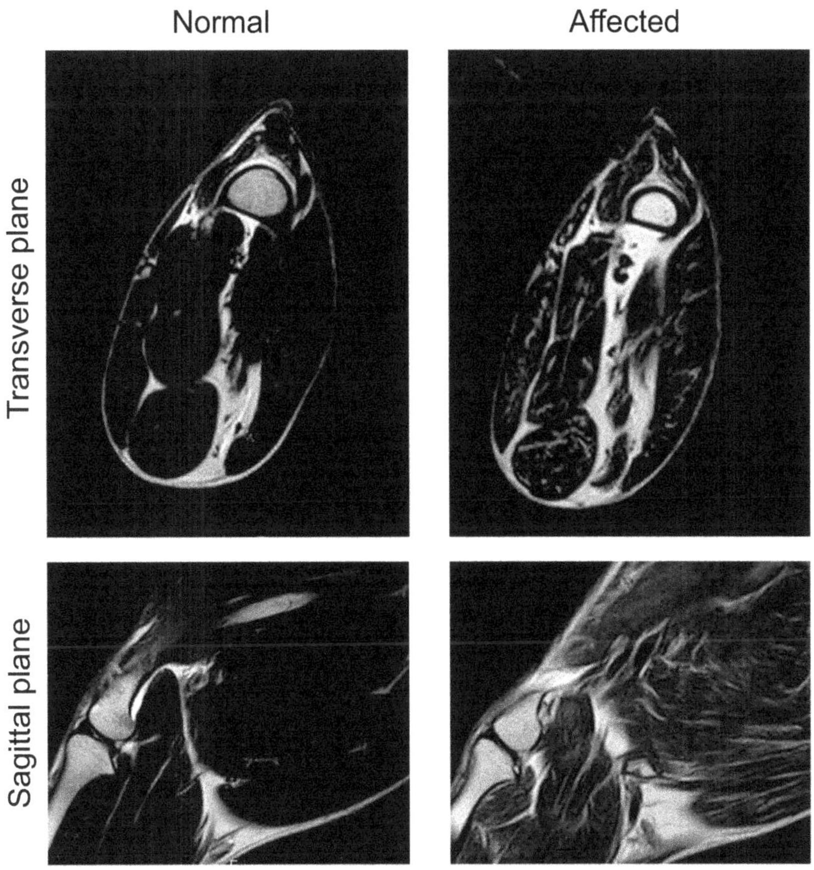

Fig. 3 Representative MRI images of the skeletal muscle from normal and affected dogs. T2-weighted MRI images of the distal thigh musculature in the transverse (top) and sagittal (bottom) planes. The normal dog illustrates the typical balance of hypointense muscle (dark tissue) and interspersed hyperintense fat. In the affected dogs, the muscles demonstrate several classic changes such as hyperintense striations or marbling, which can indicate fat, fibrosis, or a combination therein

3.5 Data Analysis

Images stored in the PACS are viewed in the DICOM viewer. In the DICOM viewer, perform the necessary measurements to describe skeletal and heart muscle size and composition (*see* **Note 8**).

4 Notes

1. T1 and T2 are the basic types of MRI images. T in T1 and T2 stands for time. T1 is the spin-lattice-relaxation time, and T2 is the transverse relaxation time.
2. The custom-made trough is a U-shaped platform made to support and position the dog during the MRI imaging. The platform has a rectangular shape with long sides and open ends. The platform is made of lucite and was manufactured at the University of Missouri Physics Machine Shop (University of Missouri, Columbia, MO, USA). The rectangular sheet of

lucite was heated to a malleable state, and the two long sides of the sheet were bent at a 90° to form the U-shaped platform. The inner distance of the U-shaped was determined according to the animal size.

3. Since inhalation anesthetic agent is known to cause sudden cardiac arrest, malignant hyperthermia, rhabdomyolysis, and hypermetabolic reaction in DMD patients [19], we have been using a continuous propofol infusion to achieve general anesthesia during the MRI imaging procedure. To remove any inhalation anesthetic gas residual, the ventilator is flushed for 30 min with oxygen before the procedure, and gas scavenging anesthetic filters are used throughout the procedure. Also, fresh soda lime is added to the ventilator before each procedure to assist in the scavenging process.
4. Dexmedetomidine was chosen as the pre-anesthetic medication because we found that it can maintain the heart rate between 80 and 100 beats per minute in most dogs, which is ideal for ECG gating.
5. Sterile water-soluble lubricant is applied to the endotracheal tube (ET) prior to laryngoscope-guided placement. ET tube position is checked by palpation, the tube is secured in place to the muzzle of the dog, and the cuff incrementally inflated until a leak cannot be heard at an inflation pressure of 20–25 cm H_2O.
6. Temperature monitoring is critical as an anesthetized dog can experience heat loss. We recommend using heat-trapping techniques, including paws wrap with latex gloves and covering the animal with blankets over the MRI flex coils.
7. If the gadolinium is administered with too much pressure in the face of the resistance of flow, this may result in leakage at the IV catheter, thus reducing the accuracy of the post-contrast scanning. The injection should flow with ease. If too much resistance is present, then the catheter should be evaluated with a flush, and the limb position may be extended slightly. Video monitoring of gadolinium administration allows the imaging technologist to observe the injection process and set the timing for the post-contrast scanning. This also allows for consistent timing between subjects.
8. Measurements of skeletal muscle size were reported as a ratio of muscle size in the sagittal and axial planes to the size of the tibia. Measurements of signal intensity were made as a region of interest (ROI) in each muscle evaluated and compared to individual control tissues. This was performed in the T2, T1, and T1 post-contrast axial sequences. Three ROIs of the semimembranosus muscle were compared to fat, marrow cavity of the femur, cranial, and caudal fascicles of the sartorius.

Spectroscopy was reported as a measurement of the peak height. Signal intensity measurements of the myocardium were performed at the plane of the papillary muscles. This was performed in T2-modified gradient echo, T2 fat-saturated modified gradient echo, T1-modified gradient echo, and the short axis T1 3D fat-saturated sequences (TE 1.3 and 1.9, both pre- and post-contrast). Three ROIs were measured in the interventricular septum, the left ventricular free wall, and papillary muscles. Three areas in the liver were used as a control.

Acknowledgments

The research on the canine DMD model in the Duan lab is currently supported by the National Institutes of Health (NS-90634 and AR-70517), Jesse Davidson Foundation-Defeat Duchenne Canada, Hope for Javier, Jackson Freel DMD Research Fund, Parent Project Muscular Dystrophy, Jett Foundation, Michael's Cause, Ryan's Quest, Solid Biosciences Inc., and the University of Missouri.

Disclosure DD is a member of the scientific advisory board for Solid Biosciences and equity holders of Solid Biosciences. DD is a member of the scientific advisory board for Sardocor Corp. In the last 3 years, the Duan lab has received research supports unrelated to this project from Solid Biosciences and Edgewise.

References

1. Dahlqvist JR, Widholm P, Leinhard OD et al (2020) MRI in neuromuscular diseases: an emerging diagnostic tool and biomarker for prognosis and efficacy. Ann Neurol 88(4): 669–681. https://doi.org/10.1002/ana.25804
2. Jungbluth H (2017) Myopathology in times of modern imaging. Neuropathol Appl Neurobiol 43(1):24–43. https://doi.org/10.1111/nan.12385
3. Mercuri E, Pichiecchio A, Allsop J et al (2007) Muscle MRI in inherited neuromuscular disorders: past, present, and future. J Magn Reson Imaging 25(2):433–440. https://doi.org/10.1002/jmri.20804
4. Aivazoglou LU, Guimaraes JB, Link TM et al (2021) MR imaging of inherited myopathies: a review and proposal of imaging algorithms. Eur Radiol. https://doi.org/10.1007/s00330-021-07931-9
5. Duan D, Goemans N, Takeda S et al (2021) Duchenne muscular dystrophy. Nat Rev Dis Primers 7:13
6. Alic L, Griffin JFT, Eresen A et al (2021) Using MRI to quantify skeletal muscle pathology in Duchenne muscular dystrophy: a systematic mapping review. Muscle Nerve 64(1):8–22. https://doi.org/10.1002/mus.27133
7. Power LC, O'Grady GL, Hornung TS et al (2018) Imaging the heart to detect cardiomyopathy in Duchenne muscular dystrophy: a review. Neuromuscul Disord 28(9):717–730. https://doi.org/10.1016/j.nmd.2018.05.011
8. McGreevy JW, Hakim CH, McIntosh MA et al (2015) Animal models of Duchenne muscular dystrophy: from basic mechanisms to gene therapy. Dis Model Mech 8(3):195–213. https://doi.org/10.1242/dmm.018424

9. Duan D (2015) Duchenne muscular dystrophy gene therapy in the canine model. Hum Gene Ther Clin Dev 26(1):57–69. https://doi.org/10.1089/humc.2015.006
10. Kornegay JN (2017) The golden retriever model of Duchenne muscular dystrophy. Skelet Muscle 7(1):9. https://doi.org/10.1186/s13395-017-0124-z
11. Thibaud JL, Monnet A, Bertoldi D et al (2007) Characterization of dystrophic muscle in golden retriever muscular dystrophy dogs by nuclear magnetic resonance imaging. Neuromuscul Disord 17(7):575–584
12. Thibaud JL, Azzabou N, Barthelemy I et al (2012) Comprehensive longitudinal characterization of canine muscular dystrophy by serial NMR imaging of GRMD dogs. Neuromuscul Disord 22(Suppl 2):S85–S99. https://doi.org/10.1016/j.nmd.2012.05.010
13. Guo LJ, Soslow JH, Bettis AK et al (2019) Natural history of cardiomyopathy in adult dogs with golden retriever muscular dystrophy. J Am Heart Assoc 8(16):e012443. https://doi.org/10.1161/JAHA.119.012443
14. Yokota T, Lu QL, Partridge T et al (2009) Efficacy of systemic morpholino exon-skipping in Duchenne dystrophy dogs. Ann Neurol 65(6):667–676. https://doi.org/10.1002/ana.21627
15. Bish LT, Sleeper MM, Forbes SC et al (2011) Long-term systemic myostatin inhibition via liver-targeted gene transfer in golden retriever muscular dystrophy. Hum Gene Ther 22(12):1499–1509. https://doi.org/10.1089/hum.2011.102
16. Bish LT, Sleeper MM, Forbes SC et al (2012) Long-term restoration of cardiac dystrophin expression in golden retriever muscular dystrophy following rAAV6-mediated exon skipping. Mol Ther 20(3):580–589. https://doi.org/10.1038/mt.2011.264
17. Barbash IM, Cecchini S, Faranesh AZ et al (2013) MRI roadmap-guided transendocardial delivery of exon-skipping recombinant adeno-associated virus restores dystrophin expression in a canine model of Duchenne muscular dystrophy. Gene Ther 20(3):274–282. https://doi.org/10.1038/gt.2012.38
18. Kornegay JN, Li J, Bogan JR et al (2010) Widespread muscle expression of an AAV9 human mini-dystrophin vector after intravenous injection in neonatal dystrophin-deficient dogs. Mol Ther 18(8):1501–1508. mt201094 [pii] 1038/mt.2010.94
19. Hayes J, Veyckemans F, Bissonnette B (2008) Duchenne muscular dystrophy: an old anesthesia problem revisited. Paediatr Anaesth 18(2):100–106. https://doi.org/10.1111/j.1460-9592.2007.02302.x

Chapter 18

Assessment of the Gene Therapy Immune Response in the Canine Muscular Dystrophy Model

Chady H. Hakim, Sandeep R. P. Kumar, Dennis Pérez-López, James Teixeira, Roland W. Herzog, and Dongsheng Duan

Abstract

The immune response is a primary hurdle in the development of gene therapy for neuromuscular diseases. Both innate and adaptive immune responses have been observed in human trials. The canine model is an excellent platform to understand immunological consequences of gene therapy. Over the last several decades, we have conducted gene replacement and gene repair therapies in the canine model of Duchenne muscular dystrophy (DMD) using adeno-associated virus (AAV)-mediated expression of the highly abbreviated micro-dystrophin gene, the larger mini-dystrophin gene, and the Cas9-based CRISPR genome editing machinery. We have evaluated the innate, humoral, and cellular immune responses to the AAV vector and the transgene product. In this chapter, we share our experience in collecting and processing of the dog blood samples for immunological assays, and our protocols for quantitative evaluation of cytokines and chemokines, antibodies, and T-cell responses.

Key words Duchenne muscular dystrophy, DMD, Canine model, Blood, Serum, Plasma, PBMCs, ELISA, Luminex, ELISpot

1 Introduction

Gene therapy is a highly powerful modality to significantly change the disease course of many neuromuscular disorders and dramatically improve the life quality of patients. The basic principle of gene therapy is to deliver a therapeutic gene to affected tissues in patients using a vehicle. Adeno-associated virus (AAV) is by far the most effective gene delivery vector. Although AAV is considerably less immunogenic compared to other viral vectors such as adenovirus, the immune response to AAV remains a major challenge to the success of gene therapy [1]. Besides the viral capsid, the therapeutic gene and its product are often foreign to the body and can induce untoward immune responses. The immune response has resulted in

Rika Maruyama and Toshifumi Yokota (eds.), *Muscular Dystrophy Therapeutics: Methods and Protocols*, Methods in Molecular Biology, vol. 2587, https://doi.org/10.1007/978-1-0716-2772-3_18,

serious adverse events and even patient death in neuromuscular disease gene therapy clinical trials [2].

Murine models are commonly used in gene therapy studies. However, mice are not ideal for studying the gene therapy immune response because mice are inbred and are housed in specific pathogen-free facilities. Canine models do not have these limitations and hence are better suited for studying the immune response.

Duchenne muscular dystrophy (DMD) is a fatal X-linked childhood muscle-wasting disease [3]. The canine DMD model was established soon after the discovery of the *DMD* gene and its gene product, dystrophin. Over the last several decades, we have explored AAV gene therapy for DMD in several canine models [4]. We have studied humoral immunity to AAV in normal and affected dogs [5] and cellular immunity to AAV following micro-dystrophin gene therapy in dystrophic dogs [6]. More recently, we characterized the immune response to the transgene product in AAV CRISPR therapy in the canine DMD model [7]. Here, we present the protocols we used to collect serum, plasma, and peripheral blood mononuclear cells (PBMC) for immunological assays. We also share the protocols to quantify serum chemokine and cytokine levels using the Luminex technology, antibody levels using the enzyme-linked immunosorbent (ELISA) assay, and the T cell response using the enzyme-linked immuno absorbent spot (ELISpot) assay.

2 Materials

2.1 Blood Drawing (Fig. 1A)

1. Personal protective equipment (PPE), including disposable gowns and/or dedicated scrubs, head covers, shoe covers, surgical masks, and gloves.
2. Color-coded BD® vacutainer blood collection tubes: Red-top tube with a silicone-coated interior for serum separation, lavender-top tube with spray-coated K2EDTA for hematology for plasma separation, green-top tube with spray-coated sodium heparin to extract peripheral blood mononuclear cell (PBMC) from blood, and light green PST™ tube with polymer gel and lithium heparin for blood chemistry (Becton, Dickinson, and Company, Franklin Lakes, NJ, USA).
3. Safety blood collection set (23G × ¾ inch or 21G × ¾ inch butterfly needle attached to a 12″ infusion line with Luer lock connector) (Greiner Bio-One, Kremsmünster, Austria).
4. BD Vacutainer one-use blood collection tube holder (Becton, Dickinson, and Company).
5. Animal hair clipper with #50 blade (Oster Professional Products, Sunbeam Products, Inc., Boca Raton, FL, USA).

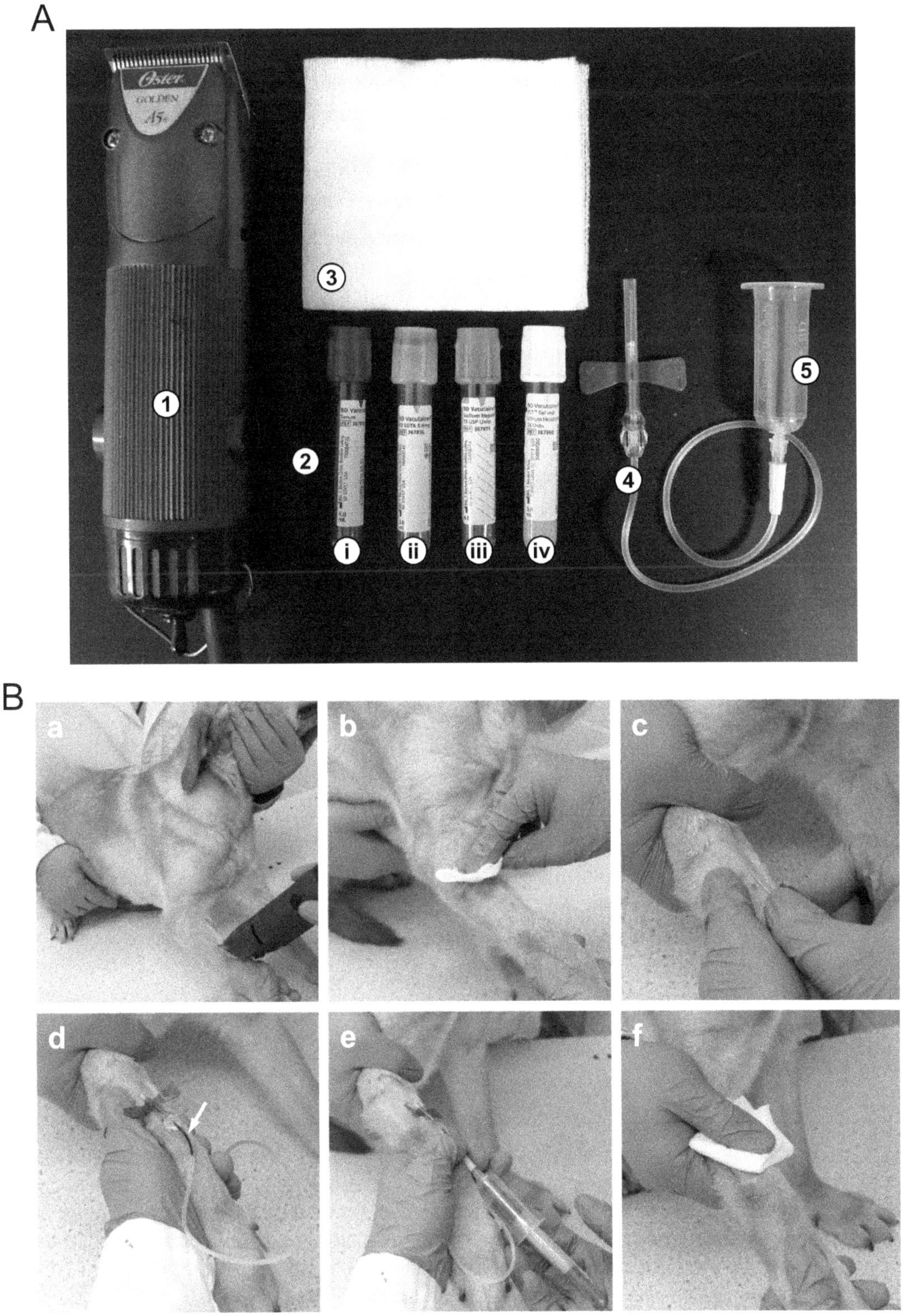

Fig. 1 Experimental materials and method for blood drawing. (**A**) Materials needed. (1) Animal hair clipper with #50 blade, (2) color-coded BD® vacutainer blood collection tubes, (i) red-top tube with a silicone-coated interior, (ii) lavender-top tube with spray-coated K2EDTA, (iii) green-top tube with spray-coated sodium heparin, and (iv) light green PST™ tube with polymer gel and lithium heparin, (3) nonwoven gauze sponges,

6. 70% Isopropyl alcohol (Walmart Stores Inc., Bentonville, AR, USA).
7. Nonwoven gauze sponges (4 inch × 4 inch) (Medline Industries Inc., Northfield, IL, USA).

2.2 Blood Sample Processing (Fig. 2A)

1. Personal protective equipment (PPE), including laboratory coat or disposable gowns, eye protection glasses, and gloves.
2. Color-coded BD® vacutainer blood collection tubes: Red-top tube with a silicone-coated interior for serum separation, lavender-top tube with spray-coated K2EDTA for hematology for plasma separation, and green-top tube with spray-coated sodium heparin to extract peripheral blood mononuclear cell (PBMC) from blood (Becton, Dickinson, and Company, Franklin Lakes, NJ, USA).
3. Eppendorf benchtop centrifuge model # 5810R with an A-4-62 rotor (Millipore Sigma, St. Louis, MO, USA).
4. Microcentrifuge tubes 2-mL Conical screw cap (Thermo Fisher Scientific, Hampton, NH, USA).
5. Penicillin-streptomycin (100X, containing 10,000 U/mL of penicillin and 10,000 μg/mL of streptomycin) (Thermo Fisher Scientific, Hampton, NH, USA).
6. L-Glutamine (200 mM, 100X) (Thermo Fisher Scientific, Hampton, NH, USA).
7. 2-Mercaptoethanol (55 mM, 1000X) (Thermo Fisher Scientific, Hampton, NH, USA).
8. Nonessential amino acid (100X) (Thermo Fisher Scientific, Hampton, NH, USA).
9. Sodium pyruvate (100 mM) (Thermo Fisher Scientific, Hampton, NH, USA)
10. Conical tube (15 and 50 mL) (Thermo Fisher Scientific, Hampton, NH, USA).
11. Fetal bovine serum (FBS) (Thermo Fisher Scientific, Hampton, NH, USA).

Fig. 1 (continued) (4) safety blood collection set (23G × ¾ inch or 21G × ¾ inch butterfly needle attached to a 12″ infusion line with Luer lock connector), and (5) BD Vacutainer one-use blood collection tube holder. (**B**) Step-by-step for blood collection. (**a**) Shave the cranial aspect of the lower forelimb. (**b**) Clean the skin on top of the cephalic vein with a soaked 70% isopropanol gauze by wiping it in a unidirectional straight motion. (**c**) Occludes the cephalic vein at the right forelimb elbow with the right thumb and then gently inserts the needle of the blood collection set into the cephalic vein. (**d**) Once the needle enters the vein, blood will appear in the needle hub (white arrow), then hold the infusion line of the blood collection set with the left thumb. (**e**) Insert the blood collection tube into the tube holder with the right hand to collect blood. (**f**) Apply pressure at the venipuncture site with dry gauze to stop the bleeding

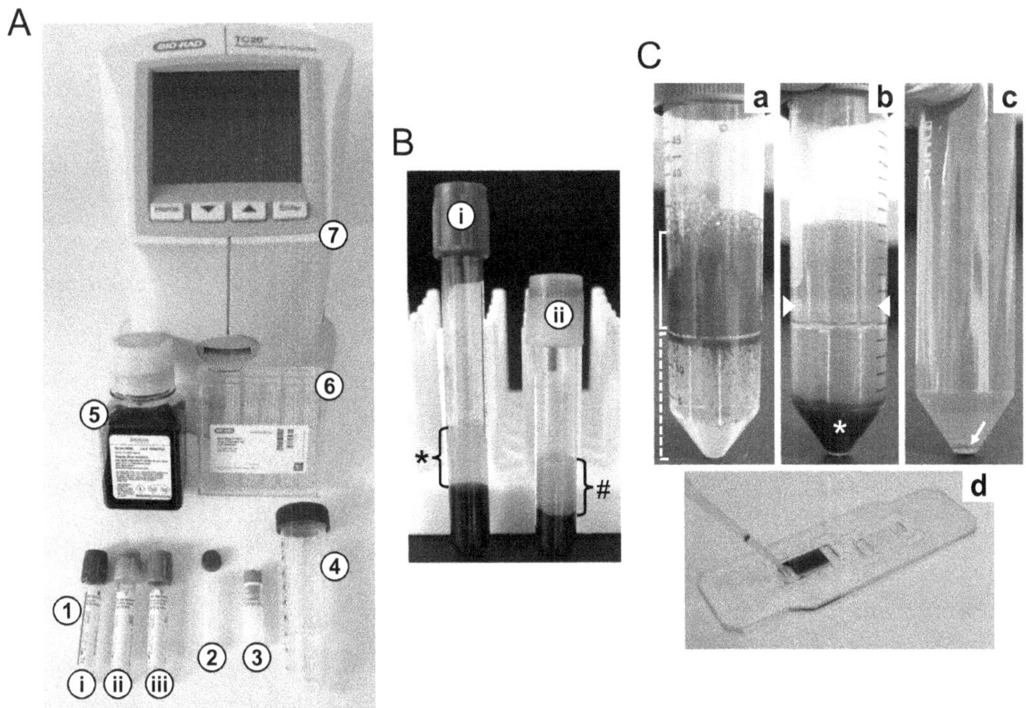

Fig. 2 Experimental materials and method for blood sample processing. (**A**) Materials needed. (1) Color-coded BD® vacutainer blood collection tubes, (i) red-top tube with a silicone-coated interior, (ii) lavender-top tube with spray-coated K2EDTA, and (iii) green-top tube with spray-coated sodium heparin, (2) microcentrifuge tubes 2 ml conical screw cap, (3) cryogenic vial, (4) stemcell SepMate™ tube, (5) trypan blue solution, (6) cell counting slides, and (7) TC20 automated cell counter. (**B**) A photograph showing the serum and plasma separation from blood cells following centrifugation. (i) Red-top tube with a silicone-coated interior for serum separation and (ii) lavender-top tube with spray-coated K2EDTA for plasma separation. *, serum (a yellowish transparent liquid). #, plasma (a yellowish transparent liquid). (**C**) A series of photographs showing the process of PBMC extraction. (**a**) A SepMate™ tube containing Ficoll-Paque density gradient medium in the bottom chamber of the tube, bellow the inset (dashed white bracket) and diluted blood sample on the upper chamber of the tube, above the inset (solid white bracket). (**b**) A SepMate™ tube following centrifugation showing the red blood cells and granulocytes are pelleted below the tube inset (*), while the PBMC layer remains above the inset (white arrow). (**c**) PBMC cells are pelleted at the bottom of the conical tube (white arrow) following centrifugation. (**d**) Resuspend PBMC cells mixed with trypan blue added to the counting slide chamber to measure the live cell count using the BioRad automated cell counter

12. RPMI medium 1640 (Thermo Fisher Scientific, Hampton, NH, USA).
13. HBSS 500 mL (Thermo Fisher Scientific, Hampton, NH, USA).
14. Ficoll-Paque PLUS (GE Healthcare Bio-Science, Uppsala, Sweden).
15. Stemcell SepMate™ tube (15 mL and 50 mL) (Stemcell Technologies Inc., Vancouver, BC, Canada).
16. Trypan blue solution (Sigma-Aldrich, Saint Louis, MO, USA).

17. TC20 automated cell counter (Bio-Rad Laboratories, Inc., Hercules, CA, USA).
18. Cell counting sliders (Bio-Rad Laboratories, Inc., Hercules, CA, USA).
19. Dimethyl Sulfoxide (DMSO) (Sigma-Aldrich, Saint Louis, MO, USA).
20. Cryogenic vial (1.8 mL) (Corning Inc., Corning, NY, USA).
21. Mr. Frosty™ freezing container (Thermo Fisher Scientific, Hampton, NH, USA).

2.3 Multiplex Luminex Cytokine Detection Assay (Fig. 3)

1. Personal protective equipment (PPE), including laboratory coat or disposable gowns, eye protection glasses, and gloves.
2. MILLIPLEX canine cytokine/chemokine magnetic bead panel kit. The kit consists of canine cytokine standard, canine cytokine quality controls 1 and 2, serum matrix, set of one 96-well plate with two sealers, assay buffer, 10X wash buffer, canine

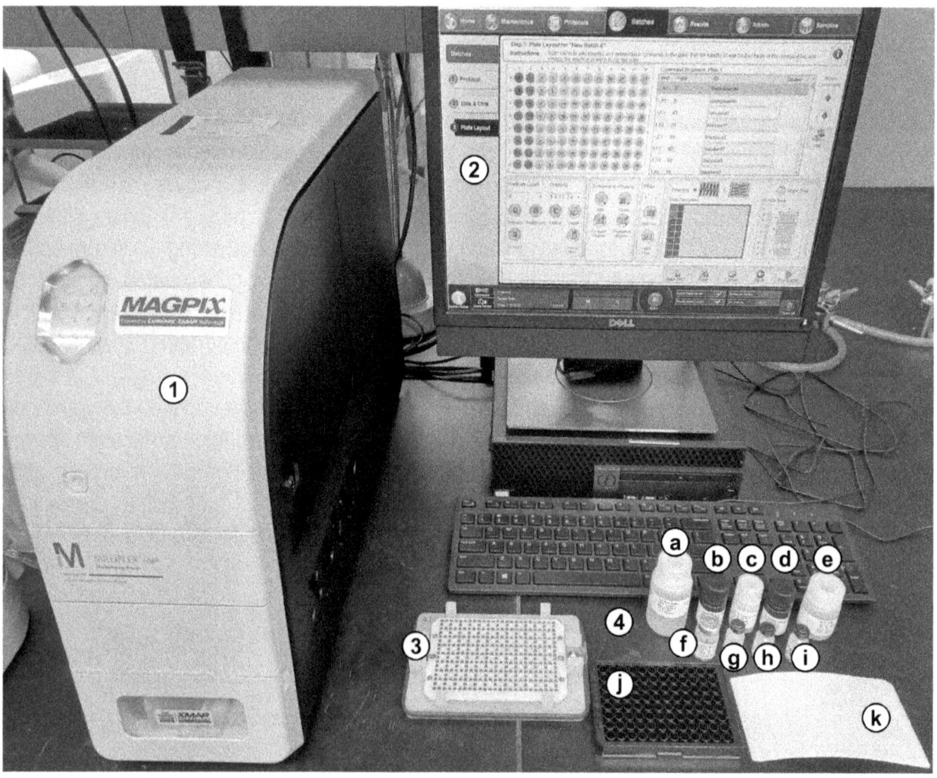

Fig. 3 Materials required for the multiplex Luminex cytokine detection assay. (1) Luminex® analyzer (MAGPIX®) system, (2) Luminex xPONENT software, (3) Magnet plate holder, (4) MILLIPLEX canine cytokine/chemokine magnetic bead panel kit: (**a**) 10X wash buffer, (**b**) premixed 13-plex beads, (**c**) canine cytokine detection antibodies, (**d**) streptavidin-phycoerythrin, (**e**) assay buffer, (**f**) serum matrix, (**g**) canine cytokine quality control 1, (**h**) canine cytokine quality control 2, (**i**) canine cytokine standard, (**j**) 96-well plate, and (**k**) plate sealer

cytokine detection antibodies, and streptavidin-phycoerythrin, and premixed 13-plex beads (Millipore Sigma, Burlington, MA, USA).

3. Orbital plate shaker (Thermo Fisher Scientific, Hampton, NH, USA).
4. Magnet plate holder (Millipore Sigma, Burlington, MA, USA).
5. MAGPIX® drive fluid (Millipore Sigma, Burlington, MA, USA).
6. Luminex® analyzer (MAGPIX®) system (Millipore Sigma, Burlington, MA, USA).
7. Luminex xPONENT 4.2 (Millipore Sigma, Burlington, MA, USA).
8. Sonicator (Thermo Fisher Scientific, Hampton, NH, USA).
9. Aluminum foil (Thermo Fisher Scientific, Hampton, NH, USA).
10. Belysa software version 1.0.19 (Millipore Sigma, Burlington, MA, USA).

2.4 Enzyme-Linked Immunosorbent (ELISA) Assay

1. Personal protective equipment (PPE), including laboratory coat or disposable gown, safety glasses or splash goggles, and gloves.
2. Canine ELISA kit (for cytokine/chemokine of interest). The kit consists of antibody-coated wells (8-well strips), antibody biotin-conjugate, standard, wash buffer, assay diluent, streptavidin-HRP, stop solution, and sealing films. (Thermo Fisher Scientific, Hampton, NH, USA).
3. Orbital plate shaker (Thermo Fisher Scientific, Hampton, NH, USA).
4. Aluminum foil (Thermo Fisher Scientific, Hampton, NH, USA).
5. ELISA plate reader (Thermo Fisher Scientific, Hampton, NH, USA).
6. Sodium carbonate (Thermo Fisher Scientific, Hampton, NH, USA).
7. Sodium biocarbonate (Thermo Fisher Scientific, Hampton, NH, USA).
8. Sodium hydroxide (NaOH) (Thermo Fisher Scientific, Hampton, NH, USA).
9. Tween-20 (Sigma-Aldrich, Inc., St. Louis, MO, USA).
10. 10X Phosphate buffered saline (PBS) (Thermo Fisher Scientific, Hampton, NH, USA).
11. Bovine serum albumin (BSA) (Thermo Fisher Scientific, Hampton, NH, USA).

12. Dog IgG whole molecule (Rockland Immunochemicals, Inc., Pottstown, PA, USA).
13. Cas9 protein (OriGene Technologies, Inc., Rockville, MD, USA).
14. Goat anti-Dog IgG (H&L)-HRP conjugated antibody (Rockland Immunochemicals, Inc., Pottstown, PA, USA).
15. SIGMAFAST OPD (o-phenylenediamine dihydrochloride) tablets (MilliporeSigma, Burlington, MA, USA).
16. Stop Solution (2N H_2SO_4) (Sigma-Aldrich, Inc., St. Louis, MO, USA).

2.5 Enzyme-Linked Immuno Absorbent Spot (ELISpot) Assay

1. Canine IFN-γ ELISpot Kit. The kit contains pre-coated canine IFN-γ microplate, biotinylated detection antibody, streptavidin conjugated to alkaline phosphatase, dilution buffer 1 and 2, 10X wash buffer, BCIP/NBT chromogen, and canine IFN-γ positive control (R&D Systems, Inc., Minneapolis, MN, USA).
2. RPMI medium (Thermo Fisher Scientific, Hampton, NH, USA).
3. Optima Grade Fetal Bovine Serum (FBS) (R&D Systems, Inc., Minneapolis, MN, USA).
4. Penicillin-streptomycin (10,000 U/mL) (100X, containing 10,000 U/mL of penicillin and 10,000 μg/mL of streptomycin) (Thermo Fisher Scientific, Hampton, NH, USA).
5. Ultra-low attachment 6-well plate (Corning Inc., Corning, NY, USA).
6. CO_2 Incubators (Thermo Fisher Scientific, Hampton, NH, USA).
7. 15-mL Centrifuge tube (Thermo Fisher Scientific, Hampton, NH, USA).
8. Trypan blue stain (Thermo Fisher Scientific, Hampton, NH, USA).
9. Cas9 protein (OriGene Technologies, Inc., Rockville, MD, USA).
10. Aluminum foil (Thermo Fisher Scientific, Hampton, NH, USA).
11. Rocking shaker (Thermo Fisher Scientific, Hampton, NH, USA).
12. Cell stimulation cocktail 500X (Thermo Fisher Scientific, Hampton, NH, USA).
13. ELISpot reader (ImmunoSpot, Shaker Heights, OH, USA).

3 Methods

3.1 Blood Drawing

The following protocol describes blood collection from the right cephalic vein (*see* **Notes 1** and **2**).

1. Follow the institutional guidelines for animal care and handling. Obtain approval from the institutional animal care and use committee (IACUC). Wear all PPE before working or handling animals. Use aseptic techniques while drawing blood from the dog.
2. Select the appropriate type of blood collection tube according to the required test and label the tube accordingly with the dog's information and the date of blood collection (*see* **Note 3**).
3. Assemble the safety blood collection set by screwing the blood collection tube holder to the Luer-lock connector at the end of the infusion line.
4. Shave the cranial aspect of the lower forelimb to allow for better visualization of the cephalic vein during the procedure (Fig. 1B.a).
5. Place the dog on the exam table in sternal recumbency with the lower forelimb extended. Clean the skin on top of the cephalic vein with a soaked 70% isopropanol gauze by wiping it in a unidirectional straight motion (Fig. 1B.b).
6. Perform the blood drawing with the help of two investigators. The first investigator first restrains the dog by standing on the dog's left side and rotates the dog's head to the left side using their left hand, then occludes the cephalic vein at the right forelimb elbow with the right thumb. The second investigator first holds the paw joint with the left hand to stabilize the limb and position the left thumb lateral to the cephalic vein to pull the skin to stabilize the vein, then gently inserts the needle of the blood collection set into the cephalic vein (*see* **Note 4**) (Fig. 1B.c). Once the needle enters the vein, blood will appear in the needle hub (Fig. 1B.d). The second investigator holds the infusion line of the blood collection set with the left thumb and inserts the blood collection tube into the tube holder with the right hand to collect blood (*see* **Note 5**) (Fig. 1B.e).
7. After the desired amount of blood is collected, the second investigator removes the collection tube from the holder, while the first investigator removes their thumb from the elbow region to the vein. Then, while pressing the entry site of the needle with dry gauze, the second investigator retracts the needle from the vein, and the first investigator applies pressure at the venipuncture site for 2 min to avoid bleeding (*see* **Notes 6** and 7) (Fig. 1B.f).

3.2 Blood Sample Processing

Follow the institutional guidelines for environmental health and safety. Obtain the necessary approval and training to process biological samples from the research institute. Wear all PPE before handling blood. Use aseptic techniques while collecting and processing blood samples.

3.2.1 Serum Separation

1. Collect whole blood in a red-top BD® vacutainer blood collection tube and allow the blood to clot by leaving the tube in an upright position and undisturbed for 15 min at room temperature.
2. Centrifuge the blood tube at 5000 RPM for 5 min. The resulting supernatant, a yellowish transparent liquid, is the serum (*see* **Note 8**) (Fig.2B).
3. Immediately transfer the serum into a clean microcentrifuge tube and store it accordingly (*see* **Note 9**).

3.2.2 Plasma Separation

1. Collect whole blood in a lavender-top BD® vacutainer blood collection tube. Immediately, mix the blood with the EDTA anticoagulant by gently inverting the tube three to five times.
2. Centrifuge the blood tube for 5000 RPM for 15 min. The resulting supernatant, a yellowish transparent liquid, is the plasma (*see* **Note 8**) (Fig.2B).
3. Immediately, transfer the plasma into a clean microcentrifuge tube and store it accordingly (*see* **Note 9**).

3.2.3 PBMC Isolation

1. Collect whole blood in a green-top BD® vacutainer blood collection tube. Immediately, mix the blood with the sodium heparin anticoagulant by gently inverting the tube three to five times
2. Prepare the media supplement mix by combining 100 mL of penicillin-streptomycin, 100 mL of L-glutamine, and 10 mL of the 2-mercaptoethanol in a 250 mL sterile bottle (*see* **Notes 10** and **11**).
3. Prepare the complete RPMI (CRPMI) by adding 5 mL of the 100X nonessential amino acids, 5 mL of the 100 mM sodium pyruvate, 50 mL of FBS, and 10.5 mL media supplement mix to 500 mL of RPMI. (*see* **Note 10**).
4. Dilute the blood sample with an equal volume of HBSS in a sterile 50 mL conical tube (*see* **Note 12**).
5. Add 15 mL of Ficoll-Paque density gradient medium into the SepMate™ tube through the center hole in the insert. Pipette the diluted blood sample slowly on the side of the SepMate™ tube. The insert in the tube separates blood from the gradient medium (*see* **Note 13**) (Fig. 2C.a).

6. Centrifuge at 1200× g for 10 min. During centrifugation, the red blood cells and granulocytes pellets below the tube insert, while the PBMC layer remains above the insert (*see* **Note 14**) (Fig. 2C.b).
7. Pour the PBMCs layer into a sterile 50-mL conical tube in one smooth motion (*see* **Note 15**). Add HBSS to bring the final volume to 50 mL. Invert the tube three to five times to wash the PBMC cells. Next, centrifuge the tube at 500× g for 5 min to pellet the PBMC cells and then discard the supernatant (Fig. 2C.c). Repeat the washing step one more time.
8. Resuspend the cells in 1 mL of CRPMI and determine the live cell count. In a separate microcentrifuge, mix an equal volume of the cell suspension and trypan blue. Add the mix into the counting slide chamber and measure the live cell count using the BioRad automated cell counter (Fig. 2C.d).
9. Dilute the cells at the desired concentration in a freezing media containing 40% CRPMI, 10% DMSO, and 40% FBS. Store the diluted cells in a cryogenic vial.
10. To freeze the cells, place the cryogenic vial immediately in a pre-chilled Mr. Frosty™ freezing container and store the container at −80 °C for 24–72 h. Following which vials should be transferred to liquid nitrogen for long-term storage.

3.3 Multiplex Luminex Cytokine Detection Assay

The following protocol describes the quantification of analytes (cytokines and/or chemokines) in canine serum or plasma samples. The protocol is performed in two days. On the first day, the sample is diluted and incubated with the detection bead overnight. On the second day, the beads are prepared for detection by the Luminex® analyzer (MAGPIX®) (*see* **Notes 16** and **17**).

Day one

1. Equilibrate the MILLIPLEX Canine cytokine/chemokine magnetic bead panel reagents to room temperature. Reconstitute the kit reagents (wash buffer, quality controls (QC1 and QC2), standard, and serum matrix) as indicated by the manufacturer, and incubate them for 10–30 min at room temperature before using them in the assay (*see* **Note 18**).
2. Dilute the sample 1:3 in the assay buffer (*see* **Notes 19** and **20**).
3. Dilute the standard fourfolds and continue with a fourfold serial dilution to obtain seven standards (*see* **Note 21**).
4. Add 200 μL of assay buffer to the 96-well plate and seal the plate with a sealing film. Shake the plate on an orbital plate shaker for 1 min at 600 rpm to wet it. Decant the assay buffer and remove any residual amount from the wells by dabbing the plate on a stack of paper towels several times.
5. Add 25 μL of assay buffer (background), the seven standards, QC1 and QC2, and sample into the designated wells. Add

25 μL of the matrix solution to the background, the seven standards, and the two QCs wells. Add 25 μL of assay buffer to the sample wells.

6. Sonicate the magnetic beads bottle for 30 s and then vortex it vigorously for 1 min. Add 25 μL of the magnetic beads to each well.
7. Seal the plate with a plate sealing film and wrap it in foil. Shake the plate on an orbital plate shaker for 16–18 h (overnight) at 600 rpm in 4 °C.

Day two

8. Continue the last hour of the shaking at room temperature using the same speed.
9. Equilibrate the Milliplex canine magnetic bead reagents to room temperature for 1 h before continuing the assay.
10. Turn on the Luminex® analyzer (MAGPIX®), and calibrate the machine according to the manufacturer's instructions. Setup the Luminex® analyzer xPONENT software with the sample and dilution information (*see* **Note 22**).
11. Unwrap the plate from the foil. Secure the plate to the magnetic plate holder and wait for 1 min to capture the beads at the bottom of the plate through the magnetic field. Remove the sealing film and decant the plate content into the sink.
12. Remove the plate from the magnetic plate holder and add 200 μL of wash buffer to each well. Seal the plate with a sealing film, cover with foil, and shake on an orbital plate shaker for 30 s at 600 rpm at room temperature.
13. Repeat steps 11 and 12 to wash the beads one more time.
14. Vortex the detection antibody for 30 s and add 25 μL of the antibody to each well. Seal the plate with a sealing film, wrap with foil, and shake on an orbital plate shaker for 1 h at 600 rpm at room temperature.
15. Remove the plate seal and add 25 μl of streptavadin to each well. Seal the plate with a sealing film, wrap with foil and shake on an orbital plate shaker for 30 min at 600 rpm at room temperature.
16. Repeat steps 11 and 12 twice to wash the beads.
17. Add 150 μL MAGPIX® drive fluid to each well. Seal the plate with a sealing film, wrap with foil and shake on an orbital plate shaker for 5 min at 600 rpm at room temperature.
18. Unwrap the plate from the foil and remove the sealing film. Load the plate into the Luminex® analyzer (MAGPIX®) and detect the bead content of each well using the Magpix analyzer with the xPONENT software. Analyze the data using the Belysa software (*see* **Note 23**).

3.4 Enzyme-Linked Immunosorbent (ELISA) Assay

The following protocol describes an enzyme-linked immunosorbent assay (ELISA) protocol using Invitrogen canine ELISA kit to detect and quantify a target cytokine and chemokine from canine serum or plasma (*see* **Note 24**).

3.4.1 ELISA Protocol to Detect Cytokine and Chemokine

1. Prepare the kit reagents (wash buffer, assay diluent, streptavidin-HRP solution, and biotin conjugate) as indicated by the manufacturer. Reconstitute the lyophilized standard and dilute according to the manufacturer guidelines to have a set of seven standards with high to low dilution.
2. Dilute the sample twofolds in the assay diluent (*see* **Notes 25** and **26**).
3. Determine the number of the 8-well strips needed for the assay and place them in the strip holder.
4. Add 100 μL of each standard and sample in the designated wells.
5. Seal the wells with the sealing films, and incubate for 2.5 h at room temperature or overnight at 4 °C on an orbital shaker.
6. Decant the wells content into the sink.
7. Add 300 μL of wash buffer to each well. Seal the wells with a sealing film, and shake gently on an orbital plate shaker for 30 s at room temperature. Remove the sealing film and decant the well content into the sink
8. Repeat step 7 three more times and at the end of the last wash, remove any residual amount from the wells by inverting the plate and tapping it onto paper towels several times.
9. Add 100 μL of biotin-conjugated antibody to each well, seal the well with sealing films and incubate for 1 h at room temperature while shaking gently on an orbital shaker.
10. Decant the well content into the sink and repeat steps 7 and 8 to wash the wells.
11. Add 100 μL of streptavidin-HRP solution to each well, seal the wells with sealing films, and incubate for 45 min at room temperature while shaking gently on an orbital shaker.
12. Decant the well content into the sink and repeat steps 7 and 8 to wash the wells.
13. Add 100 μL of TMB substrate to each well. The solution in the well will start to develop a blue color. Wrap the wells with foil and incubate for 30 min at room temperature in the dark while shaking gently on an orbital shaker.
14. Add 50 μL of stop solution to each well and tap the side of the holder gently to mix. The solution in the wells will change to yellow color.
15. Read the plate at 450 nm in an ELISA plate reader. Analyze the data according to the manufacture's guidelines (*see* **Note 27**).

3.4.2 ELISA Protocol to Detect Anti-Cas9 Total IgG in Canine Serum

The following protocol describes an enzyme-linked immunosorbent assay (ELISA) protocol using an in-house protocol to detect Cas9 total antibody from canine serum. The protocol is performed in two days. On the first day, the standard wells are coated with the dog IgG, and the sample wells are coated with the Cas9 protein. On the second day, serum samples are incubated in Cas9-coated wells, and anti-cas9 total IgG antibody is detected using Goat anti-dog IgG-HRP conjugated antibody (*see* **Note 28**).

Day 1

1. Prepare the coating buffer by dissolving 0.68 g of sodium carbonate, 3.675 g sodium biocarbonate in 475 mL of distilled water. Adjust the pH to 9.2 with 5N NaOH and bring the final volume to 500 mL using distilled water.
2. Prepare the wash buffer (0.05% Tween-20 in 1X PBS) by mixing 1 mL of Tween-20 in 2000 mL of 1X PBS.
3. Prepare the dilution buffer (6% bovine serum albumin) by dissolving 6 g of BSA in 100 mL of wash buffer.
4. Determine the desired number of 8-well ELISA strips and insert them in a strip holder.
5. Prepare the standard by diluting dog IgG whole molecule to 4000 ng/mL in coating buffer. Perform a twofold serial dilution of 4000 ng/mL standard until concentration reaches 7.8 ng/mL. This will provide an extended range of standard (4000 ng/mL–7.8 ng/mL).
6. Dilute the Cas9 protein to 25 μg/mL in coating buffer.
7. Add 50 μL of each standard to the designated standard wells, 50 μL of coating buffer to the designated blank wells, and 50 μL of Cas9 protein (25 μg/mL) in the sample wells. Each standard, blank and sample should be tested in at least duplicates.
8. Seal the plate with a sealing film and incubate it at 4 °C overnight (*see* **Note 29**).

Day 2

9. Add 200 μL of wash buffer to each well and incubate for atleast 1 min at room temperature. Decant the well content into a sink (*see* **Note 30**).
10. Repeat step 9 two more times and remove any residual wash buffer from wells by gently dabbing the plate on a stack of paper towels several times.
11. Add 200 μL of dilution buffer to each well. Seal the plate with a sealing film and incubate it for 1 h at room temperature.
12. Dilute each serum sample (1:50) in dilution buffer (*see* **Note 31**).

13. After 1 h of incubation remove the dilution buffer from each well by decanting into the sink and remove any residual amount from wells by gently dabbing the plate on a stack of paper towels several times.
14. Add 50 μL of the diluted sample in duplicate to the Cas9-coated wells. Add 50 μL of dilution buffer to each standard and blank wells. Seal the plate with a sealing film and incubate it for 1 h at 37 °C.
15. Repeat steps 9 and 10 to wash the plate.
16. Dilute goat anti-dog IgG-HRP conjugated antibody (1: 10,000) in dilution buffer and add 100 mL of it to each well (*see* **Note 31**). Seal the plate with a sealing film and incubate it for 1 h at 37 °C.
17. Repeat steps 9 and 10 to wash the plate.
18. Prepare the OPD solution by dissolving SIGMAFAST OPD tablets. First dissolve OPD tablet (wrapped in gold foil) in 20 mL distilled water. Once dissolved, add urea hydrogen peroxide/buffer tablet (wrapped in silver foil) and completely dissolve to obtain a ready-to-use buffered solution containing OPD and urea hydrogen peroxide.
19. Add 200 μL of OPD solution to each well and incubate the plate at room temperature for 10 min while protecting it from light (*see* **Note 32**).
20. Add 50 μL of stop solution (2N H_2SO_4) to each well and mix by gently tapping on the side of the plate. The solution in the wells will change to yellow color.
21. Read the plate at 492 nm in an ELISA plate reader. Analyze the data by using a four-parameter logistic curve fit (*see* **Note 27**).

3.5 Enzyme-Linked Immuno Absorbent Spot (ELISpot) Assay

The following protocol describes an enzyme-linked immuno absorbent spot (ELISpot) assay protocol using a Canine IFN-γ ELISpot kit to detect Cas9-specific T cells from canine peripheral blood mononuclear cells (PBMC) (*see* **Note 33**). However, the protocol can be adapted to detect other antigen-specific T cells.

Day 1

1. Isolate or revive previously cryopreserved PBMC.
2. Resuspend PBMC in RPMI medium supplemented with 10% FBS, 100 units/mL of penicillin, and 100 μg/mL of streptomycin and plate in an ultra-low attachment 6-well plate. Incubate the cells overnight at 37 °C with 5% CO_2 and >90% humidity.

Day 2

Preparation of ELISpot Plate

3. Take out the pre-coated ELISpot plate from the kit. Pre-wet the plate wells with 200 mL of RPMI medium supplemented with 10% FBS, 100 units/mL of penicillin, and 100 μg/mL of streptomycin.
4. Incubate the plate at room temperature for at least 30 min.

Preparation of Cells

5. During the incubation, transfer the cells into a 15 mL centrifuge tube. Pellet down the cells by centrifugation at 300 g and 4 °C for 10 min.
6. Remove the supernatant and resuspend the cell pellet with 1 mL of RPMI medium supplemented with 10% FBS, 100 units/mL of penicillin, and 100 μg/mL of streptomycin.
7. Mix 15 μL of cell suspension with 135 μL of trypan blue stain (0.04%). Perform a cell count with 10–20 μL of this cell suspension to determine the number of live cells.
8. Dilute the cells to a final concentration of 2.5×10^6 cells/mL. In case of low recovery, dilute cells to a final concentration of 1.0×10^6 cells/mL.

Cell Seeding

9. Take a wide-opening container and spray the inside with 70% ethanol, and place it inside the biosafety cabinet. Spray a stack of paper towels with 70% ethanol and place it inside the biosafety cabinet.
10. Remove RPMI medium from ELISpot plate by flicking it over the wide-opening container. Remove the leftover medium from the plate by blotting on the stack of paper towels.
11. Leave the first and last columns and rows of the ELISpot plate empty. For each sample, add 100 μL of the cell suspension to 9 wells [triplicate wells for unstimulated control, Cas9 treatment, and cell stimulation cocktail (positive control)].
12. Mix Cas9 protein at a concentration of 20 μg/mL in 1 mL of RPMI medium supplemented with 10% FBS, 100 units/mL of penicillin, and 100 μg/mL of streptomycin.
13. Dilute the 500X cell stimulation cocktail to 2X using RPMI medium supplemented with 10% FBS, 100 units/mL of penicillin, and 100 μg/mL of streptomycin.
14. Add 100 μL of RPMI medium supplemented with 10% FBS, 100 units/mL of penicillin, and 100 μg/mL of streptomycin to each unstimulated control well.

15. Add 100 μL of 20 μg/mL Cas9 protein mixture (prepared in step 14) to each Cas9 treatment well.
16. Add 100 μL of 2X cell stimulation cocktail (prepared in step 15) to each positive control well.
17. To avoid evaporation from testing wells, add 200 μL of plain RPMI medium to first and last columns and rows of the ELI-Spot plate.
18. To maintain a uniform temperature, cover the plate with aluminum foil (without wrapping) in such a way that it does not prevent gas exchange.
19. Incubate the plate for 48 h at 37 °C with 5% CO_2 and >90% humidity. Avoid disturbing the plate by repeatedly opening and closing the incubator during these 48 h (*see* **Note 34**).

Day 4

20. Prepare 1X wash buffer by mixing 50 mL of 10X wash buffer with 450 mL of distilled water.
21. Remove the plate from the incubator and flick it over a wide-opening waste container containing 10% bleach.
22. Wash the wells four times with 200 μL of 1X wash buffer.
23. Prepare detection antibody by diluting 100 μL of biotinylated detection antibody concentrate in 12 mL of buffer #1.
24. Add 100 μL of diluted detection antibody to each well and incubate overnight at 4 °C. Alternatively, the plate can be incubated at room temperature for 2 h with constant rocking.

Day 4/5

25. Wash each well with 200 μL of 1X wash buffer.
26. Prepare alkaline phosphatase-conjugated streptavidin by diluting 100 μL of streptavidin-alkaline phosphatase concentrate in 12 mL of buffer # 2.
27. Add 100 μL of diluted streptavidin-alkaline phosphatase to each well and incubate at room temperature for 2 h with constant rocking.
28. Wash the wells four times with 200 μL of 1X wash buffer.
29. Add 100 μL of BCIP/NBT substrate to each well and incubate at room temperature for 30–60 min while protecting from light. After 30 min, monitor the spot formation/development in positive control wells using a dissection microscope (*see* **Note 35**).
30. After development, decant the BCIP/NBT substrate and rinse the plate thoroughly with distilled water. Remove the underdrain plastic from the plate and wipe the bottom of the plate

with paper towels. Air-dry the plate at room temperature. Alternatively, the plate can be dried at 37 °C for 30 min (*see* **Notes 36** and **37**).

31. Analyze the plate for spot quantification with an ELISpot reader. In case an ELISpot reader is not available, spots can be counted manually under a dissection microscope. The number of spots is corrected for a million cells and is generally reported as the number of spot forming units (SFU) per 10^6 cells.

4 Notes

1. A well-trained individual should perform the venipuncture step in this procedure.
2. Besides the cephalic vein, other veins can also be used, such as the saphenous vein and the jugular vein.
3. BD offers different blood collection tubes for hematology, chemistry, coagulation, and other specialty tests. The tubes come in various volume capacities (2–10 mL) based on the amount of blood needed. They also offer BD Microtainer® for blood volume less than 0.5 mL. We use the red top tube with a silicone-coated interior for chemistry (e.g., complete comprehensive panel) and serum separation, lavender top tube with spray-coated K2EDTA for hematology (e.g., total blood count) and plasma separation, light green PST™ tube with polymer gel and lithium heparin for chemistry (e.g., cardiac troponin) and plasma separation, and green top tube with spray-coated sodium heparin to isolate PBMC.
4. Insert the needle at the most visible distal part of the vein. Use caution when inserting and advancing the needle in the vein. If a second venipuncture occurs while advancing the needle in the vein, blood may leak to the surrounding tissue resulting in formation of a hematoma. To avoid this, gently pull the needle out and apply pressure on the venipuncture site using a dry gauze until the bleeding stop. To reattempt venipuncture, insert the needle proximal to the initial venipuncture site or in a different vein.
5. Due to the vacuum pressure in the blood collection tube, the needle tip may lodge against the vessel wall, disrupting the blood flow. To overcome this hurdle, gently rotate or reposition the needle to restore blood flow.
6. To collect blood into multiple blood collection tubes, simply replace the blood-filled tube in the tube holder with the new tube. Use caution while changing the blood collection tubes to avoid disturbing the needle position in the vein.

7. Insufficient pressure on the venipuncture site may result in blood leakage to the surrounding tissue resulting in formation of a hematoma. To avoid this, hold a dry gauze tightly over the venipuncture site until the bleeding stops.
8. Use caution when moving the blood tube out of the centrifuge to avoid mixing red blood cells from the pellet with the serum or plasma. If a mix-up happens, re-centrifuge the sample as indicated.
9. If serum or plasma sample is not analyzed following the separation procedure, store immediately at −20 °C or lower. Avoid freeze-thaw as it detriments the sample components. In addition, hemolyzed, icteric, or lipemic samples can invalidate some tests.
10. All reagent preparation and the PBMC isolation should be carried out in a laminar flow hood. Equilibrate all reagents to room temperature before starting the PBMC isolation.
11. Aliquot 10.5 mL of the media supplement mix into a sterile 15-mL conical tube and store in −80 °C.
12. The amount of blood processed should be based on the total number of PBMC cells needed. In 1 mL of blood, the expected PBMC yield ranges between 0.5 and 3×10^6 cells. The SepMate™ 50 mL tube processes 4–17 mL of the whole-undiluted blood, whereas the SepMate™ 15 mL tube process 0.5–5 mL of the whole-undiluted blood.
13. While adding the density gradient medium, bubbles may be present in the lower chamber of the SepMate™ tube. However, this should not affect the PBMC isolation.
14. No need to turn off the brake setting on the centrifuge. The design of the SepMate™ tube prevents the separated layers from mixing when the centrifuge slows down.
15. The insert in the SepMate™ tube keeps the red blood cell and gradient medium separated from the isolated PBMC. However, do not hold the tube in an inverted position while pouring the PBMC layer more than 2 s.
16. The MILLIPLEX® is based on the Luminex® xMAP® technology and allows the evaluation of the cytokines and chemokines in the canine cell or tissue lysate, culture supernatant, and serum or plasma samples. Luminex uses proprietary techniques to internally color-code microspheres with two fluorescent dyes. Through precise concentrations of these dyes, 100 distinctly colored bead sets can be created, each of which is coated with a specific capture antibody. After a cytokine from a test sample is captured by the bead, a biotinylated detection antibody is introduced. The reaction mixture is then incubated

with streptavidin-PE conjugate, the reporter molecule, to complete the reaction on the surface of each microsphere. The microspheres are allowed to pass rapidly through a laser that excites the internal dyes marking the microsphere set. A second laser excites PE, the fluorescent dye on the reporter molecule. Finally, high-speed digital-signal processors identify each microsphere and quantify the result of its bioassay based on fluorescent reporter signals. The capability of adding multiple conjugated beads to each sample results in the ability to obtain multiple results from each sample. MILLIPLEX® canine cytokine/chemokine panel used in this protocol detects the following 13 analytes: GM-CSF, IFN-γ, KC-like, IP-10, IL-2, IL-6, IL-7, IL-8, IL-10, IL-15, IL-18, MCP-1, and TNF-α. 1. However, these analytes can be purchased separately or custom-prepared based on the user's need.

17. The protocol described here is a summarized version of the detailed manufacturer's protocol. We highly recommend that the user read and follow the manufacturer's guidelines and instructions in performing the assay.
18. The manufacturer instructions can be found in https://www.emdmillipore.com/US/en/product/MILLIPLEX-Canine-Cytokine-Chemokine-Magnetic-Bead-Panel-Immunology-Multiplex-Assay,MM_NF-CCYTOMAG-90K#anchor_PR.
19. Samples can be fresh or frozen. For a frozen sample, the manufacturer recommends completely thawing the sample on ice, mixing it well by vortexing, and centrifuging it to remove any particulates before dilution. We recommend not freeze/thaw a serum or plasma sample more than two times as the quality of the assay would significantly diminish.
20. When further dilution of more than threefolds is required, the manufacturer recommends using serum matrix as the diluent instead of the assay buffer.
21. According to the manufacturer's instructions, start by labeling six microcentrifuge tubes Standard #6, Standard #5, Standard #4, Standard #3, Standard #2, and Standard #1. Add 150 μl of assay buffer to each of the six tubes. Perform fourfold serial dilutions by adding 50 μL of the kit standard (standard #7) to the standard #6 tube. Mix well and transfer 50 μL of standard #6 to the standard #5 tube and mix well. Repeat the same serial dilution until standard 1.
22. The Luminex® analyzer (MAGPIX®) must be calibrated using a calibration kit and performance verification kit. The calibration and verification data are stored for 72 h in the machine. We highly recommend calibrating the machine before starting the assay to assure that the machine is clear of any error or malfunction.

23. In case the Luminex® analyzer (MAGPIX®) is not ready to analyze the plate, we recommend storing the foil-wrapped sealed pate in 4 °C for a max of 3 h and then shaking it on an orbital plate shaker while being wrapped in foil for 30 min at 600 rpm at room temperature before measurement.
24. The protocol described here is a summarized version of the detailed manufacturer's protocol. We highly recommend that the user read and follow the manufacturer's procedural guidelines and plate washing directions in the ELISA technical guide at thermofisher.com for details prior to starting the assay. We also like to point out that Thermo Fisher Scientific offers different canine cytokine and chemokine ELIS kits such as IL-10, IL-6, IL-8, TNF alpha, MCP-1, and IFN gamma.
25. Samples can be serum, plasma, cell culture supernatants, or cell culture media, and can be either a fresh or frozen sample. For a frozen sample, the manufacturer recommends completely thawing the sample on ice, mixing it well by vortexing, and centrifuging it to remove any particulates before dilution. We recommend not freeze/thaw a sample more than two times as the quality of the assay would significantly diminish. We also recommend avoiding using hemolyzed or lipemic sera in the assay.
26. We recommend that each investigator determine the optimal dilution to be used in the assay. When further dilution of more than twofolds is required, the manufacture recommends diluting the sample up to tenfolds using the assay diluent.
27. Absorbance values of the samples should be within the linear range of the standard curve.
28. Standards, Cas9 protein for coating, serum dilution, secondary antibody dilution, and OPD solution must be prepared fresh before use.
29. Alternatively, the coating can be performed by incubating the plate at 37 °C for 2 h.
30. Alternatively, an ELISA plate washer can be used.
31. We recommend that each investigator determine the optimal dilution to be used in the assay.
32. We recommend that each investigator determine the optimal incubation time for color development. Alternatively, absorbance can be measured (at 450 nm) repeatedly until the O.D. value for the highest standard is ≤ 1.0. At that time point, color development can be stopped by adding the stop solution.

33. The protocol described here is a summarized version of the detailed manufacturer's protocol. We highly recommend that the user read and follow the manufacturer's procedural guidelines
34. The incubation time of the cells with stimulant peptide/protein should be optimized. Shorter incubation time may result in minimal spot formation, while longer incubation time may cause saturation of spot formation. Therefore, leading to inaccurate assessment of results between experimental groups. An incubation of 24 h is a good starting point.
35. To prevent high background, carefully monitor the spot development after 30 min of incubation with BCIP/NBT substrate. An over-developed plate will make it difficult to accurately quantify the number of spots.
36. Plate should be rinsed thoroughly with distilled water as any residual BCIP/NBT substrate can cause over-development of the plate while drying.
37. Spot quantification cannot be performed accurately until the plate is completely dry. It is best to leave the plate for drying overnight at room temperature. Alternatively, it can be dried at 37 °C for 30–60 min.

Acknowledgments

The research on the canine DMD model in the Duan lab is currently supported by the National Institutes of Health (NS-90634 and AR-70517), Jesse Davidson Foundation-Defeat Duchenne Canada, Hope for Javier, Jackson Freel DMD Research Fund, Parent Project Muscular Dystrophy, Jett Foundation, Michael's Cause, Ryan's Quest, Solid Biosciences Inc., and the University of Missouri.

Disclosure DD is a member of the scientific advisory board for Solid Biosciences and equity holders of Solid Biosciences. DD is a member of the scientific advisory board for Sardocor Corp. In the last three years, the Duan lab has received research supports unrelated to this project from Solid Biosciences and Edgewise.

References

1. Shirley JL, de Jong YP, Terhorst C et al (2020) Immune responses to viral gene therapy vectors. Mol Ther 28(3):709–722. https://doi.org/10.1016/j.ymthe.2020.01.001
2. Philippidis A (2021) Fourth boy dies in clinical trial of Astellas' AT132. 32:19–20
3. Duan D, Goemans N, Takeda S et al (2021) Duchenne muscular dystrophy. Nat Rev Dis Primers 7:13
4. Duan D (2015) Duchenne muscular dystrophy gene therapy in the canine model. Hum Gene

Ther Clin Dev 26(1):57–69. https://doi.org/10.1089/humc.2015.006

5. Shin J-H, Yue Y, Smith B et al (2012) Humoral immunity to AAV-6, 8, and 9 in normal and dystrophic Dogs. Hum Gene Ther 23(3):287–294. https://doi.org/10.1089/hum.2011.125
6. Shin J-H, Pan X, Hakim CH et al (2013) Micro-dystrophin ameliorates muscular dystrophy in the canine model of Duchenne muscular dystrophy. Mol Ther 21(4):750–757. https://doi.org/10.1038/mt.2012.283
7. Hakim CH, Kumar SRP, Perez-Lopez DO et al (2021) Cas9-specific immune responses compromise local and systemic AAV CRISPR therapy in multiple dystrophic canine models. Nat Commun 12(1):6769. https://doi.org/10.1038/s41467-021-26830-7

Chapter 19

Use of Mesenchymal Stem Cells to Enhance the Efficacy of Gene Therapy

Hiromi Hayashita-Kinoh and Takashi Okada

Abstract

Herein, a method to use of mesenchymal stem cells (MSCs) to modulate immune response against rAAV transduction in a canine Duchenne muscular dystrophy (DMD) model is presented. The aim is to overcome the immune response against adeno-associated virus (AAV) capsid itself as well as against the AAV-derived transgene.

AAV is currently the most used viral vector because of its relative safety and high efficiency of gene transfer to nondividing cells. Since DMD is caused by a deficiency of dystrophin protein due to mutation or deletion in the dystrophin gene, dystrophin replacement therapy using AAV vectors carrying dystrophin as a therapeutic gene is an effective treatment as shown by animal experiments and clinical trials. Because DMD is a systemic disease, the amount of AAV vector required to achieve efficacy is impractically large. MSC have been used in combination with organ transplants due to their immunomodulatory effects. By using MSCs and AAVs in combination as described below, we are able to decrease the immune response to AAV capsid and the transgene as well as to reduce the dose of AAV to approximately 1/100 of the dose used in conventional clinical trials.

Key words Gene therapy, AAV vector, Mesenchymal stromal cells, Immune modulation

1 Introduction

Duchenne muscular dystrophy (DMD) is an X-linked recessive genetic disease associated with mutations in the dystrophin gene, causing progressive deterioration of skeletal and cardiac muscle. Patients with DMD develop progressive wasting of skeletal and cardiac muscle, eventually leading to early death due to cardiopulmonary dysfunction. Currently, no curative treatment for DMD has been established, but several new strategies have been developed to replace and repair the dystrophin gene, including gene supplementation therapy using viral vectors. Recombinant adeno-associated virus (rAAV) vectors have shown great promise for gene therapy due to their high safety, low toxicity, and long transgene expression period, and many clinical trials of gene therapy using

Rika Maruyama and Toshifumi Yokota (eds.), *Muscular Dystrophy Therapeutics: Methods and Protocols*,
Methods in Molecular Biology, vol. 2587, https://doi.org/10.1007/978-1-0716-2772-3_19,

rAAV have been conducted, some of which have resulted in the approval of gene therapy products [1]. Recently, however, it has been suggested that AAV vector-related toxicities such as hepatotoxicity, neurotoxicity [2], and germline transmission [3, 4] can occur when high doses of rAAV are administered systemically, and methods to reduce the dose are being investigated.

Dystrophin supplementation using rAAV has been shown to be effective in alleviating the pathology of animal models of DMD. We have previously shown that local injection of rAAV-microdystrophin into the skeletal muscle of *mdx* mice [5–7] and systemic injection of rAAV9-microdystrophin into *mdx* mice resulted in widespread expression of microdystrophin and improvement of myocardial dysfunction [8]. However, immune responses to AAV capsid proteins and transgene products hinder efficient gene therapy. To ensure effective long-term expression of the transgene, it is important to prevent immune-mediated destruction of the transduced myofibers by controlling the host immune response. Several groups, including our own, have reported that rAAV-mediated gene transfer into canine skeletal muscle elicits an immune response to rAAV or the transgene product [9–11]. Recently, a clinical gene therapy trial using rAAV2.5-minidystrophin was conducted in six patients with DMD. In this clinical trials, T-cell responses to transduced dystrophin were detected [12]. To overcome these immune responses against AAV capsid itself and/or transgenes, we previously reported that a single intravenous injection of rAAV-microdystrophin into canine fetuses successfully resulted in long-term gene expression and improved the dystrophic phenotype [13].

Mesenchymal stromal cells (MSCs) are mesoderm-derived multipotent mesenchymal cells with the potential of self-renewal. MSCs have been reported to modulate various inflammatory diseases, including graft-versus-host disease (GvHD), through their immunomodulatory effects. Furthermore, MSCs have been conditionally approved for use in Canada, New Zealand, and Japan. In addition, it has been shown in animal studies that MSCs alone can suppress inflammation in muscular dystrophy.

In this section, an effective method of administration of bone marrow-derived MSCs in combination with AAV administration to induce immune tolerance to rAAV9 vector and expression of transgene is described below.

2 Materials

2.1 AAV Preparation

2.1.1 AAV Production

1. HEK293 or 293AAV (Agilent technologies) cells.
2. Culture flasks such as T225 flask (Nunc, Thermo Fisher Scientific).

3. DMEM or DMEM/F12 supplemented with antibiotics and 10% FBS.
4. AAV Helper-Free system (Agilent technologies) contains AAV plasmids (AAV serotype 2 of Rep/Cap plasmid, marker gene expression vector with ITR sequence, pHelper). If you want to produce other AAV serotype, search the Addgene website with "AAV RepCap."
5. Microdystrophin expression plasmid or other therapeutic gene expression plasmid.
6. Transfection reagent such as PEI, PEI-Max, TransIT-Virus-GEN Transfection Reagent, or similar.
7. Purification kit such as AAVpro Purification kit (Takarabio Inc.) or ViraBind AAV Purification kit (Cell Bio Labs, Inc) or similar.
8. AAV titration kit (Takarabio, Inc.) with ITR sequence primers or microdystrophin (transgene) primers.
9. PBS, autoclaved.
10. 0.5 M EDTA, pH 8.0, autoclaved.

2.2 Animal Preparation (Intravenous Injection)

1. Using DMD dog model: Seven- to ten-week-old dogs, wild-type and $CXMD_J$ obtained from the beagle-based $CXMD_J$ breeding colony at the Middle Animal Faculty of the NCNP (Tokyo, Japan).
2. Syringes, needles
3. Anesthetic agent (Thiopental, Isoflurane), anti-allergic agent (d-Chlorpheniramine Maleate, 0.15 mg/kg body weight)

2.3 Assessment After In Vivo Gene Transduction

2.3.1 IHC

1. Slide glasses, cover glasses.
2. PBS.
3. 4% Paraformaldehyde (PFA) /PBS.
4. Triton-X100.
5. Blocking reagent, Block Ace powder (DS Pharma biomedical Co., Ltd., JAPAN) or similar.
6. Anti-Dystrophin antibody.
7. Second antibody.
8. DAPI-containing mounting reagent, Fluoromount-G™ (SouthernBiotech)or similar.

2.3.2 PBMC + AAV Re-stimulation

1. Blood collection tube (EDTA•2Na).
2. SepMate™-50 (RUO) (STEMCELL™ Technologies).
3. Density gradient media such as Lympholyte®-Mammal Cell Separation Media (CEDARLANE), or Ficoll-Paque™ (Cytiva).

4. Red blood cell lysis solution (NH_4Cl reagent) such as BD Pharm Lyse™ Lysing Buffer (BD Biosciences, #555899).
5. RPMI medium with antibiotics (penicillin and streptomycin).
6. T25 flask, 6-well or 12-well plate (both for floating cell culture).
7. PHA stock solution 1 mg/mL.
8. RNA purification reagent such as TRIzol (Thermo Fisher Scientific).
9. Reagents for reverse transcription (Superscript™ IV First-Strand Synthesis system, Thermo Fisher Scientific) and qRT-PCR.
10. Primers for 18S ribosomal RNA (QuantumRNA™ Universal 18S Internal standard), Thermo Fisher Scientific, and primers for canine IFN-γ.

3 Methods

In this section, we will focus specifically on techniques utilized for canine DMD models, $CXMD_J$, since these models are well suited to study the immune response to AAV, which is insufficient in the *mdx* mouse models of DMD.

3.1 AAV Preparation, Production, and Purification

Among various viral delivery systems, adeno-associated virus (AAV) vector is one of the most used virus vectors in these days. However, production of large amount, highly purified AAV is time-consuming and labor-intensive (*see* **Note 1**). It is also possible to purchase the necessary vectors from Vector BioLab, etc. by custom order.

0. Construct the microdystrophin expression plasmids to pAAV-LacZ vector of AAV Helper free system, pAAV-microdystrophin. Prepare 1 mg/mL solutions of plasmids.
1. Seed the HEK293 cells or 293-AAV cells (*see* **Note 2**) in 5 T225 flasks at a density of 10^7 cells per flask. Culture cells in 30 mL of DMEM or DMEM/F12 medium with 10% FBS at 37 °C in an incubator supplemented with 5% CO_2.
2. When the cells are approximately 80–90% confluent (*see* **Note 3**), set up the following transfection solution (*see* Table 1 and **Note 4**).

Mix tubes 1 and 2 and then vortex well for 30 s. Incubate at room temperature for 15 min. Fill up with DMEM to 30 mL/flask. Discard old medium from T225 flask, and then gently add 30 mL of transfection solution to each flask. Incubate the cells at the 37 °C incubator with 5% CO_2.

Table 1
Plasmid and reagent preparation for triple transfection

Tube 1	
pAAV-RepCap	0.077 μg/cm^2 culture area
pHelper	0.077 μg/cm^2
pAAV-microdystrophin or pAAV-marker gene	0.077 μg/cm^2
DMEM	7.5 mL
Tube 2	
PEI-Max	× 3 volume of transfection plasmids
DMEM with P/S	7.5 mL

3. Five days after transfection, collect the culture supernatant, then centrifuge for 2000× *g*, 15 min, at 4 °C. Then filtrate using a 0.45 μm filter (*see* **Note 5**). Wash residual cells in the flask with PBS, and then add 1/80 volume of 0.5 M EDTA.
4. Proceed with AAV purification according to the purification kit instructions.
5. Filtrate the purified AAV solution with a 0.22 μm filter, then titrate by qPCR using the AAV titration kit. Store the titrated AAV solution at −80 °C (*see* **Note 6**).

3.2 AAV and MSCs Administration Schedule (See Fig. 1)

Day 0: First administration of AAV and MSCs

Day 7: Second intravenous injection of MSCs

Day 8: Systemic injection of AAVs

Days 15, 22, 29, and 36: Obtain blood from treated dogs for blood tests

Day 36: Punch biopsy or biopsy of skeletal muscle (*see* **Note 7**)

3.3 AAV and MSCs Injection to Dogs

3.3.1 AAV Preparation for Injection

0. AAV working solution preparation must be carried out in a class IIA2 biosafety cabinet. Irradiate the workstation with ultraviolet prior to all procedures.
1. Fill the syringe with AAV solution. Use a different syringe for each AAV vector if multiple vectors are being tested. All materials that come into contact with rAAV solution (e.g., tips, tubes) should be discarded in biohazard containers as appropriate. If the AAV solution is accidentally spilled, wipe up with sodium hypochlorite solution, not with 70% ethanol.

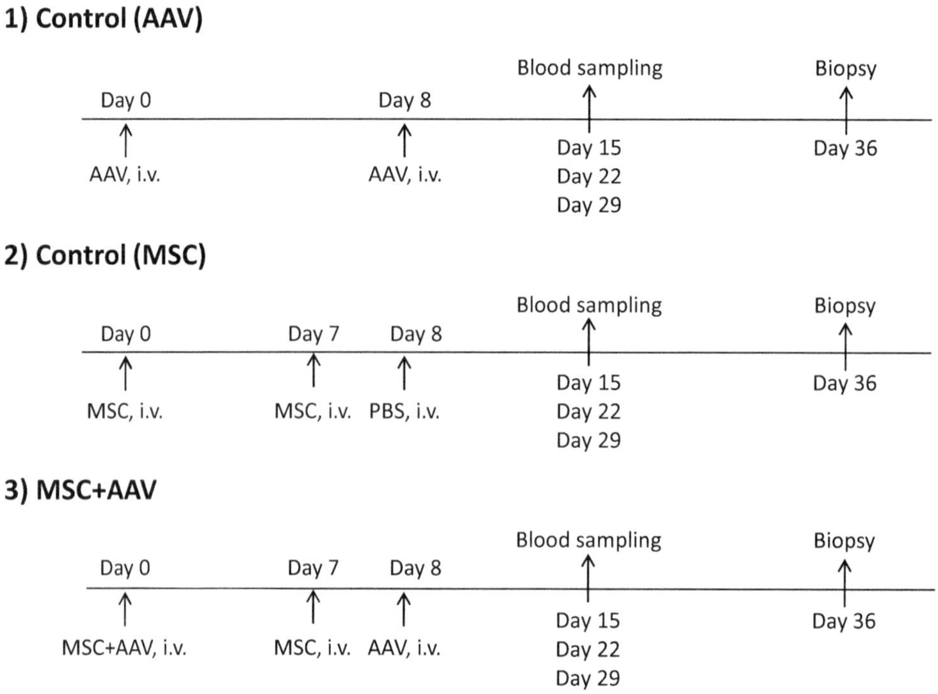

Fig. 1 AAV and MSCs administration schedule

3.3.2 MSCs Preparation for Injection

1. Bring the frozen MSCs out of the −80 °C freezer or liquid nitrogen container.
2. Thaw into a 37 °C water bath until a few ice crystals are visible.
3. Add 2 mL of PBS, and then gently suspend the cell solution pipetting up and down without bubbles.
4. Centrifuge at 1000× g for 5 min at 4 °C and resuspend in fresh PBS.
5. Count the cells, suspend with 2 mL of PBS. Fill the syringe with the MSCs solution.

3.3.3 MSCs with AAV Solution for Injection

1. When MSCs are to be administered with AAV, wash the cells with PBS, mix them with the desired amount of AAV to be injected (up to 2–5 mL).
2. Fill the syringe with mixture. Place the syringe in an icebox and transport to the site of procedure.

3.3.4 Preparation of Dogs and Administration of Vector and/or Cell Solution

0. Dogs scheduled for the procedure should fast from the day before.
1. Anesthetize the animal by intravenous injection of thiopental, maintain anesthesia with isoflurane. Monitor the SpO_2 and heart rate during procedure.

2. Shave the coat around the site of administration with an electrical shaver.
3. Secure vascular access with an indwelling needle. Flush the catheter with heparin-Na solutions.
4. Confirm that the administered dog is stable, and slowly inject the cell suspension or AAV solution. The rate of administration should not be a bolus dose. If there is any change or any abnormalities such as skin redness during administration, stop the drug administration, then take appropriate measures under the judgment of a veterinarian.
5. Remove the syringe from the catheter, turn off the anesthesia. Wait for the dog to recover from anesthesia.

3.4 Assessment After In Vivo Gene Transduction

Dissect the specimen and prepare frozen tissue blocks of skeletal muscle (*see* **Note 8**). Cut muscle cryosections at 7 μm thickness using a cryostat microtome. Stock cryosections at −80 °C before staining.

3.4.1 Immunohistochemical Staining

1. Completely air-dry the sections before beginning, but once the procedure starts, prevent sections from drying out again (*see* **Note 9**).
2. Soak in PBS to rinse the sections, and fix with 4% PFA/PBS solution for 15 min at room temperature. Rinse sections with PBS twice, then permeabilize for 5 min in PBS containing 0.1% Triton X-100. Wash sections twice with PBS. Then treat with PBS containing blocking reagent for 1 h at room temperature.
3. Incubate sections with primary antibodies diluted with blocking reagent in PBS for overnight at 4 °C (*see* **Note 10**).
4. Wash sections with PBST three times for 10 min each.
5. Incubate sections with fluorescent material-conjugated secondary antibodies diluted with blocking reagent for 30 min.
6. Wash the sections with PBST three times for 10 min each.
7. Mount the sections with DAPI-containing Fluoromount-G™. Keep away from light until the mounting medium dries.
8. Observe and take images using a fluorescence microscope.

3.4.2 T Cell Responses

Isolate Peripheral lymphocytes from 5 mL of whole blood using Lympholyte-Mammal and SepMate-50 according to the manufacture's protocol.

1. Collect 5 mL of whole blood into EDTA-2Na blood collection tube, then dilute with equal volume of PBS.
2. Add density the gradient medium to the SepMate tube. Add the diluted blood sample to the SepMate tube.
3. Centrifuge at 1200× g for 10 min at room temperature.

4. Aspirate the top plasma layer, collect enriched PBMCs to a new tube (*see* **Note 11**).
5. Wash the PBMCs with PBS, and centrifuge at 300× g for 8 min at room temperature.
6. Repeat PBS wash twice, and then suspend the cells with RPMI.
7. Count the PBMCs, and then transfer to a T25 flask. Incubate the cells at a 37 °C incubator with 5% CO_2.
8. Plate cells onto 6-well or 12-well plates with RPMI medium. Incubate with PHA as positive control or 1×10^5 v.g./cells of AAV.
9. Incubate cells for 24 h, and then isolate total RNAs from cultured lymphocytes using TRIzol or similar.
10. Synthesize first-strand cDNA using a reverse transcriptase. Measure IFN-γ expression using qRT-PCR.

4 Notes

1. If gene therapy experiments are to be performed in adult dogs, a minimum dose of 1e13vg/kg is required. Therefore, about 1e14vg is needed in most cases for preparation. To prepare this amount of vector, it is necessary to culture about 30 T225 flasks or 6 plus or more flasks for HyperFlask (1720 cm^2 culture area) scale. For purification of the AAV vector, the commercial kit described here can be used, but in the laboratory, concentration of the culture medium using TFF, ultra-centrifugal purification using cesium chloride or iodexanol, and purification using an ion-exchange column are also possible.
2. Although it is possible to use standard HEK293 cells for AAV production, the production efficiency is often lower than that of 293AAV cells.
3. When using PEI or PEI-Max for plasmid triple transfection, the cells should be close to 90% confluent. Also, the amount of medium used for transfection should be as reduced as possible for better transfection efficiency. If it is uncertain how much of the plasmid has been transfected, a small amount of GFP-expressing plasmid without ITR can be mixed with the transfection medium.
4. Note that the medium should not contain serum after transfection. Two percent FBS can be added if the cells are in bad condition, but it is not recommended to add extra protein components during vector production.
5. Some types of purification kits use only the vectors in the transfected cells. However, for many serotypes other than AAV2, about 1/2–1/3 of the total amount of vector produced

is released into the culture medium, and in this case, it would be wasteful not to purify the vector in the culture medium. TFF, ammonium sulfate precipitation, and PEG precipitation can be used to enrich the culture medium containing AAV [14].

6. If the purified AAV vector is to be stocked for a long time, it should be divided into appropriate small amounts and stored at −80 °C. It is better to use tubes with low protein binding for storage. Avoid repeated freezing and thawing.
7. When DMD dogs are being used in an experiment and you want to evaluate their running at a later time point, it is better to biopsy skeletal muscles that are not affected by running, such as the temporalis muscle (not muscles of the extremities). In $CXMD_J$, the temporalis muscle often atrophies as the disease progresses, so be careful while collecting the muscle during punch biopsy. We use disposable biopsy punches (2 to 3 mm diameter, regular type, BP-30F, KAI industries co., ltd.).
8. When creating a block of cardiac muscle, pay attention to the direction in which the muscle fiber is running. Make sure that a transverse section of the muscle can be observed during the preparation of the frozen section.
9. Immunostaining for dystrophin protein is prone to false positives, so be sure to have a negative control (no primary antibody, only secondary antibody).
10. The primary antibody reaction can be done for 1 h at 37 °C, but the resulting signal is often not very good.
11. If the PBMC layer contains visible red blood cells, treat with a hemolytic agent to remove them.

References

1. Mendell JR, Al-Zaidy S, Shell R, Arnold WD, Rodino-Klapac LR, Prior TW, Lowes L, Alfano L, Berry K, Church K et al (2017) Single-dose gene-replacement therapy for spinal muscular atrophy. N Engl J Med *377*:1713–1722
2. Hinderer C, Katz N, Buza EL, Dyer C, Goode T, Bell P, Richman LK, Wilson JM (2018) Severe toxicity in nonhuman primates and piglets following high-dose intravenous administration of an adeno-associated virus vector expressing human SMN. Hum Gene Ther *29*:285–298
3. Arruda VR, Fields PA, Milner R, Wainwright L, De Miguel MP, Donovan PJ, Herzog RW, Nichols TC, Biegel JA, Razavi M et al (2001) Lack of germline transmission of vector sequences following systemic administration of recombinant AAV-2 vector in males. Mol Ther *4*:586–592
4. Favaro P, Downey HD, Zhou JS, Wright JF, Hauck B, Mingozzi F, High KA, Arruda VR (2009) Host and vector-dependent effects on the risk of germline transmission of AAV vectors. Mol Ther *17*:1022–1030
5. Yoshimura M, Sakamoto M, Ikemoto M, Mochizuki Y, Yuasa K, Miyagoe-Suzuki Y, Takeda S (2004) AAV vector-mediated microdystrophin expression in a relatively small percentage of mdx myofibers improved the mdx phenotype. Mol Ther *10*:821–828
6. Yuasa K, Sakamoto M, Miyagoe-Suzuki Y, Tanouchi A, Yamamoto H, Li J, Chamberlain JS, Xiao X, Takeda S (2002) Adeno-associated virus vector-mediated gene transfer into dystrophin-deficient skeletal muscles evokes

enhanced immune response against the transgene product. Gene Ther *9*:1576–1588
7. Bostick B, Shin JH, Yue Y, Duan D (2011) AAV-microdystrophin therapy improves cardiac performance in aged female mdx mice. Mol Ther *19*:1826–1832
8. Shin JH, Nitahara-Kasahara Y, Hayashita-Kinoh H, Ohshima-Hosoyama S, Kinoshita K, Chiyo T, Okada H, Okada T, Takeda S (2011) Improvement of cardiac fibrosis in dystrophic mice by rAAV9-mediated microdystrophin transduction. Gene Ther *18*:910–919
9. Ohshima S, Shin JH, Yuasa K, Nishiyama A, Kira J, Okada T, Takeda S (2009) Transduction efficiency and immune response associated with the administration of AAV8 vector into dog skeletal muscle. Mol Ther *17*:73–80
10. Yuasa K, Yoshimura M, Urasawa N, Ohshima S, Howell JM, Nakamura A, Hijikata T, Miyagoe-Suzuki Y, Takeda S (2007) Injection of a recombinant AAV serotype 2 into canine skeletal muscles evokes strong immune responses against transgene products. Gene Ther *14*:1249–1260
11. Zhu J, Huang X, Yang Y (2009) The TLR9-MyD88 pathway is critical for adaptive immune responses to adeno-associated virus gene therapy vectors in mice. J Clin Invest *119*:2388–2398
12. Mendell JR, Campbell K, Rodino-Klapac L, Sahenk Z, Shilling C, Lewis S, Bowles D, Gray S, Li C, Galloway G et al (2010) Dystrophin immunity in Duchenne's muscular dystrophy. N Engl J Med *363*:1429–1437
13. Hayashita-Kinoh H, Yugeta N, Okada H, Nitahara-Kasahara Y, Chiyo T, Okada T, Takeda S (2015) Intra-amniotic rAAV-mediated microdystrophin gene transfer improves canine X-linked muscular dystrophy and may induce immune tolerance. Mol Ther *23*:627–637
14. Tomono T, Hirai Y, Okada H, Miyagawa Y, Adachi K, Sakamoto S, Kawano Y, Chono H, Mineno J, Ishii A et al (2018) Highly efficient ultracentrifugation-free chromatographic purification of recombinant AAV serotype 9. Mol Ther Method Clin Dev *11*:180–190

Chapter 20

Exon-Skipping for a Pathogenic *COL6A1* Variant in Ullrich Congenital Muscular Dystrophy

Sara Aguti, Fady Guirguis, Carsten Bönnemann, Francesco Muntoni, Véronique Bolduc, and Haiyan Zhou

Abstract

Single nucleotide variants that alter splice sites or splicing regulatory elements can lead to the skipping of exons, retention of introns, or insertion of pseudo-exons (PE) into the mature mRNA transcripts. When translated, these changes can disrupt the function of the synthesized protein. Splice-switching antisense oligonucleotides (ASOs) are synthetic, modified nucleic acids that can correct these aberrant splicing events. They are currently in active clinical development for a number of conditions and have been approved by regulatory agencies for the treatment of neuromuscular disorders such as Duchenne muscular dystrophy and spinal muscular atrophy. We have previously reported that splice-switching ASOs effectively skip a pathogenic PE that causes Ullrich congenital muscular dystrophy (UCMD). This erroneous PE insertion is caused by a deep-intronic variant located within intron 11 of *COL6A1* (c.930+189 C>T). Here, we describe the detailed protocols and workflow that our labs have used to assess the efficacy of ASOs to skip this PE in vitro. The protocols include designing ASOs; isolating, culturing, and transfecting fibroblasts; extracting RNA and protein; and validating splicing correction at the mRNA and protein levels using quantitative reverse transcription PCR (qRT-PCR) and western blot assays, respectively.

Key words Antisense oligonucleotide, Exon skipping, Deep intronic mutation, Pseudo-exon, Pre-mRNA splicing, Skin fibroblasts, COL6, Congenital muscular dystrophy

1 Introduction

Collagen VI (COL6) is an extracellular matrix protein that forms microfibrillar networks, which support various organs and tissues such as skeletal muscles and lungs [1, 2]. In there, COL6 plays multiple critical roles that range from providing mechanical support to regulating cellular processes in nearby cells. The *COL6A1*, *COL6A2*, and *COL6A3* genes encode the three major α(VI) polypeptide chains—α1(VI), α2(VI), and α3

(VI) [1]. COL6 monomers are formed by the intertwining of these three chains, which assemble through their triple-helical domains in a 1:1:1 ratio. The monomers dimerize, and the dimers form tetramers intracellularly. The tetramers get secreted into the extracellular space, where they assemble to form the aforementioned microfibrillar networks.

Deleterious variants in any of COL6's three major genes (*COL6A1*, *COL6A2*, or *COL6A3*) underlie a group of congenital and early-onset muscular dystrophies known as collagen VI-related dystrophies, which affect at least 0.9 per 100,000 individuals in certain populations. Collagen VI-related dystrophies encompass a spectrum of disease severity that range from the mild Bethlem to the severe Ullrich congenital muscular dystrophy (UCMD). They manifest with varying degrees of distal joint hyperlaxity, proximal joint contractures, loss of ambulation, muscle weakness, and respiratory insufficiency [3].

Single nucleotide variants that alter precursor messenger RNA (pre-mRNA) splicing make up one of the most common types of variants that underlie UCMD [3–6]. Splicing is a complex, highly regulated process that removes introns from pre-mRNA to create the continuous sequence of coding exons in the mature mRNA. Splicing relies on the accurate recognition of splicing regulatory elements (SREs) and the 5′ splice donor (SD) and 3′ splice acceptor (SA) sites by the splicing machinery—the spliceosome. The splice sites define the exon-intron junctions, and the SREs are *cis*-acting exonic and intronic splicing enhancers and silencers (ESE, ISE, ESS, and ISS) (Fig. 1). Intronic variants can disrupt or create SREs, or create new active splice sites. These changes most commonly lead to intronic retention or pseudo-exon (PE) insertion into the mature mRNA, which if translated can negatively impact the physiological function of the synthesized protein [7]. Recently, we identified a recurrent dominant-negative PE-inserting variant within intron 11 of *COL6A1* (c.930+189C>T), which causes UCMD. This de novo C>T substitution creates a novel SD site and activates an upstream cryptic SA site within intron 11 resulting in the inclusion of a 72 nucleotide-long PE located between these two cryptic splice sites. The 24 amino acid residues encoded by this PE get incorporated within and disrupt the amino terminus of the collagen α1(VI) chain's triple-helical domain. The α1(VI) chains that contain this 24 amino acid-long insertion likely retain their ability to assemble into COL6 monomers, dimers, and tetramers. However, these chains are thought to disrupt the proper assembly and physiological function of COL6 microfibrils in the extracellular space in a dominant-negative manner [4–7]. Correcting this aberrant splicing event to skip the PE has the therapeutic potential to reduce disease severity.

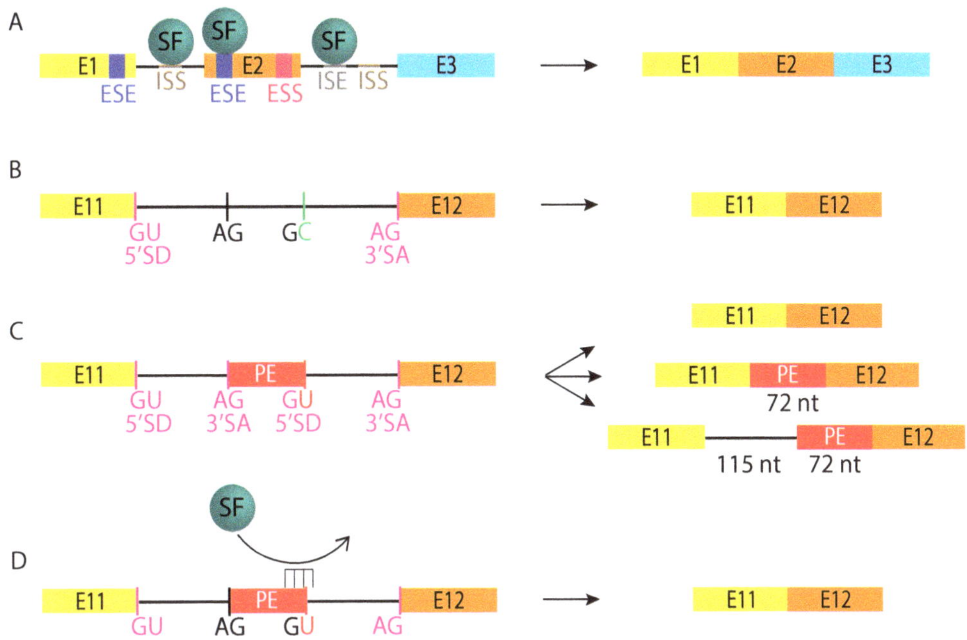

Fig. 1 Schematic diagram of simplified splicing mechanisms. (**a**) A generic diagram of pre-mRNA made up of three exons (E1, E2, and E3) and two introns (lines between exons) with exonic and intronic splicing regulatory elements (ESE, ESS, ISE, and ISS). Splicing factors (SF) recognize the regulatory elements, then the introns get spliced out, and the continuous coding mRNA sequence is produced. (**b**) The segment of *COL6A1* pre-mRNA that contains exons 11 and 12 (E11 and E12) and intron 11. Intron 11 is flanked by a 5′ splice donor (SD) and a 3′ splice acceptor (SA) sites. In the absence of splice-altering variants, intron 11 gets spliced out, and E11 and E12 are joined in the resulting transcripts. (**c**) Pre-mRNA transcribed from *COL6A1* gene with the c.930+189C>T variant (C>U in RNA). This variant creates a novel 5′SD site and activates a previously dormant 3'SA site located 72 nucleotides (nt) upstream of the novel SD site. When these two new sites are recognized by the spliceosome, the 72-nt-long pseudo-exon (PE) gets inserted between E11 and E12 in the mature mRNA. When only the novel 5'SD site is recognized, both the 72-nt-long PE and a 115-nt-long region upstream of it get inserted into the mature mRNA. However, this latter transcript is out-of-frame and gets degraded without being translated. When neither of the new sites is recognized, the wild-type mature transcript is produced. (**d**) Splice-switching ASOs sterically block the spliceosome from recognizing the novel 5′SD site and result in the correct splicing out of the entire intron 11 from the mature transcripts

Pre-mRNA splicing can be modulated using antisense oligonucleotides (ASOs), which are single-stranded, synthetic nucleic acids that typically range in length from 18 to 30 bases. Splice-switching ASOs recognize and hybridize with their target pre-mRNA sequences through sequence complementarity [8, 9]. By binding to their target sequences, ASOs sterically block the splicing factors from recognizing the splice sites and/or SREs located at these sequences [10]. Depending on the splice sites and SREs that the ASOs target, pre-mRNA sequences get either included in (exon inclusion) or excluded from (exon skipping) the mature mRNA transcript (Fig. 1) [7].

Splice-switching ASOs are chemically modified to have therapeutically favorable pharmacokinetic and pharmacodynamic properties. These properties include increased resistance to nucleases, enhanced target binding affinity, reduced immunostimulatory potential, and favored spliceosome blocking activity over RNase H degradation activity. ASO modifications can either be changes to the oligonucleotide's backbone or sugar component. For example, phosphorothioate (PS) and phosphorodiamidate morpholino (PMO) ASOs have a phosphorothioate or a phosphorodiamidate backbone, respectively, instead of the phosphodiester backbone. Additionally, the deoxyribose and ribose sugars are replaced by a morpholine ring in PMOs, and the sugar's 2′ position has 2′-O-methyl or 2′-O-methoxyethyl substitutions in 2′-OMe and 2′-MOE ASOs [8–10].

Multiple ASOs have been approved by the US Food and Drug Administration (FDA) for different conditions. Five of these drugs are splice-switching ASOs for the treatment of neuromuscular disorders (Eteplirsen, Golodirsen, Viltolarsen, and Casimersen for Duchenne muscular dystrophy; Nusinersen for spinal muscular atrophy) [11, 12]. Considering ASOs' potential of becoming a safe and effective therapeutic intervention to correct the aberrant splicing events that underlie UCMD, we tested and demonstrated the efficacy of different ASO chemistries to skip the aforementioned *COL6A1* intron 11 pathogenic PE in vitro at the mRNA and protein levels. Additionally, we showed that the ASOs improved the quality of the extracellular matrix synthesized by the treated cells [5, 6]. We describe here how to design ASOs; isolate, culture, and transfect fibroblasts; and use molecular readouts to assess the ASOs' exon-skipping efficacy in vitro at the RNA and protein levels (Fig. 2).

2 Materials

2.1 ASO Design

1. Human Splicing Finder (https://www.genomnis.com/access-hsf).
2. RNA fold web server (http://rna.tbi.univie.ac.at/cgi-bin/RNAWebSuite/RNAfold.cgi).
3. OligoAnalyzer™ Tool to analyze GC %, Tm and self-complementarity (https://eu.idtdna.com/calc/analyzer).
4. Off-target sequence detection (https://gggenome.dbcls.jp/en/hg19/1/).

2.2 ASO Preparation

1. ASOs are synthesized commercially (e.g., Eurogentec Ltd, Gene Tools LLC).
2. DNase- and RNase-free water (not DEPC-treated).

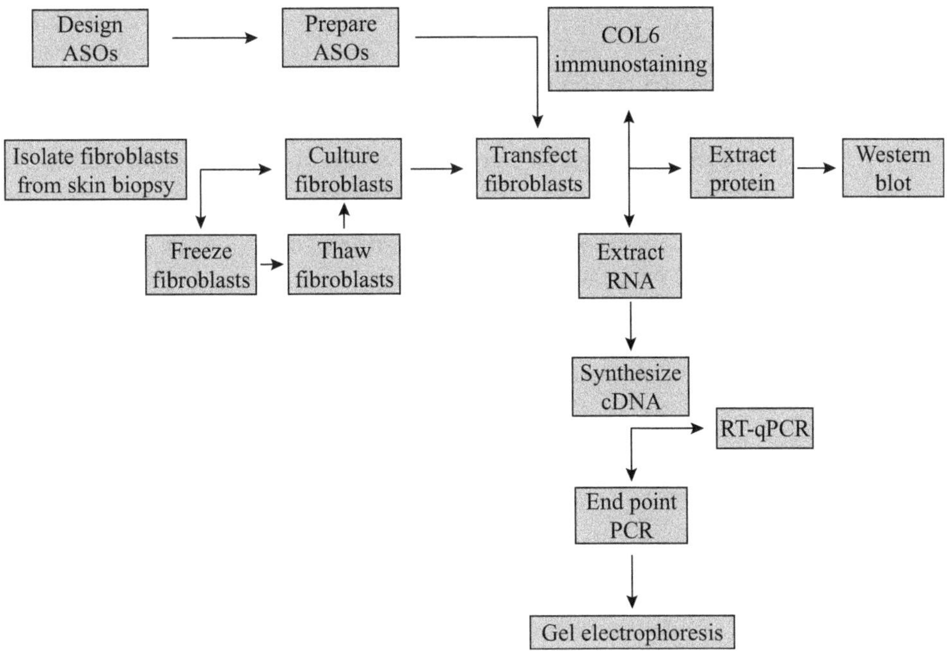

Fig. 2 Procedure overview for designing and testing splice-switching ASOs

2.3 Fibroblast Culture

1. Culture media and supplements. Growth medium of human fibroblast cells is composed of Dulbecco's Modified Eagle Medium (DMEM), 10% fetal bovine serum (FBS), 1% GlutaMAX, and 1% penicillin-streptomycin.
2. Trypsin 2.5% (for fibroblast isolation from skin biopsy) and 0.05% trypsin–EDTA (for trypsinization in subculture).
3. Collagenase Type IV solution.
4. Phosphate-buffered saline (PBS), 1x.
5. Bijou tubes.
6. 60 mm sterile tissue culture dishes and T12.5, T75, and T175 flasks with vented cap.
7. Sterile scalpels.
8. Incubator set at 37 °C and 5% CO_2.
9. Automated cell counter, cell counting slide, trypan blue.
10. Cryoprotective agent (e.g., DMSO).
11. Isopropanol chamber or small polystyrene box.
12. Sterile cryogenic storage vials.
13. 70% ethanol.

2.4 Fibroblast Transfection

1. Lipofectamine 2000 (L2K) transfection reagent and Opti-Minimal essential Medium (Opti-MEM) for transfections with non-PMO ASOs.

2. Medium with reduced serum and no antibiotic (DMEM + 5% FBS) and Endo-Porter PEG transfection reagent (available at Gene Tools, LLC) for transfections with PMO ASOs (*see* **Note 1**).
3. 6 or 24 well plates, or 35 mm diameter culture dishes.
4. Sterile 1.5 mL Eppendorf tubes.
5. Target-specific and scrambled control ASOs and PMOs from Subheading 2.2 (Table 1).

2.5 RNA Extraction and cDNA Synthesis

1. ASO-treated and control fibroblast cultures from Subheading 2.4.
2. RNA isolation kit (e.g., RNeasy Mini Kit).
3. β-mercaptoethanol (β-ME), ice, and 70–100% ethanol.
4. RNase-free water.
5. 1.5 mL and 0.2 mL RNase-free Eppendorf tubes.
6. NanoDrop spectrophotometer.
7. Reagents for reverse transcription (e.g., high-capacity RNA-to-cDNA™ Kit or Superscript IV).
8. Thermocycler.

2.6 Polymerase Chain Reaction (PCR) and Gel Electrophoresis

1. cDNA template from Subheading 2.5.
2. Taq Polymerase System and set of primers (*COL6A1*: forward: 5′-TAC CAG GGA ATG AAG GGA GA-3′ and reverse: 5′-GTC CTT GAA TGC CGT CAA AC-3′).
3. 0.2 mL PCR tube.
4. Thermocycler.
5. Tris–borate–EDTA 1× (TBE) buffer.
6. 1.5% (w/v) agarose gel.
7. DNA gel stain, loading buffer, and DNA ladder.
8. Gel imaging system.

2.7 Quantitative Reverse Transcription PCR (qRT-PCR)

1. cDNA template from Subheading 2.5.
2. qPCR assays (Table 2).
3. Takyon ROX Probe qPCR kit.
4. 96-well real-time PCR plate and MicroAmp™ Optical Adhesive Film.
5. Real-time PCR thermal cycler.

2.8 Protein Extraction and Western Blotting

1. Plates of ASO-treated and control fibroblast cultures from Subheading 2.4.
2. PBS, 1×.
3. Cell lysis buffer (e.g., RIPA).

Table 1
List of ASOs tested in this study

ASO ID	Sequence (5'->3')	Chemistry	ASO ID	Sequence (5'->3')	Chemistry
ASO-1	GGCGGUGGAGCGGGUCUGCAGGACA	2'-OMe	**2'OMe-SA**	GUGGAGCGGGUCUGCAGGACACGUG	2'-OMe
ASO-2	CGAGGGGCGGUGGAGCGGGUCUGCA	2'-OMe	**2'OMe-PEX1**	AGAUGGAGGGGACGGCGAGG	2'-OMe
ASO-3	GACGGCGAGGGGCGGUGGAGCGGGU	2'-OMe	**2'OMe-PEX2**	GGCUGUCCUUGUCCUCCCAGAUGGA	2'-OMe
ASO-4	GAGGGGACGGCGAGGGGCGGUGGAG	2'-OMe	**2'OMe-PEX3**	GGCUGUCCUUGUCCUUCCAG	2'-OMe
ASO-5	AGAUGGAGGGGACGGCGAGGGGCGG	2'-OMe	**2'OMe-PEX4**	GUGCCUGGGUGGCUGUCCUU	2'-OMe
ASO-6	CUUCCAGAUGGAGGGGACGGCGAGG	2'-OMe	**2'OMe-SD1**	AGGCACCUUUGCUGGGUGCCUGGGU	2'-OMe
ASO-7	UUGUCCUUCCAGAUGGAGGGGACGG	2'-OMe	**2'OMe-SD2**	UGAAAGUGACACAGGCACCUUUGCU	2'-OMe
ASO-8	UGUCCUUGUCCUUCCAGAUGGAGGG	2'-OMe	**PMO-SA**	AGGACACCTGGTCTCCTGTCAGGGA	PMO
ASO-9	GUGGCUGUCCUUGUCCUUCCAGAUG	2'-OMe	**PMO-PEX1**	GCTGTCCTTGTCCTTCCAGATGGAG	PMO
ASO-10	CCUGGGUGGCUGUCCUUGUCCUUCC	2'-OMe	**PMO-PEX2**	GTGGCTGTCCTTGTCCTTCCAGATG	PMO
ASO-11	GGGUGCCUGGGUGGCUGUCCUUGUC	2'-OMe	**PMO-PEX3**	GTGCCTGGGTCGCTGTCCTTGTCCT	PMO
ASO-12	UUGCUGGGUGCCUGGGUGGCUGUCC	2'-OMe	**PMO-PEX4**	TTGTCCTTCCAGATGGACGGGAC	PMO
ASO-13	CACCUUUGCUGGGUGCCUGGGUGGC	2'-OMe	**PMO-SD**	TGAAAGTGACACAGGCACCTTTGCT	PMO
ASO-14	ACAGGCGCCUUUGCUGGGUGCCUGG	2'-OMe	**ASO-5**	AGATGGAGGGGACGGCGAGGGGCGG	2'-MOE
ASO-15	GUGACACAGGCGCCUUUGCUGGGUG	2'-OMe	**ASO-6**	CTTCCAGATGGAGGGGACGGCGAGG	2'-MOE
ASO-Scr	ACATGCAGGCCAGGCGGACG	2'-MOE	**ASO-5/6**	AGATGGAGGGGACGGCGAGG	2'-MOE

Table 2
List of primers and probes used in amplifying total and mutant-specific *COL6A1* transcripts by real-time PCR

	Forward (5′->3′)	Reverse (5′->3′)	Probe	Product size (bp)
COL6A1 Total	CCGACTGCGC TATCAAGAA	AATCAGGTAC TTATTCTCC TTCAGGT	probe #17 from ROCHE Universal Probe Library (cat.num 04686900001)	77
***COL6A1* Mutant**	TACCAGGGAA TGAAGGGAGA	CCTGGAGCCC TTTGCTG	/56-FAM/ATCTGGAAG/ZEN/ GACAAGGACAGCCAC/ 3IABkFQ/	123
PGK1	CAGCTGCTGGG TCTGTCAT	GCTGGCTCGGC TTTAACC	probe #67 from ROCHE Universal Probe Library (cat.num 04688660001)	103

4. Protease inhibitors (e.g., Roche complete).
5. Bucket and ice.
6. Cell lifters (optional).
7. 1.5 mL Eppendorf tubes.
8. Microcentrifuge.
9. Protein assay kit (e.g., BCA assay).
10. Incubator (or oven), set at 37 °C.
11. Loading buffer (e.g., ThermoFisher NuPAGE).
12. β-mercaptoethanol.
13. Western blot electrophoresis and transfer materials and reagents (e.g., ThermoFisher NuPAGE system), including precast running gel (e.g., NuPAGE Bis-Tris 4–12%), running buffer (e.g., MOPS), transfer buffer (e.g., NuPAGE transfer), sponges, filter paper, and polyvinylidene difluoride (PVDF) membrane.
14. Methanol.
15. Tris-buffered saline (TBS), 10×.
16. Tween-20.
17. Dry milk.
18. Primary antibodies (anti-Colα1(VI)-PE, and anti-Colα1(VI)) (*see* **Note 2**) and a loading control (e.g., anti-tubulin).
19. Secondary antibodies, horseradish peroxidase (HRP)-conjugated.
20. Enhanced chemiluminescence reagents (e.g., Thermo Fisher Pierce ECL immunoblotting substrate).
21. Bio-Imager (e.g., Bio-Rad ChemiDoc).

2.9 Immunofluorescence Staining for Extracellular COL6

1. 22 mm diameter glass coverslip precoated with collagen type I (alternatively, eight-chamber glass slides could be used).
2. Fibroblast cultures from Subheading 2.4.
3. L-ascorbic acid (e.g., Alpha Laboratories).
4. PBS, 1×.
5. Paraformaldehyde.
6. Goat serum.
7. Mouse anti-collagen type VI antibody (e.g., Sigma-Aldrich MAB1944).
8. Goat anti-mouse secondary antibody (e.g., Thermofisher A11029; goat anti-mouse Alexa 488).
9. 4′,6-diamidino-2-phenylindole (DAPI).
10. Hydromount.
11. Fluorescence microscope.

3 Methods

In this section, we will describe all the basic procedures, following the workflow from the design of ASOs to the assessment of their efficiency in skipping the PE from the mature *COL6A1* transcripts at the RNA and protein levels (Fig. 2).

3.1 ASO Design

1. Designing ASOs capable of skipping the PE from the *COL6A1* mRNA can be achieved using one of two different approaches (Fig. 3):
   - *RNA-tiling approach to target the mis-spliced PE and its flanking sequences.* Start to design the first ASO (20–25-bp long) at the 5′ flanking sequence of the PE (~20-bp upstream of the PE). The ASO's sequence should be complementary to the target pre-mRNA sequence. Then, move down by constant increments (e.g., 5 bp) from this ASO's 5′ end and start designing the following ASO. Repeat these steps until you have designed enough overlapping ASOs that cover the whole length of the PE and its 5′ and 3′ flanking sites (Fig. 3b).
   - *Targeting splicing enhancer sequences to sterically block the binding of splicing factors.* To predict the locations of splice enhancer motifs, on the Human Splicing Finder website, paste the target sequence (sequence of the PE plus 50 bp upstream and downstream of the PE), and select "Analyze a sequence." Once the splice enhancer motifs' locations have been predicted, design the ASO overlapping these sequences. The yellow line represents the ratio between

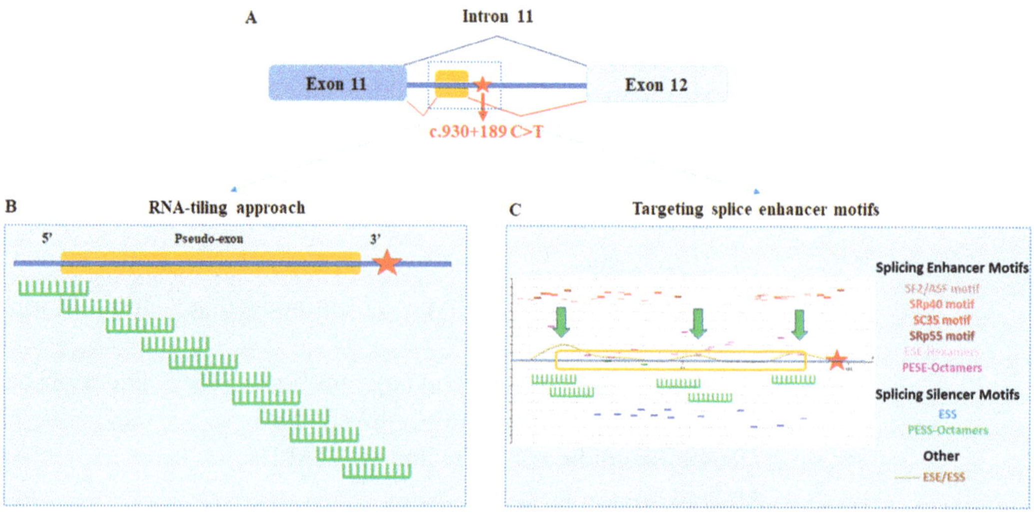

Fig. 3 Schematic representations of the *COL6A1* intron 11 containing the deep intronic variant and two approaches used for exon-skipping ASO design. (**a**) Representation of the *COL6A1* deep intronic variant (c.930+189 C>T, red star) and the resulting intronic segment of intron 11 that is included into the mature mRNA as a PE (yellow bar). (**b**) Illustration of the RNA-tiling approach. The ASO sequences are designed to cover all the PE and its 5′ and 3′ flanking sites. (**c**) Illustration of an alternative ASO design approach, based on splicing motifs' predictions. The green arrows point to regions rich in exon splice enhancer motifs

splicing enhancer and silencer motifs (ESE/ESS): when its value is above zero, this means that the splice enhancer motifs are more abundant compared to the splice silencer motifs in that particular region. Therefore, these regions are potential target regions for ASO design (Fig. 3c).

2. The optimal length of ASO sequences is between 20 and 25 bases. Calculate the GC content of the ASO sequence, using OligoAnalyzer™ Tool, which should be between 40% and 60%.
3. Predict the secondary structure of the target pre-mRNA sequence. Use RNA fold web server. On the website, paste the pre-mRNA sequence (at least 200 bp of length centered around the PE). Try avoiding closed regions of the secondary structure of the target pre-mRNA for ASO design; target open or semi-open regions.
4. Check for the presence of off-target sequences. Access the GGgenome page, paste the sequence complementary to ASO sequences and choose the genome or RNA sequence database of the species of interest. Search for both strands and enter 0 for number of mismatched/gaps.
5. The scrambled ASO, which is a nontargeting ASO, is designed with the same base compositions as the designed ASO but with a sequence that is not complementary to any pre-mRNA.

3.2 ASO Preparation

1. Before reconstituting ASOs that are shipped lyophilized, briefly centrifuge tubes at maximum speed to collect the ASOs at the bottom of the tube. Working in a laminar flow hood, add sterile DNase- and RNase-free water to obtain a stock concentration of 100 μM. For PMOs, reconstitute to 1 mM.
2. Prepare working dilutions in sterile DNase- and RNase-free water (e.g., 20 μM, 10 μM, 5 μM, 2.5 μM).

3.3 Establishment of Primary Skin Fibroblast Culture

1. Prepare complete growth medium with 10% FBS + 1% GlutaMAX + 1% penicillin-streptomycin (P/S) (450mL DMEM + 50 mL FBS + 5 mL GlutaMAX + 5 mL P/S) and warm in a water bath for 30 min before use.
2. Aseptically, collect skin sample in a bijou tube with growth medium. Keep at 4 °C until ready to process.
3. Cut skin into small pieces in a 60-mm petri dish, and then transfer into a 15 mL centrifuge tube.
4. Add 2 mL of trypsin 2.5% and incubate at 37 °C for 15 min (*see* **Note 3**).
5. Add 8 mL of warm growth medium and centrifuge at 400× g for 5 min.
6. Discard the supernatant, resuspend the pellet in 2 mL collagenase type IV solution, and incubate at 37 °C for 90 min (*see* **Note 4**).
7. Add 8 mL of growth medium and centrifuge at 400× g for 5 min.
8. Discard the supernatant, resuspend the pellet in 4 mL of growth medium and transfer to a T12.5 flask and incubate at 37 °C.
9. Approximately two days later, some fibroblasts should become visible. Add 1–2 mL growth medium to the original medium (*see* **Note 5**).
10. When most of the cells are attached (after 4–5 days), discard the medium, wash once with 4 mL PBS, and add 4 mL of growth medium.
11. After another week, cells should be confluent. Discard the medium, add 1 mL of 0.05% Trypsin–EDTA and incubate for 3 min in the incubator. When cells appear detached, add one volume (1 mL) of pre-warmed growth medium to inactivate trypsin.
12. Transfer cell suspension to a falcon tube and centrifuge for 5 min at 500 × *g*. Then discard the supernatant and resuspend the pellet in 20 ml of growth medium. Split the cell suspension into two T75 flasks for further culturing (10 mL in each flask) (*see* **Note 6**).

3.4 Subculture of Skin Fibroblasts

1. Pre-warm growth medium in a 37 °C water bath for 30 min before use.
2. Thaw a frozen vial in a 37 °C water bath.
3. Add 1 volume of growth medium to 1 volume of thawed cells in a falcon tube, mix gently, centrifuge at 500× g for 5 min, and remove the supernatant containing the freezing medium.
4. Resuspend the cell pellet in growth medium. Transfer into culture dish or flask (10 mL of medium for one T75 flask).
5. Incubate flask at 37 °C/5% CO_2 and change medium every 3 days.
6. Passage cells when they reach 80–90% confluency. Wash culture dish or flask twice with PBS. Add pre-warmed 0.05% trypsin–EDTA solution (e.g., 3 mL in T75 or 4 mL in T175) and leave flask or dish in the incubator for 3 min.
7. When cells appear detached, add one volume of pre-warmed growth medium (e.g., 3 mL in T75 or 4 mL in T175).
8. Transfer the suspension to a falcon tube and centrifuge for 5 min at 500× g. Then, discard the supernatant and resuspend the pellet in pre-warmed growth medium. Split the cell suspension into flasks or culture dishes.

3.5 Preparation of Fibroblasts for Transfection

1. Seed the cells in 6-well plates (or 35 mm diameter dishes) at 2×10^5 cells in 2 mL growth medium per well, a seeding density to reach 90% confluency on the following day (Fig. 4a). Alternatively, seed 3×10^4 cells in 50 μL growth medium per well in 24-well plates (*see* **Note** 7).
2. For the transfection experiment, three controls are needed:
   - Blank control (untreated patient's skin fibroblasts).
   - Mock control (patient's skin fibroblasts treated with transfection reagent only).
   - Scrambled control (patient's skin fibroblasts treated with scrambled ASO).

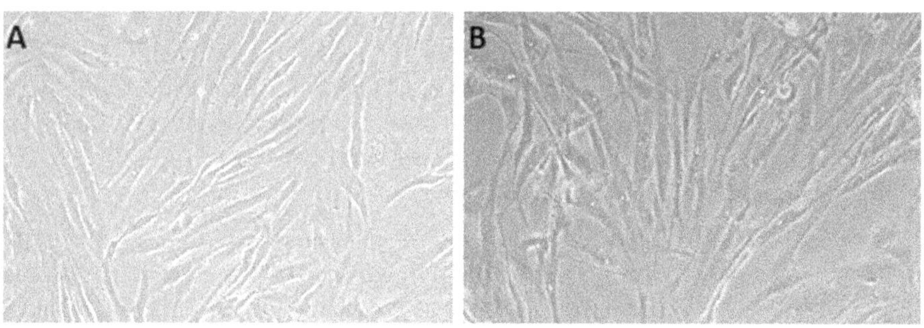

Fig. 4 (**a**) Optimal seeding density for the transfection experiment obtained 24 h after seeding 2×10^5 fibroblasts in 6-well plate cells. (**b**) Representative image of cultured skin fibroblasts 24 h after transfection

3.6 Transfection of Fibroblasts with Non-PMO ASOs (2′-OMe and 2′-MOE) (See Note 8)

1. On the day of the transfection experiment, first discard the medium from the 6-well plates and add 1 ml of Opti-MEM to each well. Incubate plates at 37 °C/5% CO_2 while preparing transfection mixtures.
2. Prepare the transfection reagents mixture in sterile 1.5 mL tubes. Two different mixtures are needed for each sample.
   - For ASOs, tube A contains 99 μL Opti-MEM and 1 μL ASO at the desired concentration (e.g., 1 μL ASO at 20 μM to obtain a 20 nM final concentration). For the mock control, tube A contains 100 μL Opti-MEM only.
   - Tube B contains 95 μL Opti-MEM and 5 μL lipofectamine 2000 (L2K).
3. Mix the solutions of tubes A and B together and incubate the final transfection mix for 20 min at room temperature (RT).
4. Add 800 spi1;μL Opti-MEM to the transfection mix to obtain a final volume of 1 mL (Fig. 5a).
5. Discard the Opti-MEM from the 6-well plate and add 1 mL transfection mix to each well (Fig. 5b).
6. Incubate the plate for 24 h at 37 °C/5% CO_2 (*see* **Note 9**).
7. After 24 h, take the plate out of the incubator to stop the transfection and proceed to the next step. If the transfection is followed by RNA extraction, *see* Subheading 3.8. If the transfection is followed by protein studies, *see* Subheadings 3.11 and 3.12.

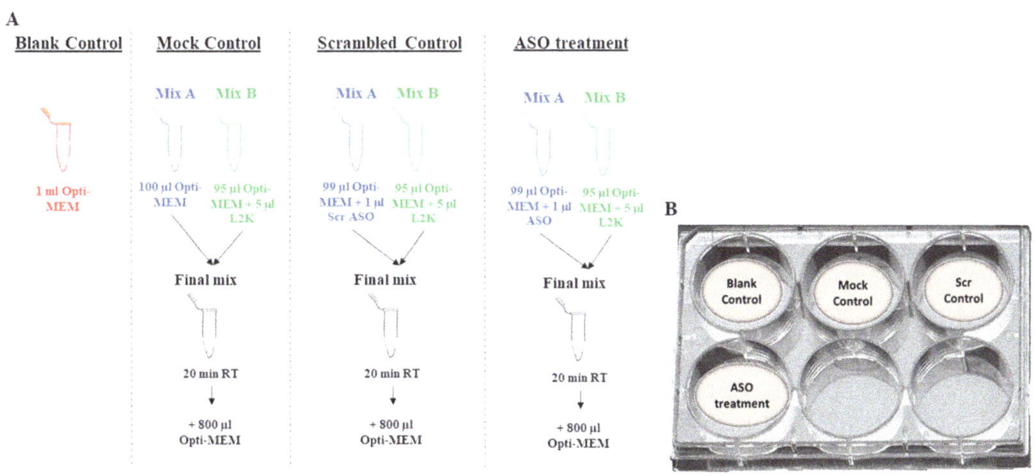

Fig. 5 Schematic representation of (**a**) the different transfection mixes and (**b**) the 6-well plate layout. L2k: Lipofectamine 2000. Scr: Scrambled

3.7 Transfection of Fibroblasts with PMOs (See Note 10)

1. On the day of the transfection experiment, discard the medium from the 24-well plates and add 300 μL of DMEM containing 5% FBS.
2. Add the PMO to the wells by pipetting directly into the medium, and gently swirl the plate to distribute the PMO solution.
3. Add Endo-Porter (2.4 μL per 300 μL of medium for a final concentration of 8 μM of Endo-Porter) by pipetting directly into the medium. Swirl plate immediately after each addition of Endo-Porter.
4. Incubate the plate at 37 °C/5% CO_2.
5. After 48–72 h, take the plate out of the incubator to stop the transfection and proceed to the next step. If the transfection is followed by RNA extraction, *see* Subheading 3.8. If the transfection is followed by protein studies, *see* Subheadings 3.11 and 3.12.

3.8 RNA Extraction and cDNA Synthesis

1. Before starting the RNA extraction, spray the working environment with RNaseZap™ to minimize RNase contamination.
2. Visually inspect each well under a microscope to verify cell confluency and viability before harvesting (Fig. 4b) (*see* **Note 11**).
3. Aspirate the medium from the 6-well or 24-well plate and wash the cells once with PBS (e.g., 1 mL of PBS for 6-well and 500 μL for 24-well plate).
4. Add to each well 350 μL of lysis solution composed of RLT buffer (from QIAGEN kit) and β-mercaptoethanol (β-ME) (10 μL β-ME per 1 mL RLT buffer).
5. Collect lysed cells by scraping them using a cell scraper and transfer them to RNase-DNase-free 1.5-mL tubes.
6. Add one volume of 70% ethanol, mix and transfer to an RNA extraction column.
7. Follow the manufacturer's instructions and elute RNA in the final step in 30 μL of RNase-DNase-free water (*see* **Note 12**).
8. Measure the concentration and quality of extracted total RNA using NanoDrop Spectrophotometer. The absorbance at 260 nm (A260) is used to calculate the concentration.
9. Perform first-strand cDNA synthesis using 50 ng to 1 μg of total RNA and random hexamer primers. Follow the manufacturer's instructions to complete the reverse transcription reaction.

3.9 Polymerase Chain Reaction (PCR) and Gel Electrophoresis

1. Perform PCR according to standard procedures. Add 1 μL cDNA (~100 ng of cDNA) in a 25 μL PCR reaction with 500 pmol of each primer, 200 μM of dNTPs, 1.5 mM $MgCl_2$, 1 unit of Taq polymerase, and 1× PCR buffer and use DNase/RNase-free water to top-up the volume to 25 μL (*see* **Note 13**).
2. The PCR amplification program consists of initial denaturation at 95 °C for 3 min followed by 30 cycles of denaturation at 95 °C for 30 s, annealing at 60 °C for 30 s and extension at 72 °C for 30 s, and a final extension step at 72 °C for 10 min.
3. Verify a small amount of the PCR product (~5 μL) by electrophoresis on an agarose gel with SYBR Safe DNA stain and visualize under a UV transilluminator. The gel concentration depends on the number of base pairs of the PCR product. With the PCR product generated using the primers listed above, use 1.5–2.5% agarose gel.

3.10 Real-Time Quantitative PCR

1. Reconstitute the assays in molecular biology-grade water as recommended by the manufacturer (*see* **Note 14**).
2. Determine the number of samples and the number of assays to be used in the experiment: a minimum of two samples (control and experimental) and two assays (housekeeping and target gene) are required, but additional samples and targets can be added.
3. Calculate the volumes needed for a 10-μL reaction (e.g., 4 μL of diluted cDNA, 2 μL of 5× buffer, 0.5 μL of 20× assay, 3.5 μL of water).
4. Prepare the master mix solution by combining all components except for the cDNA. The minimal volume required for the master mix depends on the desired number of samples and of technical replicates (i.e., three replicates).
5. Dilute each cDNA in molecular biology-grade water 1/10 to 1/20 (depending on the concentration of the cDNA from Subheading 3.8. The volume required depends on the number of technical replicates and assays.
6. Distribute master mix solution in wells across the plate, and add the diluted cDNA.
7. Seal the plate with an optical adhesive cover, briefly spin, and run in the real-time PCR machine with standard methods (*see* **Note 15**).
8. At the end of the run, record the cycle threshold (Ct) values to determine the relative expression, using the $2^{-\Delta\Delta Ct}$ method.

3.11 Protein Extraction and Western Blotting

1. Visually inspect fibroblast culture plates under a microscope to verify cell confluency before processing protein extraction (*see* **Note 16**).
2. Prepare a 50× protease inhibitor stock solution by dissolving one tablet of Roche complete Protease inhibitor cocktail in 1 mL of cold water. Keep on ice (*see* **Note 17**).
3. Chill on ice four series of 1.5-mL tubes (four tubes for each well). In the first series of tubes, add a volume of the 50× protease inhibitor solution that is equivalent to 1/50th the volume of medium in the well (e.g., 6 μL of 50× protease inhibitor if transfection is done in 24-well plates using 300 μL of medium). The second, third, and fourth series of tubes will be used in steps 10, 11, and 13.
4. Prepare a working solution of the cell lysis buffer by adding protease inhibitors (to a final concentration of 1×) to RIPA buffer. Mix by inversion, briefly spin, place on ice, and proceed immediately to the next step.
5. Transfer a plate of cells from the incubator to ice.
6. Collect the medium from the first well and transfer the medium to a tube containing the protease inhibitor solution (prepared in step 3), mix the tube by inversion, return to ice. Immediately wash the first well with cold PBS 1×.
7. Discard PBS and add cell lysis solution (prepared in step 4) directly to the well. (e.g., for a 24-well plate, use 35 μL per well; for a 6-well plate, use 200 μL).
8. Repeat steps 6 and 7 for the remaining wells. Proceed quickly.
9. For each well, scrape cells using a cell lifter or a pipette tip, and pipet up and down a few times, avoiding the creation of bubbles.
10. Collect lysates in the empty prechilled 1.5 mL tubes (the second series of tubes from step 3) and incubate on ice for 30 min.
11. During this 30-min incubation period, centrifuge the first series of tubes (in which the medium was collected) at 4 °C and 15,000× g for 10 min. Transfer the supernatant to a new, prechilled, 1.5 mL tube (the third series of tubes from step 3) and discard the pellet. This constitutes the medium fraction.
12. When the incubation of the cell lysates is complete, centrifuge at 15,000× g for 20 min, at 4 °C.
13. Transfer the supernatant to a new, prechilled, 1.5 mL tube (the fourth series of tubes from step 3) and discard the pellet. This constitutes the cell fraction.
14. Determine protein concentration for all samples with a protein assay according to manufacturer's instructions (e.g., BCA Protein Assay Kit).

15. Dilute all protein samples to the same concentration (we generally dilute protein samples to 0.5 or 1 mg/mL) in NuPAGE LDS sample buffer (final concentration of 1×) and β-mercaptoethanol (final concentration of 5%) (*see* **Note 18**).
16. When ready to use, heat the samples in a thermomixer at 95 °C for 5 min. Vortex at maximum speed for 5 s, and briefly centrifuge.
17. Assemble the gel electrophoresis apparatus according to manufacturer's instructions.
18. Load the samples into the gel. Per well, load approximatively 5 μg for cell lysates protein samples, or 15 μg for medium protein samples.
19. Run the gel at 200 V for 50 min, or until dye front reaches the bottom of the gel.
20. Prepare transfer buffer by adding methanol to the transfer buffer to reach a final concentration of 5% methanol.
21. Uncast gel, trim sides, and wet in transfer buffer. Activate a PVDF in methanol for 15 s and wet in transfer buffer. Assemble the blot and the blotting apparatus according to manufacturer's instructions.
22. Run transfer at 25 V for 1.5 h.
23. Prepare a solution of 1× TBS containing 0.05% tween (TBS-T).
24. Prepare blocking buffer: 5% milk (w/v) in TBS-T.
25. When transfer is complete, retrieve the PVDF membrane and wash it in 1× TBS solution.
26. Incubate the membrane in blocking buffer for 1 h at RT with gentle agitation.
27. Dilute one primary antibody in the blocking buffer (1:50 for Colα1(VI)-PE, 1:7500 for Colα1(VI), or 1:6000 for tubulin). Incubate the membrane in this primary antibody solution overnight at 4 °C with gentle agitation (*see* **Note 19**).
28. Wash the membrane in TBS-T three times for 5 min each, with agitation.
29. Dilute the appropriate secondary antibody in the blocking solution (1:10,000 of goat anti-rabbit-HRP, or 1:10,000 of goat anti-mouse-HRP). Incubate the membrane in the secondary antibody solution for 2 h at RT with gentle agitation.
30. Wash the membrane in TBS-T three times for 5 min each, with agitation.
31. Incubate the membrane in the ECL solution according to manufacturer's instructions and visualize using the imager (Fig. 6).

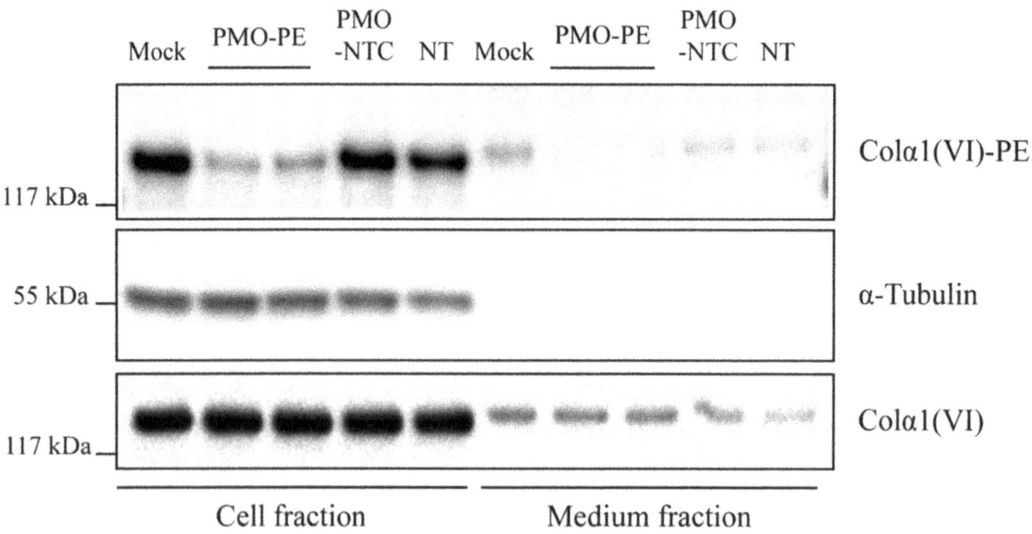

Fig. 6 Protein quantification of PE-containing collagen α1(VI) chains in skin fibroblast cultures after ASO treatment. Representative western blots of proteins isolated from cell lysates (cell fraction) or secretome (medium fraction) show that treatment with PE-targeting PMOs (PMO-PE) reduces the levels of pseudo-exon-containing α1(VI) chains (Colα1(VI)-PE), compared to negative control PMO (PMO-NTC), vehicle (mock) treatment, or no treatment (NT). Total levels of collagen α1(VI) chains (Colα1(VI)) are unchanged. Alpha-tubulin was used as a loading control in the cell fraction. PE: Pseudo-exon. NTC: Nontargeting control. NT: Non-treated

3.12 Immunofluorescence Staining for Extracellular COL6

1. Place a 22 mm diameter precoated coverslip into a 6-well plate.
2. Seed the cells at a concentration of 2×10^5 cells per well.
3. Transfect the cells the following day as described in Subheading 3.6 or 3.7.
4. 24 h after transfection, remove the transfection medium and add 2-ml pre-warmed growth medium containing 50 μg/mL L-ascorbic acid to the wells.
5. After a minimum of 24 h and up to 4 days of treatment with L-ascorbic acid, remove the medium and wash once with PBS (*see* **Note 20**).
6. Fix the cells on the coverslip with 600 μL of 4% pre-warmed paraformaldehyde for 10 min at RT. Rinse the coverslips thoroughly with PBS (*see* **Note 21**).
7. Discard the PBS and incubate the cells for 1 h at RT with the blocking solution (5% goat serum diluted in PBS).
8. Discard the blocking solution and incubate the cells with primary antibody (Sigma-Aldrich MAB1944, 1:2,000 diluted in the blocking solution) for 1 h at RT.

9. Rinse the coverslips three times with PBS for 5 min, and then incubate the cells with secondary antibody (Thermofisher A11029; goat anti-mouse Alexa 488, 1:500 diluted in the blocking solution) for 1 h at RT (*see* **Note 22**).
10. Rinse the coverslips three times with PBS for 5 min and then mount the coverslips with Hydromount containing DAPI to stain the nuclei.
11. Leave the slides to dry in the dark at RT for at least 1 h before the image acquisition under a fluorescence microscope.

4 Notes

1. Endo-Porter is used for in vitro transfection of PMO.
2. Antibody Colα1(VI)-PE is a gift from Raimund Wagener, University of Cologne; antibody Colα1(VI) is a gift of Mon-Li Chu, Thomas Jefferson University.
3. Ideally skin biopsy should be processed as soon as possible, or within five days when stored at 4 °C. Sample incubation in trypsin can be left at 37 °C for up to 1 h.
4. Sample incubation in collagenase can be left at 37 °C for up to 2 h.
5. Do not aspirate the medium to avoid losing unattached cells.
6. After the establishment of the first passage, always freeze down some vials of cells at low passage before starting any experiments. Cells can be stored at -80 °C for a short period, then should be stored in the vapor phase of liquid nitrogen for long-term storage.
7. Penicillin and streptomycin should not be used in the medium during transfection experiments. This is because lipofectamine increases cell permeability, which may increase the amount of antibiotics delivered into the cells, resulting in cytotoxicity.
8. The protocol described is an example of 2′-MOE and 2′-OMe ASOs when using lipofectamine 2000 as the transfection reagent.
9. The time of incubation can vary from 6 to 24 h.
10. The protocol described is an example of PMO when using Endo-Porter as the transfection reagent.
11. If cells are detached, it may indicate that the ASO is toxic.
12. Alternatively, add 500 μL of Trizol to the wells after the PBS wash, and lyse cells by pipetting up and down. Transfer the lysate to an Eppendorf tube and incubate at room temperature

for 5 min. Follow manufacturer's instructions to extract and precipitate RNA. Resuspend the final pellet in a small volume of water (e.g., 15 μL).

13. To avoid amplifying DNA, primers are designed to target exon-exon junctions. These primers are specific for the amplification of cDNA since they cannot bind to genomic DNA, which contains introns between exons.
14. When an assay is used for the first time, a dilution curve of a control sample should be run to ensure that the assay's efficiency is between 90% and 110%.
15. The annealing/elongation temperature can be decreased, if needed, for an underperforming assay.
16. Fibroblasts should be fully confluent before collecting proteins. In addition, L-ascorbic acid should be added to the culture medium at least 24 h before protein collection.
17. This stock solution can be aliquoted and stored at −20 °C until use.
18. Protein samples (medium and cell fractions) and protein samples diluted in the loading buffer can be stored at −80 °C until ready to proceed.
19. To probe for both Colα1(VI)-PE and tubulin antibodies using a single membrane, we recommend cutting the membrane in half at approximately the 100-kDa mark. The top half of the membrane is then incubated with the Colα1(VI)-PE primary antibody solution, and the bottom half, with the tubulin primary antibody solution. To probe for both Colα1(VI)-PE and Colα1(VI) antibodies, we recommend running two duplicate gels instead.
20. L-ascorbic acid is used to allow collagen VI secretion into the extracellular space. Incubation with L-ascorbic acid should be at least 24 h, but in some experiments, longer incubation periods might be needed.
21. Paraformaldehyde should be pre-warmed at 37 °C to preserve the intact morphology of the cells. After the incubation with paraformaldehyde, fixed coverslips can be kept in PBS 1× in the fridge for a few weeks until the following staining.
22. Collagen VI chain-specific antibodies are available. These include anti-collagen type VI α-1 chain antibody (17023-1-AP, Proteintech) and anti-collagen type VI α-3 chain antibody, clone 3C4 (MAB1944, Sigma-Aldrich). The generic anti-collagen type VI antibody, clone VI-26 (MAB3303, Sigma-Aldrich), recognizes all α-1, α-2, and α-3 chains.

Acknowledgments

This work was supported by the Muscular Dystrophy UK (grant 17GRO-PG36-0168), Wellcome Trust (grant 204841/Z/16/Z), the Intramural Research Program of the NIH NINDS, and the National Institutes of Health Research Biomedical Research Centre (NIHR BRC) at Great Ormond Street Hospital for Children NHS Foundation Trust and University College London. We thank Raimund Wagener and Mon-Li Chu for providing collagen VI-specific antibodies.

References

1. Cescon M, Gattazzo F, Chen P, Bonaldo P (2015) Collagen VI at a glance. J Cell Sci 128(19):3525–3531
2. Theocharis AD, Skandalis SS, Gialeli C, Karamanos NK (2016) Extracellular matrix structure. Adv Drug Deliv Rev 97:4–27
3. Bonnemann CG (2011) The collagen VI-related myopathies: muscle meets its matrix. Nat Rev Neurol 7(7):379–390
4. Cummings BB, Marshall JL, Tukiainen T, Lek M, Donkervoort S, Foley AR et al (2017) Improving genetic diagnosis in Mendelian disease with transcriptome sequencing. Sci Transl Med 9(386):eaal5209
5. Aguti S, Bolduc V, Ala P, Turmaine M, Bonnemann CG, Muntoni F et al (2020) Exon-skipping oligonucleotides restore functional collagen VI by correcting a common COL6A1 mutation in Ullrich CMD. Mol Ther Nucleic Acids 21:205–216
6. Bolduc V, Foley AR, Solomon-Degefa H, Sarathy A, Donkervoort S, Hu Y et al (2019) A recurrent COL6A1 pseudoexon insertion causes muscular dystrophy and is effectively targeted by splice-correction therapies. JCI Insight 4(6):e124403
7. Vaz-Drago R, Custodio N, Carmo-Fonseca M (2017) Deep intronic mutations and human disease. Hum Genet 136(9):1093–1111
8. Havens MA, Hastings ML (2016) Splice-switching antisense oligonucleotides as therapeutic drugs. Nucleic Acids Res 44(14):6549–6563
9. Scoles DR, Minikel EV, Pulst SM (2019) Antisense oligonucleotides: a primer. Neurol Genet 5(2):e323
10. Rinaldi C, Wood MJA (2018) Antisense oligonucleotides: the next frontier for treatment of neurological disorders. Nat Rev Neurol 14(1):9–21
11. Gagliardi M, Ashizawa AT (2021) The challenges and strategies of antisense oligonucleotide drug delivery. Biomedicine 9(4):433
12. Roberts TC, Langer R, Wood MJA (2020) Advances in oligonucleotide drug delivery. Nat Rev Drug Discov 19(10):673–694

Part V

Genome Editing/CRISPR

Chapter 21

CRISPR-Cas9 Correction of Duchenne Muscular Dystrophy in Mice by a Self-Complementary AAV Delivery System

Yu Zhang, Rhonda Bassel-Duby, and Eric N. Olson

Abstract

Duchenne muscular dystrophy (DMD) is a fatal neuromuscular disorder, caused by mutations in the *DMD* gene coding dystrophin. Applying clustered regularly interspaced short palindromic repeats (CRISPR) and CRISPR-associated proteins (CRISPR-Cas) for therapeutic gene editing represents a promising technology to correct this devastating disease through elimination of underlying genetic mutations. Adeno-associated virus (AAV) has been widely used for gene therapy due to its low immunogenicity and high tissue tropism. In particular, CRISPR-Cas9 gene editing components packaged by self-complementary AAV (scAAV) demonstrate robust viral transduction and efficient gene editing, enabling restoration of dystrophin expression throughout skeletal and cardiac muscle in animal models of DMD. Here, we describe protocols for cloning CRISPR single guide RNAs (sgRNAs) into a scAAV plasmid and procedures for systemic delivery of AAVs into a DMD mouse model. We also provide methodologies for quantification of dystrophin restoration after systemic CRISPR-Cas9–mediated correction of DMD.

Key words Gene editing, sgRNA, Adeno-associated virus, DMD mouse model, Dystrophin

1 Introduction

Duchenne muscular dystrophy (DMD) is a severe monogenic muscle disease, caused by mutations in the dystrophin gene located on the X chromosome [1, 2]. Dystrophin maintains muscle membrane integrity by linking the dystroglycan complex with the actin cytoskeleton [3, 4]. Absence of dystrophin in skeletal and cardiac muscle causes muscle degeneration and myocardial fibrosis, followed by respiratory and cardiac failure and, ultimately, premature death. There is no effective treatment for this debilitating disease. Numerous therapeutic approaches have been attempted to mitigate DMD disease progression, including corticosteroid supplementation, antisense oligonucleotide injection, and micro-dystrophin gene replacement [5–7]. However, none of these approaches enable

Rika Maruyama and Toshifumi Yokota (eds.), *Muscular Dystrophy Therapeutics: Methods and Protocols*,
Methods in Molecular Biology, vol. 2587, https://doi.org/10.1007/978-1-0716-2772-3_21,

correction of the underlying genetic mutation nor restore endogenous dystrophin expression. At present, the lack of transformative treatment emphasizes the need for development of new therapeutic approaches for DMD.

Gene editing by clustered regularly interspaced short palindromic repeats (CRISPR) and CRISPR-associated proteins (CRISPR-Cas) offers the ability to precisely and efficiently modify the genome and correct disease-causing mutations [8–10]. With this approach, Cas9 nuclease is directed to a specific genomic locus through interaction with a single guide RNA (sgRNA) that anneals to complementary DNA sequences [8–12]. If the target DNA sequence is immediately followed by a protospacer adjacent motif (PAM), the RNA-guided nuclease generates DNA double-stranded breaks (DSBs). Depending on the cell type, CRISPR-Cas9-induced DSBs can be repaired by three distinct DNA repair pathways, including homology-directed repair (HDR), classical nonhomologous end joining (C-NHEJ), and microhomology-mediated end joining (MMEJ). HDR is confined to proliferating cells and requires a DNA repair template. Although HDR generates a precise modification at the target locus, this pathway is not feasible for correcting *DMD* mutations in vivo since skeletal and cardiac muscles are post-mitotic. Alternatively, DNA DSBs can be repaired by C-NHEJ when there is no sequence microhomology present at the breakage point, or by MMEJ when there are 2–25 base pairs of microhomology on each side of the DSB [13, 14]. The latter two DNA repair pathways occur in both proliferating and post-mitotic cells and can be adapted to correct *DMD* mutations in vivo.

There are several aspects of DMD that make it suitable for CRISPR-Cas9 therapeutic gene editing [15–17]. First, the central rod domain of the dystrophin protein is mainly composed of spectrin-like repeats. In-frame deletion of this modular structure converts DMD to Becker muscular dystrophy (BMD), which is a relatively mild form of muscular dystrophy. Second, the location of the *DMD* gene on the X chromosome only requires gene editing of a single mutant allele in affected boys, precluding potential off-target mutagenesis of the wild-type allele. Third, unlike toxic gain-of-function or dominant negative mutations, which require near-complete correction or elimination of the mutant allele, DMD is caused by a loss-of-function mutation of the dystrophin gene. Studies in patients with BMD have estimated that ~20% of normal levels of dystrophin protein could significantly slow disease progression [18]. Therefore, gene editing of a relatively small fraction of the mutant allele is sufficient to confer therapeutic benefits.

Using adeno-associated virus (AAV) as a delivery tool, we and others performed CRISPR-Cas9-mediated gene editing in animal models of DMD [19–30]. C-NHEJ or MMEJ-induced insertions and deletions (INDELs) reframed or skipped the mutant *Dmd* exon and subsequently restored the open reading frame of the dystrophin gene. Gene-edited animals restored dystrophin expression and demonstrated functional improvement in skeletal muscle and heart. Moreover, sustained dystrophin expression can be observed in DMD mice for 12–18 months after a single systemic dose of AAV9-encoded Cas9 [23, 28].

However, several studies demonstrated that single-stranded AAV (ssAAV)-delivered CRISPR sgRNA was a rate limiting factor for in vivo gene editing, and there was a preferential depletion of CRISPR sgRNA compared to Cas9 nuclease [23, 26]. To address this issue, we adapted the self-complementary AAV (scAAV) system to deliver CRISPR sgRNAs for in vivo gene editing [25, 31]. scAAV has a mutation in one end of the inverted terminal repeat (ITR) and carries a double-stranded viral genome [32]. This property allows scAAV to bypass second-strand synthesis, which is a rate-limiting step for gene expression from ssAAV [33, 34]. Moreover, double-stranded scAAV is more resistant to DNA degradation, thereby sustaining more copies of stable episomes after viral transduction [35, 36]. DMD mice receiving a systemic injection of scAAV-sgRNAs showed higher gene editing and dystrophin restoration compared to ssAAV-treated cohorts [35, 36]. Thus, CRISPR-Cas gene editing components delivered by scAAV can be deployed to correct diverse DMD mutations and offer the prospect of a potential gene therapy for the permanent correction of other genetic diseases.

In this chapter, we describe protocols for cloning CRISPR sgRNAs into a scAAV plasmid and procedures for systemic delivery of AAVs into a DMD mouse model. We also provide methodologies for quantification of dystrophin restoration after systemic CRISPR-Cas9–mediated correction of DMD.

2 Materials

2.1 Reagents for Cloning CRISPR sgRNAs into scAAV Plasmid

1. Oligonucleotides (Integrated DNA Technologies).
2. NEBuffer 2 (10×) (New England Biolab).
3. Tango Buffer (10×) (Thermo Fisher Scientific).
4. Dithiothreitol (DTT) (Thermo Fisher Scientific).
5. ATP (10 mM) (Thermo Fisher Scientific).
6. FastDigest Esp3I (Thermo Fisher Scientific).
7. FastDigest BpiI (Thermo Fisher Scientific).
8. T4 DNA ligase (New England Biolab).
9. One Shot Stbl3™ Chemically Competent *E. coli* (Thermo Fisher Scientific).
10. S.O.C. Medium (Thermo Fisher Scientific).

11. Chloramphenicol (Sigma-Aldrich).
12. Ampicillin (Sigma-Aldrich).
13. Glycerol (Sigma-Aldrich).
14. Qiaprep Spin Miniprep Kit (QIAGEN).
15. NucleoBond PC 500 Maxiprep Kit (MACHEREY-NAGEL).
16. AhdI (New England Biolab).
17. CutSmart Buffer (New England Biolab).
18. Standard DNA gel electrophoresis equipment and reagents.

2.2 Reagents for Injecting AAV CRISPR-Cas9 into DMD Mice

1. Ultrafine insulin syringe (Becton, Dickinson and Company).
2. Sodium chloride solution 0.9% in water (Sigma-Aldrich).

2.3 Reagents for Skeletal and Cardiac Muscle Dystrophin Western Blot Analysis

1. Liquid nitrogen.
2. Bessman Tissue Pulverizer (Repligen).
3. Protein lysis buffer (10% SDS, 62.5 mM Tris–HCl pH 6.8, 1 mM EDTA, protease inhibitor).
4. Pellet Pestle and Motor (Fisher Scientific)..
5. 1-mL syringe (Becton, Dickinson and Company).
6. 22 and 27 gauge needles (Becton, Dickinson and Company).
7. Pierce BCA Protein Assay Kit (Thermo Fisher Scientific).
8. 4× Laemmli Sample Buffer (Bio-Rad).
9. β-mercaptoethanol (Sigma-Aldrich).
10. 4–20% Criterion™ TGX™ Precast Protein Gel (Bio-Rad).
11. Protein ladder (Bio-Rad).
12. Running buffer (25 mM Tris–HCl, 192 mM glycine, 0.1% SDS, pH 8.3).
13. PVDF membrane (Bio-Rad)
14. Transfer buffer (25 mM Tris–HCl, 192 mM glycine, 0.1% SDS, 20% methanol, pH 8.3).
15. Tris-buffered saline (TBS, pH 7.5).
16. Blocking buffer (5% w/v nonfat dry milk, 1× TBS, 0.1% Tween-20).
17. Wash buffer (0.1% Tween-20 in TBS).
18. Mouse anti-dystrophin antibody (MANDYS8, Sigma-Aldrich).
19. Mouse anti-vinculin antibody (V9131, Sigma-Aldrich).
20. Goat anti-mouse IgG (H+L)-HRP antibody (Bio-Rad).
21. Western Blotting Luminol Reagent (Santa Cruz).

2.4 Reagents for Skeletal and Cardiac Muscle Dystrophin Immunostaining

1. Gum tragacanth (Sigma-Aldrich).
2. Tissue-Tek® O.C.T. tissue freezing compound (SAKURA).
3. Liquid nitrogen.
4. Isopentane (Sigma-Aldrich).
5. Hydrophobic pen.
6. Phosphate-buffered saline (PBS, pH 7.4) (Sigma-Aldrich).
7. Triton X-100 (Sigma-Aldrich).
8. M.O.M.® (Mouse on Mouse) Immunodetection kit containing blocking reagent, protein concentrate, and biotinylated anti-mouse IgG reagent (Vector Laboratories).
9. Mouse anti-dystrophin primary antibody (MANDYS8, Sigma-Aldrich).
10. Rabbit anti-laminin Laminin primary antibody (Sigma-Aldrich, L9393).
11. Fluorescein Avidin D, FITC (Vector Laboratories).
12. Goat anti-rabbit IgG (H+L) secondary antibody conjugated with Alexa Fluor® 555 (Thermo Fisher Scientific).
13. DAPI (Sigma-Aldrich).
14. Shandon Immu-mount (Fisher Scientific).

3 Methods

3.1 Generating scAAV with Multiplexed CRISPR sgRNAs

Cloning CRISPR sgRNAs into the scAAV vector relies on two steps of Golden Gate Assembly. The first step is to clone the annealed sgRNA oligonucleotides into three donor plasmids using Esp3I (BsmBI)-mediated Golden Gate Assembly. The donor plasmids do not carry AAV ITRs, which prevents potential sequence recombination during assembly. The second step is to subclone the U6, H1, and 7SK sgRNA expression cassettes from each donor plasmid into the scAAV vector using BpiI (BbsI)-mediated Golden Gate Assembly.

3.1.1 Ordering and Annealing CRISPR sgRNA Oligonucleotides

1. Order CRISPR sgRNA oligonucleotides with adaptor sequences for the donor plasmid:

$$\text{Forward oligonucleotides}: 5' - \text{CACCG(N)}_{20} - 3'$$

$$\text{Reverse oligonucleotides}: 5' - \text{AAAC(N)}_{20}\text{C} - 3'$$

2. Anneal CRISPR sgRNA oligonucleotides as indicated in Table 1 and incubate in a thermocycler: (95 °C for 5 min, cool down to 25 °C at 5 °C/min). Dilute the annealed oligonucleotides 1:200 in water.

Table 1
CRISPR sgRNA oligonucleotide annealing reaction

Components	Amount
Forward oligonucleotides (100 μM)	1 μL
Reverse oligonucleotides (100 μM)	1 μL
NEBuffer 2 (10×)	1 μL
ddH_2O	7 μL
Total	10 μL

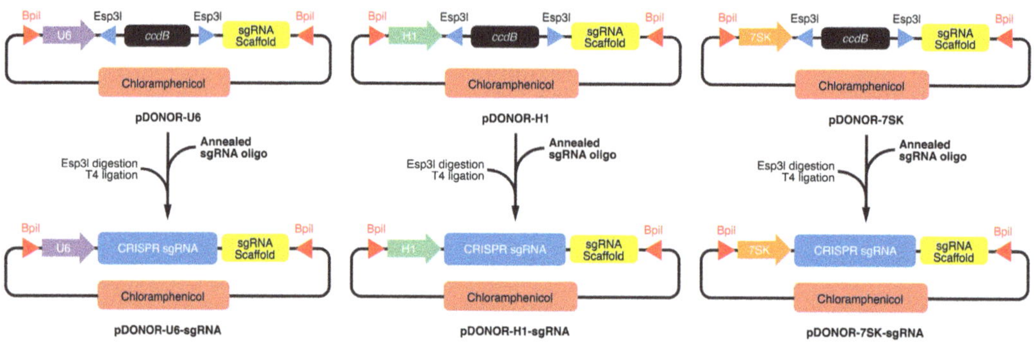

Fig. 1 Assembly of CRISPR sgRNAs into donor plasmids. Annealed CRISPR sgRNA oligos are cloned into three independent donor plasmids using Esp3I-mediated Golden Gate Assembly method. Each donor plasmid contains a unique RNA polymerase III promoter, U6, H1, and 7SK. ccdB is a negative selection gene to prevent unassembled plasmid from being transformed into *E. coli*. Chloramphenicol is the antibiotic resistance gene. Blue triangle depicts Esp3I restriction enzyme recognition site. Red triangle depicts BpiI restriction enzyme recognition site

3.1.2 Golden Gate Assembly of CRISPR sgRNAs into Donor Plasmids

There are three independent donor plasmids with each carrying a unique RNA polymerase III promoter (U6, H1, and 7SK) to drive sgRNA expression (Fig. 1). Each donor plasmid carries a negative selection gene *ccdB* and a chloramphenicol resistance gene (*see* **Note 1**). CRISPR sgRNA must be cloned into these donor plasmids separately.

1. Set up three independent Golden Gate Assembly reactions as indicated in Table 2 and incubate the reaction mixture in a thermocycler: (37 °C for 5 min followed by 21 °C for 5 min (6 cycles), 65 °C for 10 min to denature the restriction enzyme and ligase, cool down to 4 °C).
2. Transform 50 μL of One Shot Stbl3™ Chemically Competent *E. coli* with 5 μL of assembled mixture. Incubate on ice for 30 min followed by heat shock at 42 °C for 30 s.
3. Recover the Stbl3™ *E. coli* in 450 μL of S.O.C. medium and place on a shaker for 45 min (300 RPM, 37 °C).

Table 2
Golden Gate Assembly of CRISPR sgRNAs into donor plasmids

Components	Amount
Donor plasmid (U6 or H1 or 7SK) (100 ng/μL)	1 μL
Annealed sgRNA oligonucleotides (1:200 diluted)	2 μL
Tango buffer (10×)	2 μL
DTT (10 mM)	1 μL
ATP (10 mM)	1 μL
FastDigest Esp3I restriction enzyme	1 μL
T4 DNA ligase	1 μL
ddH_2O	11 μL
Total	20 μL

4. Plate the Stbl3™ *E. coli* on LB agar plate containing 25 μg/mL of chloramphenicol and incubate at 37 °C overnight.
5. The next day, pick three colonies and culture each colony in 5 mL of LB broth with 25 μg/mL of chloramphenicol at 37 °C overnight.
6. Isolate donor plasmid from bacterial culture using Qiaprep Spin Miniprep Kit according to the manufacturer's protocol.
7. Confirm assembly of CRISPR sgRNA into donor plasmid by Sanger sequencing using primer:

   5′ – GTATGTTGTGTGGAATTGTGAG – 3′

8. Donor plasmids assembled with CRISPR sgRNA can be stored at −20 °C.

3.1.3 Golden Gate Assembly of CRISPR sgRNAs into scAAV Vector

The second round of Golden Gate Assembly aims to subclone CRISPR sgRNAs from each donor plasmid into the scAAV vector (Fig. 2). The scAAV plasmid carries a negative selection gene *ccdB* and an ampicillin resistance gene (*see* **Note 1**). The fully assembled scAAV plasmid carries three copies of the same CRISPR sgRNA driven by three independent RNA polymerase III promoters U6, H1, and 7SK.

1. Set up Golden Gate Assembly reactions as indicated in Table 3 and incubate the reaction mixture in a thermocycler: (37 °C for 5 min followed by 21 °C for 5 min (6 cycles), 65 °C for 10 min to denature the restriction enzyme and ligase, cool down to 4 °C).

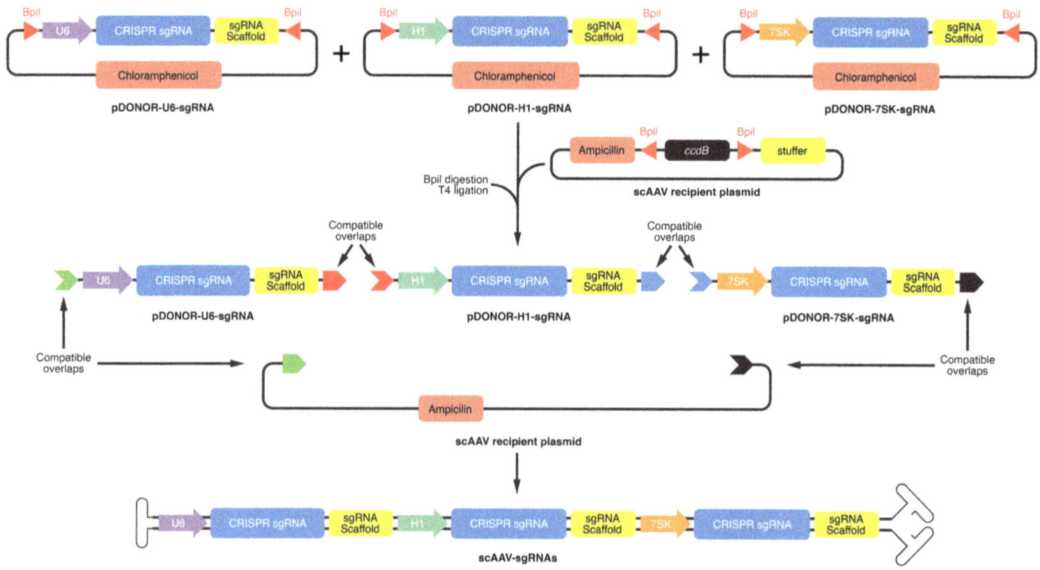

Fig. 2 Assembly of CRISPR sgRNAs into scAAV plasmid. A CRISPR sgRNA is cloned into three independent donor plasmids, digested with BpiI, and cloned into scAAV plasmid using Golden Gate Assembly method. ccdB is a negative selection gene to prevent unassembled plasmid from being transformed in *E. coli*. Chloramphenicol and ampicillin are antibiotic resistance genes used in donor plasmids and scAAV plasmid, respectively. Red triangle indicates the BpiI restriction enzyme recognition site

Table 3
Golden Gate Assembly of CRISPR sgRNAs into scAAV plasmid

Components	Amount
Donor plasmid 1 (U6-sgRNA) (50 ng/μL)	1 μL
Donor plasmid 2 (H1-sgRNA) (50 ng/μL)	1 μL
Donor plasmid 3 (7SK-sgRNA) (50 ng/μL)	1 μL
scAAV plasmid (50 ng/μL)	1 μL
Tango buffer (10×)	2 μL
DTT (10 mM)	1 μL
ATP (10 mM)	1 μL
FastDigest BpiI restriction enzyme	1 μL
T4 DNA ligase	1 μL
ddH_2O	10 μL
Total	20 μL

2. Transform 50 μL of One Shot Stbl3™ Chemically Competent *E. coli* (*see* **Note 2**) with 5 μL of assembled mixture. Incubate on ice for 30 min followed by heat shock at 42 °C for 30 s.
3. Recover the Stbl3™ *E. coli* in 450 μL of S.O.C. medium and place on a shaker for 45 min (300 RPM, 37 °C).
4. Plate the Stbl3™ *E. coli* on LB agar plate containing 50 μg/mL of ampicillin and incubate at 37 °C overnight.
5. The next day, pick several colonies and culture each colony in 5 mL of LB broth with 50 μg/mL of ampicillin at 37 °C for 12–14 h (*see* **Note 3**).
6. Isolate the scAAV plasmid from a bacterial culture using Qiaprep Spin Miniprep Kit according to the manufacturer's protocol.
7. Confirm assembly of three copies of the CRISPR sgRNA into scAAV plasmid by Sanger sequencing using primers (*see* **Note 4**):

U6 forward primer 5′ – GAGGGCCTATTTCCCATGATTCC – 3′

H1 forward primer 5′ – TGTCGCTATGTGTTCTGGGAAATCA – 3′

8. Perform AhdI restriction digestion of sequence confirmed plasmid to ensure that ITRs of scAAV plasmid are maintained without significant recombination (Fig. 3) (*see* **Note 5**).
9. For scAAV plasmid with correctly assembled CRISPR sgRNAs and intact ITRs, generate glycerol stock from the corresponding bacterial culture by mixing 500 μL of the bacterial culture with 500 μL of 50% glycerol. Store at −80 °C for long-term storage.
10. Set up a large-scale bacterial culture from the glycerol stock and purify at least 500 μg of the scAAV plasmid using a maxiprep kit.
11. Perform AhdI restriction digestion of plasmid from maxiprep to ensure that ITRs of scAAV plasmid are maintained without significant recombination.
12. Produce scAAV with multiplexed CRISPR sgRNAs (*see* **Note 6**).

3.2 In Vivo CRISPR-Cas9 Gene Editing of DMD Mice Using scAAV

In vivo CRISPR-Cas9-mediated correction of *Dmd* mutation in mice requires systemic injection of two AAVs into postnatal DMD mice. The first AAV is a single-stranded AAV encoding *Streptococcus pyogenes* Cas9 (*Sp*Cas9), and the second AAV is a double-stranded scAAV expressing three copies of the CRISPR sgRNA.

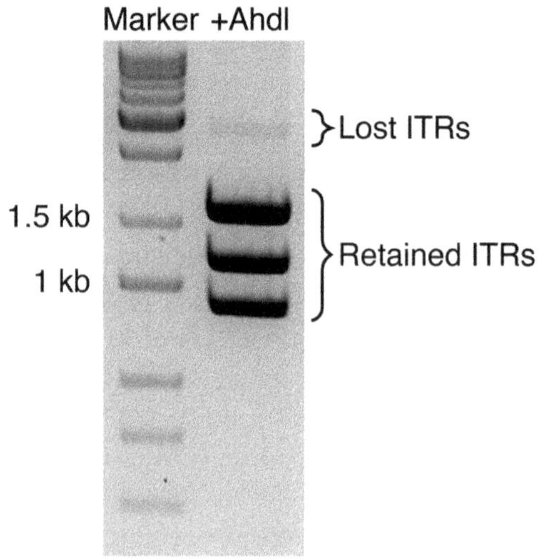

Fig. 3 AhdI restriction digestion of the scAAV plasmid. The scAAV plasmid contains three AhdI recognition sites, which are located in the two ITR regions and the ampicillin resistant gene. Restriction digestion of the scAAV plasmid using AhdI enzyme generates three bands after 1.5% agarose gel electrophoresis. The upper band above the three AhdI-digested bands indicates loss of ITR. Typically, a scAAV plasmid with less than 10% ITR recombination is recommended to be used for AAV packaging

1. Weigh postnatal day 4 (P4) DMD mice and calculate total amount of AAVs used for injection. For each mouse, 1×10^{14} vg/kg of ssAAV-*Sp*Cas9 and 1×10^{14} vg/kg of scAAV-sgRNAs is recommended to achieve efficient in vivo gene editing.
2. Prepare AAV injection mixture by diluting ssAAV-*Sp*Cas9 and scAAV-sgRNAs with saline to achieve total volume of 100 μL.
3. Inject AAV mixtures intraperitoneally (IP) into P4 DMD mice using an insulin syringe.
4. Muscle harvested from CRISPR gene edited DMD mice can be used to quantify dystrophin restoration at 4 weeks post-IP injection.

3.3 Quantification of Dystrophin Restoration in DMD Mice After Systemic AAV CRISPR Gene Editing

Many assays can be used to assess dystrophin restoration in DMD mice after systemic administration of AAV CRISPR gene editing components. Western blot analysis is performed to quantitatively measure the percentage of dystrophin protein restoration in skeletal muscles and heart. Immunostaining is performed to quantify the percentage of myofibers or cardiomyocytes expressing dystrophin protein.

3.3.1 Skeletal and Cardiac Muscle Dystrophin Western Blot Analysis

1. Euthanize mouse and harvest skeletal muscles and hearts from AAV CRISPR gene edited DMD mice (*see* **Note 7**).
2. Freeze muscle samples in liquid nitrogen. Pulverize frozen muscle samples into fine powder using liquid nitrogen cooled tissue pulverizer (*see* **Note 8**).
3. Lyse pulverized muscle samples in 250 μL lysis buffer. Use a Pellet Pestle Motor to homogenize the protein lysate for 20 s.
4. Pass the protein lysate through a 22-gauge needle for ten times. Repeat with a 27-gauge needle for ten times.
5. Measure protein concentration using Pierce BCA Protein Assay Kit according to the manufacturer's protocol.
6. Prepare protein sample by mixing 50 μg of total protein with 4× Laemmli sample buffer (containing 5% β-mercaptoethanol) to achieve a final volume of 20 μL. Boil protein samples at 95 °C for 5 min. Store the protein sample at −80 °C if not using it immediately.
7. Load 50 μg of total protein in 4–20% 10-well or 18-well Criterion™ TGX™ Precast Protein Gel. Run the SDS-PAGE at 80 V for 30 min followed by 130 V for 2 h.
8. Transfer the protein to PVDF membrane using the wet transfer method (*see* **Note 9**).
9. Block the PVDF membrane in blocking buffer at room temperature for 1 h.
10. Incubate the PVDF membrane with mouse anti-dystrophin antibody (1:1000) at 4 °C overnight.
11. Wash the PVDF membrane with TBST buffer three times.
12. Incubate the PVDF membrane with goat anti-mouse IgG (H+L)-HRP antibody (1:10,000) at room temperature for 1 h. Wash the PVDF membrane with TBST buffer three times.
13. Develop the PVDF membrane using Western Blotting Luminol Reagent according to the manufacturer's protocol. Detect dystrophin protein using either X-ray film or digital imaging.
14. Use stripping buffer to remove the antibodies from the PVDF membrane.
15. Re-block the PVDF membrane in blocking buffer at room temperature for 1 h.
16. Incubate the PVDF membrane with mouse anti-vinculin antibody (1:1000) at room temperature for 1 h, followed by washing the PVDF membrane with TBST buffer three times.
17. Incubate the PVDF membrane with goat anti-mouse IgG (H+L)-HRP antibody (1:10,000) at room temperature for 1 h. Wash the PVDF membrane with TBST buffer three times.

18. Develop the PVDF membrane using Western Blotting Luminol Reagent according to the manufacturer's protocol. Detect vinculin protein using either X-ray film or digital imaging.

3.3.2 Skeletal and Cardiac Muscle Dystrophin Immunostaining

1. Embed skeletal muscles of AAV CRISPR edited DMD mice in a 1:2 volume mixture of gum tragacanth and tissue-freezing compound. Submerge heart in tissue-freezing compound.
2. Snap freeze the embedded skeletal muscles and heart in supercooled isopentane to −155 °C and store at −80 °C for long-term storage.
3. Perform cryosectioning of skeletal muscles and heart (eight-micron transverse sections) using a cryostat.
4. Air dry the cryosections at room temperature for 20 min.
5. Use a hydrophobic pen to draw a water-repellent barrier around the muscle sections.
6. Rehydrate and delipidate muscle sections with 1% Triton X-100 in phosphate buffered saline (PBS; pH 7.4) for 30 min.
7. Wash muscle sections with PBS three times (5 min each wash).
8. Block muscle sections with M.O.M IgG Blocking Reagent (two drops of M.O.M Blocking Reagent stock solution in 2.5 mL of PBS) at room temperature for 1 h.
9. Wash muscle sections with PBS twice (2 min each wash).
10. Equilibrate muscle sections with M.O.M diluent (600 μL of M.O.M Protein Concentrate stock solution to 7.5 mL of PBS) at room temperature for 5 min.
11. Incubate muscle sections with mouse anti-dystrophin primary antibody (1:500) and rabbit anti-laminin primary antibody (1:500) dissolved in M.O.M diluent at room temperature for 1 h (*see* **Note 10**).
12. Wash muscle sections with PBS twice (2 min each wash).
13. Incubate muscle sections with M.O.M biotinylated anti-mouse IgG reagent (1:250) dissolved in M.O.M diluent at room temperature for 10 min.
14. Wash muscle sections with PBS three times (5 min each wash).
15. Incubate muscle sections with Avidin D conjugated with FITC (1:250), goat anti-rabbit IgG secondary antibody conjugated with Alexa Fluor 555 (1:500), and DAPI (1:250) dissolved in M.O.M diluent at room temperature for 30 min.
16. Wash muscle sections with PBS three times (5 min each wash).
17. Mount the slides and detect dystrophin expression using a fluorescent microscope.

4 Notes

1. *ccdB* is a lethal gene that targets *E. coli* DNA gyrase. Propagation of the original donor plasmid and scAAV plasmid requires *ccdB*-resistant *E. coli* strain.
2. scAAV plasmid contains ITRs, which is prone to DNA recombination. It is recommended to use recombination defective *E. coli* strain to propagate AAV plasmid.
3. Culturing *E. coli* transformed with AAV plasmid should not exceed over 14 h. This is because AAV ITRs are unstable in *E. coli* and plasmids that lose the ITRs have a replication advantage in transformed cells.
4. U6 forward sequencing primer primes at the U6 promoter and can be used to confirm CRISPR sgRNA cloned after the U6 and H1 promoters. H1 forward sequencing primer primes at the H1 promoter and can be used to confirm CRISPR sgRNA cloned after the H1 and 7SK promoters.
5. AhdI cuts at two ITRs and the ampicillin resistant gene, which digests scAAV plasmid into three fragments. Loss of AAV ITRs results in an undigested upper band (Fig. 3). Plasmids with significant ITR recombination cannot be used to package AAV. Typically, scAAV plasmid with less than 10% ITR recombination is recommended to be used for AAV packaging.
6. We recommend packaging scAAV using a Viral Core facility or commercial vendor to ensure that both AAV purity and titer meets standard in vivo application.
7. Muscles from one side of the limb can be used for biochemical analysis, such as genomic DNA, RNA, and protein analysis. Muscles from the contralateral limb can be used for histological analysis, such as immunostaining, hematoxylin, and eosin staining. Diaphragm and heart can be cut into two parts for biochemical and histological analysis.
8. Pulverized muscle samples can be divided into three parts for isolating genomic DNA (20% of total sample), RNA (40% of total sample), and protein (40% of total sample).
9. We recommend using the wet transfer method for transferring large molecular weight proteins. Keep transfer buffer and apparatus in the cold room. The recommended transfer condition is 100 V for 1.5 h.
10. Alternatively, incubate muscle sections with mouse anti-dystrophin primary antibody (1:1500) and rabbit anti-laminin primary antibody (1:1500) dissolved in M.O.M diluent at 4 °C overnight.

Acknowledgments

We thank J. Cabrera for graphics and H. Li and E. Sanchez-Ortiz for providing insight to this manuscript. This work was supported by the NIH (grants HL130253), the Senator Paul D. Wellstone Muscular Dystrophy Specialized Research Center (grant P50 HD 087351), and the Robert A. Welch Foundation (grant 1-0025 to E. N.O.).

References

1. Hoffman EP, Brown RH Jr, Kunkel LM (1987) Dystrophin: the protein product of the Duchenne muscular dystrophy locus. Cell 51(6):919–928
2. Koenig M, Hoffman EP, Bertelson CJ et al (1987) Complete cloning of the Duchenne muscular dystrophy (DMD) cDNA and preliminary genomic organization of the DMD gene in normal and affected individuals. Cell 50(3):509–517
3. Gao QQ, McNally EM (2015) The dystrophin complex: structure, function, and implications for therapy. Compr Physiol 5(3):1223–1239. https://doi.org/10.1002/cphy.c140048
4. Guiraud S, Aartsma-Rus A, Vieira NM et al (2015) The pathogenesis and therapy of muscular dystrophies. Annu Rev Genomics Hum Genet 16:281–308. https://doi.org/10.1146/annurev-genom-090314-025003
5. Quattrocelli M, Zelikovich AS, Jiang Z et al (2019) Pulsed glucocorticoids enhance dystrophic muscle performance through epigenetic-metabolic reprogramming. JCI Insight 4(24). https://doi.org/10.1172/jci.insight.132402
6. Lim KR, Maruyama R, Yokota T (2017) Eteplirsen in the treatment of Duchenne muscular dystrophy. Drug Des Devel Ther 11:533–545. https://doi.org/10.2147/DDDT.S97635
7. Duan D (2018) Systemic AAV micro-dystrophin gene therapy for Duchenne muscular dystrophy. Mol Ther 26(10):2337–2356. https://doi.org/10.1016/j.ymthe.2018.07.011
8. Cong L, Ran FA, Cox D et al (2013) Multiplex genome engineering using CRISPR/Cas systems. Science 339(6121):819–823. https://doi.org/10.1126/science.1231143
9. Jinek M, Chylinski K, Fonfara I et al (2012) A programmable dual-RNA-guided DNA endonuclease in adaptive bacterial immunity. Science 337(6096):816–821. https://doi.org/10.1126/science.1225829
10. Mali P, Yang L, Esvelt KM et al (2013) RNA-guided human genome engineering via Cas9. Science 339(6121):823–826. https://doi.org/10.1126/science.1232033
11. Ran FA, Cong L, Yan WX et al (2015) In vivo genome editing using Staphylococcus aureus Cas9. Nature 520(7546):186–191. https://doi.org/10.1038/nature14299
12. Zetsche B, Gootenberg JS, Abudayyeh OO et al (2015) Cpf1 is a single RNA-guided endonuclease of a class 2 CRISPR-Cas system. Cell 163(3):759–771. https://doi.org/10.1016/j.cell.2015.09.038
13. Iyer S, Suresh S, Guo D et al (2019) Precise therapeutic gene correction by a simple nuclease-induced double-stranded break. Nature 568(7753):561–565. https://doi.org/10.1038/s41586-019-1076-8
14. Gallagher DN, Haber JE (2018) Repair of a site-specific DNA cleavage: old-school lessons for Cas9-mediated gene editing. ACS Chem Biol 13(2):397–405. https://doi.org/10.1021/acschembio.7b00760
15. Chemello F, Bassel-Duby R, Olson EN (2020) Correction of muscular dystrophies by CRISPR gene editing. J Clin Invest 130(6):2766–2776. https://doi.org/10.1172/JCI136873
16. Min YL, Bassel-Duby R, Olson EN (2019) CRISPR correction of Duchenne muscular dystrophy. Annu Rev Med 70:239–255. https://doi.org/10.1146/annurev-med-081117-010451
17. Zhang Y, Long C, Bassel-Duby R et al (2018) Myoediting: toward prevention of muscular dystrophy by therapeutic genome editing. Physiol Rev 98(3):1205–1240. https://doi.org/10.1152/physrev.00046.2017
18. Hoffman EP, Kunkel LM, Angelini C et al (1989) Improved diagnosis of Becker muscular dystrophy by dystrophin testing. Neurology 39(8):1011–1017. https://doi.org/10.1212/wnl.39.8.1011
19. Amoasii L, Hildyard JCW, Li H et al (2018) Gene editing restores dystrophin expression in a canine model of Duchenne muscular

dystrophy. Science 362(6410):86–91. https://doi.org/10.1126/science.aau1549

20. Amoasii L, Li H, Zhang Y et al (2019) In vivo non-invasive monitoring of dystrophin correction in a new Duchenne muscular dystrophy reporter mouse. Nat Commun 10(1):4537. https://doi.org/10.1038/s41467-019-12335-x
21. Amoasii L, Long C, Li H et al (2017) Single-cut genome editing restores dystrophin expression in a new mouse model of muscular dystrophy. Sci Transl Med 9(418). https://doi.org/10.1126/scitranslmed.aan8081
22. Bengtsson NE, Hall JK, Odom GL et al (2017) Muscle-specific CRISPR/Cas9 dystrophin gene editing ameliorates pathophysiology in a mouse model for Duchenne muscular dystrophy. Nat Commun 8:14454. https://doi.org/10.1038/ncomms14454
23. Hakim CH, Wasala NB, Nelson CE et al (2018) AAV CRISPR editing rescues cardiac and muscle function for 18 months in dystrophic mice. JCI Insight 3(23). https://doi.org/10.1172/jci.insight.124297
24. Long C, Amoasii L, Mireault AA et al (2016) Postnatal genome editing partially restores dystrophin expression in a mouse model of muscular dystrophy. Science 351(6271):400–403. https://doi.org/10.1126/science.aad5725
25. Min YL, Chemello F, Li H et al (2020) Correction of three prominent mutations in mouse and human models of Duchenne muscular dystrophy by single-cut genome editing. Mol Ther 28(9):2044–2055. https://doi.org/10.1016/j.ymthe.2020.05.024
26. Min YL, Li H, Rodriguez-Caycedo C et al (2019) CRISPR-Cas9 corrects Duchenne muscular dystrophy exon 44 deletion mutations in mice and human cells. Sci Adv 5(3): eaav4324. https://doi.org/10.1126/sciadv.aav4324
27. Nelson CE, Hakim CH, Ousterout DG et al (2016) In vivo genome editing improves muscle function in a mouse model of Duchenne muscular dystrophy. Science 351(6271): 403–407. https://doi.org/10.1126/science.aad5143
28. Nelson CE, Wu Y, Gemberling MP et al (2019) Long-term evaluation of AAV-CRISPR genome editing for Duchenne muscular dystrophy. Nat Med 25(3):427–432. https://doi.org/10.1038/s41591-019-0344-3
29. Tabebordbar M, Zhu K, Cheng JKW et al (2016) In vivo gene editing in dystrophic mouse muscle and muscle stem cells. Science 351(6271):407–411. https://doi.org/10.1126/science.aad5177
30. Moretti A, Fonteyne L, Giesert F et al (2020) Somatic gene editing ameliorates skeletal and cardiac muscle failure in pig and human models of Duchenne muscular dystrophy. Nat Med 26(2):207–214. https://doi.org/10.1038/s41591-019-0738-2
31. Zhang Y, Li H, Min YL et al (2020) Enhanced CRISPR-Cas9 correction of Duchenne muscular dystrophy in mice by a self-complementary AAV delivery system. Sci Adv 6(8):eaay6812. https://doi.org/10.1126/sciadv.aay6812
32. McCarty DM, Fu H, Monahan PE et al (2003) Adeno-associated virus terminal repeat (TR) mutant generates self-complementary vectors to overcome the rate-limiting step to transduction in vivo. Gene Ther 10(26): 2112–2118. https://doi.org/10.1038/sj.gt.3302134
33. Ferrari FK, Samulski T, Shenk T et al (1996) Second-strand synthesis is a rate-limiting step for efficient transduction by recombinant adeno-associated virus vectors. J Virol 70(5): 3227–3234
34. Fisher KJ, Gao GP, Weitzman MD et al (1996) Transduction with recombinant adeno-associated virus for gene therapy is limited by leading-strand synthesis. J Virol 70(1): 520–532
35. Ren C, Kumar S, Shaw DR et al (2005) Genomic stability of self-complementary adeno-associated virus 2 during early stages of transduction in mouse muscle in vivo. Hum Gene Ther 16(9):1047–1057. https://doi.org/10.1089/hum.2005.16.1047
36. McCarty DM (2008) Self-complementary AAV vectors; advances and applications. Mol Ther 16(10):1648–1656. https://doi.org/10.1038/mt.2008.171

Chapter 22

Preparation of NanoMEDIC Extracellular Vesicles to Deliver CRISPR-Cas9 Ribonucleoproteins for Genomic Exon Skipping

Kei Watanabe, Peter Gee, and Akitsu Hotta

Abstract

The CRISPR-Cas9 system has quickly become the standard tool for genome editing. To deliver this system to target cells, adeno-associated virus (AAV) vectors are commonly used. In fact, AAV vectors have been utilized to deliver the CRISPR-Cas9 system to induce genomic exon skipping and restore the dystrophin protein in various Duchenne muscular dystrophy model animals. Despite the high transduction efficiency, AAV vector-mediated delivery has several limitations, such as the packaging size, prolonged overexpression of Cas9, immunogenicity against the AAV capsid, and the risk of integrating a part of the AAV genomic sequence into the host cell. To overcome these issues, we have recently engineered a transient delivery system utilizing VSV-G pseudotyped extracellular vesicles (EVs) termed NanoMEDIC (nanomembrane-derived extracellular vesicles for the delivery of macromolecular cargo). NanoMEDIC utilizes an HIV-derived Gag protein to package Cas9 protein and gRNA into EVs. The Cas9 and Gag proteins are fused to a heterodimerizer and conditionally dimerized by the addition of an inducible chemical ligand to recruit Cas9 protein into EVs. sgRNA is packaged into EVs through an HIV-derived RNA packaging signal and is subsequently released by two self-cleaving ribozymes. Utilizing these features, NanoMEDIC can achieve highly efficient packaging of the Cas9 protein and gRNA for genome editing into a variety of target cells and in vivo. Here, we describe a step-by-step protocol, including the gRNA-expressing vector construction and large-scale NanoMEDIC production, for in vivo genome editing.

Key words Genome editing, Duchenne muscular dystrophy, CRISPR-Cas9, Extracellular vesicle, Virus-like particle, Exon skipping, Gene therapy

1 Introduction

1.1 CRISPR-Cas9 Genome Editing Therapy

Clustered regularly interspaced short palindromic repeat (CRISPR)-associated protein 9 (Cas9) enables efficient genome editing in human cells, giving it the potential to treat genetic disorders [1, 2]. For example, genome editing by CRISPR-Cas9 has been demonstrated for exon skipping therapy of Duchenne muscular dystrophy (DMD) in several preclinical animal models [3–6].

Rika Maruyama and Toshifumi Yokota (eds.), *Muscular Dystrophy Therapeutics: Methods and Protocols*,
Methods in Molecular Biology, vol. 2587, https://doi.org/10.1007/978-1-0716-2772-3_22,

1.2 Duchenne Muscular Dystrophy (DMD)

DMD is a genetic disorder caused by loss-of-function mutations in the dystrophin gene on the X chromosome. The dystrophin gene encodes the dystrophin protein, which connects cytoskeletal actin fibers and the cellular membrane through the formation of the dystroglycan complex. The loss of dystrophin protein leads to fragile skeletal muscle, such that DMD patients undergo progressive muscle weakness. Notably, an exon skipping approach to modulate the mRNA splicing pattern is applicable to restore the protein reading frame of dystrophin. Antisense oligonucleotide drugs (i.e., Eteplirsen and Viltolarsen) have been approved in multiple countries; however, the effect of antisense oligonucleotides is transient. The induction of exon skipping at the genomic level has an advantage in terms of longer therapeutic effect. Genomic exon skipping is possible with CRISPR-Cas9 technology by disrupting the splice acceptor or donor site (or sometimes by splicing an enhancer motif) via an indel mutation or base modification.

1.3 In Vivo Delivery Approaches

To induce genome editing in vivo, the intracellular delivery of Cas9/gRNA is critical, since the target genomic DNA is protected by multiple biological layers. Until now, various delivery technologies have been reported to deliver CRISPR-Cas9 in animal models, including adeno-associated virus (AAV) vectors. The AAV vector is an attractive vehicle for gene therapy applications due to its high transduction efficiency and low toxicity; however, it has some concerns in the context of genome editing. For example, the expression from AAV vectors typically persists for years in nondividing cells [7], and the prolonged expression of the Cas9 nuclease may enhance the risk of off-target mutagenesis [8] or the insertion of the AAV vector backbone into the host genome [9].

To avoid these issues, the delivery of CRISPR-Cas9 and gRNA as a ribonucleoprotein (RNP) complex is ideal compared to DNA-based vectors, as it is rapidly degraded in cultured cells. Moreover, RNPs do not carry the risk of insertional mutagenesis [10, 11] and can avoid cellular inhibitory RNA [12].

1.4 Extracellular Vesicle (EV)-Based Delivery Technology

Extracellular vesicles (EVs) are small (typically 30–300 nm in diameter) lipid bilayer particles released from the cell membrane. Exosomes (30–150 nm in size) and microvesicles (150–300 nm in size) are the predominant EVs in a narrow sense, but enveloped virus particles or virus-like particles (VLPs) can also be considered EVs in a broad sense. In general, natural EVs have a limited capacity to package and transduce intracellular components compared to viral vectors. To accelerate the particle release and transduction efficiency, a few viral components, such as envelope or capsid proteins, can be utilized to artificially engineer EVs. These proteins are encapsulated into a nano-capsule to mimic the viral entry machinery and enter the target cells efficiently. Yet they do not completely recapitulate a viral vector, since nano-capsules lack most viral genomic nucleic acids and do not replicate in target cells. Thus, EVs offer a transient but efficient delivery vehicle for extracellular macromolecules. For Cas9/gRNA delivery, several EV systems have been reported, including CasP LV [13], NanoBlades [14], VEsiCas [15], and Gesicle [16]. By overexpressing Cas9 and gRNA in EV producer cells with viral protein components, EVs containing Cas9 RNP are released into the culture media. However, in most EV systems, efficient Cas9 RNP packaging is still challenging.

1.5 NanoMEDIC System

Recently, we developed a novel EV delivery system termed NanoMEDIC (nanomembrane-derived extracellular vesicles for the delivery of macromolecular cargo) [17]. NanoMEDIC utilizes two unique mechanisms to actively package Cas9 protein and gRNA into EVs. For the Cas9 packaging, FRB (FKBP12-rapamycin binding domain) is fused with Cas9, and FKBP12 (FK506-binding protein) is fused with HIV Gag. When a heterodimerization ligand, AP21967 (rapamycin analogue), is added into EV producer cells, dimerization between FRB and FKBP12 is induced [18, 19], and Cas9 protein is actively packaged into the EVs by interacting with HIV Gag. Gag protein is composed of three domains: MA (matrix), CA (capsid), and NC (nucleocapsid). The MA domain at the N'-terminus interacts with the plasma membrane, and the CA domain promotes self-assembly to form the core structure of EVs. sgRNA flanked by self-cleaving ribozymes is fused with an HIV packaging signal to interact with the NC domain of Gag for packaging into EVs. After the packaging, gRNA is liberated after ribozyme cleavage [20, 21]. Despite its complicated packaging mechanisms, the production of NanoMEDIC is straightforward and almost identical to conventional lentiviral vector production procedures (Fig. 1).

For CRISPR genome editing in cultured cells, many commercial reagents (such as lipofection reagents) or instruments (such as electroporators) are available for common cell lines. However, NanoMEDIC is useful for hard-to-transfect cells, such as nondividing cells or primary cells with limited proliferation capacity. In fact, NanoMEDIC has been shown to have robust activity in a variety of cell types, including skeletal muscle cells or neurons differentiated from human-induced pluripotent stem cells (iPSCs). NanoMEDIC also works in mice by intramuscular injection, although large-scale production is recommended. For large-scale production of NanoMEDIC, we adapted a suspension culture of HEK293T cells and transfected the NanoMEDIC components into a large cell number of producer cells (1.2×10^9 cells) by flow electroporation using the MaxCyte electroporation platform.

In this protocol paper, we describe (i) how to construct a gRNA-expressing plasmid for the target, (ii) how to produce NanoMEDIC at small (10–15 mL) and large (480 mL) scale, and (iii) how to assess genome editing outcomes.

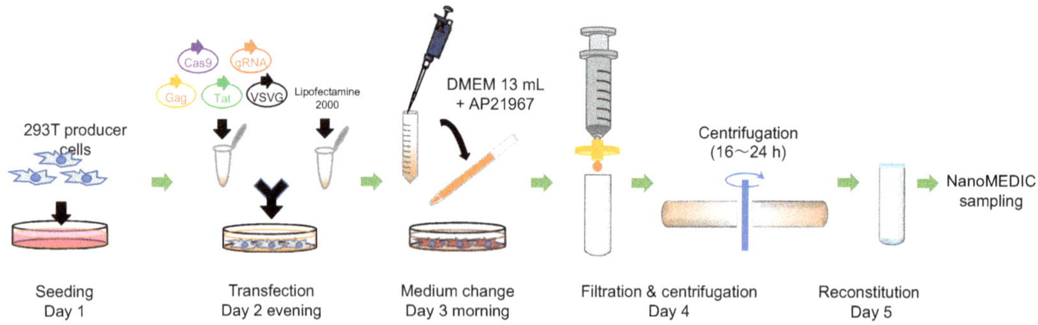

Fig. 1 Schematic of the experimental procedures for the NanoMEDIC production. On day 1, HEK293T producer cells are seeded on a 10-cm dish. On the evening of day 2, five plasmid DNAs (express FRB-Cas9, Ribozyme-gRNA-Ribozyme, FKBP12-Gag, Tat, and VSV-G, respectively) are transfected into the producer cells. On day 3, the media is exchanged for fresh DMEM containing AP21967 to induce the heterodimerization of FRB-Cas9 and FKBP12-Gag. On day 4, the supernatant that contains NanoMEDIC is harvested. The supernatant is filtrated with a 0.45-μm pore syringe filter, and centrifugation is performed for at least 16 h. On day 5, after the centrifugation, the supernatant is removed, and the NanoMEDIC pellet is reconstituted by adding chilled-HBSS and incubated at 4 °C for 2–4 h. After suspending firmly, the NanoMEDIC solution is transferred into several screw cap tubes for the subsequent inoculation experiment or storage at −80 °C

2 Materials

2.1 Construction of CRISPR-gRNA Vectors for NanoMEDIC

1. Design multiple Cas9/gRNA targeting sites with a NGG PAM sequence. If necessary, utilize web-based gRNA design tools, such as CRISPOR or CRISPRdirect. If possible, pick several gRNA sequences within your target region, since the genome editing efficiency varies with the sequence.

   $5'$-$N_1N_2N_3N_4N_5N_6N_7N_8N_9N_{10}N_{11}N_{12}N_{13}N_{14}N_{15}N_{16}$-$N_{17}N_{18}N_{19}N_{20}$NGG-$3'$

   Design Primer #1 by copying the first 20 nucleotides of the sequence from the designed gRNA sequence above. Primer #1 (41-mer) is composed of a 20-mer of the gRNA-targeting region (in blue) and a part (21-mer) of the gRNA scaffold sequence (in green) as shown below:

   - Primer #1 (41-mer):

     $5'$-$N_1N_2N_3N_4N_5N_6N_7N_8N_9N_{10}N_{11}N_{12}N_{13}N_{14}N_{15}$-$N_{16}N_{17}N_{18}N_{19}N_{20}$ GTTTTAGAGCTATGCTGGAAA-$3'$

2. Design Primer #3. Primer #3 (see below) is composed of a 15-mer overhang with the $5'$-end of the EagI site in the pENTR-AmCyan vector, reverse complementary sequence of the first 6 bp of the gRNA (in small letters), a 37-mer hammerhead ribozyme, and the gRNA target sequence of 20-mer (in blue). Note: The small letter sequence region ($n_6n_5n_4n_3n_2n_1$) must be a reverse complement of the first six bases of the gRNA target sequence to form a secondary structure that is identified by the hammerhead ribozyme for self-cleavage of the gRNA part (Fig. 2).

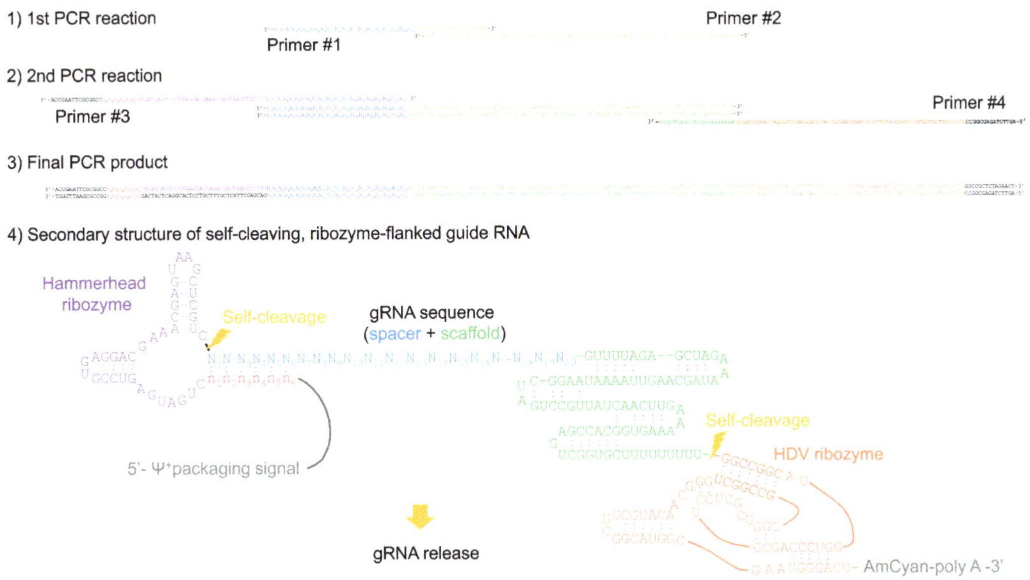

Fig. 2 Design and construction of gRNA flanked by two ribozyme sequences. First, the target sequence of the gRNA is determined, and the spacer sequence (in blue) is inserted into Primer #1. The first PCR reaction is performed using Primer #1 and Primer #2. The gRNA scaffold sequence from SpCas9 is indicated in green letters. Next, the second PCR reaction is performed to amplify the first PCR product with Primer #3 and Primer #4, which respectively contain a hammerhead ribozyme (in purple) and HDV ribozyme (in orange). Finally, the PCR product after the second PCR reaction is inserted into a pENTR-AmCyan plasmid vector and subsequently cloned into a PL-5LTR-RGR-AmCyan-A vector. Once the gRNA flanked by the two ribozymes is expressed from the PL-5LTR-RGR vector, the two ribozymes (purple and orange) form the secondary structure for self-cleavage, as shown. These structures enable gRNA to release from the extra sequences that are needed for packaging into the NanoMEDIC particle

- Primer #3 (79-mer):

  5′-ACCGAATTCGCGGCC$n_6n_5n_4n_3n_2n_1$CTGATGA GTCCGTGAGGACGAAACGAGTAAGCTCGTC $N_1N_2N_3N_4N_5N_6N_7N_8N_9N_{10}N_{11}N_{12}N_{13}N_{14}N_{15}N_{16}N_{17}$- $N_{18}N_{19}N_{20}$ -3′

3. Alongside universal Primer #2 and #4 below, order Primer #1 and #3 for custom oligo-DNA synthesis. Primer #2 contains the gRNA scaffold sequence from SpCas9, and Primer #4 contains a 15-mer overhang with the 3′-end of the EagI site in the pENTR-AmCyan vector, 68-mer of HDV ribozyme, and a part (22-mer) of the gRNA scaffold sequence. For all primers, no chemical modification is necessary. Column purification grade (or cloning grade) is sufficient.

   - Primer #2 (108-mer):

     5′-AAAAAAAAGCACCGACTCGGTG CCACTTTTTCAAGTTGATAACGGACTAG CCTTATTTTAACTTGCTATGCTGTTTCCAGCATAG CTCTAA-3′

- Primer #4 (105-mer):

  5′-AGTTCTAGAGCGGCCGTCCCATTCGCCATG CCGAAGCATGTTGCCCAGCCGGCGCCAGCGAGG AGGCTGGGACCATGCCGGCCAAAAAAAAAG CACCGACTCGGT-3′
- AmCyan reverse primer (5′-GTGTGGAACTGGCATCTG TAGT-3′)
- EagI restriction enzyme (NEW ENGLAND BioLabs, Cat # R0505S)
- KOD-Plus-Neo (TOYOBO, Cat # KOD-401X5)
- In-Fusion HD cloning kit (Clontech/Takara Bio USA Inc., Cat # 639650)
- LR Clonase II (Thermo Fisher, Cat # 11791020)
- Agarose S (WAKO, Cat # 312-01193)
- Wizard® SV Gel and PCR Clean-Up System (Promega, Cat # A9281)
- LB media (Sigma, Cat # L7275-500TAB)
- Agar powder (for LB agar plate) (Nacalai Tesque, Cat # 01028-85)
- Ampicillin (Nacalai Tesque, Cat # 19769-64)
- Kanamycin (Nacalai Tesque, Cat # 19757-01)

2.2 Small-Scale Transfection by Lipofection

- Cell culture

  HEK293T cells or 293T cells (i.e., ATCC CRL-3216)

  Do not use HEK293 cells, as this cell line does not express SV40 large T antigen and does not permit episomal amplification of a plasmid DNA with SV40 replication origin. The 293FT cell line (Thermo Fisher), Lenti-X 293T cell line (Takara Bio), and Gesicle Producer 293T cell line (Takara Bio) may also be utilized.

  - TrypLE Select (1×) (Gibco, Cat #12563-011)
  - DMEM (High Glucose) (Nacalai Tesque, Cat # 08459-64)
  - Fetal Bovine Serum (FBS; biosera, Cat # FB-1365/500)

    Note: A lot check of the FBS should be performed before use for good cell growth, plasmid DNA transfection efficiency, and NanoMEDIC or lentivirus particle production. Avoid using inappropriate lots or expired lots.
  - D-PBS (−) (1×) (Nacalai Tesque, Cat # 14249-24)
  - Cell strainer 40-μm Nylon (Falcon, Cat # 352340)
- Transfection
  - Lipofectamine 2000 (Thermo Fisher, Cat #11668500)
  - Opti-MEM (Thermo Fisher, Cat # 31985070)

- Chemical supplement
  - A/C Heterodimerizer (AP21967) (Takara Clontech, Cat # Z5057N). The stock solution should be dissolved in ethanol to a concentration of 500 μM. Store in a shading tube (to protect from light) at −20 °C.
- DNA plasmids
  - pENTR-AmCyan (Addgene ID: 138481)

    The Gateway (GW) entry vector is used to insert the gRNA sequence flanked by two ribozyme elements. Subsequently, the insert sequences are transferred to a PL-5LTR-GW-A vector by the Gateway LR reaction.
  - PL-5LTR-GW-A (Addgene ID: 138480)

    The Gateway destination vector is used to express the ribozyme-flanked gRNA from the Gateway entry vector after the Gateway LR reaction.
  - pHLS-EF1a-FKBP12-Gag^{HIV}-A (Addgene ID: 138476)

    This construct expresses Gag protein of HIV but does not contain the Pol region, which include protease, reverse transcriptase, and integrase. FKBP12, chemically induced to dimerize with FRB domain, is fused to the N-terminus of the Gag protein.
  - pHLS-EF1a-FRB-SpCas9-A (Addgene ID: 138477)

    FRB, chemically induced to dimerize with FKBP12 domain, is fused to the N-terminus of SpCas9.
  - pcDNA3.1-Tat^{HIV} (Addgene ID: 138478)

    This plasmid encodes HIV Tat protein, which is important for driving the expression of gRNA from the HIV LTR promoter.
  - pVSV-G (Addgene ID: 138479)

    Glycoprotein from vesicular stomatitis virus (VSV-G) is known to bind to low-density lipoprotein receptor (LDLR) and induces endocytosis. VSV-G also induces the release of VLPs from early endosomes.
  - PL-5LTR-RGR(DMD#1)-AmCyan-A (Addgene ID: 138482)

    A positive control vector expressing ribozyme element-flanked gRNA. This gRNA sequence was designed to target the splicing acceptor of exon 45 in the human DMD (Dystrophin) gene.

2.3 Large-Scale Transfection by Flow Electroporation

- LV-MAX™ Lentiviral Production System (Thermo Fisher Scientific, Cat # A35684)

  Note: This kit contains the Viral Production Cells derived from 293F cells and the cells are adapted for suspension culture with LV-MAX™ Production Medium. Since 293F cells do not express SV40 large T antigen, it is recommended to transduce the cells with an integrating vector with SV40 large T antigen expression.
- LV-MAX™ Production Medium (Thermo Fisher Scientific, Cat # A3583401)
- Nalgene single-use PETG Erlenmeyer flasks with filtered caps, 1,000 mL (Thermo Fisher Scientific, Cat # 4115-1000)
- MaxQ 2000 CO_2 Plus, Digital CO_2 Resistant Shaker (Thermo Fisher Scientific, Cat#88881101)8–10% CO_2, 37 °C incubator for the orbital shaker
- MaxCyte STX instrument (Cat # STX)
- CL-2 RUO Processing Assembly (MaxCyte, Cat# SCL-2)
- MaxCyte Electroporation (EP) buffer, 500 mL (MaxCyte, Cat# EPB-5)
- 5 mL syringe with Luer-lock tip (Terumo, Cat # SS-05LZ)
- 50 mL syringe with Luer-lock tip (Terumo, Cat # SS-50LZ)
- Benzonase endonuclease, purity grade I (≥99%) (Merck Millipore, Cat# 71205-3CN)
- $MgCl_2$ solution (Stock solution: 500 mM. Final concentration: 5 mM)

2.4 Concentration by Centrifugation

- Colletion and concentration of NanoMEDIC for a small-scale production:
  - 10 mL Syringe with Luer-lock tip (Terumo, Cat # SS-10LZ)
  - 0.45 μm pore Minisart NML syringe filters (Sartorius, Cat # 16555K)
- Note: Do not use a pore size of 0.20 μm or smaller for the NanoMEDIC filtration, as it will significantly reduce the yield.
  - 15 mL, Open-top thin-wall ultra-clear open-top tube, 16 × 96 mm – 50Pk (Beckman Coulter, Cat # 361706).
  - Swinging-Bucket rotor JS-24.15 (6 × 15 mL, k-factor: 376, Beckman Coulter, Cat # 362396) or equivalent rotor. A swing rotor is recommended over an angled rotor, as it makes locating precipitated NanoMEDIC in the bottom of a centrifugation tube easier.
  - Avanti JXN-30 Centrifuge (Beckman Coulter) or equivalent ultra-centrifuge machine.
  - HBSS (Hanks' Balanced Salt Solution) (Thermo, Cat # 14175095).

- Harvest and concentration of NanoMEDIC for large-scale production.
  - Stericup Quick Release-HV Sterile Vacuum Filtration System, 500 mL process volume, 0.45 μm pore size, Low protein binding PVDF membrane (Merck Millipore, Cat # S2HVU05RE)
- Note: Do not use a pore size of 0.20 μm or smaller for the NanoMEDIC filtration, as it will significantly reduce the yield.
  - 38.5 mL, open-top thin-wall ultra-clear tube, 25 × 89 mm – 50Pk (Beckman Coulter, Cat # 344058)
  - Swinging-Bucket rotor JS-24.38 (6 × 38 mL, k-factor: 334, Beckman Coulter, Cat # 360743) or equivalent rotor.
  - Avanti JXN-30 Centrifuge (Beckman Coulter) or equivalent ultra-centrifuge machine.
  - HBSS (Hanks' Balanced Salt Solution) (Thermo, Cat # 14175095)

2.5 Inoculation

48-well or 96-well tissue culture plate

2.6 Assessing Genome Editing Efficiency by Sanger Sequencing and TIDE/ICE Analysis

- MonoFas genome DNA extraction kit (animos, Cat # A06-0002 A)
- PrimeSTAR GXL DNA polymerase (Clontech/Takara Bio USA Inc., Cat # R050B)
- Mupid-exU electrophoresis system (Mupid, Cat # EXU-1)
- TapeStation4200 DNA analyzer (Agilent Technologies, Cat # G2991BA)
- A PC with internet connection

3 Methods

3.1 Designing CRISPR-gRNA for NanoMEDIC Delivery

For genomic exon skipping, a splice acceptor or splice donor site of the target exon needs to be destroyed by indels induced by CRISPR-Cas9/gRNA. Hence, the gRNA target sequence should be designed around the splice acceptor or splice donor site. At the same time, we found some variations in genome editing and exon skipping efficiencies depending on the gRNA design. Therefore, multiple gRNAs should be tested for the optimal induction of exon skipping.

In general, avoid specific sequences that may form stable secondary structures, extreme (>70–80%) GC content, or polynucleotide stretches (i.e., more than five T's). Additionally, the number of possible off-target sites should be considered when designing gRNA. If possible, choose a gRNA sequence that resembles fewer sequences in the human genome to decrease the possibility of off-target editing. Web-based free search programs (such as CRISPOR, CRISPR direct, and so on) can calculate the number of potential off-target sites to exclude gRNA with highly similar off-target sites or too many potential off-target sites.

3.2 Construction of NanoMEDIC gRNA Vector (Fig. 2)

As shown in Fig. 2, two PCR reactions are necessary to construct a gRNA with a scaffold sequence flanked by two ribozyme sequences (ribozyme-gRNA-ribozyme: RGR). The final PCR product will be cloned into the pENTR-AmCyan vector digested by EagI to obtain pENTR-RGR-AmCyan. Lastly, the RGR cassette with AmCyan cDNA will be cloned into the PL-5LTR-GW-A vector by a LR Clonase II reaction to obtain the PL-5LTR-RGR-AmCyan-A vector.

Perform the first PCR reaction using Primers #1 and #2 with KOD-Plus-Neo polymerase or other high-fidelity DNA polymerases. This is an overlap extension PCR or fusion PCR reaction that links the gRNA target and gRNA scaffold sequences, so no other template DNA is needed.

Reagent	Volume (μL)
Ultrapure water	34
10× KOD -Plus -Neo Buffer	5
dNTPs	5
$MgCl_2$	3
Primer #1, 10 μM	1
Primer #2, 10 μM	1
KOD -Plus -Neo Enzyme	1
Total	50

Step (3 steps)	Temperature (°C)	Time (sec)	Cycle
Pre-denature	94	10	1
Denaturation Annealing Extension	98 60 68	10 30 10	35 cycles
Hold	4	∞	1

1st PCR product: 131 bp

NNNNNNNNNNNNNNNNNNNNGTTTTAGAG
CTATGCTGGAAACAGCATAGCAAGTTAAAATAAGG
CTAGTCCGTTATCAACTTGAAAAAGTGGCACCGAG
TCGGTGCTTTTTTTTTGAATTCAAACCCGGGC

1. Digest pENTR-AmCyan plasmid DNA with the EagI restriction enzyme for an hour. Run on a 1% agarose gel electrophoresis and cut out the 3,053 bp band. Perform a gel extraction and quantify the concentration of purified DNA.
2. Perform the second PCR reaction with the first PCR product and Primers #3 and #4. This step adds the ribozyme sequences on both ends of the gRNA.

Reagent	Volume (μL)
Ultrapure water	33.5
10× KOD -Plus -Neo Buffer	5
dNTPs	5
$MgCl_2$	3
Primer #3, 10 μM	1
Primer #4, 10 μM	1
PCR Product 1	0.5
KOD -Plus -Neo Enzyme	1
Total	50

Step	Temperature (°C)	Time (sec)	Cycle
Pre-denature	94	10	1
Denaturation Annealing Extension	98 60 68	10 30 10	35 cycles
Hold	4	∞	1

3. After the second PCR reaction, purify the PCR product by running on a 2% agarose gel and using a gel extraction kit.
4. Insert the second PCR product into EagI-digested pENTR-AmCyan plasmid DNA by In-Fusion reaction.

In-Fusion reaction	Volume (μL)
5× In-Fusion HD	1
PCR2 product	x (final 10 ng)
pENTR-AmCyan digest by EagI	x (final 50–100 ng)
Total	5

5. Incubate the reaction mixture at 50 °C for 15–60 min.
6. After the In-Fusion reaction, transform 1 μL of the reaction mixture with DH5α chemically competent *E. coli*.

7. The pENTR-RGR-AmCyan vector is Kanamycin resistant. Streak the transformed *E. coli* on an LB agar plate with Kanamycin for selection.
8. The following day, pick some *E. coli* colonies, incubate them in LB media with Kanamycin and culture overnight.
9. Extract plasmid DNA using a mini-prep kit and perform restriction enzyme digestion using BamHI. The expected band sizes are 3015 bp and 268 bp.
10. Additionally, perform Sanger sequencing to confirm the gRNA sequence using the AmCyan reverse primer (5′-GTGTGGAACTGGCATCTGTAGT-3′).
11. Once gRNA and the surrounding sequences (ribozyme part) are confirmed, perform a Gateway LR clonase reaction to insert the RGR construct in the pENTR-RGR-AmCyan vector into the PL-5LTR- GW-A vector to construct the PL-5LTR-RGR-AmCyan-A vector.

LR clonase reaction	Volume (μL)
TE buffer (pH 8.0)	Fill to 5 μL
PL-5LTR-GW-A	x (final 100 ng)
pENTR-RGR-AmCyan	x (final 100 ng)
5× LR ClonaseII reaction mix	1
Total	5

12. Incubate at 25 °C for 1 h.
13. Add 0.5 μL of proteinase K solution (2 μg/μL) and incubate for 10 min at 37 °C to stop the LR clonase reaction.
14. After the LR reaction, transform DH5α chemical competent cells with 1 μL of the reaction mixture.
15. The final product, PL-5LTR-RGR-AmCyan-A vector, is ampicillin resistant. Therefore, use a LB agar plate with ampicillin for the selection.
16. The following day, pick some *E. coli* colonies, inoculate them in 2-mL LB media with ampicillin, and culture overnight.
17. Extract plasmid DNA from each clone using a mini-prep kit and perform restriction enzyme digestion using BamHI. The expected band patterns are 5178 bp, 838 bp, and 268 bp.
18. Once successful cloning of the gRNA is confirmed, make a frozen *E. coli* stock.
19. Use an endotoxin-free Midi-prep to prepare high-purity plasmid DNA for the NanoMEDIC manufacturing (five kinds of plasmids, including the one with your gRNA). After

isopropanol precipitation of the plasmid DNA, use a smaller volume of nuclease-free ultrapure water to dissolve the plasmid DNA at a high concentration, ideally 2,000–3,000 ng/μL.

3.3 Lipofection Method for NanoMEDIC Production

Before starting the NanoMEDIC production, HEK293T cells should be cultured for at least one week after thawing. Passage every 2–3 days to make sure cells are at logarithmic growth phase.
Day 1: Seeding HEK293T cells for transfection.

1. Prepare semi-confluent HEK293T cells in a tissue culture-grade 10-cm dish.

   Note: Do not use over-confluent HEK293T cells, as transfection efficiency will be compromised.

2. Aspirate the media, wash the cells with 10-mL PBS (pre-warmed to room temperature), and aspirate again.
3. Add 1 mL of TrypLE Select (diluted one to five times in PBS) to the side wall of the plate and rock the plate gently to cover the cells with the solution.
4. Then aspirate the TrypLE solution immediately without disturbing the cells and place the plate in a 37 °C incubator for about 3 min.
5. Tap the plate after the incubation to make sure all the cells are detached from the plate.
6. Add 10 mL of culture medium and dissociate the cells by pipetting up and down against the bottom plate to ensure single cell separation.
7. If you find cell clumps are remaining, transfer the cell suspension to a 50 mL tube with a 40 μm cell strainer on top to remove the clumps.
8. Count the cell number and seed 3×10^6 cells per 10-cm plate.

   Note: The seeding density of HEK293T cells is critical for optimizing the transfection efficiency and achieving good NanoMEDIC production. If necessary, optimize the seeding cell density to make sure the cells are almost confluent at the time of the NanoMEDIC harvest.

Day 2 Evening: Transfection

1. On the following day, prepare a 5 mL Eppendorf tube (tube A) and 1.5 mL Eppendorf tube (tube B) per NanoMEDIC sample.
2. Add 1 mL of Opti-MEM media to both tube A and tube B.
3. For tube A, add 50 μL of lipofectamine 2000, mix two to three times gently and then incubate tube B for 5 min at room temperature.

   For tube B, add the following amount of each plasmid and mix:

(a) 10 μg pHLS-EF1a-FKBP12-GagHIV-A

(b) 10 μg pHLS-EF1a-FRB-SpCas9-A

(c) 2 μg pcDNA3.1-TatHIV

(d) 1 μg pVSV-G

(e) 10 μg PL-5LTR-RGR-AmCyan-A

4. Transfer the solution in Tube B to Tube A and mix up and down by pipetting a few times. Incubate the mixed solution (A&B) for 20 min at room temperature.
5. After incubation, add the mixed lipofection solution (A&B) to the HEK293T cells in a dropwise manner. Be careful not to dispense with too much force, otherwise the cells will detach.
6. Incubate the cells in a 37 °C, 5% CO_2 incubator overnight.

Day 3 (Morning Time): Media Change

1. Change the media the next morning (about 15–18 h after the transfection).
2. Pre-warm an aliquot of DMEM + 10% FBS media (12 mL for one 10 cm dish) at 37 °C in water-bath to minimize cold shock to the producer cells.
3. Add AP21967 to the pre-warmed media at a final concentration of 300 nM. Discard the old media containing the lipofectamine mix and add the pre-warmed fresh media with 300-nM AP21967 (for 10 mL of media, add 6 μL of 500-μM AP21967 solution).

   Note: Add AP21967 to the pre-warmed media right before the medium change. The AP21967 solution should be kept on a cold block, because the reagent is in 100% ethanol and prone to evaporation. Spin down the tube before use.
4. Incubate in a 37 °C, 5% CO_2 incubator until harvesting.

   Note: After transfection of the NanoMEDIC plasmids, NanoMEDIC particles are produced from the producer cells. Although NanoMEDIC particles are not pathogenic, they should be treated as infectious material under proper biosafety procedures. For example, use a biosafety cabinet, not clean bench. Also use double gloves when handling NanoMEDIC and discard the outer gloves regularly. All the waste should be disinfected by autoclaving, hypochlorous acid treatment, or 70% ethanol wiping.

Day 4 (Evening Time): Harvest NanoMEDIC

1. The next evening (2 days from the transfection), harvest NanoMEDIC from a 10 cm dish by using a 10 mL syringe with a Luer lock. Pull the plunger to insert 1–2 mL of air into the

syringe before drawing up the NanoMEDIC solution. This will help push the NanoMEDIC solution from the syringe completely at a later step.

2. Attach a 0.45 μm syringe filter to the Luer lock syringe. Note: For larger batches of NanoMEDIC (several 10 cm dishes or more), a vacuum filter unit can also be used to remove cellular debris from the NanoMEDIC solution.
3. Dispense the syringe into a clean 15 mL open-top thin-wall ultra-clear centrifugation tube (Beckman). Make sure to fill more than 10 mL of the 15-mL tube, otherwise the tube might crash during the centrifugation. Note: If immediate centrifugation is not feasible, you may transfer the content to a 15 mL or 50 mL conical tube with a screw cap and store at 4 ° C for a few days. Do not freeze at this stage, as serum protein will precipitate after the freeze-thaw cycle. A portion of unconcentrated NanoMEDIC can be set aside at 4 °C to check the efficiency of the concentration if necessary.
4. Centrifugation rotor and buckets should be cooled inside of a centrifuge machine or in a refrigerator at 4 °C. Before the centrifugation, pre-centrifuge with a low speed at 4 °C for around 30 min to cool the inside of the machine.
5. Adjust the balance between a pair of 15 mL tubes using a scale or a balance. Place the tube into a bucket and close the screw cap. It is recommended to apply grease on the screw cap so that it turns smoothly.
6. Set the 15 mL tubes into chilled buckets and carefully set onto the swing rotor.

   Note: If the buckets are not properly set onto the rotor, they may fall during the centrifugation, which risks a serious accident. Read the manual of the centrifuge machine and rotor before the experiment and be familiar with the procedures.
7. Centrifuge the harvested NanoMEDIC solution at 8,800 × *g* for at least 16 h at 4 °C to spin down NanoMEDIC.

   Note: NanoMEDIC can be centrifugated at a higher centrifugation speed (Max RCF 92,000 × *g*) for 2 h at 4 °C, similar to the VSV-G pseudotyped lentiviral vector. However, we observed better yields of NanoMEDIC with low-speed overnight centrifugation.

Day 4 (Morning Time): Resuspension of NanoMEDIC

1. The following day, stop the centrifugation and carefully take out the tubes from the buckets. Do not disturb the tubes, because the NanoMEDIC pellet is soft and fragile. Gently aspirate the supernatant using a 10 mL pipette and directly discard into a waste bottle with bleach.

2. Tilt the tube and aspirate any remaining media on the side wall of the tube using a 1,000 μL pipette. You might see a white pellet at the bottom of the centrifuge tube. This pellet is mainly composed of cellular debris, and NanoMEDIC itself is too small to see. Hence, there is no problem if you see a clear pellet so long as you can locate the position of the NanoMEDIC pellet at the bottom of the tube (in the case of an angled rotor, the pellet should be located on the side wall). If necessary, mark the location of the pellet by a pen.
3. Add 100 μL of chilled HBSS (PBS can be used alternatively) to the pellet. Make sure the entire pellet is soaked by HBSS solution. Cover the tube with parafilm and incubate the standing tube at 4 °C for 2–4 h to loosen the pellet. You may leave the tube overnight if desired.
4. Using a 200-μL pipette set at 50–80 μL volume, suspend the precipitated NanoMEDIC by pipetting up and down at least 20 times. Make sure to wash the side walls, since NanoMEDIC is distributed around the pellet, although it is not visible. Try not to introduce bubbles. The total volume of NanoMEDIC per tube usually ends up being 110–120 μL with some residual media in the tube.
5. Transfer the suspension into several screw cap tubes (i.e., 10–60 μL per tube). If used within a week, store at 4 °C. If not, freeze at −80 °C or −150 °C in a deep freezer (do not use a liquid nitrogen tank for biosafety). Avoid the freeze-thaw cycle more than two times, as this can significantly compromise infectivity.

3.4 Large-Scale Suspension Culture and Flow Electroporation Method

Day 1: Preparation of suspension cells for flow electroporation

1. Prepare LV-MAX Viral Production cells as a suspension culture with LV-MAX Production Medium, as described in the LV-MAX Lentiviral Production System manual. Alternatively, HEK293T cells might be adapted for the suspension culture with LV-MAX Production Medium by gradually switching the media composition.
2. Passage LV-MAX Viral Production cells the day before the electroporation to a density of 1.5–2 × 10^6 cells/mL in two 1000 mL Erlenmeyer flasks. The culture volume should be around 250 mL per flask. For passaging, it is not necessary to centrifuge the cells and aspirate the media. Instead, simply diluting the cells in fresh LV-MAX Production Medium is sufficient.
3. During the cell culture at 37 °C in 5% CO_2, the flask should be shaken at the suggested rpm based on the LV-MAX instruction manual. We used 100 rpm for a 19 mm shaking diameter.

4. The cell number should double by the following day to 3–4 × 10^6 cells/mL. For a total culture of 500 mL, the total cell number should reach 1.5–1.6 × 10^9 cells.

Day 2: Flow Electroporation (Fig. 3)

1. Centrifuge LV-MAX Viral Production cells at 200 × *g* for 10 min with 50 mL tubes (total 20 tubes). Larger centrifuge tubes can be used if available.
2. Aspirate the media or decant the centrifuge tube. Be careful not to disrupt the cell pellet.
3. Resuspend 1.4–1.5 × 10^9 cells in 100 mL of MaxCyte EP buffer for washing. 1–2 × 10^9 cells can be resuspended in 50–100 mL of MaxCyte EP buffer to ensure sufficient buffer exchange. If serum is present, then the cells must be washed two times.
4. Take an aliquot of the cells and dilute them in MaxCyte EP buffer at a 1:10 ratio. Count the cells using a cell counter. The cell viability should be greater than 95% by trypan blue staining.
5. Centrifuge the cells at 200 × *g* for 10 min.
6. Aspirate the MaxCyte buffer or decant the centrifuge tube. Be careful not to disrupt the cell pellet.
7. Resuspend the cell pellet in MaxCyte buffer to a cell density of 0.8–1.0 × 10^8 cells/mL. The volume of the cell pellet should be considered when adding MaxCyte buffer to the final volume. For 1.2 × 10^9 cells, resuspend the cells in a total volume of 12.8 mL MaxCyte buffer.
8. Prepare five plasmid DNAs as listed below. The total DNA concentration in the final electroporation volume is 268 μg/mL. The concentration of each plasmid is listed below as well as the amount of DNA to add for 12.8 mL of electroporation volume. Mix the cell mixture with the plasmid DNAs well using a serological pipette at least ten times. This is important, because the DNA plasmids may be viscous and a homogenous mixture of cells and DNA is desired.
   (a) 1,038 μg PL-5LTR-RGR-AmCyan-A (gRNA expressing vector); 81 μg/mL
   (b) 1,038 μg pHLS-EF1a-FKBP12-GagHIV-A; 81 μg/mL
   (c) 1,038 μg pHLS-EF1a-FRB-SpCas9-A; 81 μg/mL
   (d) 208 μg pcDNA1-TatHIV; 16 μg/mL
   (e) 51 μg pVSV-G; 4 μg/mL
9. Spray the CL-2 Processing Assembly package with 70% ethanol and place the package inside a biosafety cabinet. There are three labeled bags connected to the CL-2 Processing Assembly: "Sample Bag," "Collection Bag," and "Air Bag."

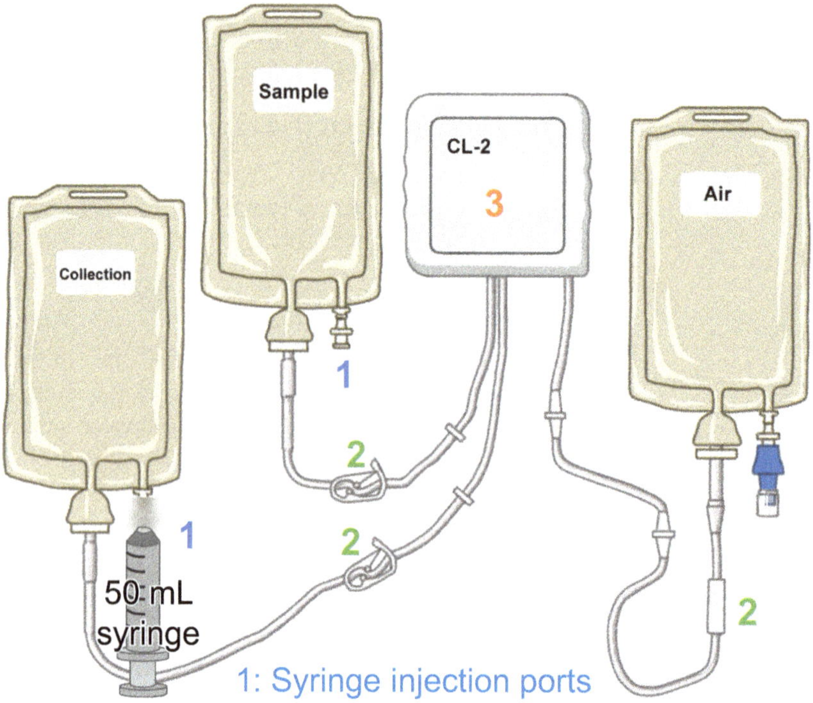

Fig. 3 MaxCyte CL-2 Processing Assembly for flow electroporation and instrument setup for large-scale production of NanoMEDIC. The MaxCyte CL-2 Processing Assembly is designed for flow electroporation and made up of several components connected by tubing. There is one main electroporation chamber and three bags: a Sample Bag for adding cells and DNA before the electroporation; a Collection Bag where the cells will flow after the electroporation; and an Air Bag, which is used by a peristaltic pump to push the sample from the Sample bag to the electroporation chamber and from the electroporation chamber to the Collection Bag. The Sample Bag and Collection Bag each have a syringe injection port (No. 1 in blue) to connect syringes to load samples or to draw samples out, respectively. The tubing connecting each bag has clamps (No. 2 in green) that remain closed while loading the cells and DNA plasmids into the Sample Bag and are opened for the electroporation. After the electroporation is complete, the clamps are closed to prevent the electroporated cells from flowing back into the tubing

10. Open the package and wipe down the plunger connection ports of the sample and collection bags with 70% ethanol.
11. Close the clamps for the Sample Bag and Collection Bag to prevent samples from leaking after loading.
12. Remove a plunger from a 50 mL Luer lock syringe and insert the barrel into the Sample Bag.
13. Position the bag and syringe vertically with the open end of the syringe barrel pointed up.

14. Pour the cell and DNA mixture into the syringe and allow the mixture to flow into the Sample Bag by gravity. No need to re-insert the plunger as it is not stable.
15. Once all the cells loaded, remove the syringe barrel from the Sample Bag.
16. Using a new 5 mL Luer lock syringe, pull the plunger back so that approximately 5 mL of air is taken up into the syringe.
17. Connect the new 5 mL Luer lock syringe to the Sample Bag via the syringe injection port. The 5 mL of air can be pushed into the sample bag to push any residual sample remaining in the syringe injection port and prevent the cells in the Sample Bag from sticking to the surface. The syringe from the Sample Bag can be removed, and the syringe connection port will automatically reseal. Alternatively, the syringe can remain attached to the Sample Bag connection port during electroporation. The presence of the syringe does not affect the cells or the electroporation.

    Note: An alternative option to loading the cell and DNA mixture into the Sample Bag is to attach a blunt cannula (or blunt 17 Gauge syringe) to the 50 mL Luer lock syringe and take up the cell and DNA mixture with 5 mL of air. Remove the blunt cannula, attach the Luer lock syringe to the Sample Bag and inject the cell and DNA mix altogether with 5 mL of air into the Sample Bag. The 50 mL syringe can be removed from the Sample Bag or remain attached to the connection port.
18. Prepare the Collection Bag by inserting approximately 5 mL of air and attaching a 50 mL syringe for correcting the electroporated cells later. Using a new 50 mL Luer lock syringe, pull the plunger inside a biosafety cabinet. Pull the plunger back so that approximately 5 mL of clean sterilized air is taken up into the syringe. Then connect the syringe into the plunger connection port of the Collection Bag (not the Sample Bag) and push in the 5 mL of air into the bag. Keep the syringe attached to the Collection Bag, as the syringe will be used to collect the electroporated cells from this bag.

    Note: 5 mL of air prevents the cells from sticking to the bag and the volume does not have to be precise.
19. Carry the sample-loaded CL-2 Processing Assembly, with one 50-mL syringe attached the Collection Bag and an optional syringe attached to the Sample bag, to the MaxCyte STX instrument.
20. Turn on the STX laptop and log into the computer.
21. Open the STX software by clicking the icon on the desktop.
22. Move the electroporation port switch on the STX to open the ports for the CL-2 and close the static electroporation port.

23. Place the Sample Bag in the top left tray.
24. Place the Collection Bag with 50 mL syringe on the top right tray.
25. Place the bag labeled Air Bag in the bottom left holder of the MaxCyte STX.
26. In the STX software, select Optimization Energy 9 protocol.

    Note: You may use Optimization Energy 8 protocol, if you observe significant cell damage.
27. Next, select CL-2 from the processing assembly from the drop-down menu.
28. Follow the step-by-step instructions to load the tubing for the Sample Bag, Collection Bag, and Air Bag into the correct positions.
29. Open the clamps on the Sample Bag, Collection Bag, and Air Bag.
30. Prime the tubing by pressing the "Prime" button in the STX software to fill the line with cells from the Sample Bag to just before the electroporation port.
31. Press the "Start" button for the electroporation.
32. The instrument will charge energy for 40–60 s and afterward start to electroporate the cells. 100 mL of concentrated cells and DNA can be electroporated in less than 30 min.
33. After the electroporation is finished, press the "Empty" button in the MaxCyte software, which will push any residual cells in the tubing from the electroporation chamber to the Collection Bag.
34. Close all clamps to keep the cells in the Collection Bag.
35. Detach the tubing of the CL-2 from the STX and then carry the CL-2 Processing Assembly to the clean bench.
36. Tilt the Collection Bag at an angle with the Plunger connection port and 50 mL syringe pointing down so that the cells will collect near the port with the syringe. Slowly pull back the 50 mL plunger of the syringe attached to the Collection Bag to draw up the cells as much as possible. It is relatively easy to recover the cells and unnecessary to wash the Collection Bag with MaxCyte buffer to recover any residual cells.
37. Transfer the cells from the syringe into a 50 mL centrifuge tube.
38. Add 64 units/mL of Benzonase to the tube and $MgCl_2$ to a final concentration of 5 mM. **Note**: A final $MgCl_2$ concentration of 1–2 mM is also feasible.
39. Incubate the centrifuge tube at 37 °C for 40 min to remove excess untransfected plasmid DNA by benzonase treatment and assist with the membrane recovery after electroporation.

40. After incubation, resuspend the electroporated cells with 480 mL of LV-MAX Production Medium (pre-warmed to 37 °C) at a density of 2.5×10^6 cells/mL. Divide 480 mL of the resuspended cells into two 1,000 mL Nalgene single-use PETG Erlenmeyer flasks with filtered caps; each flask will contain 240 mL of the resuspended cells.
41. Add 144 μL of 500-μM AP21967 solution to each flask (240-mL media) at a final concentration of 300 nM.
42. Place the Erlenmeyer flasks in the 37 °C 5% CO_2 incubator on an orbital shaker set at 100 rpm shaking speed (for MaxQ 2000 CO_2 Plus, Digital CO_2 Resistant Shaker, orbit: 19 mm).

Day 4 (Evening Time): Harvest and concentration by centrifugation

1. Two days after electroporation, transfer the cells and culture media into centrifuge tubes and centrifuge at 200 × *g* for 10 min to pellet producer cells.
2. Transfer the supernatant (containing NanoMEDIC) to new centrifuge tubes and centrifuge at 500 × *g* for 10 min in order to remove any debris.
3. Transfer the supernatant that contains NanoMEDIC to a 0.45-μm pore vacuum filter to further remove cell debris. Note: After centrifugation and filtration, approximately 450–460 mL of the original 480 mL is remaining.
4. Transfer 38 mL of the NanoMEDIC solution into a clean 38.5 mL open-top thin-wall ultra-clear centrifugation tube (Beckman). With the SJ-24.38 rotor, up to six tubes can be processed at once (38 mL × 6 = 228 mL). Only half of the NanoMEDIC supernatant can be concentrated at one time with the above rotor, however. Store the remaining NanoMEDIC-containing supernatant at 4°C until the next day for further centrifugation. Do not freeze NanoMEDIC at this stage. A small aliquot of unconcentrated NanoMEDIC can be set aside at 4°C for checking the efficiency of the concentration if necessary.
5. Centrifugation rotor and buckets should be cooled inside of the centrifuge machine or in a refrigerator at 4°C. Before the centrifugation, pre-centrifuge at 8,800 × *g*, 4 °C for around 30 min to cool the inside of the centrifuge machine.
6. Adjust the balance between a pair of 38.5 mL tubes using a scale or a balance. Place the tube into the bucket and screw the cap shut. It is recommended to apply grease on the cap so that it turns smoothly.
7. Set the 38.5 mL tubes into chilled buckets and carefully set the buckets onto the swing rotor. Note: If the buckets are not properly set onto the rotor, they may fall during the

centrifugation, which risks a serious accident. Read the manual of the centrifuge machine and rotor before the experiment and be familiar with the procedures.

8. Centrifuge the harvested NanoMEDIC solution at 8,800 × *g* for at least 16 h at 4 °C to spin down NanoMEDIC.
9. After the centrifugation, discard the supernatant and resuspend the NanoMEDIC pellet using 80–100 μL of chilled HBSS per tube. Make sure the entire pellet is soaked by HBSS solution. Cover the tube with parafilm and incubate the standing tube at 4 °C for 2–4 h to loosen the pellet. You may leave the tube overnight if desired.
10. Using a 200 spi1;μL pipette set at 50–80 μL volume, suspend the precipitated NanoMEDIC by pipetting up and down at least 20 times. Make sure to wash the side walls, since NanoMEDIC is distributed around the pellet, although it is not visible. Try not to introduce bubbles. The total volume of NanoMEDIC per tube usually ends up being 110–120 μL with some residual media in the tube.
11. After finishing the concentration of the first batch of NanoMEDIC particles, repeat steps 5–10 for the remaining NanoMEDIC supernatant from step 4.
12. At the end, you should be able to harvest around 1,200–1,400 μL of NanoMEDIC solution from the 12 centrifugation tubes.

3.5 Inoculation of NanoMEDIC

1. Prepare target cells for NanoMEDIC inoculation one day before the experiment. If you would like to use fresh NanoMEDIC, seed the target cells on the day of the centrifugation. To check the infectivity of NanoMEDIC, we recommend using HEK293T cells as a control, in addition to your target cells of interest.
2. Seed 50,000 cells per well for a 48-well plate. Make sure to prepare an additional well for the non-infection control. The seeding cell density can be optimized for different cell types.
3. Add NanoMEDIC solution into one of the seeded wells. Usually, 10 μL of concentrated NanoMEDIC should be enough to detect robust genome editing via NHEJ without severe cellular toxicity. Unlike the lentiviral vector, the addition of polybrene is not recommended for NanoMEDIC inoculation.
4. Incubate the cells at 37 °C in a 5% CO_2 incubator.
5. Change the media the next day (at least more than 12 h post-inoculation) if necessary.
6. Culture the cells 2–3 days or longer. Harvest the cells by trypsinization. If desired, some portion of the cells can be expanded for downstream applications or as a backup.

3.6 Assessing Genome Editing Efficiency by Sanger Sequencing and TIDE/ICE Analysis (Fig. 4)

1. Extract genomic DNA from the harvested cells (NanoMEDIC inoculated and non-inoculated control) using a genomic DNA extraction kit (i.e., MonoFas).
2. PCR amplify the target region of the gRNA. Make sure to PCR amplify the non-inoculated control as well. Run the PCR products on an agarose gel and extract the expected DNA band.
3. Perform Sanger sequencing of the PCR product. If NanoMEDIC-mediated genome editing is successful, the electropherogram of the sequence shows a disturbance from the Cas9 cutting site, typically 3–4 bp upstream from the PAM sequence.

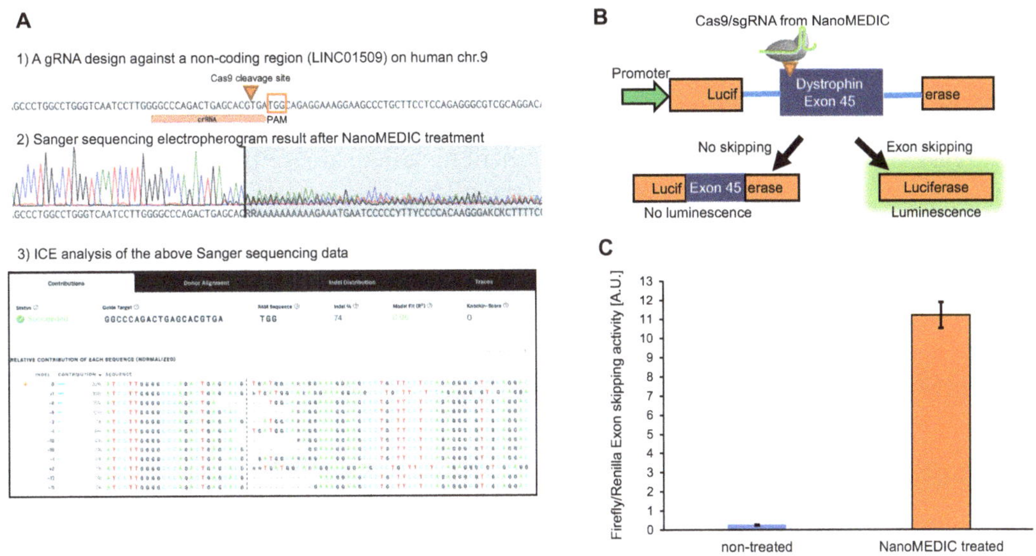

Fig. 4 Expected outcome of indel analysis and DMD exon skipping activity measurement. (**a**) NanoMEDIC packaging gRNA against a non-coding region (LINC01509) on human chr.9 and SpCas9 were inoculated to HEK293T cells. Three days after the inoculation of 20 μL of concentrated NanoMEDIC, genomic DNA was harvested, and Sanger sequencing was performed. The electropherogram of the Sanger sequencing showed a disturbance at the Cas9 cleavage site. To quantify the indel percentage of the Sanger sequencing data, ICE analysis was performed using ab1 electropherogram files from NanoMEDIC-treated and non-treated samples. The ICE analysis showed 74% of indels at the target locus. (**b**) To detect the exon skipping activity induced by Cas9/gRNA packaged into NanoMEDIC, we constructed an exon skipping reporter vector based on the luciferase assay system. In this reporter vector, firefly luciferase cDNA is split into two exons, and exon 45 from human Dystrophin gene is inserted between them. If exon 45 is successfully skipped by Cas9/gRNA, full-length firefly luciferase protein is expressed, and the exon skipping efficiency can be detected as luminescence. For one well of a 96-well plate, we seeded 50,000 cells per well, and the exon skipping reporter vector (pPV-EF1a-Luc2(V323I)-hDMD-Ex45[+]-iP-A, 100 ng) and Renilla luciferase vector (pRL-TK, 20 ng) were co-transfected into the cells using Lipofectamine 2000 reagent (1 μL). On the next day, 20 μL of concentrated NanoMEDIC solution packaged with the gRNA targeting the exon 45 on the human dystrophin gene was added into the well. (**c**) A luciferase assay was performed 3 days after the NanoMEDIC inoculation. The Firefly luciferase value was divided by the Renilla luciferase value to normalize the transfection efficiency. The NanoMEDIC-treated well showed much higher Firefly/Renilla Exon skipping activity compared to the non-treated well

4. For detailed information of the indel pattern and proportion, TIDE or ICE analysis can be performed using the ab1 files from the Sanger sequencing.

   TIDE: https://tide.nki.nl/

   ICE: https://ice.synthego.com/#/

4 Notes

4.1 CRISPR Genome Editing Target

The target DNA sequence editable by CRISPR-Cas9 can also be targeted by NanoMEDIC editing. If editing of your gRNA by NanoMEDIC delivery is not observed (but control gRNA is observed), check the DNA sequence to make sure your gRNA is active in producer HEK293T cells. If it is not active, design a new gRNA. If it is active, double check the NanoMEDIC production procedures. In rare cases, the target sequence interferes with the secondary structure of the ribozyme, such that the gRNA target may have difficultly functioning in NanoMEDIC. If so, it is recommended to design a new gRNA.

4.2 Condition of Producer Cells

1. The efficiency of the NanoMEDIC production directly depends on the condition of the producer cell (HEK293T or LV-MAX Viral Production cells). Do not conduct the experiment unless the producer cells are healthy.
2. To keep the producer cells healthy, be careful with the daily cell handling. Passage every 2–3 days at the semi-confluent stage (when there is still some space between cells). Never overgrow the cells, as this would significantly reduce the transfection efficiency and NanoMEDIC productivity. Uneven seeding of the cells or insufficient cell dissociation could also cause regional overgrowth, which should be avoided. It is important to add solution (i.e., PBS or DMEM) gently and slowly along the side wall of the dish, as HEK293T cells are easy to detach. Cold PBS may cause cell detachment as well. Add enough PBS for washing to remove residual FBS before trypsinization. Perform enough trypsinization and suspend well to avoid cell clumps. After seeding onto a dish or plate, make sure the cells are distributed evenly. Do not touch the seeded dish or plate until the cell attach.

4.3 Transfection (Quality of Plasmid)

The purity and quality of the plasmids can largely affect the transfection efficiency and cell viability of the producer cells. Therefore, an endotoxin-free midi- or maxi-prep kit is recommended for the plasmid DNA purification. At the final step of the plasmid DNA purification, TE buffer is recommended to elute the plasmid DNA pellet in some purification kits. However, we found TE buffer reduces the transfection efficiency of the electroporation, presumably due to the chelate effect of EDTA. Therefore, we recommend eluting plasmid DNA in nuclease-free ultrapure water and maintaining a high DNA concentration (i.e., 2,000–3,000 ng/μL) to minimize carrying over of impurities (including EDTA) to the electroporation.

- Transfection scale(production scale)

3. When no concentration is needed, NanoMEDIC can also be produced in 6- or 12-well plates. In this case, the amount of plasmid DNA and reagent for the NanoMEDIC production needs to be calculated according to the surface ratio. Although the total amount of NanoMEDIC produced in 6- or 12-wells is less than in a 10 cm dish and the NanoMEDIC titer can be low (i.e., 100 times) compared with concentrated titers, you may test more than six kinds of NanoMEDIC side by side without the concentration step.

4.4 Alternative Concentration Methods

For the NanoMEDIC concentration, there are other two methods you may incorporate: centrifugation at higher speed (93,000 × *g*) for 3 h or applying a Lenti-X concentrator (Clontech, Cat# 631231).

- 3 h centrifugation at 93,000 × *g*

This is a common concentration condition for VSV-G pseudotyped lentiviral vectors. With this high speed but short time, we find the pellet is much tighter, so the pellet needs to be dissolved more thoroughly. In our hands, the VLP yield was better with a lower speed centrifugation, presumably due to the gentler condition.

- Lenti-X concentrator

Lenti-X is a PEG-based precipitation method for nanoparticles. For 9 mL of filtered NanoMEDIC solution from a 10 cm plate, add 3 mL of chilled Lenti-X Concentrator. Invert several times to mix well and incubate at 4 °C for 30 min to overnight. Centrifuge at >1,500 × *g* for 45 min at 4 °C. Aspirate the media and resuspend the pellet in HBSS or PBS. For inoculation, use multiple doses to avoid cellular toxicity due to the contaminated Lenti-X reagent.

4.5 Storage of NanoMEDIC

- Storage for a few days

If you will use NanoMEDIC within a few days, store it at 4 °C and do not freeze.

- Storage for more than 1 week

If you will store NanoMEDIC for more than 1 week, freeze it at −80 °C. It is recommended to aliquot the NanoMEDIC into several tubes at an appropriate volume (i.e., 30 μL each) to avoid multiple freeze-thaw cycles.

4.6 *In Vivo* Application

We directly injected 50 μL of concentrated NanoMEDIC into the gastrocnemius muscle of luciferase exon skipping reporter mice (17). By delivering two gRNAs (one targeting *DMD* exon45 splice acceptor and another targeting *DMD* exon45 splice donor), we observed 6.9 ± 1.4% exon skipping.

4.7 Other

- Confirmation of the packaging of Cas9 protein

Western blotting can be utilized to confirm Cas9 protein packaging into NanoMEDIC. Add a mixture of 4× Laemmli sample buffer and DTT (final concentration of DTT is 50 mM) to NanoMEDIC and heat inactivate at 95 °C for 5 min to prepare the loading sample for western blotting. Use an anti-Cas9 antibody (Active motif, Cat # 61577) or anti-DmrC antibody (TaKaRa bio, Cat # 635091) for the detection. Anti-HIV p24 antibody (Abcam, Cat # ab9071) or anti-VSVG antibody (Merck, Cat # V5507-2ML) can be used as the positive control.

Acknowledgments

We are grateful to Dr. Yusuke Kojima, Dr. Peter Karagiannis, Dr. Eman Taha, and Joseph Lee for critical reading of the manuscript. This research was supported in part by AMED grants (JP19ek0109293, JP21bm0104001, JP21bm0804005, JP21im0210115), the T-CiRA Join Research program by Takeda Pharmaceutical Company, and an Intramural Research Grant (No. 28-6) for Neurological and Psychiatric Disorders of NCNP.

References

1. Jinek M, Chylinski K, Fonfara I, Hauer M, Doudna JA, Charpentier E A programmable dual-RNA-guided DNA endonuclease in adaptive bacterial immunity. Science 337:816–821
2. Knott GJ, Doudna JA CRISPR-Cas guides the future of genetic engineering. Science 361: 866–869
3. Li HL, Fujimoto N, Sasakawa N, Shirai S, Ohkame T, Sakuma T et al (2015) Precise correction of the dystrophin gene in duchenne muscular dystrophy patient induced pluripotent stem cells by TALEN and CRISPR-Cas9. Stem Cell Rep 4(1):143–154
4. Long C, Amoasii L, Mireault AA, McAnally JR, Li H, Sanchez-Ortiz E et al (2016) Postnatal genome editing partially restores dystrophin expression in a mouse model of muscular dystrophy. Science 351(6271):400–403
5. Nelson CE, Hakim CH, Ousterout DG, Thakore PI, Moreb EA, Castellanos Rivera RM et al (2016) *In vivo* genome editing improves muscle function in a mouse model of Duchenne muscular dystrophy. Science 351(6271):403–407
6. Amoasii L, Hildyard JCW, Li H, Sanchez-Ortiz E, Mireault A, Caballero D et al (2018)

Gene editing restores dystrophin expression in a canine model of Duchenne muscular dystrophy. Science 362(6410):86–91

7. Buchlis G, Podsakoff GM, Radu A, Hawk SM, Flake AW, Mingozzi F et al (2012) Brief report Factor IX expression in skeletal muscle of a severe hemophilia B patient 10 years after AAV-mediated gene transfer. Blood 119: 3038–3041
8. Fu Y, Foden JA, Khayter C, Maeder ML, Reyon D, Joung JK et al (2013) High-frequency off-target mutagenesis induced by CRISPR-Cas nucleases in human cells. Nat Biotechnol 31(9):822–826
9. Nelson CE, Wu Y, Gemberling MP, Oliver ML, Waller MA, Bohning JD et al (2019) Long-term evaluation of AAV-CRISPR genome editing for Duchenne muscular dystrophy. Nat Med 25(3):427–432
10. Xu H, Wang B, Ono M, Kagita A, Fujii K, Sasakawa N et al (2019) Targeted disruption of HLA genes via CRISPR-Cas9 generates iPSCs with enhanced immune compatibility. Cell Stem Cell 24(4):566–578.e7
11. Kim S, Kim D, Cho SW, Kim J, Kim JS (2014) Highly efficient RNA-guided genome editing in human cells via delivery of purified Cas9 ribonucleoproteins. Genome Res 24(6):1012–1019
12. Kagita A, Lung MSY, Xu H, Kita Y, Sasakawa N, Iguchi T et al (2021) Efficient ssODN-mediated targeting by avoiding cellular inhibitory RNAs through precomplexed CRISPR-Cas9/sgRNA ribonucleoprotein. Stem Cell Rep 16(4):985–996
13. Choi JG, Dang Y, Abraham S, Ma H, Zhang J, Guo H et al (2016) Lentivirus pre-packed with Cas9 protein for safer gene editing. Gene Ther 23(7):627–633
14. Mangeot PE, Risson V, Fusil F, Marnef A, Laurent E, Blin J et al (2019) Genome editing in primary cells and *in vivo* using viral-derived Nanoblades loaded with Cas9-sgRNA ribonucleoproteins. Nat Commun 10(1):1–15
15. Montagna C, Petris G, Casini A, Maule G, Franceschini GM, Zanella I et al (2018) VSV-G-enveloped vesicles for traceless delivery of CRISPR-Cas9. Mol Ther Nucleic Acids 12: 453–462
16. Campbell LA, Coke LM, Richie CT, Fortuno LV, Park AY, Harvey BK (2019) Gesicle-mediated delivery of CRISPR/Cas9 ribonucleoprotein complex for inactivating the HIV Provirus. Mol Ther 27(1):151–163
17. Gee P, Lung MSY, Okuzaki Y, Sasakawa N, Iguchi T, Makita Y et al (2020) Extracellular nanovesicles for packaging of CRISPR-Cas9 protein and sgRNA to induce therapeutic exon skipping. Nat Commun 11(1):1–18
18. Putyrski M, Schultz C (2012) Protein translocation as a tool: the current rapamycin story. FEBS Lett 586:2097–2105
19. Bayle JH, Grimley JS, Stankunas K, Gestwicki JE, Wandless TJ, Crabtree GR (2006) Rapamycin analogs with differential binding specificity permit orthogonal control of protein activity. Chem Biol 13(1):99–107
20. Yoshioka S, Fujii W, Ogawa T, Sugiura K, Naito K (2015) Development of a mono-promoter-driven CRISPR/Cas9 system in mammalian cells. Sci Rep 5:1–8
21. Ferré-D'Amaré AR, Scott WG (2010) Small self-cleaving ribozymes. Cold Spring Harb Perspect Biol 2:a003574
22. Nikolic J, Belot L, Raux H, Legrand P, Gaudin Y, Albertini AA (2018) Structural basis for the recognition of LDL-receptor family members by VSV glycoprotein. Nat Commun 9(1):1–12

Chapter 23

Restoration of Dystrophin Expression in Mdx-Derived Muscle Progenitor Cells Using CRISPR/Cas9 System and Homology-Directed Repair Technology

Yue Jin, Yan Shen, Il-man Kim, Neal L. Weintraub, Mark Hamrick, and Yaoliang Tang

Abstract

Duchenne muscular dystrophy (DMD) is a progressive myopathy caused by mutations in genes encoding dystrophin proteins that ultimately lead to depletion of myogenic progenitor cells (MPCs). Several approaches have been used to correctly express the *dystrophin* gene in induced pluripotent stem cells (iPSCs), including deletion of mutated exon 23 (ΔEx23) by clustered regularly interspaced short palindromic repeat (CRISPR)/CRISPR-associated gene 9 (Cas9)-mediated gene editing technology. However, this approach is labor-intensive due to individual colony picking and genotyping to verify allelic modification. Here, we present a protocol to restore the function of the *dystrophin* gene by using homology-directed repair (HDR)-based CRISPR/Cas9 and inducing myogenic program of reprogrammed iPSCs from Mdx mice by inducible muscle-specific transcription factor MyoD.

Key words Duchenne muscular dystrophy, iPSC-derived myoprogenitors, CRISPR-Cas9-HDR gene editing

1 Introduction

Duchenne muscular dystrophy (DMD) is a progressive myopathy caused by mutations in genes encoding dystrophin proteins that ultimately lead to depletion of myogenic progenitor cells (MPCs) [6, 8]. Several approaches have been used to correct *dystrophin* gene expression in iPSCs [5], including deletion of mutated exon 23 (ΔEx23) by clustered regularly interspaced short palindromic repeats (CRISPR)/CRISPR-associated 9 (Cas9) gene-editing techniques [2]. However, this approach is labor- intensive due to inefficiency. Here, we designed two guide RNAs (gRNAs) to flank mutated exon23 after Cas9-mediated double-stranded breaks (DSBs). Next, we incorporated the homology arms at the 5′- and 3′-ends of the cleavage site into a floxed elongation factor-1 (EF-1)

Rika Maruyama and Toshifumi Yokota (eds.), *Muscular Dystrophy Therapeutics: Methods and Protocols*, Methods in Molecular Biology, vol. 2587, https://doi.org/10.1007/978-1-0716-2772-3_23,

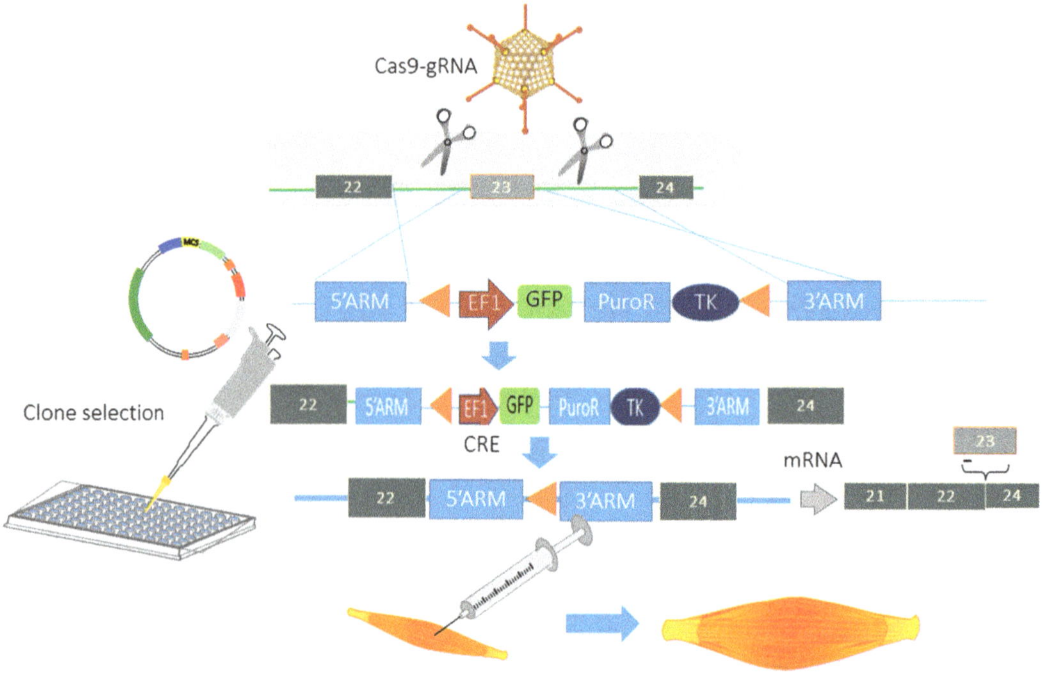

Fig. 1 Deletion of exon 23 of dystrophin using a combination of CRISPR/Cas9 and HDR. (**a**) Schematic diagram of CRISPR/Cas9 and HDR-mediated deletion of exon 23. Step1: Cas9 nuclease targets intron 22 and intron 23 via two gRNAs. Double-strand breaks (DSBs) of Cas9 result in the excision of mutant exon 23. Step 2: By HDR, a portion of the open reading frame (ORF) is replaced by a selection cassette containing GFP and puromycin resistance markers. Step 3: The triple-tagged cassette flanked by loxP sites can be excised from the genome by Cre mRNA

promoter-driven triple-tagged selection cassette (LoxP + Insulator Sequence + EF-1 + GFP-T2A-Puro-P2A-HSV-TK-PolyA + Insulator Sequence + LoxP) in one vector. The vector was transfected into iPSCs for HDR-based targeting of exon23. EF-1 promoter-driven GFP and puromycin (Puro) selection markers promote selection of HDR-modified iPSCs, followed by removal of the floxed selection cassette using Cre recombinase mRNA (Fig. 1). Finally, ganciclovir treatment was used to kill cells without excised cassettes by herpes simplex virus type I thymidine kinase (HSV-tk) negative selection. We used an inducible vector system for myogenic-specific transcription factor (MyoD) to induce muscle differentiation of iPSCs reprogrammed from Mdx mice.

2 Materials

- Dmd^{mdx} mice (The Jackson Laboratory).
- Collagenase IV (Worthington Biochemical Corporation, LS004189).

- Dispase (STEMCELL Technologies, 7923).
- CytoTune™ iPS 2.0 Sendai Reprogramming kit (Thermo Fisher).
- Mouse recombinant Leukemia Inhibitory Factor (LIF) (ESGRO® Supplement, Millipore).
- MEK/GSK-3 inhibitor supplement (MEK/GSK-3 Inhibitor Supplement, Millipore).
- TVP solution (PBS with 1% chicken serum, 0.025% trypsin, and 1.27 mM EDTA).
- Immunocompromised B10ScSn.Cg-Prkdc^{scid} Dmd^{mdx}/J (The Jackson Laboratory)
- Plasmid lenti-CRISPRv2-blast (Addgene plasmid no.83480, a gift from Dr. Mohan Babu).
- Plasmid lentiGuide-Hygro-iRFP670 (Addgene plasmid no. 99377, a gift from Dr. Kristen Brennand).
- Plasmid lenti-TRE-VP64-mouse-MyoD-T2A-dsRedExpress2 (Addgene, 60625, a gift from Dr. Charles Gersbach).
- Packaging plasmid psPAX2 (Addgene plasmid no. 12260, a gift from Dr. Didier Trono), and envelope plasmid pMD2.G (Addgene plasmid no. 12259, a gift from Dr. Didier Trono).
- Quick Ligation Kit (NewEngland BioLabs).
- 293FT cells (Thermo Scientific).
- Blasticidin (Sigma).
- Hygromycin B (Invitrogen).
- Puromycin (MP Biomedicals).
- Ganciclovir (Thermo Fisher).
- Doxycycline (Fisher BioReagents).
- Carbenicillin (Millipore).
- X-GaL (MilliporeSigMa).
- Homologous-directed recombination targeting vector MCS1-LoxP-EF1α-GFP-T2A-Puro-P2A-hsvTK-pA-LoxP-MCS2 (System Biosciences, HR210-PA-1).
- 5′ and 3′ homology arms specific to the *Dmd* gene Exon23 (BIOMATIK).
- Electroporation transfection system (Nero Transfection System, Invitrogen).
- Cre recombinase mRNA (130-101-113, Miltenyi Biotec).
- Lipofectamine™ MessengerMAX™ Transfection Reagent (Thermo Fisher).
- QIAamp DNA Blood Mini Kit (QIAGEN).
- T100TM Thermal Cycler (BIO-RAD).

- PrimeSTAR Max Premix (Takara).
- QIAEX II Gel Extraction Kit (QIAGEN, 20051).
- TOPO® TA Cloning® Kits (Invitrogen).
- QIAprep Spin Miniprep Kit (QIAGEN).
- RNAzol (Molecular Research Center, INC).
- CFX96TM Real-Time System (BIO-RAD).
- PowerUP SYBR Green Master Mix (Thermo Scientific, A25742).
- Dystrophin antibody (Thermo Fisher).
- Alexa555-conjugated goat-anti-rabbit antibody (Thermo Fisher).
- HardSet Antifade Mounting Medium with DAPI (Vector Laboratories).

3 Methods

3.1 Isolation of Skin Fibroblasts from Adult Dmd ^{mdx} Mice

1. Mince adult mouse tails from Dmd^{mdx} mice into 1 mm pieces
2. Digest the minced tails with Dulbecco's minimal essential medium/Ham's F12 (DMEM/F12) medium containing 0.1% collagenase and 1 U/mL of dispase for 2 h at 37 °C in a 5% CO_2 incubator
3. Mount the tissue explants on gelatin-coated plates and culture at fibroblast culture medium (Dulbecco modified Eagle medium [DMEM] with 10% fetal calf serum [FBS], 100 IU/mL penicillin G, 100 μg/mL streptomycin, 0.25 μg/mL amphotericin B, 2 mM L-glutamine, 0.10 mM nonessential amino acids, and 0.05 mM β-mercaptoethanol) at 37 °C and 5% CO_2 for 10 d

3.2 Generation of iPSCs from Dmd^{mdx} Skin Fibroblasts

1. Plate 5×10^4 skin fibroblasts per well on a 6-well plate two days prior to infection.
2. Add Sendai vector containing KOS, hc-Myc, and hKlf4 at MOI of 5:5:3 mix to cells.
3. Replace media with fresh fibroblast medium 24 h after infection.
4. Change to mouse ES cell culture medium (Dulbecco modified Eagle medium [DMEM] with 10% fetal calf serum [FBS], 100 IU/mL penicillin G, 100 μg/mL streptomycin, 0.25 μg/mL amphotericin B, 2 mM L-glutamine, 0.10 mM nonessential amino acids, and 0.05 mM β-mercaptoethanol, 1× mouse recombinant Leukemia Inhibitory Factor (LIF), 1× MEK/GSK-3 inhibitor supplement).

5. Pick putative iPSC colonies with 10 μL sterile pipette tips by morphological selection criteria [2].
6. Dissociate mouse iPSC colonies with TVP solution (PBS with 1% chicken serum, 0.025% trypsin, and 1.27 mM EDTA) and seed onto a 0.1% gelatin-coated dish containing mouse ES cell culture medium.
7. Change the media every other day.

3.3 Design and Construction of Guide Sequences to Completely Delete Exon 23 of DMD Gene [4]

1. Design two pairs of gRNA oligos targeting intron flanking dystrophin exon 23 using the online free program (http://crispor.tefor.net/crispor.py). The designed pairs are as follows:
   - i22sense: 5′-CACCGTTAAGCTTAGGTAAAATCAA-3′
   - i22anti-sense: 5′-AAACTTGATTTTACCTAAGCTTAAC-3′
   - i23sense: 5′-CACCGAGTAATGTGTCATACCTTCT-3′
   - i23 anti-sense: 5′-AAACAGAAGGTATGACACATTACTC-3′
2. Phosphorylate and anneal each pair of oligos at 37 °C for 30 min, at 95 °C for 5 min, and then ramp down to 25 °C at 5 °C/min
   - 1 μL Oligo sense (100 μM)
   - 1 μL Oligo anti-sense (100 μM)
   - 1 μL 10× T4 Ligation Buffer (NEB)
   - 6.5 μL ddH_2O
   - 0.5 μL T4 PNK (NEB M0201S)
   - Total 10 μL
3. Digest and dephosphorylate lenti-CRISPRv2-blast and lenti-Guide-Hygro-iRFP670 plasmids using BsmB1/Esp3I
   - 5 μg lentiCRISPRv2-blast or lenti-Guide-Hygro-iRFP670
   - 3 μL BsmBI (Esp3I, NEB)
   - 3 μL CutSmart™ Buffer (NEB)
   - 6 μL 10× FastDigest Buffer
   - 0.6 μL 100 mM DTT
   - X μL ddH_2O
   - Total 60 μL
4. Run the digested plasmids in 0.8% agarose gel, cut off the larger bands, and extract DNAs using QIAquick Gel extraction kit and elute DNAs in ddH_2O.
5. Dilute annealed oligos in ddH_2O at a dilution of 1:200.

6. Establish ligation reactions using the Quick Ligation Kit by cloning annealed oligos into the vectors (i22 for plenti-CRISPRv2-blast and i23 for plenti-Guide-Hygro-iRFP670
   - X μL BsmBI/Esp3I digested plasmid (50 ng)
   - 1 μL diluted oligo duplex
   - 5 μL 2× Quick Ligase Buffer (NEB)
   - X μL ddH_2O
   - Subtotal 10 μL
   - 1 μL Quick Ligase (NEB)
   - Total 11 μL
7. Transformation into SURE 2 Supercompetent cells.
8. Lentiviral vector packaging:
   8.1 Co-transfect plentiCRISPR V2-Blast-gRNAi22 and plentiGuide-Hygro-RFP670-gRNAi23 into 293FT cells respectively using lipofactaimine 3000 with packaging plasmid psPAX2 and envelope plasmid pMD2.G.
   8.2 Collect lentiviral supernatant 48 h after the initial plasmid transfections for two consecutive days.
   8.3 Purify the lentiviral vectors with PEG6000 (8.5% final concentration) and NaCl (0.4-M final concentration) to the 0.45 μM syringe filtered supernatant as previously reported [3, 7].
9. Generate Cas9/gRNA-modified miPSC:
   9.1 Use mouse ES media containing lenti-CRISPR V2-blast-gRNAi22, lenti-Hygro-iRFP670-gRNAi23, and control (empty vector: lenti-CRISPR V2-blast and lenti-Hygrp-iRFP670) with 8 μg/mL polybrene to infect miPSC.
   9.2 Select infected miPSCs in mES medium containing 2.5 μg/mL blasticidin and 100 μg/mL hygromycin B for 72 h after infection until large resistant colonies are visible.

3.4 Construction of Homologous-Directed Recombination (HDR) Targeting Vector System [4]

A ~1 kb 5′ homology arm and a ~0.8 kb 3′ homology arm were synthesized and cloned into a homologous-directed recombination targeting system at 5′ and 3′ multiple cloning sites separately.

DMD3′HDR-Forward	5′-CGATCCCGTGCCACCTT-3′
DMD3′HDR-Reverse	5′-GGGAAGGAAATATGGCAGAAATTAAACA-3′
DMD5′HDR-Forward	5′-ACATGTCTTATCAGTCAAGAGATCA-3′
DMD5′HDR-Reverse	5′-CGAGAAGCGTTCAGAGGAAA-3′

3.5 Generation of Stable Cell Lines with DMD-Exon 23 Deletion

1. Transfect 1.5 μg of homologous-directed recombination targeting vector into 1×10^5 Cas9/gRNA-infected iPSCs by electroporation at 1200 V 40 ms (one pulse)
2. Select the HDR-integrated iPSCs in mES medium containing 1 μg/mL puromycin for 48 h
3. Remove LoxP flanked selection cassette with Cre recombinase mRNA using Lipofectamine™ MessengerMAX™ Transfection Reagent.
4. Treat the iPSCs with 5 μg/mL ganciclovir for 48 h to kill cassette-bearing cells, which express the *hsv-tk* gene.

3.6 Differentiation of Mouse iPS Cells into Myogenic Progenitor Cells (MPC) by Lenti-TRE-VP64-MyoD-T2A-dsRed Vector

1. Prepare lenti-TRE-VP64-mouse-MyoD-T2A-dsRedExpress2-PuroR vectors by co-transfecting the plasmid lenti-TRE-VP64-mouse-MyoD-T2A-dsRedExpress2-PuroR with packaging plasmid psPAX2 and envelope plasmid pMD2.G into 293FT cells, and collect lentiviral supernatant 48 h after initial plasmid transfection for two consecutive days.
2. Infect miPSCs with medium containing lenti-TRE-VP64-mouse-MyoD-T2A-dsRed-Puro and 8 μg/mL Polybrene.
3. Select stable miPSCs infected with lenti-TRE-VP64-mouse-MyoD-T2A-dsRed-Puro in mES medium containing 5 μg/mL puromycin for 48 h after infection and maintaining until large resistant colonies became visible.
4. Culture cells in mES media.
5. Induce *MyoD* transgene expression by supplementing the growth medium with 3 μg/mL doxycycline for at least 10 days.
6. Replace the media with fresh growth media supplemented with doxycycline every 2 days.

3.7 MPC Transplantation [5]

1. To determine whether cell transplantation with MPCs of $\text{iPSC}^{\text{Cas9-}\Delta\text{Exo23-HDR-CRE}}$ restores dystrophin protein expression in the muscle of Mdx mice, inject equal amounts of MPCs from $\text{iPSC}^{\text{Cas9-Ctrl}}$ or $\text{iPSC}^{\text{Cas9-}\Delta\text{Exo23-HDR-CRE}}$ intramuscularly into the TA muscle of 8-week-old MDX/SCID mice.
2. Anesthetize male immunodeficient Prkdcscid Dmd^{mdx} mice with ketamine/xylazine.
3. Inject 30 μL suspended MPCs in DMEM containing 5×10^5 MPCs from $\text{iPSC}^{\text{Cas9-Ctrl}}$ or $\text{iPSC}^{\text{Cas9-}\Delta\text{Exo23-HDR-CRE}}$ intramuscularly into the tibialis anterior (TA) muscle of each 8-week-old mouse.

3.8 Immunofluorescent Tissue Staining to Detect Dystrophin Expression in Dmd^{mdx} Mouse Model

1. Isolate TA muscles from immunodeficient Dmd^{mdx} mouse 4 weeks after injection of dox-induced MPCs from iPSC $^{Cas9\text{-}Ctrl}$ or iPSC $^{Cas9\text{-}\Delta Exo23\text{-}HDR\text{-}CRE}$.
2. Embed TA muscle in OCT compound and cut into 5 μm sections following snap frozen.
3. Fix the tissue with 4% paraformaldehyde and block the tissue with 5% goat serum for 1 h.
4. Incubate tissue sections with rabbit anti-dystrophin antibody (1:50).
5. Stain with the Alexa555-conjugated goat-anti-rabbit antibody (1:400).
6. Mount the sections using HardSet Antifade Mounting Medium with DAPI.
7. Image tissue staining on a Zeiss 780 upright confocal microscope as shown in Fig. 2, four weeks after cell transplantation. We observed restored dystrophin protein in the muscles injected with MPCs from iPSC$^{Cas9\text{-}\Delta Exo23\text{-}HDR\text{-}CRE}$, but not in controls; TA muscle from wild type C57BL/6 mice was used as a positive control for dystrophin staining. Regenerated muscle fibers from donor MPCs showed variable fiber sizes and were relatively small compared to the size of muscle cells from wide type mice (C57BL/6).

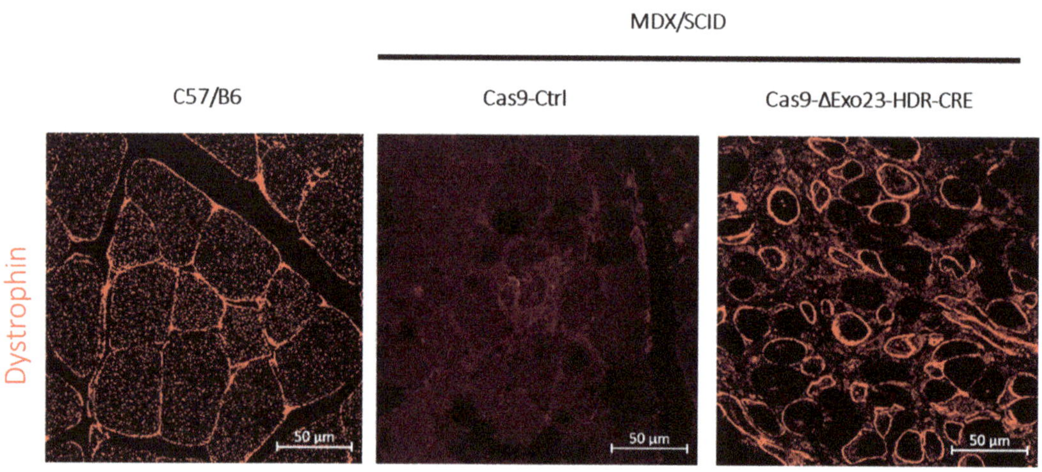

Fig. 2 Restoration of dystrophin expression in vivo. Immunofluorescence analysis of dystrophin expression in TA muscles of C57BL/6 and MDX/SCID mice at 4 weeks after injection of Dox-induced MPCs from iPSC $^{Cas9\text{-}Ctrl}$ and iPSC Cas9-ΔExo23-HDR-CRE (scale bar = 50 μm)

4 Notes

1. The lentiviral vector-based Cas9 system uses double-cut sgRNAs and HDR-based donor vectors to integrate EF-1 promoter-driven GFP, puromycin resistance genes, and HSV-TK cassettes in iPSCs. This results in a high level of HDR-mediated precision substitution. We previously showed that double cleavage of double gRNAs is inefficient in deleting mutant exon23 by nonhomologous end joining (NHEJ) repair [2]. Compared to direct deletion of mutant Exon23 from iPSC^{Mdx} based on two gRNAs, which has the disadvantage of extremely low efficiency and cumbersome clone selection and genotyping process, HDR-mediated CRISPR/Cas9 gene editing strategy by integrating multiple gene selection tools, including GFP, puromycin resistance genes and herpes simplex virus thymidine kinase (HSV-TK) driven by EF-1 promoter in HDR design, improves the efficiency of selecting correctly edited iPSC clones by utilizing a triple selection strategy, including FACS or the use of antibiotics [4].
2. In design of the HDR donor DNA, the 1 kb 5′ homology arm (HA) is identical to the sequence around the DSB created by the i22 gRNA, and the 0.8 kb 3′ HA is identical to the sequence around the DSB created by the i23 gRNA.
3. In this study, we chose the EF-1 promoter-driven donor HDR reporter system over the promoterless donor HDR reporter system because the *DMD* gene is only expressed in muscle progenitors and mature muscle cells [1], not in undifferentiated pluripotent stem cells; therefore, the promoterless HDR reporter system did not allow the selection of iPSCs, in which exon 23 was correctly replaced. Since the HDR donor has two LoxP sites on both sides, this makes possible to remove HDR donors by Cre/LoxP recombination. With our cellular system, we achieved an efficient deletion of exon 23 [4].
4. It should be noted that when designing HDR donor DNAs, it is necessary to mutate the protospacer adjacent motif (PAM) sequences (5′-NGG-3′, in which "N" represents any nucleotide base) of 5′-HA and 3′-HA to prevent insertion of HDR to be deleted by the activated CRISPR/Cas9 system.

Acknowledgments

This work was supported by the American Heart Association (AHA) Transformational Project Award 18TPA34170104 and the National Institutes of Health (NIH) R01HL146481 (to I.K.); NIH R01HL134354 (to N.L.W. and Y.T.); as well as AHA Grant-in-Aid 16GRNT31430008 and NIH R01HL086555 (to Y.T.).

References

1. Jelinkova S, Fojtik P, Kohutova A, Vilotic A, Marková L, Pesl M, Jurakova T, Kruta M, Vrbsky J, Gaillyova R et al (2019) Dystrophin deficiency leads to genomic instability in human pluripotent stem cells via NO synthase-induced oxidative stress. Cells 8:53
2. Jin Y, Shen Y, Su X, Weintraub N, Tang Y (2019) CRISPR/Cas9 technology in restoring dystrophin expression in iPSC-derived muscle progenitors. J Vis Exp
3. Jin Y, Shen Y, Su X, Cai J, Liu Y, Weintraub NL, Tang Y (2020a) The small GTPases Rab27b regulates mitochondrial fatty acid oxidative metabolism of cardiac mesenchymal stem cells. Front Cell Dev Biol 8:209–209
4. Jin Y, Shen Y, Su X, Weintraub NL, Tang Y (2020b) Effective restoration of dystrophin expression in iPSC (Mdx)-derived muscle progenitor cells using the CRISPR/Cas9 system and homology-directed repair technology. Comput Struct Biotechnol J 18:765–773
5. Jin Y, Shen Y, Weintraub NL, Tang Y (2020c) Using iRFP genetic labeling technology to track tumorogenesis of transplanted CRISPR/Cas9-edited iPSC in skeletal muscle. Methods Mol Biol 2126:73–83
6. Su X, Shen Y, Jin Y, Jiang M, Weintraub N, Tang Y (2019) Purification and transplantation of myogenic progenitor cell derived exosomes to improve cardiac function in duchenne muscular dystrophic mice. J Vis Exp
7. Su X, Shen Y, Weintraub NL, Tang Y (2020) Imaging and tracking stem cell engraftment in ischemic hearts by near-infrared fluorescent protein (iRFP) labeling. Methods Mol Biol 2150: 121–129
8. Su X, Shen Y, Jin Y, Weintraub NL, Tang YL (2021) Identification of critical molecular pathways involved in exosome-mediated improvement of cardiac function in a mouse model of muscular dystrophy. Acta Pharmacologica Sinica 42:529–535

Part VI

Small Molecules

Chapter 24

Effects of Glucocorticoids in Murine Models of Duchenne and Limb-Girdle Muscular Dystrophy

Michelle Wintzinger, Karen Miz, Allen York, Alexis R. Demonbreun, Jeffery D. Molkentin, Elizabeth M. McNally, and Mattia Quattrocelli

Abstract

In vivo testing of glucocorticoid steroids in dystrophic mice offers important insights in benefits and risks of those drugs in the pathological context of muscular dystrophy. Frequency of dosing changes the spectrum of glucocorticoid effects on muscle and metabolic homeostasis. Here, we describe a combination of non-invasive and invasive methods to quantitatively discriminate the specific effects of intermittent (once-weekly) versus mainstay (once-daily) regimens on muscle fibrosis, muscle function, and metabolic homeostasis in murine models of Duchenne and limb-girdle muscular dystrophies.

Key words Glucocorticoid steroids, *mdx* mice, *Sgcg* mice, Intermittent dosing, Hydroxyproline, Grip strength, Treadmill assay, In situ force analysis, Glycemia

1 Introduction

Muscular dystrophies are genetically inherited diseases inducing chronic degeneration and functional impairment of muscle [1]. Duchenne muscular dystrophy is caused by mutations in the gene encoding dystrophin and is genetically modeled in mice by the *mdx* mutation [2]. Limb-girdle muscular dystrophies encompass a heterogeneous group of subtypes based on the mutated gene. The subtype R5 (previously known as type 2C) is caused by mutations in the gene encoding γ-sarcoglycan (*Sgcg*) and is genetically modeled in mice by *Sgcg* knockout [3]. For both *mdx* and *Sgcg* strains, the *DBA/2J* genetic background augments the dystrophic pathophysiology progression in muscle as compared to the traditional *C57BL/10* background [4]. Glucocorticoid steroids like prednisone are standard of care for Duchenne muscular dystrophy. Glucocorticoids help manage the chronic inflammation of degenerating muscles, prolonging the ambulatory age in these patients [5]. However, chronic daily intake of these drugs induces untoward side effects,

Rika Maruyama and Toshifumi Yokota (eds.), *Muscular Dystrophy Therapeutics: Methods and Protocols*,
Methods in Molecular Biology, vol. 2587, https://doi.org/10.1007/978-1-0716-2772-3_24,

including metabolic syndrome [6]. Moreover, the present glucocorticoid regimens used for Duchenne are not indicated for other muscular dystrophies [7]. Several approaches are being pursued to minimize and possibly revert the side effects of once-daily glucocorticoid dosing. We have recently shown that intermittent once-weekly dosing of glucocorticoids improved repair, fibrosis, and function in both *mdx* and *Sgcg* models of muscular dystrophy [8, 9]. Importantly, opposite to the mainstay once-daily dosing, the intermittent regimen did not induce the typical metabolic side effects associated with chronic glucocorticoid dosing in mice, including hyperglycemia, wasting, and weakness [10]. Here, we review methods and examples of assessments of glucocorticoid dosing in *mdx* and *Sgcg* mice from the same DBA/2J genetic background.

2 Materials

2.1 Injectable Glucocorticoid Solution

1. Prednisone stock: Weigh and dilute prednisone powder (Sigma #P6254) in DMSO in sterile conditions to a final concentration of 5 mg/mL. Prepare aliquots for single use (avoid freeze-thaw) and store at −20 °C. *See* **Note 1** for other glucocorticoids.
2. Vehicle stock: 100% DMSO aliquoted and stored as prednisone stock.
3. Sterile physiological solution: This will be used to dilute drug or vehicle solutions prior to injection.
4. Insulin syringes.

2.2 Hydroxyproline Dosing Solutions

1. Mortar-pestle or cryomill to cryo-pulverize flash-frozen muscle tissue.
2. Glass pyrex tubes with anti-evaporation screwcap.
3. 6 M HCl
4. 1 M Hydroxyproline standard: Dissolve powder to 1 M concentration in 6 M HCl, prepare aliquots for single use (avoid freeze-thaw) and store at −20 °C.
5. 7% Chloramine T solution: Dissolve 700 mg chloramine T in 10 mL ultrapure water. Store at 4 °C in the dark for no longer than ~3 weeks.
6. Acetate citrate buffer: Combine 57 g anhydrous sodium acetate, 33.4 g citric acid monohydrate, 435 mL of 1 M NaOH, 385 mL of 100% 2-propanol, and ultrapure water up to 1 L. Store at 4 °C in the dark.
7. Ehrlich's reagent: Resuspend 3 g 4-dimethylaminobenzaldehyde in 10 mL 100% ethanol and add 675 μL sulfuric acid in a

fume hood with careful drop-by-drop pipetting (highly exergonic reaction). Prepare right prior to use.

8. Solution A: 20% chloramine T solution, 80% acetate citrate buffer. Prepare right prior to use. Prepare at least 100 μL per sample, including all standards.
9. Solution B: 18.75% Ehrlich's reagent, 81.25% 2-propanol. Prepare right prior to use. Prepare at least 1.5 mL per sample, including all standards.
10. Oven, thermal bath, plate reader.

2.3 Grip and Treadmill Assays

1. Grip strength meter with single sensor for mice (maximum 1 kg capacity) with standard pull bars.
2. Open Treadmill with stimulus assembly, ~10% uphill incline, individual mouse lanes, drive motor to adjust speed (3–100 m/min range), exercising belt with grip-prone texture.

2.4 Force-Frequency Analysis in Tibialis Anterior Muscles

1. In situ force analysis equipment and sensor
2. 4-0 U.S.P silk braided surgical suture
3. Isoflurane-based inhalation anesthesia system
4. Precision scale, digital caliper

2.5 Measurement of Body Weight and Glycemia

1. Scale
2. Glucometer with test strips

3 Methods

All analyses should be performed blinded to treatment cohorts.

Here, we report protocols and example data from parallel analyses in two background-matched genetic models of muscular dystrophy, *DBA/2J-mdx* mice and *DBA/2J-Sgcg* mice.

Four-month-old *DBA/2J-mdx* males were obtained from the JAX colony #013141. For this strain, it is appropriate to use only males as the X-linked Duchenne muscular dystrophy primarily affects boys. Use at least five mice per cohort.

Four-month-old *DBA/2J-Sgcg* mice (homozygous null) were obtained from the original mutant strain [3] after at least ten backcrossing generations in the *DBA/2J* background. For this strain, randomize equal numbers of males and females across treatment cohorts. Use at least five mice per cohort.

See **Note 2** for considerations regarding age of test mice.

3.1 Glucocorticoid Regimens

1. Record mouse weight immediately before injection.
2. Prepare the injectable dose in a final volume of 50 μl sterile physiological solution. Add the appropriate volume of prednisone stock to a final dose of 1 mg/kg, for example, 4 μL (5 μg/μL, total of 20 μg) for a 20 g mouse.
3. Prepare a vehicle dose with the corresponding volumes of vehicle stock and sterile physiological solution.
4. Load one sterile, single-use insulin syringe per dose, making sure no air is introduced.
5. Restrain mice to expose abdomen and inject dose intraperitoneally.
6. Frequency of dosing: **once-daily regimen** consists of 1x prednisone dose/day and is controlled by once-daily vehicle injection. **Once-weekly regimen** consists of 1x prednisone dose (day 1) followed by 6× vehicle doses (day 2–6) per week. It is controlled by once-daily vehicle injection.
7. Time of dosing: Inject mice at the start of their subjective day-time, that is, within the first two hours of lights-on in their housing space. This circadian time presents a critical window to maximize the effects induced by exogenous glucocorticoids through the glucocorticoid receptor [11, 12].
8. Conduct regimens for a total of 12 weeks (*see* **Note 3**). Collect tissues at 24 h after last prednisone dose.

3.2 Quantitation of Anti-fibrotic Effects Through Hydroxyproline Dosing

1. Cryo-pulverize flash-frozen muscle and collect 50 mg cryo-powder in a glass pyrex tube.
2. Add 500 μL of 6 M HCl per sample, cap tightly.
3. Prepare the hydroxyproline standards into glass pyrex tubes with the following amounts of 1 M hydroxyproline stock: 0 μL (0 μmol), 5 μL (5 μmol), 10 μL (10 μmol), and 20 μL (20 μmol). Adjust volume for each standard to 500 μL with 6 M HCl, cap tightly.
4. Incubate overnight at 110 °C in oven (or thermal block) in fume hood to perform acid hydrolysis. An oven is recommended to ensure thermal homogeneity and allow the tubes to be incubated while standing into a Pyrex glass container.
5. After overnight incubation, bring samples to room temperature, set thermal block or water bath at 58 °C.
6. Transfer 10 μL of sample supernatant, paying attention to avoiding small tissue particles, into new labeled tubes.
7. Keeping tubes at room temperature, add 150 μL of 100% 2-propanol per sample, vortex and leave cap open.
8. Add 75 μl Solution A, vortex, leave cap open, and incubate at room temperature for 10 min.

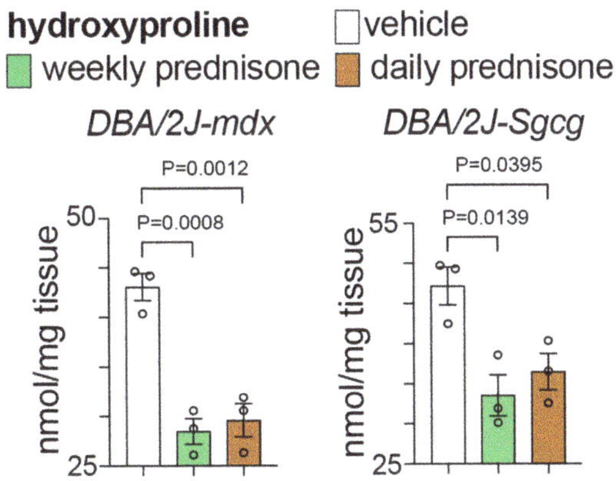

Fig. 1 Compared to vehicle, both weekly and daily prednisone treatments improved fibrosis (decreased hydroxyproline content) in quadricep muscle of dystrophic mice, as quantitated here through hydroxyproline dosing. $N = 3$♂/group (*mdx*), (1♂ + 2♀)/group (*Sgcg*). 1-way ANOVA with Sidak's multi-comparison test

9. Add 1 mL Solution B, vortex, close caps, and incubate at 58 °C for 45 min.
10. Transfer all tubes on ice until ready for plate reader analysis.
11. Spin at 5000 × *g* for 5 min at 4 °C.
12. Transfer 100 μL per replicate (≥3 replicates for each standard and sample are recommended) into wells of a transparent 96-well plate suitable for absorbance analyses at plate reader. Alternatively, use spectrophotometer with appropriate amounts for the dedicated cuvettes.
13. Measure absorbance at 558 nm.
14. Normalize hydroxyproline amount to initial cryopowder weight (e.g., nmol/mg).
15. Figure 1 shows example data of glucocorticoid regimen effects on hydroxyproline content, used as quantitative measure of fibrotic collagen deposition, in skeletal muscle of *DBA/2J-mdx* mice and *DBA/2J-Sgcg* mice after a 12-week-long treatment.

3.3 Quantitation of Regimen-Specific Effects on Grip Strength and Treadmill Performance

1. For grip strength: Record force developed by each mouse through bilateral grasp of forelimbs while pulling the mouse away from the metal grid.
2. Hold the tail and pull the animal away with a constant, gentle movement, keeping the mouse parallel to the work surface.
3. Repeat the grip force measurements three times per mouse, resting the mouse for 1 min between pulls, and record the highest number.

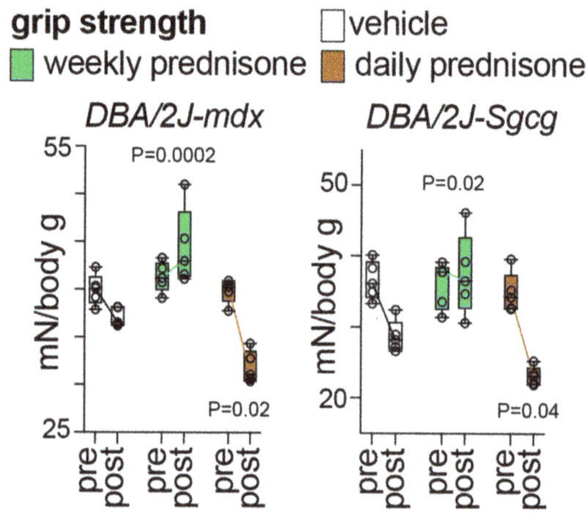

Fig. 2 Compared to vehicle, weekly prednisone improved the decline in weight-normalized grip strength in both *mdx* and *Sgcg* mice. Daily prednisone exacerbated the loss of strength. $N = 5$♂/group (mdx), (3♂ + 2♀)/group (Sgcg). 2-way ANOVA with Dunnet's multi-comparison test versus vehicle

4. Normalize grip strength to body weight, recorded immediately before grip assay.
5. Figure 2 shows example data of glucocorticoid regimen effects on grip strength. In both *mdx* and *Sgcg* mice, grip strength declined with disease progression. Once-weekly prednisone counteracted the strength decline, while this was exacerbated by once-daily prednisone.
6. For treadmill assay: Place mice in individual treadmill lanes, start the conveyor belt at 3 m/min and increase speed by 1 m/min every minute (1m/min^2 acceleration).
7. Induce mice to continue running by a shock grid at the end of the conveyor until they display evident signs of exhaustion, for example, 30 s on the shock grid without successful efforts to start running again.
8. Record distance (m) and total time (s) until exhaustion.
9. Analyze treadmill assay performance as either time-to-exhaustion, or distance-to-exhaustion, or weight-normalized work (kg*m^2*s($^{-2}$), Joule units).
10. Figure 3 shows example data of glucocorticoid regimen effects on treadmill performance. In both *mdx* and *Sgcg* mice, treadmill performance declined with disease progression. Once-weekly prednisone partially counteracted the decline in running performance, while this was exacerbated by once-daily prednisone.

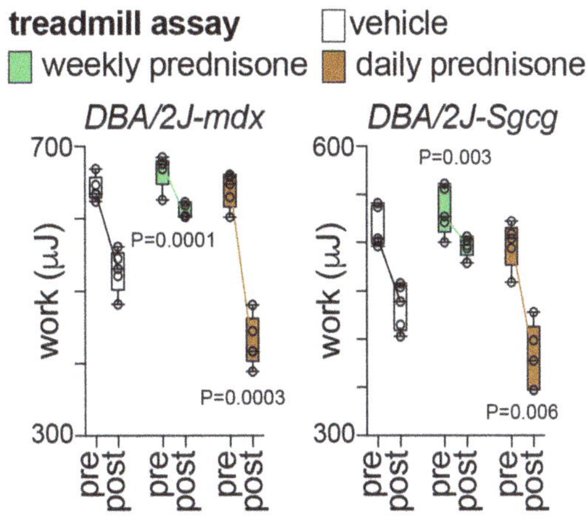

Fig. 3 Compared to vehicle, weekly prednisone improved the decline in treadmill performance until exhaustion in dystrophic mice. Daily prednisone exacerbated the loss of running performance. $N = 5$♂/group (mdx), (3♂ + 2♀)/group (Sgcg). 2-way ANOVA with Dunnet's multi-comparison test versus vehicle

11. It is recommended to measure grip strength and treadmill performance at baseline (e.g., one day before first injection) and at treatment endpoint (e.g., 24 h after last prednisone dose) to quantitate relative trends compared to baseline.
12. If assaying both grip strength and treadmill in the same day, perform grip assay before treadmill assay to avoid possible effects of fatigue. Allow mice to walk freely in the cage for 30 min between grip and treadmill assays.
13. Keep operator-dependent variables (e.g., operator, instruments, room conditions, time-of-day) for both assays constant across cohorts and time points.
14. Avoid collecting tissues immediately after a treadmill assay. It is recommended to collect tissues at least 24 h after the last treadmill assay.

3.4 Quantitation of Regimen-Specific Effects on Muscle Force

1. Anesthetize mouse with 1.5% isoflurane, lay the mouse supine on the assay platform of the Aurora in situ system, block the knee with the dedicated clamp.
2. Ensuring tendons and muscles are not damaged, remove the hindlimb skin to uncover the *tibialis anterior* muscle.
3. Detach the *tibialis anterior* muscle from the tibia by sliding and expanding forceps between muscle and bone.
4. Secure the distal tendon to the force probe through surgical suture, gently sever the tendon, and separate the muscle from the tibia. The *tibialis anterior* muscle should be linked to the force probe on the distal side and to the knee region on the proximal side.

5. Adjust platform and probe to ensure that the probe is in line with the knee.
6. Place two electrical probes into the leg, one at the distal end of the *tibialis anterior* and one under the knee cap near the sciatic nerve.
7. Run test pulses to adjust muscle tension to the equilibrium point. The equilibrium point is the muscle tension at the initial resting point to which the muscle returns to after test pulses of electricity are administered. Record the muscle length in millimeter at equilibrium with a digital caliper (L_0).
8. Run the force-frequency test through tetanus stimulations at increasing frequency, that is, from 25 Hz to 200 Hz with intervals of 25 Hz. Pause 1 min between tetani. Record force in N for each isometric contraction (P_0).
9. Repeat procedure and test on contralateral *tibialis anterior* muscle.
10. Proceed to euthanasia, collect and record each *tibialis anterior* muscle with a precision scale.
11. Calculate specific force (N/mm^2) for each tetanus frequency as (P_0 N)/[(muscle mass mg/1.06 mg/mm^3)/L_f mm]. 1.06 mg/mm^3 is the mammalian muscle density. $L_f = L_0*0.6$, where 0.6 is the muscle to fiber length ratio in *tibialis anterior* muscle [13]. The specific force is often converted to and reported as N/cm^2 units.
12. Figure 4 shows example data of glucocorticoid regimen effects on muscle force at end of treatments. Compared to vehicle in both *mdx* and *Sgcg* mice, once-weekly prednisone improved force generation, while once-daily prednisone decreased it.

3.5 Quantitation of Regimen-Specific Effects on Glycemia and Weight

1. To record weights, place mouse on a scale in a suitable container to avoid escaping or jumping. Record weight to the nearest tenth of a gram when the mouse is not moving.
2. To record glycemia, fast mice for 2 h [14], clip the tail end and measure the blood drop with a glucometer.
3. Record glycemia and weights at baseline (e.g., one day before first injection) and at treatment endpoint (e.g., 24 h after last prednisone dose) to quantitate relative trends compared to baseline.
4. Record glycemia and weights at the same time of day throughout treatment.
5. Analyze glycemia and body weight trends as absolute values or relative trends to baseline.
6. Figure 5 shows example data of glucocorticoid regimen effects on glycemia. Compared to vehicle in both *mdx* and *Sgcg* mice,

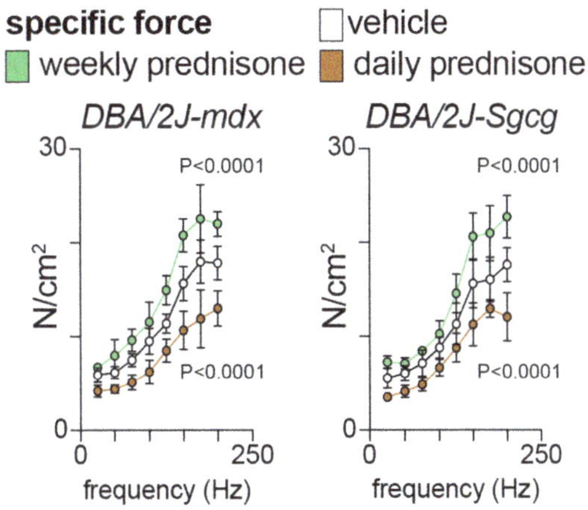

Fig. 4 At endpoint, in situ force-frequency analysis with tibialis anterior muscles showed gain of force with weekly prednisone and loss of force with daily prednisone compared to vehicle. $N = 3♂$/group (mdx), $(2♂ + 1♀)$/group (Sgcg). 2-way ANOVA with Dunnet's multi-comparison test versus vehicle

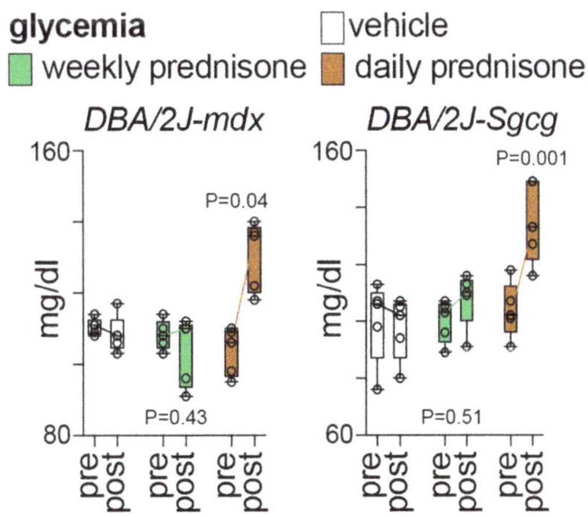

Fig. 5 Compared to vehicle, weekly prednisone did not induce changes in glycemia. Daily prednisone induced hyperglycemia in dystrophic mice, correlating with the known pro-hyperglycemic effects of this regimen. $N = 5♂$/group (mdx), $(3♂ + 2♀)$/group (Sgcg). 2-way ANOVA with Dunnet's multi-comparison test versus vehicle

once-daily prednisone (dysmetabolic regimen) increased glycemia while once-weekly prednisone did not induce significant changes.

7. Figure 6 shows example data of glucocorticoid regimen effects on body weight. Compared to vehicle in both *mdx* and *Sgcg* mice, once-daily prednisone (pro-wasting regimen) decreased body weight, while once-weekly prednisone did not induce significant changes. *See* **Note 4**.

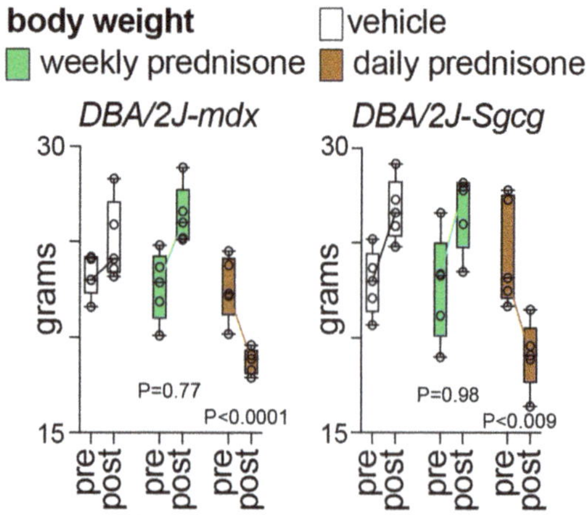

Fig. 6 Compared to vehicle, weekly prednisone did not induce changes in body weight curves, while daily prednisone induced loss of body weight, correlating with the known pro-wasting effects of this glucocorticoid regimen. $N = 5$♂/group (mdx), (3♂ + 2♀)/group (Sgcg). 2-way ANOVA with Dunnet's multi-comparison test versus vehicle

4 Notes

1. Here, we use prednisone as 1 mg/kg, as this dose successfully discriminated the effects diverging the once-weekly from the once-daily regimens in both dystrophic mouse lines [8–10]. Other glucocorticoids, dosages, and intermittence intervals should be tested to evaluate differences/similarities with analyses and trends discussed here.
2. Here, we are reporting methods and data related to glucocorticoid treatments in mice of 4 months of age at start. This age-at-start is convenient for 12-week-long regimens. Indeed, treated mice will be 4 months old at start and ~7 months old at endpoint. This age window encompasses onset and progression of pathophysiology in both *DBA/2J-mdx* and *DBA/2J-Sgcg* mouse lines [15], without the overt muscle degeneration seen at later stages. If treatments are conducted at an earlier or later age, analyses should take into account the pre-symptomatic and advanced stages of disease progression in these mice.
3. A 12-week-long treatment enables for stabilization of the divergent regimen-specific changes, with the additional consideration of ages at start and endpoint. Especially for once-daily glucocorticoid regimens, special attention should be paid when protracting these regimens longer than 12 weeks or if conducting them in mice older than ~7 months of age. Indeed, the wasting effects of this regimen tend to be of particular burden to dystrophic mice at advanced stages of disease, when dystrophic muscle degeneration is already prominent per se.

4. Mice are more susceptible to glucocorticoid-driven atrophy than humans, likely because of the higher preponderance of type-II myofibers in their muscles [16]. Weight curves offer an immediate albeit imperfect proxy estimate of overall wasting. If MRI-based systems to estimate lean and fat masses are available, these provide a better estimation of the dysmetabolic burden of once-daily versus once-weekly prednisone. Using MRI-based systems, it is indeed possible to simultaneously quantitate changes in adiposity versus lean mass (absolute or relative) over time.

Acknowledgments

The *DBA/2J-Sgcg* mice were donated by Dr. Elizabeth McNally (Northwestern University). This work was supported by DK121875, DK121875, HL158531, DK130908, NS047726, HL061322 and AG078174 (NIH), Start-up funds (CCHMC), Trustee Award (CCHMC), Heart Institute Translational Funds (CCHMC) and CuSTOM pilot grant (CCHMC).

References

1. Emery AE (2002) The muscular dystrophies. Lancet 359(9307):687–695. https://doi.org/10.1016/S0140-6736(02)07815-7
2. Bulfield G, Siller WG, Wight PA et al (1984) X chromosome-linked muscular dystrophy (mdx) in the mouse. Proc Natl Acad Sci U S A 81(4):1189–1192
3. Hack AA, Ly CT, Jiang F et al (1998) Gamma-sarcoglycan deficiency leads to muscle membrane defects and apoptosis independent of dystrophin. J Cell Biol 142(5):1279–1287
4. Hammers DW, Hart CC, Matheny MK et al (2020) The D2.mdx mouse as a preclinical model of the skeletal muscle pathology associated with Duchenne muscular dystrophy. Sci Rep 10(1):14070. https://doi.org/10.1038/s41598-020-70987-y
5. McDonald CM, Henricson EK, Abresch RT et al (2018) Long-term effects of glucocorticoids on function, quality of life, and survival in patients with Duchenne muscular dystrophy: a prospective cohort study. Lancet 391(10119):451–461. https://doi.org/10.1016/S0140-6736(17)32160-8
6. Quattrocelli M, Zelikovich AS, Salamone IM et al (2020) Mechanisms and clinical applications of glucocorticoid steroids in muscular dystrophy. J Neuromuscul Dis. https://doi.org/10.3233/JND-200556
7. Walter MC, Reilich P, Thiele S et al (2013) Treatment of dysferlinopathy with deflazacort: a double-blind, placebo-controlled clinical trial. Orphanet J Rare Dis 8:26. https://doi.org/10.1186/1750-1172-8-26
8. Quattrocelli M, Salamone IM, Page PG et al (2017) Intermittent glucocorticoid dosing improves muscle repair and function in mice with limb-girdle muscular dystrophy. Am J Pathol 187(11):2520–2535. https://doi.org/10.1016/j.ajpath.2017.07.017
9. Quattrocelli M, Barefield DY, Warner JL et al (2017) Intermittent glucocorticoid steroid dosing enhances muscle repair without eliciting muscle atrophy. J Clin Invest 127(6):2418–2432. https://doi.org/10.1172/JCI91445
10. Quattrocelli M, Zelikovich AS, Jiang Z et al (2019) Pulsed glucocorticoids enhance dystrophic muscle performance through epigenetic-metabolic reprogramming. JCI Insight 4(24). https://doi.org/10.1172/jci.insight.132402
11. Lamia KA, Papp SJ, Yu RT et al (2011) Cryptochromes mediate rhythmic repression of the glucocorticoid receptor. Nature 480(7378):

552–556. https://doi.org/10.1038/nature10700

12. Rizzini L, Levine DC, Perelis M et al (2019) Cryptochromes-mediated inhibition of the CRL4(Cop1)-complex assembly defines an evolutionary conserved signaling mechanism. Curr Biol 29(12):1954–1962 e1954. https://doi.org/10.1016/j.cub.2019.04.073
13. Burkholder TJ, Fingado B, Baron S et al (1994) Relationship between muscle fiber types and sizes and muscle architectural properties in the mouse hindlimb. J Morphol 221(2):177–190. https://doi.org/10.1002/jmor.1052210207
14. Carper D, Coue M, Laurens C et al (2020) Reappraisal of the optimal fasting time for insulin tolerance tests in mice. Mol Metab:101058. https://doi.org/10.1016/j.molmet.2020.101058
15. Roberts NW, Holley-Cuthrell J, Gonzalez-Vega M et al (2015) Biochemical and functional comparisons of mdx and Sgcg(-/-) muscular dystrophy mouse models. Biomed Res Int 2015:131436. https://doi.org/10.1155/2015/131436
16. Sali A, Guerron AD, Gordish-Dressman H et al (2012) Glucocorticoid-treated mice are an inappropriate positive control for long-term preclinical studies in the mdx mouse. PLoS One 7(4):e34204. https://doi.org/10.1371/journal.pone.0034204

Chapter 25

High-Throughput Screening to Identify Modulators of Sarcospan

Cynthia Shu, Ekaterina Mokhonova, and Rachelle H. Crosbie

Abstract

High-throughput screening enables the discovery of disease-modifying small molecules. Here, we describe the development of a scalable, cell-based assay to screen for small molecules that modulate sarcospan for the treatment of Duchenne muscular dystrophy. We detail the hit validation pipeline, which includes secondary screening, gene/protein quantification, and an in vitro membrane stability assay.

Key words Duchenne muscular dystrophy, High-throughput screening, Sarcospan, Drug discovery

1 Introduction

Sarcospan (SSPN) is a 25-kDa transmembrane protein expressed in skeletal and cardiac muscle that interacts with three major laminin-binding complexes in muscle: the dystrophin-glycoprotein complex (DGC), the utrophin-glycoprotein complex (UGC), and the α7β1D-integrin complex. These complexes form a critical link between the cytoskeleton and the extracellular matrix (ECM) [1–8].

Mutations in the genes encoding for membrane complex proteins result in over 30 types of muscular dystrophy (MD) that cause muscle weakness and degeneration. The most common and lethal form of MD is Duchenne muscular dystrophy (DMD). DMD is caused by mutations in the dystrophin gene (*DMD*) on the X chromosome that result in little or no production of dystrophin [9]. Dystrophin is a vital protein for protecting muscle fibers from contraction induced injury [9]. It connects the cytoskeleton of a muscle fiber to the ECM through the cell membrane [10–16]. The DGC may also play a role in cell signaling by interacting with proteins that send and receive chemical signals. Loss of dystrophin from the cell membrane causes loss of the entire DGC from the sarcolemma. Without DGC, muscle cells are damaged and, over time, are replaced with scar tissue and fat in a process called fibrosis.

Rika Maruyama and Toshifumi Yokota (eds.), *Muscular Dystrophy Therapeutics: Methods and Protocols*,
Methods in Molecular Biology, vol. 2587, https://doi.org/10.1007/978-1-0716-2772-3_25,

DMD is one of the most frequent genetic conditions affecting approximately 1 in 3500–5000 male births worldwide. Onset of DMD symptoms occurs in early childhood, usually between ages 3 and 5. Patients lose the ability to walk at approximately 10 years of age, but with corticosteroid treatment, the initial age of wheelchair reliance is extended to 13 years of age [17]. Approximately 90% of patients over 18 years of age develop dilated cardiomyopathy and respiratory dysfunction, leading to premature death in the third decade of life [18, 19].

Currently, the only Food and Drug Administration approved treatments for DMD are corticosteroids and exon skipping therapies. In clinical trials, corticosteroids extended the age of loss of ambulation by several years, and exon skipping therapies slowed disease progression, but neither are not considered a cure [20, 21]. Although significant progress has been made in the development of DMD therapies, there remains a need for treatments that can improve the quality of life and life expectancy of individuals affected by DMD.

Investigations using several transgenic mouse models demonstrate that overexpression of sarcospan in the mouse model for Duchenne muscular dystrophy ameliorates pathology and restores muscle cell binding to laminin [22–24]. Sarcospan improves cell surface expression of the dystrophin- and utrophin-glycoprotein complexes as well as α7β1 integrin. SSPN modulation of integrin signaling is required for extracellular matrix attachment and muscle force development [24, 25]. SSPN overexpression also addresses cardiac and pulmonary complications, which are the leading causes of death in DMD patients [26].

For therapeutic purposes, small molecules are the most commonly used approach to treat diseases due to ease of delivery and reduced risk for immune responses relative to viral or cell-based therapies. In this chapter, we describe the development of high-throughput screening to identify modulators of sarcospan (workflow summarized in Fig. 1) [27]. New modulators of sarcospan are predicted to benefit patients with DMD and other forms of muscular dystrophies caused by deficits in muscle cell adhesion.

2 Materials

2.1 Molecular Cloning of Sarcospan Reporter Plasmids

1. Genomic DNA (Bioline).
2. Phusion High-Fidelity DNA Polymerase (New England Biolabs).
3. Primers used for reporter construct cloning and sequencing (Table 1).
4. PureLink HiPure Plasmid DNA Purification kit (Life Technologies).

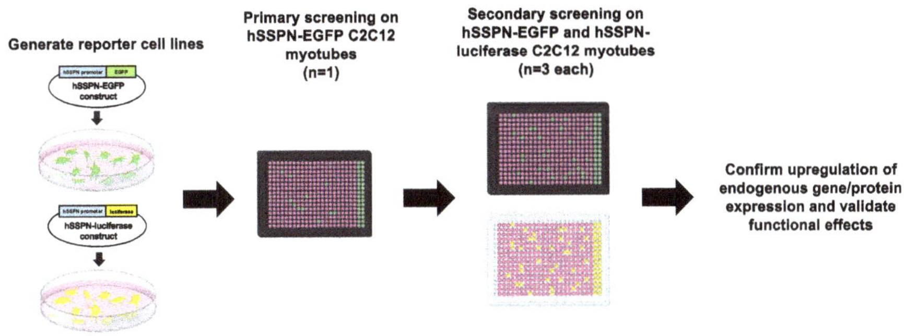

Fig. 1 Workflow for high-throughput screening for sarcospan modulators. (*1*) Generate promoter-reporter C2C12 cells using fluorescence and luminescence reporter constructs. (*2*) Conduct primary screening on hSSPN-EGFP C2C12 myotubes followed by secondary screening on both reporter cell lines to validate hits. (*3*) Confirm upregulation of endogenous gene and protein expression before conducting functional effects

Table 1
Primers used for reporter construct cloning and sequencing

Primer	Sequence (5′ → 3′)	Location
Universal forward primer	GTGTAGATCTCAGGTGGGTGTCC TGGTATAA	2 kb upstream of hSSPN TSS
EGFP construct reverse primer	GTGTAAGCTTCTCCTCCCCGCAC TCCTT	Exon 1 of hSSPN
Luciferase construct reverse primer	GTGTAAGCTTGCTCCTCCCCGCAC TCCTT	Exon 1 of hSSPN
EGFP sequencing forward primer 1	ATAACCGTATTACCGCCATGCATTA	25 bp upstream of hSSPN insert
Luciferase sequencing forward primer 1	CAGAACATTTCTCTGGCCTAACTGG	35 bp upstream of hSSPN insert
Universal sequencing forward primer 2	CTCTAAGTGCTACTGAGTAGAGGTA	600 bp within hSSPN insert
Universal sequencing forward primer 3	CAGCCACTTGGAGAC TGAGGAGAGA	1200 bp within hSSPN insert

5. NEBuffer 3.1 (New England Biolabs).
6. Zymoclean Gel DNA Recovery Kit (Zymo Research).
7. pmEGFP-1 plasmid (Addgene).
8. pGL4.17 Luc2 plasmid (Promega).
9. T4 DNA ligase (Invitrogen).
10. One Shot TOP10 chemically competent *E. coli* (Thermo Fisher).
11. LB broth (MP Biomedicals): 10 g tryptone, 5 g yeast extract, 10 g NaCl per liter final concentration.

12. Bacto Dehydrated Agar (BD): 1.5% final concentration in LB broth.
13. PureLink Quick Plasmid Miniprep (Life Technologies).
14. Plasmid Maxi Kit (Qiagen).

2.2 Generation of Stable Reporter Cell Lines

1. C2C12 cell line.
2. C2C12 growth medium: 20% FBS in DMEM.
3. C2C12 differentiation medium: 2% horse serum in DMEM.
4. G418.
5. Lipofectamine 3000 Transfection Reagent (Thermo Fisher Scientific).
6. Bright-Glo Luciferase Assay System (Promega).

2.3 High-Throughput Screening

1. FluoroBrite DMEM (Gibco).
2. Insulin Transferrin Selenium.
3. EL406 Combination Washer Dispenser (Biotek).
4. Multidrop 384 (Thermo Fisher Scientific).
5. BioMek FX Automated Workstation (Beckman Coulter Life Sciences).
6. ImageXpress Confocal High-Content Imaging System (Molecular Devices).
7. MetaXpress High-Content Image Acquisition and Analysis Software (Molecular Devices).
8. 384-well black, clear bottom plate, with lid (Greiner).
9. 70% ethanol, USP grade.
10. Hoechst.

2.4 Secondary Screening

1. Bright-Glo Luciferase Assay System (Promega).
2. Envision plate reader (PerkinElmer).

2.5 Culture of Dystrophin-Deficient Myoblasts for Hit Validation

1. Dystrophin-deficient murine and human cell line [28, 29].
2. H2K growth medium: 20% HI-FBS, 2% L-glutamine, 2% chicken embryo. extract, 1% penicillin-streptomycin, and 20 U/mL of fresh interferon gamma in DMEM. Filter to sterilize.
3. H2K differentiation medium: 5% horse serum, 2% L-glutamine, and 1% penicillin-streptomycin in DMEM. Keep sterile.
4. Gelatin (Sigma Aldrich). 0.01% final concentration in DIW. Filter to sterilize.

5. Matrigel (Corning). 0.1 mg/ml final concentration in DMEM. Keep sterile.
6. Skeletal muscle growth medium: Skeletal Muscle Growth Medium Kit (Promocell) and 1% penicillin-streptomycin. Keep sterile.
7. Skeletal muscle differentiation medium: Skeletal Muscle Differentiation Medium Kit (Promocell) and 1% penicillin-streptomycin. Keep sterile.

2.6 Validation of Gene and Protein Expression

1. Trizol.
2. Chloroform.
3. Ethyl Alcohol, 200 proof.
4. Direct-zol RNA Miniprep Plus RNA extraction kit (Zymo Research).
5. NanoDrop 1000 (Thermo Fisher Scientific).
6. iScript cDNA synthesis kit (Bio-Rad).
7. SsoFast EvaGreen Supermix (Bio-Rad).
8. Primers for mouse sarcospan and beta-actin (Table 2).
9. TaqMan Gene Expression Assay (Thermo Fisher Scientific).
10. QuantStudio 5 Real-Time PCR System (Thermo Fisher Scientific).
11. RIPA buffer (Thermo Fisher Scientific) containing Halt protease and phosphatase inhibitor.
12. DC Protein Assay (Bio-Rad).
13. Laemmli sample buffer: 5×; 15% sodium dodecyl sulfate, 50% glycerol, 0.33 M Tris-HCl, 25% β-mercaptoethanol, 1% bromophenol blue in DIW pH 6.8.

Table 2
Primers used for gene expression analysis of murine cells

Primer	Accession number	Sequence (5′ → 3′)	Location	Amplicon length	Amplification efficiency
SSPN F	NM_010656.3	TGCTAGTCAGAGATACTCCGTTC	Exons 1–2	103 bp	94%
SSPN R		GTCCTCTCGTCAACTTGGTATG			
ACTB F	NM_007393.5	TCCTGACCCTGAAGTACCCCAT	Exons 1–2	131 bp	104%
ACTB R		CTCGGTGAGCAGCACAGGGT			

14. Running buffer: 50 mM MOPS, 50 mM Tris base, 3.5 mM SDS, and 0.8 mM EDTA in DIW pH to 8.3.
15. Transfer buffer: 200 mM glycine, 25 mM Tris base, 20% methanol in DIW pH 8.3.
16. Tris-buffered saline (TBST): 25 mM Tris, 100 mM NaCl, 0.2% Tween-20, pH to 7.4.
17. Blocking buffer: 5% non-fat dried milk in TBST.
18. 4–12% Tris-glycine or Bis-Tris polyacrylamide gels (Life Technologies).
19. Nitrocellulose membrane.
20. Primary antibodies
    - SSPN N-terminus (sc-393187, Santa Cruz Biotechnology)
    - SSPN extracellular loop 2 (EC2) (LS-C747357, LifeSpan Biosciences)
    - GAPDH (Mab374, Millipore)
21. Secondary antibody
    - Goat anti-mouse IgG HRP (Abcam)
    - Goat anti-rabbit IgG HRP (Abcam)
22. Modified RIPA buffer: 1% Triton X-100, 0.05% DOC, 0.05% SDS, 10 mM DTT and Halt protease and phosphatase inhibitors in RIPA.
23. Modified TBST: TBS with 0.05% Tween-20.
24. Tetramethylbenzidine solution.
25. SuperSignal West Pico Chemiluminescent Substrate (Thermo Fisher Scientific).
26. Autoradiography films.
27. Film developer.
28. 96-well Nunc MaxiSorp plate (Invitrogen).
29. Recombinant human SSPN standard (Novus Biologicals).
30. Microplate reader.
31. Sonicator Sonic Dismembrator Model 100 (Fisher Scientific).

2.7 Membrane Stability Assay

1. Osmotic shock solution: 5 mM HEPES, 5 mM KCl, 1 mM $MgCl_2$, 5 mM NaCl, 1.2 mM $CaCl_2$, and 1 mM glucose. Add 26, 50, 173, 192 mM sucrose to reach osmolarities of 50, 80, 100, 280, and 300 mosmol.
2. VAPRO vapor pressure osmometer (Wescor Inc.).
3. Creatine kinase assay (Sekisui Diagnostics).
4. Microplate reader.

3 Methods

3.1 Molecular Cloning of SSPN Reporter Plasmids

1. Amplify a 2-kb region encompassing the SSPN promoter from human genomic DNA using Phusion High-Fidelity DNA Polymerase with the primers indicated in Table 1. The primers contain leader sequences and restriction sites for BglII (AGATCT) or HindIII (TTCGAA).
2. Purify PCR products using the PureLink HiPure Plasmid DNA Purification kit and digest with BglII and HindIII in NEBuffer 3.1.
3. Electrophorese 2-kb digested PCR products on agarose gels. Excise and purify using the Zymoclean Gel DNA Recovery Kit.
4. Digest plasmids with BglII and HindIII and ligate to PCR products using T4 DNA ligase.
5. Linearize plasmid constructs with BglII, which digests the region upstream of the SSPN promoter.
6. Use purified and linearized plasmids to transform One Shot TOP10 chemically competent *E. coli* grown on agar containing the appropriate antibiotic.
7. Confirm presence of SSPN promoter construct by colony PCR of individual colonies.
8. Inoculate positive colonies in liquid culture overnight.
9. Purify plasmids from bacterial cultures using the PureLink Quick Plasmid Miniprep and sequence using the primers listed in Table 2 to confirm the presence and accuracy of the SSPN promoter region.
10. Select bacterial cultures for large cultures and collected for plasmid purification using the Plasmid Maxi Kit.

3.2 Generation of Stable Reporter Cell Lines

1. Culture C2C12 myoblasts in C2C12 growth medium at 37 °C with 5% CO_2.
2. Transfect with 5 μg of linearized hSSPN-EGFP or hSSPN-luciferase plasmid using Lipofectamine 3000 Transfection Reagent.
3. After 24 h, remove transfection reagent and replace with C2C12 growth medium containing 800 μg/mL of G418 for 4 weeks to generate stable cell lines expressing reporter protein under control of the human SSPN promoter.
4. Validate expression of GFP-tagged protein using fluorescence microscopy and luciferase-tagged protein using the Bright-Glo Luciferase Assay System.

3.3 High-Throughput Screening

1. For primary screening, use an $n = 1$ well per compound.
2. Seed hSSPN-EGFP myoblasts at 500 cells per well in 50 μL of growth medium in 384-well black, clear bottom microplates using a Multidrop 384.
3. After 72 h, the cells should be 90–100% confluent. Using an EL406 combination washer dispenser, replace growth medium with 50 μL of C2C12 differentiation medium.
4. Incubate for 48 h and aspirate medium, leaving a residual volume of 10 μL, and add 30 μL of fresh C2C12 differentiation medium.
5. Using a Biomek Fx, add 0.5 μL of small molecule in DMSO or DMSO alone (for vehicle and positive control wells, respectively).
6. To ensure proper mixing of the DMSO, add an additional 50 μL of C2C12 differentiation medium to all wells except the positive control wells, which instead received 50 μL of medium containing insulin transferrin selenium (ITS) to reach a final concentration of 1% ITS. The final concentration of drug in each treated well is 5.5 μM in 0.55% DMSO and 0.55% DMSO only for vehicle and positive control wells.
7. Incubate for 48 h and replace medium with 50 μL of Fluorobrite DMEM.
8. Image each plate using ImageXpress Micro Confocal High Content Imaging System and determine fluorescence intensity of imaged cells using a custom module analysis in MetaXpress Analysis software. Analysis settings are as follows: top hat (size, 12; filter shape: circle), adaptive threshold (source: top hat; minimum width, 10; maximum width, 800; intensity above local background, 500), filter mask (filter type: minimum area filter; minimum value, 500).
9. Use Robust strictly standardized mean difference (SSMD*) to assess plate quality and for hit selection. $\mathrm{SSMD}^* = X_P - X_N / 1.4826\sqrt{s_P^2 + s_N^2}$, where XP, XN, SP, and SN are the medians and median absolute deviations of the positive and negative controls, respectively [30]. For plate quality, SSMD* ≥ 1 indicates a good quality moderate positive control.

3.4 Secondary Screening

1. Conduct secondary screening of hits from primary screen on both reporter cell lines ($n = 3$ wells per line).
2. Using Multidrop 384 seed 500 hSSPN-EGFP or hSSPN-luciferase myoblasts per well in 50 μL of C2C12 growth medium into 384-well black, clear bottom microplates (for hSSPN-EGFP) or white, clear bottom microplates (for hSSPN-luciferase).
3. Repeat screening protocol from Subheading 3.3.

4. For hSSPN-luciferase cells, assay as described in following steps.
5. Allow plates to equilibrate to RT before using an EL406 combination washer dispenser to aspirate medium and a Multidrop 384 to add 60 μL of a mixture containing equal parts of Bright-Glo Luciferase Assay System reagent and C2C12 differentiation medium.
6. After a 3-min incubation at RT, measure luminescence signal.
7. Calculate relative luminescence units (fold change of signal from treated cells over vehicle treated cells).
8. Determine which compounds reproducibly increased SSPN reporter expression in both cell lines.
9. The compounds that increase expression in all replicates are considered "confirmed hits," which then must be further validated to increase endogenous target gene and protein expression as described in the following sections.

3.5 Culture of Dystrophin-Deficient Myoblasts for Hit Validation

3.5.1 Gene Expression Analysis

1. For proliferation of conditionally immortalized H2K *mdx* myoblasts, coat flasks with 0.01% gelatin for 30 min at 37 °C.
2. Aspirate gelatin and add cells in H2K growth medium. Culture cells at 33 °C with 5% CO_2.
3. For differentiation of H2K *mdx* myoblasts, coat a plate with 0.1 mg/mL of Matrigel for 1 h at 37 °C.
4. Aspirate Matrigel and add cells in the H2K growth medium.
5. Upon reaching 90% confluence, change medium to H2K differentiation medium. Culture cells at 37 °C with 5% CO_2.

3.5.2 Culture of Dystrophin-Deficient Human Myoblasts

1. For proliferation of human myoblasts, culture cells in Skeletal muscle growth medium at 37 °C with 5% CO_2.
2. For differentiation of human myoblasts, allow cells to reach 100% confluence and replace growth medium with Skeletal muscle differentiation medium. Culture cells at 37 °C with 5% CO_2.

3.6 Validation of Gene and Protein Expression

3.6.1 Gene Expression Analysis

1. Treat mouse and/or human myotubes for 48 h with 0–50 μM of confirmed hits.
2. Lyse cells in Trizol and purify RNA using the Direct-zol RNA Miniprep Plus RNA extraction kit.
3. Determine RNA concentrations using a NanoDrop and add 750 ng of RNA into a 20 μL reaction of the iScript cDNA synthesis kit with the following cycling conditions: 25 °C for 5 min, 42 °C for 30 min, 85 °C for 5 min.
4. For mouse gene expression, set up a reaction containing Sso-Fast EvaGreen Supermix, 400 nM of each primer listed in Table 2, and cDNA corresponding to 37.5 ng RNA.

5. Run the following reaction using the QuantStudio 5 Real-Time PCR System: 55 °C for 2 min, 95 °C for 2 min, 40 cycles of 95 °C for 10 s, and 62 °C for 30 s.
6. For qPCR of human samples, use TaqMan assays to quantify SSPN (assay ID Hs01025520m_1) and ACTB (assay ID Hs01060665_g1) with the following reaction conditions: 50 °C for 2 min, 95 °C for 10 min, 40 cycles of 95 °C for 15 s, and 62 °C for 1 min.
7. Analyze data using the ddCT method and normalize to reference gene ACTB with vehicle-treated samples serving as the calibrator (relative expression of vehicle control = 1).

3.6.2 Immunoblotting

1. Treat mouse and/or human myotubes for 48 h with confirmed hits.
2. Lyse cells in RIPA buffer (Thermo Fisher Scientific) and rock for 1 h at 4 °C (*see* **Note 1**).
3. Centrifuge at 1000 RPM for 30 min at 4 °C and collect supernatant.
4. Determine protein concentration using the DC protein assay and normalize all samples to 2 mg/mL in RIPA and 1× Laemmli sample buffer.
5. Heat samples to 95 °C for 2 min before loading 40 μg onto 4–12% Tris-glycine or Bis-Tris polyacrylamide gel and run for 2 h at 100 volts at RT.
6. Transfer to a nitrocellulose membrane for 2 h at 100 volts at 4 °C (*see* **Note 2**).
7. Block nitrocellulose membranes with blocking buffer for 1 h at RT.
8. Add SSPN N-terminus (1:200) and GAPDH (1:10,000) primary antibodies diluted in fresh blocking buffer and incubate on a rocker overnight at 4 °C.
9. Following three 10 min TBST washes, incubate the membranes in goat anti-mouse IgG HRP (1:5000 for SSPN, 1: 10,000 for GAPDH in 5% milk) for 1 h at RT on a rocker.
10. Following three 10 min TBST washes, incubate the membranes in SuperSignal West Pico Chemiluminescent for 5 min at RT on an orbital shaker and process membranes to determine luminescence signal.
11. Normalize target protein to loading control GAPDH with vehicle-treated cells serving as the calibrator sample (relative protein levels of vehicle control = 1) (*see* **Note 3**).

3.6.3 ELISA to Quantify SSPN Protein Levels

1. Lyse treated DMD myotubes with modified RIPA (*see* **Note 1**) buffer on ice.
2. Sonicate three times, 10s/each with the power output 2–3 (scale "0–10") at the probe tip (*see* **Note 2**).
3. Centrifuge lysate at 12,000 × *g* for 10 min at 4 °C.
4. Collect supernatant and determine protein concentration by DC protein assay (*see* **Note 3**).
5. For preparation of ELISA plate, coat a 96-well Nunc MaxiSorp plate with 100 μL of 1 μg/mL SSPN N-terminus antibody per well overnight at 4 °C.
6. Wash three times with modified TBST.
7. Block for 1 h at RT in 1% bovine serum albumin and wash three times with 1× TBST.
8. Add 100 μL of recombinant human SSPN standard (*see* **Note 4**) and lysates (*see* **Note 5**) to the plate, incubate overnight at 4 °C, and wash three times with 1X TBST.
9. Add 100 μL of 400 ng/mL SSPN EC2 antibody to each well, incubate for 2 h at RT, and wash three times with 1× TBST.
10. Add 100 μL of Goat anti-rabbit IgG HRP (1:10,000) to each well, incubate for 2 h at RT, and wash three times with 1× TBST.
11. Add 100 μL of tetramethylbenzidine to each well and incubate for 30–40 min at RT. Protect from light during incubation.
12. Add 100 μL of stop solution to each well and measure absorbance at 450 nm.
13. Calculate the concentration of SSPN protein for each sample using the standard curve and normalize to protein concentration and vehicle-treated controls.

3.7 Membrane Stability Assay

1. Add sucrose to osmotic shock base solution to reach osmolarities of 50, 80, 100, 280, and 300 mosmol.
2. Determine actual osmolarity using a osmometer.
3. Culture dystrophin-deficient myotubes in a 24-well plate.
4. After 48 h of treatment, subject myotubes to 20 min of osmotic shock at 37 °C using 28.5–223.5 milliosmole (mosmol) solutions.
5. Collect the supernatant, centrifuge at 1000 RPM for 5 min to pellet cell debris, and collect supernatant into new tube. Keep on ice or freeze at −80 °C for later use.
6. Trypsinize adherent cells, centrifuge at 1000 RPM for 5min, remove supernatant, and resuspend pellet in 50 μL DIW before lysing cells by three freeze-thaw cycles using liquid nitrogen. Keep on ice or freeze at −80 °C for later use.

7. Measure creatine kinase (CK) levels in both the supernatant and lysate fractions. In a 96-well plate, add 4 μL of each sample and 140 μL of reagent per well in triplicate.
8. Calculate the units per liter (U/L) as follows: (mOD/min)(total volume in mL)(dilution factor)/(6.22 M^{-1} cm^{-1})(light path in cm)(sample volume in mL).
9. The percent CK release is calculated as follows: $CK_{extracellular}/(CK_{extracellular} + CK_{intracellular}) * 100$.

4 Notes

Section: Generation of Stable Reporter Cell Lines

1. At this stage, do not let C2C12 myoblasts reach over 80% confluency, or they will lose their ability to differentiate.
2. The recommended concentration of G418 for antibiotic selection of C2C12 cells containing reporter constructs is between 500 and 4000 μg/mL. We found that 800 μg/mL G418 was sufficient to induce death of non-transfected C2C12 cells after one week. This concentration may vary between batches of cells. Therefore, we recommend conducting a kill curve experiment using a broad range of G418 to determine the optimal concentration for your cells.
3. After generation of stable reporter cells lines, validate expression of the reporter proteins in both myoblasts and myotubes. This ensures the reporter is being expressed after differentiation and that the myoblasts retain their ability to differentiate.

Section: High-Throughput Screening

1. For all solutions used in screening, make 20% excess to account for dead volume lost in media dispensing machines.
2. For the positive control, make a 1.8% ITS solution. Adding 50 μL of the 1.8% ITS to 40 μL of volume in wells leads to a final concentration of 1% ITS in 90 μL of media.

Section: Secondary Screening

1. Ensure cells are fully lysed after addition of the Bright-Glo Luciferase Assay System reagent to achieve full reporter detection.

Section: Immunoblotting

1. For every lysis, use RIPA buffer with freshly added protease and phosphatase inhibitors.

2. We typically use 0.45 μM pore size nitrocellulose membrane. In some cases, when amount of the SSPN in analyzed samples is very low, using 0.2 μM pore sizes nitrocellulose membrane may be beneficial.
3. If normalization to total protein is required, stain the membrane with Ponceau s or other commercially available *Total Protein Stain reagent after transfer and capture image of total protein before blocking the membrane. Ponceau s can be washed off with PBS or TBS before blocking. Commercially available reagents may require using appropriate destaining solutions before proceeding to the blocking step.*

Section: ELISA to Quantify SSPN Protein Levels

1. For every lysis, use mRIPA buffer with freshly added protease and phosphatase inhibitors. Lyse cells in mRIPA buffer without DTT as DTT may interfere with DC protein assay.
2. Keep samples on ice between sonication rounds. Quickly rinse with ddH_2O and wipe sonicator tip between samples to eliminate sample carryover.
3. After determination of protein concentration, add DTT to the samples to a final concentration 10 mM.
4. Standard curves generated using recombinant human SSPN (rhSSPN) titration curve. Make serial dilutions of rhSSPN in mRIPA with DTT. Concentrations of the standard protein for the curve depend on the level of protein of interest (sarcospan) in the analyzed samples. We typically use 7-point 1:2 serial dilution starting with 0.5 ng/mL to 0.008 ng/mL of rhSSPN. For the samples with lower concentrations of sarcospan, we use 7-point 1:2 serial dilution starting with 0.2 ng/mL to 0.003 ng/mL of rhSSPN.
5. Run three to four technical replicates of each sample. If volume of material is limited, try serial dilutions of the samples and chose the dilution of the samples that would not compromise ELISA signal. We use samples diluted 1:4 or 1:6 in mRIPA buffer.

Acknowledgments

This work was supported by the Muscle Cell Biology, Pathogenesis, and Therapeutics Training Grant (NIH T32 AR065972); Pilot and Feasibility Seed Grant program (NIH/NIAMS P30 AR057230); UCLA Department of Integrative Biology & Physiology Eureka Scholarship; Muscular Dystrophy Association (MDA 274143; Venture Philanthropy Program); NIH NIAMS (R01 AR048179); and NIH NHLBI (R01 HL126204).

References

1. Marshall JL, Crosbie-Watson RH (2013) Sarcospan: a small protein with large potential for Duchenne muscular dystrophy. Skelet Muscle 3(1):1. https://doi.org/10.1186/2044-5040-3-1
2. Crosbie RH, Heighway J, Venzke DP, Lee JC, Campbell KP (1997) Sarcospan, the 25-kDa transmembrane component of the dystrophin-glycoprotein complex. J Biol Chem 272 (50):31221–31224. https://doi.org/10.1074/jbc.272.50.31221
3. Barresi R, Moore SA, Stolle CA, Mendell JR, Campbell KP (2000) Expression of gamma-sarcoglycan in smooth muscle and its interaction with the smooth muscle sarcoglycan-sarcospan complex. J Biol Chem 275 (49):38554–38560. https://doi.org/10.1074/jbc.M007799200
4. Crosbie RH, Lebakken CS, Holt KH, Venzke DP, Straub V, Lee JC, Grady RM, Chamberlain JS, Sanes JR, Campbell KP (1999) Membrane targeting and stabilization of sarcospan is mediated by the sarcoglycan subcomplex. J Cell Biol 145(1):153–165. https://doi.org/10.1083/jcb.145.1.153
5. Crosbie RH, Lim LE, Moore SA, Hirano M, Hays AP, Maybaum SW, Collin H, Dovico SA, Stolle CA, Fardeau M, Tomé FM, Campbell KP (2000) Molecular and genetic characterization of sarcospan: insights into sarcoglycan-sarcospan interactions. Hum Mol Genet 9 (13):2019–2027. https://doi.org/10.1093/hmg/9.13.2019
6. Parvatiyar MS, Marshall JL, Nguyen RT, Jordan MC, Richardson VA, Roos KP, Crosbie-Watson RH (2015) Sarcospan regulates cardiac isoproterenol response and prevents duchenne muscular dystrophy-associated cardiomyopathy. J Am Heart Assoc 4(12):e002481. https://doi.org/10.1161/JAHA.115.002481
7. Parvatiyar MS, Brownstein AJ, Kanashiro-Takeuchi RM, Collado JR, Dieseldorff Jones KM, Gopal J, Hammond KG, Marshall JL, Ferrel A, Beedle AM, Chamberlain JS, Renato Pinto J, Crosbie RH (2019) Stabilization of the cardiac sarcolemma by sarcospan rescues DMD-associated cardiomyopathy. JCI Insight 5(11):e123855. https://doi.org/10.1172/jci.insight.123855
8. Gibbs EM, Marshall JL, Ma E, Nguyen TM, Hong G, Lam JS, Spencer MJ, Crosbie-Watson RH (2016) High levels of sarcospan are well tolerated and act as a sarcolemmal stabilizer to address skeletal muscle and pulmonary dysfunction in DMD. Hum Mol Genet 25 (24):5395–5406. https://doi.org/10.1093/hmg/ddw356
9. Lovering RM, Porter NC, Bloch RG (2005) The muscular dystrophies: from genes to therapies. Phys Ther 85(12):1372–1388
10. Petrof BJ, Shrager JB, Stedman HH, Kelly AM, Sweeney HL (1993) Dystrophin protects the sarcolemma from stresses developed during muscle contraction. Proc Natl Acad Sci U S A 90(8):3710–3714. https://doi.org/10.1073/pnas.90.8.3710
11. Ervasti JM, Campbell KP (1993) A role for the dystrophin-glycoprotein complex as a transmembrane linker between laminin and actin. J Cell Biol 122(4):809–823. https://doi.org/10.1083/jcb.122.4.809
12. Ervasti JM, Campbell KP (1991) Membrane organization of the dystrophin-glycoprotein complex. Cell 66(6):1121–1131. https://doi.org/10.1016/0092-8674(91)90035-W
13. Ervasti JM, Kahl SD, Campbell KP (1991) Purification of dystrophin from skeletal muscle. J Biol Chem 266(14):9161–9165. https://doi.org/10.1016/S0021-9258(18)31565-5
14. Matsumura K, Ervasti JM, Ohlendieck K, Kahl SD, Campbell KP (1992) Association of dystrophin-related protein with dystrophin-associated proteins in mdx mouse muscle. Nature 360(6404):588–591. https://doi.org/10.1038/360588a0
15. Ervasti JM, Ohlendieck K, Kahl SD, Gaver MG, Campbell KP (1990) Deficiency of a glycoprotein component of the dystrophin complex in dystrophic muscle. Nature 345 (6273):315–319. https://doi.org/10.1038/345315a0
16. Ibraghimov-Beskrovnaya O, Ervasti JM, Leveille CJ, Slaughter CA, Sernett SW, Campbell KP (1992) Primary structure of dystrophin-associated glycoproteins linking dystrophin to the extracellular matrix. Nature 355(6362):6696–6702. https://doi.org/10.1038/355696a0
17. Koeks Z, Bladen CL, Salgado D, van Zwet E, Pogoryelova O, McMacken G, Monges S, Foncuberta ME, Kekou K, Kosma K, Dawkins H, Lamont L, Bellgard MI, Roy AJ, Chamova T, Guergueltcheva V, Chan S, Korngut L, Campbell C, Dai Y, Wang J, Barišić N, Brabec P, Lähdetie J, Walter MC, Schreiber-Katz O, Karcagi V, Garami M, Herczegfalvi A, Viswanathan V, Bayat F, Buccella F, Ferlini A, Kimura E, van den Bergen JC, Rodrigues M, Roxburgh R, Lusakowska A, Kostera-Pruszczyk A, Santos R, Neagu E, Artemieva S, Rasic

VM, Vojinovic D, Posada M, Bloetzer C, Klein A, Díaz-Manera J, Gallardo E, Karaduman AA, Oznur T, Topaloğlu H, El Sherif R, Stringer A, Shatillo AV, Martin AS, Peay HL, Kirschner J, Flanigan KM, Straub V, Bushby K, Béroud C, Verschuuren JJ, Lochmüller H (2017) Clinical outcomes in duchenne muscular dystrophy: a study of 5345 patients from the TREAT-NMD DMD global database. J Neuromuscul Dis 4 (4):293–306. https://doi.org/10.3233/JND-170280

18. Ryder S, Leadley RM, Armstrong N, Westwood M, de Kock S, Butt T, Jain M, Kleijnen J (2017) The burden, epidemiology, costs and treatment for Duchenne muscular dystrophy: an evidence review. Orphanet J Rare Dis 12 (1):79. https://doi.org/10.1186/s13023-017-0631-3
19. Nigro G, Comi LI, Politano L, Bain RJ (1990) The incidence and evolution of cardiomyopathy in Duchenne muscular dystrophy. Int J Cardiol 26(3):271–277. https://doi.org/10.1016/0167-5273(90)90082-g
20. Schnell F, Donoghue C, Dworzak J, Charleston J, Frank D, Wilton S (2017) Development of a validated Western blot method for quantification of human dystrophin protein used in Phase 700 II and III clinical trials of eteplirsen for the treatment of Duchenne muscular dystrophy (DMD). Neurology 88(16 Supplement):5.105
21. Moxley RT 3rd, Pandya S, Ciafaloni E, Fox DJ, Campbell K (2010) Change in natural history of Duchenne muscular dystrophy with long-term corticosteroid treatment: implications for management. J Child Neurol 9:1116–1129. https://doi.org/10.1177/0883073810371004
22. Marshall JL, Holmberg J, Chou E, Ocampo AC, Oh J, Lee J, Peter AK, Martin PT, Crosbie-Watson RH (2012) Sarcospan-dependent Akt activation is required for utrophin expression and muscle regeneration. J Cell Biol 197 (7):1009–1027. https://doi.org/10.1083/jcb.201110032
23. Mamsa H, Stark RL, Shin KM, Beedle AM, Crosbie RH (2022) Sarcospan increases laminin-binding capacity of α-dystroglycan to ameliorate DMD independent of Galgt2. Hum Mol Genet. 31(5):718–732. https://doi.org/10.1093/hmg/ddab276
24. Peter AK, Marshall JL, Crosbie RH (2008) Sarcospan reduces dystrophic pathology: stabilization of the utrophin-glycoprotein complex. J Cell Biol 183(3):419–427. https://doi.org/10.1083/jcb.200808027
25. Marshall JL, Chou E, Oh J, Kwok A, Burkin DJ, Crosbie-Watson RH (2012) Dystrophin and utrophin expression require sarcospan: loss of α7 integrin exacerbates a newly discovered muscle phenotype in sarcospan-null mice. Hum Mol Genet 21(20):4378–4393. https://doi.org/10.1093/hmg/dds271
26. Shu C, Parfenova L, Mokhonova E, Collado JR, Damoiseaux R, Campagna J, John V, Crosbie RH (2020) High-throughput screening identifies modulators of sarcospan that stabilize muscle cells and exhibit activity in the mouse model of Duchenne muscular dystrophy. Skelet Muscle 10(1):26. https://doi.org/10.1186/s13395-020-00244-3
27. Shu C, Kaxon-Rupp AN, Collado JR, Damoiseaux R, Crosbie RH (2019) Development of a high-throughput screen to identify small molecule enhancers of sarcospan for the treatment of Duchenne muscular dystrophy. Skelet Muscle 9(1):32. https://doi.org/10.1186/s13395-019-0218-x
28. Zhang XD (2011) Illustration of SSMD, z score, SSMD*, z* score, and t statistic for hit selection in RNAi high-throughput screens. J Biomol Screen 16(7):775–785. https://doi.org/10.1177/1087057111405851
29. Morgan JE, Beauchamp JR, Pagel CN, Peckham M, Ataliotis P, Jat PS, Noble MD, Farmer K, Partridge TA (1994) Myogenic cell lines derived from transgenic mice carrying a thermolabile T antigen: a model system for the derivation of tissue-specific and mutation-specific cell lines. Dev Biol 162(2):486–498. https://doi.org/10.1006/dbio.1994.1103
30. Mamchaoui K, Trollet C, Bigot A, Negroni E, Chaouch S, Wolff A, Kandalla PK, Marie S, Di Santo J, St Guily JL, Muntoni F, Kim J, Philippi S, Spuler S, Levy N, Blumen SC, Voit T, Wright WE, Aamiri A, Butler-Browne G, Mouly V (2011) Immortalized pathological human myoblasts: towards a universal tool for the study of neuromuscular disorders. Skelet Muscle 1(1):34. https://doi.org/10.1186/2044-5040-1-34

Chapter 26

Identifying FDA-Approved Drugs that Upregulate Utrophin A as a Therapeutic Strategy for Duchenne Muscular Dystrophy

Christine Péladeau and Bernard J. Jasmin

Abstract

Duchenne muscular dystrophy (DMD) is a neuromuscular disease caused by mutations and deletions within the DMD gene, which result in a lack of dystrophin protein at the sarcolemma of skeletal muscle fibers. The absence of dystrophin fragilizes the sarcolemma and compromises its integrity during cycles of muscle contraction, which, progressively, leads to reductions in muscle mass and function. DMD is thus a progressive muscle-wasting disease that results in a loss of ambulation, cardiomyopathy, respiratory impairment, and death. Although there is presently no cure for DMD, recent advances have led to many promising treatments. One such approach entails increasing expression of a homologous protein to dystrophin, named utrophin A, which is endogenously expressed in both healthy and DMD muscle fibers. Upregulation of utrophin A all along the sarcolemma of DMD muscle fibers can, in part, compensate for the absence of dystrophin. Over the years, our laboratory has focused a significant portion of our efforts in identifying and characterizing drugs and small molecules for their ability to target utrophin A and cause its overexpression. As part of these efforts, we have recently developed a novel ELISA-based high-throughput drug screen, to identify FDA-approved drugs that increase the expression of utrophin A in muscle cells in culture as well as in dystrophic mice. Here, we describe our overall strategy to identify and characterize several FDA-approved drugs that upregulate utrophin A expression and provide details on all experimental approaches. Such strategy has the potential to lead to the rapid development of novel therapeutics for DMD.

Key words Duchenne muscular dystrophy, FDA-approved drugs, Disease, Utrophin, Muscle, Drug screen, ELISA

1 Introduction

Duchenne muscular dystrophy (DMD) is a genetic neuromuscular disorder caused by mutations or deletions in the dystrophin gene, thereby preventing the production of a functional dystrophin protein in skeletal muscle of afflicted patients. Dystrophin normally accumulates at the sarcolemma of all muscle fibers where it ensures maintenance of myofiber integrity throughout repeated muscle

Rika Maruyama and Toshifumi Yokota (eds.), *Muscular Dystrophy Therapeutics: Methods and Protocols*, Methods in Molecular Biology, vol. 2587, https://doi.org/10.1007/978-1-0716-2772-3_26,

contractions. A lack of dystrophin, as seen in DMD muscle, results in cycles of muscle fiber degeneration and regeneration. Eventually, the muscle's ability to regenerate its fibers fails, and muscle beds become infiltrated with fat and connective tissues causing a major loss in muscle strength and function.

In both healthy and DMD skeletal muscles, utrophin A is preferentially expressed at myotendinous (MTJ) and neuromuscular (NMJ) junctions [1–5]. Because of its sequence identity and functional similarity to dystrophin, utrophin A is an attractive target for the treatment of DMD. This concept involves stimulating via different means, including pharmacological and genetic approaches [6–19], expression of utrophin A throughout dystrophic muscle fibers such that its expression is no longer restricted to the MTJ and NMJ. In this context, several studies have demonstrated that increasing utrophin A expression all along the sarcolemma of dystrophic muscle fibers is an effective approach to compensate for the lack of dystrophin [20]. Thus, many preclinical studies have focused over the years on developing strategies to counteract the detrimental effects of DMD by upregulating the endogenous levels of utrophin A in dystrophic muscle [17, 18, 21–30].

Among the approaches to upregulate utrophin A levels, pharmacological interventions appear particularly promising for the treatment of DMD because it offers several clear advantages. For example, this therapeutic approach is applicable to all DMD patients, regardless of the dystrophin mutation. Moreover, a pharmacological approach stimulating utrophin A expression would also likely be efficient in positively impacting all muscles throughout the body. Finally, its ease of administration would also be another key advantage as it would be similar to taking medication orally.

Our overall strategy in our laboratory has focused on capitalizing on the availability of FDA-approved drugs that are clinically used for other diseases. Such a repurposing approach may accelerate considerably the development and implementation of effective therapeutics for DMD while reducing toxicological concerns. As part of our ongoing work, we have recently developed a novel, ELISA-based, high-throughput drug screen to identify FDA-approved drugs that stimulate the expression of utrophin A in skeletal muscle cells [23]. Hits from the screen were then comprehensively tested in multiple systems for their ability to increase utrophin A. Here, we describe our overall strategy and technical approaches that recently resulted in the identification and characterization of several FDA-approved drugs that increase utrophin A expression, thus paving the way for the use of novel therapeutics against DMD.

2 Materials

2.1 In-Cell ELISA High-Throughput Drug Screen

1. In-Cell ELISA Colorimetric Detection Kit (Thermo Fisher Scientific, Massachusetts, USA)
2. 384-well microplates
3. Screenwell FDA-approved drug library V2 (Enzo Life Sciences, Cederlane, Ontario, Canada)
4. Phosphate-buffered saline: (PBS, 10×): 80 g NaCl, 2 g KCl, 14.4 g Na_2HPO_4 2.4 g KH_2PO_4 dissolved in 1 L ddH_2O
5. Tris-buffered saline: (TBS, 20×): Dissolve 48 g of Tris and 176 g of NaCl in 800 mL of double distilled H_2O (ddH_2O). Adjust pH to 7.5 and complete the volume up to 1 L with ddH_2O
6. 4% formaldehyde: 2.75 mL of 16% methanol-free formaldehyde in 8.25 mL of 1× TBS
7. C2C12 myoblasts (American Type Culture Collection, Virginia, USA)
8. Complete growth media: Dulbecco's Modified Eagle Medium (DMEM), 10% fetal bovine serum (FBS), 1% penicillin-streptomycin
9. Utrophin A antibody (1:500; Novocastra, Leica biosystems, Concord, ON, Canada)
10. Synergy H1 microplate reader

2.2 Confirmation Western Blotting

1. C2C12 myoblasts (American Type Culture Collection)
2. Complete growth media: DMEM, 10% fetal bovine serum (FBS), 1% penicillin-streptomycin
3. 6-well culture plates
4. Cell scrapers
5. Urea extraction buffer: 7 M urea, 2 M thiourea, 4 M 3-[(3-cholamidopropyl) dimethylammonio]-1-propanesulfonate hydrate (CHAPS), 100 mM 1,4-dithiothreitol (DTT), and 125 mM Tris-HCL at pH 6.8. Dissolve and complete volume to 10 mL.
6. CB-X Protein Assay kit (G-Biosciences, Missouri, USA)
7. Loading buffer (2×): 430 μL of 70% glycerol (dissolved in ddH_2O), 300 μL of 20% sodium dodecyl sulfate (SDS) (dissolved in ddH_2O), 150 μL of 1 M Tris (pH 6.8), 100 μL of 2-mercaptoethanol, 20 μL of bromophenol blue solution
8. Tris-glycine (10×): 30 g Tris, 144 g glycine, 10 g SDS. Dissolve in 1 L of ddH_2O

9. Resolving gel (7%): 14.9 mL of ddH_2O, 7.0 mL of 30% acrylamide mix, 7.5 mL of 1.5 M Tris (pH 8.8), 300 μL of 10% SDS solution (dissolved in ddH_2O), 300 μL of 10% ammonium persulfate (dissolved in ddH_2O), 24 μL of *N,N,N′,N-′*-Tetramethylethylenediamine (TEMED)
10. Stacking gel (5%): 5.5 mL of ddH_2O, 1.3 mL of 30% acrylamide mix, 1.0 mL of 1.0 M Tris (pH 6.8), 80 μL of 10% SDS solution (dissolved in ddH_2O), 80 μL of 10% ammonium persulfate (dissolved in ddH_2O), 8 μL of TEMED
11. 0.45 μm nitrocellulose membrane
12. Whatman paper
13. Transfer Buffer (1×): 6 g Tris, 28.8 g glycine, 0.4 g SDS, 400 mL methanol. Complete the volume to 2 L with ddH_2O
14. 1× PBS containing 0.02% Tween-20 (PBST)
15. 20% Glycerol solution: 20% glycerol, 50 mM Tris at pH 7.4 in 1 L of ddH_2O
16. 5% Blocking milk solution: 10 g powdered milk, 20 mL of 10× PBS, 200 μL Tween-20. Dissolve in ddH_2O and complete the volume to 200 mL
17. 1% milk solution: 0.5 g powdered milk, 5 mL of 10× PBS, 50 μL Tween-20. Dissolve in ddH_2O and complete the volume to 50 mL
18. Ponceau solution (10×): 1 g of ponceau, 9.2 mL of trichloroacetic acid. Complete to 50 mL with ddH_2O
19. Utrophin A antibody (Novocastra)
20. Goat anti-mouse HRP-conjugated IgG secondary antibody (Jackson Immuno Research, Bar Harbor, USA)
21. Anti-β-Actin Antibody (C4) HRP (Santa-Cruz, Dallas, Texas, USA)
22. Chemiluminescent HRP detection reagent

2.3 FDA-Approved Drug Treatment in Dystrophic Mice

2.3.1 Drug Treatment in Mice

1. C57BL/10ScSn-Dmdmdx/J (mdx) mice and C57BL/10 (wild-type) mice (The Jackson Laboratory, Bar Harbor, USA)
2. 1 mL insulin syringes
3. Sterile saline

2.3.2 Mice Muscle Dissection and Protein Extraction

1. 70% ethanol diluted in ddH_2O
2. Euthanyl solution (65 mg/ml)
3. Cryotubes
4. Fine scissors and tweezers for dissections

5. Optimal cutting temperature (OCT) media (Thermo Fisher Scientific)
6. Tissue OCT plastic cryomold
7. Isopentane solution (2-methylbutane)
8. Kimwipes
9. Stainless steel tissue BioPulvarizer cooled on dry ice and a standard hammer
10. Small spatula
11. Pulsing vortex mixer (VWR, Mississauga, Ontario, Canada)

2.3.3 Immunofluorescence

1. Glass microscope slides (length: 75 mm, width: 25 mm)
2. M.O.M Immunodetection kit (Vector Laboratories, Burlington, ON, Canada)
3. Phosphate-buffered saline: (PBS, 10×): 80 g NaCl, 2 g KCl, 14.4 g Na_2HPO_4 2.4 g, KH_2PO_4 dissolved in 1 L ddH_2O. Create a 1 L 1× PBS solution by diluting 100 mL of 10× PBS with 900 mL of ddH_2O.
4. Utrophin A antibody (Novocastra)
5. Laminin antibody (Sigma-Aldrich, Oakville, Canada)
6. Texas Red conjugated streptavidin antibody (Vector laboratories)
7. Goat anti-rabbit Alexa Fluor 488 IgG secondary antibody (Thermo Fisher Scientific)
8. Vectashield containing DAPI staining (Vector Laboratories)
9. Rectangular glass coverslip
10. Fluorescent microscope

3 Methods

3.1 Drug Identification by an In-Cell ELISA High-Throughput FDA-Approved Drug Screen

1. Using the 384-well microplate, load 5 μl of each drug provided from the FDA-approved drug library, or phosphate-buffered saline as a control. Load each drug at a concentration based on doses used in the clinic by patients as listed in the Drug Bank database [31] (*see* Fig. 1). Perform this in three replicates.
2. Spin down the microplates at 1000 rpm for 5 min before seeding the cells.
3. Seed 2000 cells per well in 45 μL complete growth media, directly in the 384-well plate containing the drugs and incubate overnight in a 37 °C incubator at 5% CO_2 (*see* **Note 1**).
4. On the next day, remove the media and add 50 μL of 4% formaldehyde to fix the cells. Incubate for 15 min at room

	1	2	3	4	5	6	7	8	9	10	11	12	13	14	15	16	17	18	19	20	21	22	23	24
A	1	2	3	4	5	6	7	8	9	10	11	12	13	14	15	16	257	273	289	305	321	337	353	389
B	17	18	19	20	21	22	23	24	25	26	27	28	29	30	31	32	258	274	290	306	322	338	354	390
C	33	34	35	36	37	38	39	40	41	42	43	44	45	46	47	48	259	275	291	307	323	339	355	391
D	49	50	51	52	53	54	55	56	57	58	59	60	61	62	63	64	260	276	292	308	324	340	356	392
E	65	66	67	68	69	70	71	72	73	74	75	76	77	78	79	80	261	277	293	309	325	341	357	393
F	81	82	83	84	85	86	87	88	89	90	91	92	93	94	95	96	262	278	294	310	326	342	358	394
G	97	98	99	100	101	102	103	104	105	106	107	108	109	110	111	112	263	279	295	311	327	343	359	395
H	113	114	115	116	117	118	119	120	121	122	123	124	125	126	127	128	264	280	296	312	328	344	360	396
I	129	130	131	132	133	134	135	136	137	138	139	140	141	142	143	144	265	281	297	313	329	345	361	397
J	145	146	147	148	149	150	151	152	153	154	155	156	157	158	159	160	266	282	298	314	330	346	362	398
K	161	162	163	164	165	166	167	168	169	170	171	172	173	174	175	176	267	283	299	315	331	347	363	399
L	177	178	179	180	181	182	183	184	185	186	187	188	189	190	191	192	268	284	300	316	332	348	364	400
M	193	194	195	196	197	198	199	200	201	202	203	204	205	206	207	208	269	285	301	317	333	349	365	401
N	209	210	211	212	213	214	215	216	217	218	219	220	221	222	223	224	270	286	302	318	334	350	366	SAL
O	225	226	227	228	229	230	231	232	233	234	235	236	237	238	239	240	271	287	303	319	335	351	367	SAL
P	241	242	243	244	245	246	247	248	249	250	251	252	253	254	255	256	272	288	304	320	336	352	368	SAL

C2C12 Cells

Drugs
Saline

Fig. 1 FDA-approved drug plate layout. Each well is seeded with 2000 C2C12 cells/well and treated with a different FDA-approved drug or saline as a control for 24 h

temperature in a fume hood. Remove formaldehyde and wash the microplate twice with 50 μL of 1× TBS.

5. Using the In-Cell ELISA Colorimetric Detection Kit and provided solutions: Add 50 μL of 1× permeabilization buffer (0.11 mL of Thermo Scientific Surfact-Amps X-100 in 11 mL 1× TBS). Incubate for 15 min at room temperature. Remove the permeabilization buffer and wash twice with 1× TBS.
6. Add 50 μL of quenching solution (0.38 mL of 30% H_2O_2 in 11 mL 1× TBS). Incubate for 20 min at room temperature. Remove the quenching solution and wash twice with 1× TBS.
7. Add 50 μL of blocking buffer. Incubate for 30 min at room temperature. Remove the blocking buffer.
8. Add 50 μL of diluted primary utrophin A antibody at a dilution of 1:500, in a solution of 50% blocking buffer and 50% 1× wash buffer (7.5 mL of 20× TBS, 1.5 mL of Surfact-Amps 20 Detergent in 142 mL of water). Seal the microplate with the provided plate sealer and incubate overnight at 4 °C.
9. Remove the primary antibody and wash three times with 1× wash buffer. Add 50 μL of the provided HRP conjugate (30 μL of HRP conjugate in 12 mL of 1× wash buffer). Remove the HRP conjugate and wash the microplate three times with 1× wash buffer.
10. Add 50 μL of TMB substrate to the microplate. Incubate the substrate at room temperature until desired blue color has been achieved (protect from light throughout this incubation by covering the microplate with aluminum foil). Stop the reaction within 15 min by adding 50 μL of TMB stop solution. Using a microplate reader, measure the absorbance at a wavelength of 450 nm within 30 min of stopping the reaction.
11. Wash the microplate twice with ddH_2O. Add 50 μL of the provided whole-cell stain. Incubate for 5 min at room

temperature. Remove stain and wash three times with ddH_2O until all excess stain is removed.

12. Add 50 μL of elution buffer. Incubate for 10 min at room temperature. Measure the absorbance at 615 nm.
13. To calculate the amount of utrophin A protein relative to the number of cells in each condition, divide the readings from "**step 10**" by the readings from "**step 12**."
14. Repeat this drug screen three times.
15. Take the average of the relative ratios of the saline controls to use as a limit to find hits, that is, the drugs with a higher ratio than the saline control are considered hits. Average the ratios of the drug-treated cells and compare the expression of utrophin A in saline-treated cells versus drug-treated cells. Since it has been shown that a twofold increase of utrophin A in muscle is sufficient to improve the dystrophic pathology [20], hits that demonstrate over a twofold increase than the saline control should be picked for further in vivo assessment in dystrophic mdx mice.

3.2 Confirming Drug-Induced Utrophin A Upregulation by Western Blotting

1. Subsequent confirmation of the effects of the drugs is performed by treating C2C12 myoblasts with each drug for a period of 24 h. Using a sterile 6-well-plate, seed 75,000 C2C12 cells in 2 mL of DMEM complete growth media in each well. Create a dose curve by treating the cells with different concentrations of the FDA-approved drug identified as a hit (including the dose used in the screen). For example, if the drug induced an increase of utrophin A at 1 μM in the screen, treat the cells with at least one lower dose (0.1 μM) and one higher dose (5 μM). This will confirm the activation of utrophin A in the C2C12 cells and will permit you to determine if the drug is having a dose-dependent effect. In parallel, add the same volume of saline in a different well as a control. Treat the cells for 24 h and incubate at 37 °C, 5% CO_2.
2. On the following day, remove the media and wash every well with 1 mL of 1× PBS. Remove the PBS and add 150 μL of urea extraction buffer. Harvest the cells using a cell scraper. Transfer the cell lysates to microcentrifuge tubes. Leave the samples on ice for 30 min and vortex every 10 min. Spin the lysate samples at 13,000 rpm at 4 °C for 10 min. Keep the supernatant for western blotting analysis.
3. Perform the protein quantification by following the directions of the CB-X Protein Assay kit. Based on this quantification, prepare samples in order to load 20 μg of protein on an SDS-polyacrylamide (PAGE) gel (detailed below). Add 2× loading buffer to the sample at a ratio of 1:1. Heat the samples at 70 °C for 10 min.

4. To prepare the 7% precast gel. Mix the resolving gel ingredients (ddH_2O, 30% acrylamide mix, 1.5M Tris (pH 8.8) and 10% SDS). Add the 10% ammonium persulfate and TEMED at the moment of casting the gel to polymerize the gel. Pour the liquid gel between two glass plates (1 mm spacing). Leave 0.5cm of space below the insertion of the electrophoresis comb. Immediately cover the gel with 2 mL of 100% ethanol.
5. When the resolving gel is polymerized, remove the ethanol, and cast the stacking gel over the resolving gel. Insert the comb slowly to reduce production of bubbles. This gel can be kept overnight at 4 °C, rapped in wet paper towel, and sealed with plastic wrap.
6. The next day, unwrap the gel, remove the comb, and set it up into a standard protein electrophoresis apparatus. Pour 1× Tris-glycine buffer in between the electrodes to the top. Verify for leakage.
7. Load the samples in each well of the gel and include an appropriate pre-stained protein molecular weight marker (*see* **Note 2**). Electrophorese at 50 V, until the samples run past the stacking gel, then boost the voltage to 100 V. Run until the loading dye is released from the bottom of the gel.
8. Carefully remove the gel from the glass plates. Transfer the gel to a plastic container and submerge the gel with a 20% glycerol solution. Incubate at room temperature, for 30 min, on a rocking platform. Replace the glycerol solution by 1× transfer buffer.
9. Set up the transfer using a standard transfer apparatus. In a large plastic container and using the provided cassette, create a transfer sandwich. At the bottom is the black side of the cassette, add a sponge, two Whatman papers, the gel, a 0.45 μm nitrocellulose membrane, two more Whatman papers (all trimmed to the size of the sponge), another sponge, and the other side of the cassette. Keep all the components of the transfer sandwich submerged in 1× transfer buffer at all times. Ensure that there are no bubbles. Close the cassette tightly with the clamp. Place the transfer sandwich in the apparatus and put in an ice box. Run the transfer at 4 °C overnight at a constant amperage equivalent to 45–50 V (constant amps).
10. The next morning, increase the power to constant amperage equivalent to 100 volts (approximately 300 mA) and run for an additional 2 h changing the ice pack every 1 h maximum.
11. After the transfer is completed, carefully separate the nitrocellulose membrane from the gel and place the membrane in a plastic container. Proceed with ponceau staining using a 1× ponceau solution (diluted in 1× PBS) covering the membrane.

Place the container on a rocking platform for 5 min. Remove ponceau solution and take away the excess stain by adding ddH_2O.

12. Observe the whole protein staining to ensure equal loading of the samples. Remove the ponceau staining by washing the membrane with PBST. Remove the PBST.
13. Add 5% blocking milk solution and incubate at room temperature on a rocking platform for 60 min. Remove the blocking solution.
14. Dilute the utrophin A antibody at a concentration of 1:500 in 1% milk solution. Cover the membrane with the diluted antibody and incubate at room temperature for 2 h or at 4 °C overnight on a rocking platform.
15. Wash the membrane three times 10 min, at room temperature with 1× PBST.
16. Dilute a goat anti-mouse HRP conjugated antibody at 1:2500 in 1% milk solution (or another appropriate anti-mouse conjugated secondary antibody). Add the antibody to cover the membrane and incubate 1 h at room temperature on a rocking platform.
17. Wash three times, 10 min with 1× PBST.
18. Cover the membrane with a chemiluminescent HRP detection reagent. Expose the blot to film or in a chemiluminescent detection imaging system.
19. Rinse the membrane in 1× PBST.
20. Dilute a β-actin HRP antibody at 1:10,000 in 1% milk solution. Add the dilution over the membrane and incubate 1 h at room temperature on a rocking platform.
21. Wash three times, 10 min with 1× PBST. Detect and expose with a chemiluminescent HRP detection reagent as described above.
22. Identify the utrophin A band (running approximately at 395 kDa) by aligning with the protein molecular weight marker. Normalize the utrophin A band intensity to β-actin (or other appropriate housekeeping genes). Compare the normalized utrophin A band intensity of the drug-treated cells versus the saline-treated cells and perform statistical analysis in order to determine significance (see example in Fig. 2).

3.3 FDA-Approved Drug Treatment in Dystrophic Mice

3.3.1 Mdx and Wild-Type Mice Drug Treatment

1. The drug dose used for the in vivo experiment should be based on the doses used in the clinic by patients as listed in the Drug Bank [31] or as used previously in other animal studies. Under a laminar flow cabinet, dilute the drug in 100–200 μL of sterile saline, freshly prepared daily in a sterile vial (*see* **Note 3**).

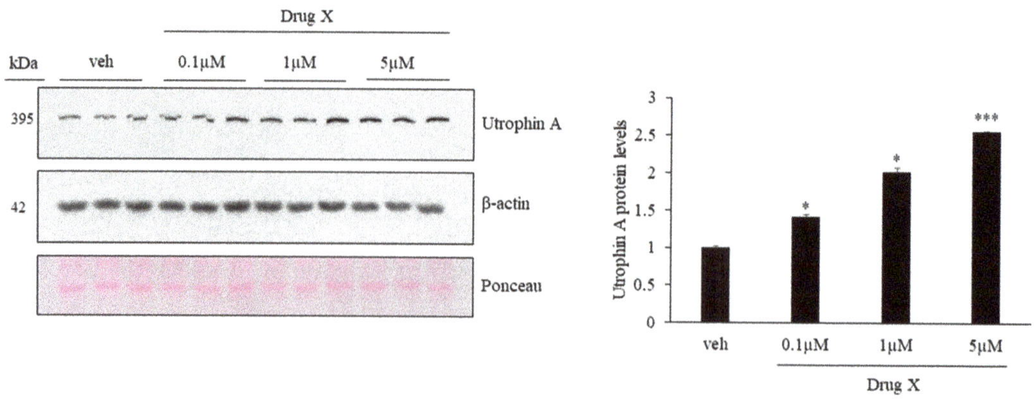

Fig. 2 Representative western blot and quantification demonstrating the level of utrophin A protein expression following a 24 h drug X- or vehicle treatment at different doses, in C2C12 cells

2. The mdx and wild-type mice need to be housed individually, and the colonies need to be maintained in a proper animal facility. All animal studies need to be performed under ethical regulation with the use of an approved animal protocol, working within the animal facility and with veterinary as well as animal care staff.
3. For a period of 4 weeks [18, 21–25, 32, 33], treat eight to ten 6-week-old mdx and wild-type mice for each condition (drug- or saline-treated) (*see* **Note 4**). To evaluate if the study is sufficiently powered to detect an effect on utrophin A expression, calculate the sample size [34]. Scruff each mouse (either mdx or wild-type) carefully to stabilize any movement. Using a sterile insulin syringe, treat the mouse with the drug or saline control daily, by intraperitoneal injections. Alternate the injection site every day.
4. Weigh each mouse weekly by placing the mouse in a high-walled container on a standard scale (*see* **Note 5**).
5. At the end of the treatment, euthanize the mice by injecting 0.5 mL of a stock euthanyl solution of 65 mg/ml, intraperitoneally and by performing cervical dislocation.

3.3.2 Mdx and Wild-Type Mice Dissections

1. Clean the dissection tools with 70% ethanol. Dry the tools and place them on a clean surface. Lay out a clean mat to perform the dissections.
2. Place the mouse on its back. With the use of sharp fine scissors and tweezers, make an incision in the skin to expose the abdomen. Cut along the bottom of the rib cage to expose the ribs and abdominal cavity. Push down or cut the liver to further expose the diaphragm. Carefully cut down the middle of the diaphragm splitting it into two halves (left and right). Hold the first half of the diaphragm with tweezers and release it from the

rib cage by cutting as closely to the ribs as possible. Transfer the first half of the diaphragm to a labeled cryotube. Freeze the tissue immediately in the liquid nitrogen.

3. Remove the second half of the diaphragm as described above. In a tissue OCT cryomold, add one layer of OCT (*see* **Note 6**). Remove all bubbles. Take the diaphragm and cut two flat pieces of liver at the same size. Sandwich the diaphragm between the two pieces of liver and place it in the mold over the OCT. Cover the sandwich with another layer of OCT and freeze in melting isopentane. When the OCT is hardened, quickly transfer it to a cryotube and freeze it in liquid nitrogen. This tissue will be used for histological analyses.
4. Once the diaphragm is removed you will see the heart under the ribs. With tweezers, hold the heart and cut the surrounding vasculature. Wipe down the excess blood with a kimwipe and place it in a cryotube. Freeze in liquid nitrogen.
5. To access the leg's lower skeletal muscles, make a light incision around the ankle area to cut through the skin and not through the tendons. Make a second incision in the skin from the ankle toward the knee on the inside of the mouse's leg. Use the tweezers to gently pull the skin up to expose the musculature of the lower leg.
6. At the front of the lower leg, you will see the tibia bone and the tibialis anterior (TA) muscle at the right of the tibia. Cautiously cut along both sides of the TA and cut the distal TA tendon. Detach the TA at the distal end with the tweezers and carefully trim the proximal tendon as close to the knee as possible. Perform this procedure to isolate the TA muscle from both limbs.
7. Transfer one of the TA muscles in a labeled cryotube and immediately freeze the tissue in the liquid nitrogen for future protein extraction.
8. For histological analysis of the TA muscle, add one layer of OCT in the cryomold, as described above. Place the TA in the mold over the OCT and cover it with another layer of OCT. Freeze in melting isopentane. When the OCT is hardened, transfer it to a labelled cryotube and freeze it immediately in liquid nitrogen.
9. For a more thorough evaluation of utrophin A expression in skeletal muscles, dissect other skeletal muscles of the lower limbs such as the extensor digitorum longus (EDL), the soleus, the gastrocnemius and the plantaris, using the same approach described above.

3.3.3 Muscle Tissue Protein Extractions

1. Leave a tissue pulvarizer to cool on dry ice for 15 min and a small spatula in liquid nitrogen.
2. Take out the frozen muscle tissues (prepared for protein analysis only) from the liquid nitrogen and keep them on dry ice. The pulvarizer contains a flat base on which to place the tissue and also contains a separate metal cylinder that will be used to crush the frozen tissue. Clean the pulvarizer's base and cylinder with 100% ethanol. Dry them with a kimwipe. Take the muscle with clean tweezers and place it on the cold pulvarizer. Place the metal cylinder over the tissue. With two to three sharp blows with a hammer onto the cylinder, break down the muscle tissue into a powder. Use the cooled spatula to scoop up the muscle powder and transfer it to a 1.5 mL microcentrifuge tube. Freeze at −80 °C until further use.
3. Add 200–500 μL of urea extraction buffer to the muscle powder (200 μL for EDL, soleus, and plantaris and 500 μL for diaphragm, TA, gastrocnemius, and heart).Vortex the samples for 30 min at room temperature (pulse settings: 30 s, with 10 s pause intervals).
4. Spin the samples at 20,000 × *g* for 15 min at 4 °C. Keep the supernatant.
5. Perform the protein quantification by following the directions of the CB-X Protein Assay kit. Based on this quantification, prepare samples in order to load 5–10 μg of protein on a 7% SDS-PAGE gel (described above). Add 2× loading buffer to each sample at a ratio of 1:1. Heat the samples at 70 °C for 10 min.
6. Perform the western blot as described in Subheading 3.2.

3.3.4 Utrophin A Expression at the Muscle Fiber Sarcolemma

1. Place the OCT-embedded TA and diaphragm muscles on dry ice (*see* **Note 7**). Avoid any freeze-thaw.
2. Using a histology cryostat, cut 10 μm cross-sections of each sample and place them carefully on a clean microscopy slide. One slide should contain at least three muscle cross-sections from wild-type mice, three sections of saline-treated mice, and three others from drug-treated mice (*see* Fig. 3). Keep the slides at −80 °C for storage.
3. Sections should be prepared for immunofluorescence analysis using the M.O.M Immunodetection kit: Block 1 h with the provided M.O.M Mouse Ig Blocking reagent by carefully adding 1 mL of the solution on the slide to cover the muscle sections. Wash the slide three times for 5 min by placing the slides in a slide holder and dipping them in a small container with 1× PBS solution. Dispose of the used 1× PBS. Add 1 mL

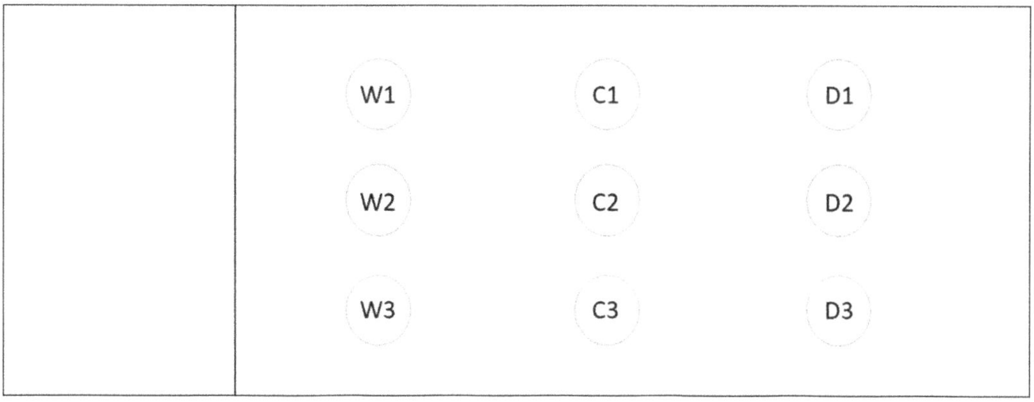

Fig. 3 Microscopy slide layout of muscle cross-sections. W = wild-type, C = mdx saline-treated control, D = mdx drug-treated

of the M.O.M Diluent solution B and incubate at room temperature for 5 min. Remove solution by tipping the slide on its side.

4. Co-stain with the utrophin A (1:200) and the laminin (1:500) antibodies (*see* **Note 8**), diluted in the M.O.M Diluent solution B. Add 1 mL to the slides and incubate at room temperature for 30 min. Wash the slides three times for 5 min with 1× PBS.
5. Incubate the muscle cross-sections with the provided M.O.M Biotinylated Anti-mouse IgG Reagent (10 μL in 2.5 mL of M. O.M Diluent solution B) for 10 min. Wash the slides three times for 5 min with 1× PBS.
6. Co-stain with a Texas Red-conjugated streptavidin antibody (1:500) and a goat anti-rabbit Alexa Fluor 488 antibody (1: 500) diluted in M.O.M Diluent solution B. Incubate at room temperature for 10 min. Cover the slides with a dark container or aluminum foil to protect from light. Wash the slides three times for 5 min with 1× PBS.
7. Mount the slides with Vectashield containing DAPI staining by adding approximately 60 μL of Vectashield solution on the slide. Carefully apply a coverslip over the stained muscle cross-sections. Seal around the coverslip with clear nail polish.
8. Visualize the utrophin A stained mouse muscle cross-sections using a fluorescent microscope.
9. Count the fibers containing sarcolemmal staining of utrophin A normalized to the number of total fibers. Compare the utrophin A sarcolemmal localization of mdx drug-treated mice vs mdx saline-treated mice (*see* Fig. 4).

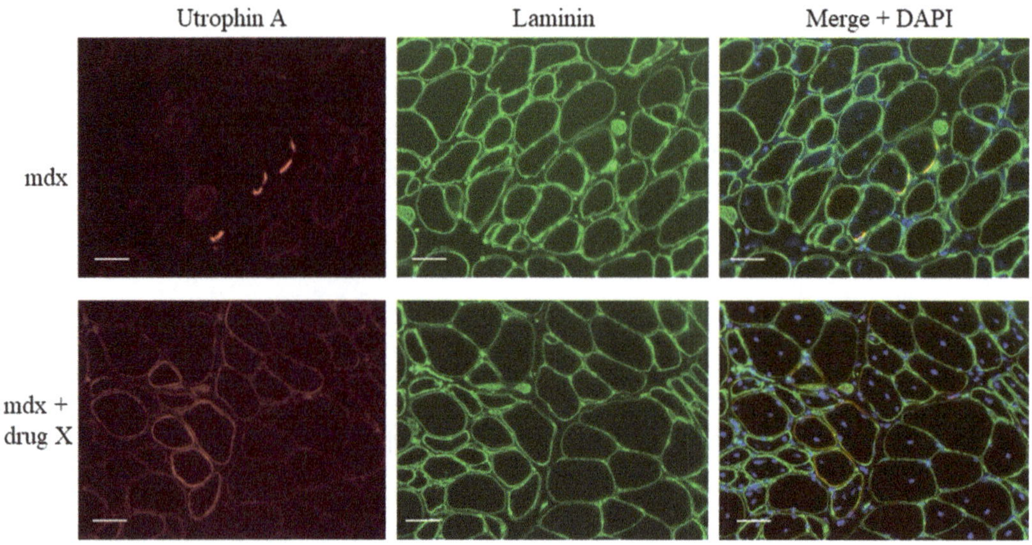

Fig. 4 Representative examples of cross-sections obtained from TA mdx mice treated with Drug X or vehicle control (saline) for 4 weeks were immunostained with utrophin A (Red), laminin (green) highlighting the sarcolemma, and DAPI (blue) staining the nuclei. Scale bars 50 μm

4 Notes

1. If using a 96 well plate, seed 10,000 cells in 100 μL DMEM and double volumes for the screen.
2. Utrophin A is a large 395 kDa protein [35]; make sure to use a protein molecular weight marker for analysis of high molecular weight proteins.
3. Do not exceed IP injection volumes over 200 μL in mice. A higher volume could be harmful.
4. Mdx mice start developing a dystrophic phenotype between 3 and 6 weeks of age [36].
5. Mdx mice have an overall larger weight than wild-type mice.
6. If you do not have access to a cryomold, you can create one by cutting a 1 mL pipette tip in half to create a boat-like shape.
7. Avoid any freeze-thaw of the embedded OCT samples. This will damage the muscle samples and will render them unusable for histological analysis.
8. Laminin staining is used to highlight the muscle fiber sarcolemma.

Acknowledgments

We are indebted to many current and previous members of the Jasmin lab who have made important contributions in this research program, leading us to perform this chemical screen. We would like to give special thanks to Nadine Adam for her contribution in editing the manuscript. We are also grateful to the Canadian Institutes of Health Research, the Association Française contre les Myopathies, the Muscular Dystrophy Association (USA), and Jesse's Journey for their generous financial support over the years.

References

1. Baby SM, Bogdanovich S, Willmann G et al (2010) Differential expression of utrophin-A and -B promoters in the central nervous system (CNS) of normal and dystrophic mdx mice. Brain Pathol 20:323–342
2. Burton EA, Tinsley JM, Holzfeind PJ et al (1999) A second promoter provides an alternative target for therapeutic up-regulation of utrophin in Duchenne muscular dystrophy. Proc Natl Acad Sci U S A 96:14025–14030
3. Dennis CL, Tinsley JM, Deconinck AE et al (1996) Molecular and functional analysis of the utrophin promoter. Nucleic Acids Res 24: 1646–1652
4. Nguyen TM, Ellis JM, Love DR et al (1991) Localization of the DMDL gene-encoded dystrophin-related protein using a panel of nineteen monoclonal antibodies: presence at neuromuscular junctions, in the sarcolemma of dystrophic skeletal muscle, in vascular and other smooth muscles, and in proliferating brain cell lines. J Cell Biol 115:1695–1700
5. Weir AP, Burton EA, Harrod G et al (2002) A- and B-utrophin have different expression patterns and are differentially up-regulated in mdx muscle. J Biol Chem 277:45285–45290
6. Nakamura A, Takeda S. Mammalian models of Duchenne muscular dystrophy: pathological characteristics and therapeutic applications. https://www.hindawi.com/journals/bmri/2011/184393/
7. Goyenvalle A, Seto JT, Davies KE et al (2011) Therapeutic approaches to muscular dystrophy. Hum Mol Genet 20:R69–R78
8. Durbeej M, Campbell KP (2002) Muscular dystrophies involving the dystrophin-glycoprotein complex: an overview of current mouse models. Curr Opin Genet Dev 12: 349–361
9. Chakkalakal JV, Thompson J, Parks RJ et al (2005) Molecular, cellular, and pharmacological therapies for Duchenne/Becker muscular dystrophies. FASEB J 19: 880–891
10. Bogdanovich S, Perkins KJ, Krag TOB et al (2004) Therapeutics for Duchenne muscular dystrophy: current approaches and future directions. J Mol Med 82:102–115
11. Gregorevic P, Blankinship MJ, Allen JM et al (2004) Systemic delivery of genes to striated muscles using adeno-associated viral vectors. Nat Med 10:828–834
12. Gregorevic P, Blankinship MJ, Allen JM et al (2008) Systemic microdystrophin gene delivery improves skeletal muscle structure and function in old dystrophic mdx mice. Mol Ther 16:657–664
13. Rodino-Klapac LR, Janssen PML, Shontz KM et al (2013) Micro-dystrophin and follistatin co-delivery restores muscle function in aged DMD model. Hum Mol Genet 22:4929–4937
14. Guiraud S, Davies KE (2017) Pharmacological advances for treatment in Duchenne muscular dystrophy. Curr Opin Pharmacol 34:36–48
15. Ramos J, Chamberlain JS (2015) Gene therapy for Duchenne muscular dystrophy. Expert Opin Orphan Drugs 3:1255–1266
16. Odom GL, Gregorevic P, Allen JM et al (2008) Microutrophin delivery through rAAV6 increases lifespan and improves muscle function in dystrophic dystrophin/utrophin-deficient mice. Mol Ther 16:1539–1545
17. Guiraud S, Squire SE, Edwards B et al (2015) Second-generation compound for the modulation of utrophin in the therapy of DMD. Hum Mol Genet 24:4212–4224
18. Tinsley JM, Fairclough RJ, Storer R et al (2011) Daily treatment with SMTC1100, a novel small molecule utrophin upregulator, dramatically reduces the dystrophic symptoms in the mdx mouse. PLoS One 6:e19189

19. Squire S, Raymackers JM, Vandebrouck C et al (2002) Prevention of pathology in mdx mice by expression of utrophin: analysis using an inducible transgenic expression system. Hum Mol Genet 11:3333–3344
20. Tinsley J, Deconinck N, Fisher R et al (1998) Expression of full-length utrophin prevents muscular dystrophy in mdx mice. Nat Med 4: 1441–1444
21. Péladeau C, Adam NJ, Jasmin BJ (2018) Celecoxib treatment improves muscle function in mdx mice and increases utrophin A expression. FASEB J 32:5090–5103
22. Péladeau C, Ahmed A, Amirouche A et al (2016) Combinatorial therapeutic activation with heparin and AICAR stimulates additive effects on utrophin A expression in dystrophic muscles. Hum Mol Genet 25:24–43
23. Péladeau C, Adam N, Bronicki LM et al (2020) Identification of therapeutics that target eEF1A2 and upregulate utrophin A translation in dystrophic muscles. Nat Commun 11:1990
24. Al-Rewashdy H, Ljubicic V, Lin W et al (2015) Utrophin A is essential in mediating the functional adaptations of mdx mouse muscle following chronic AMPK activation. Hum Mol Genet 24:1243–1255
25. Ljubicic V, Miura P, Burt M et al (2011) Chronic AMPK activation evokes the slow, oxidative myogenic program and triggers beneficial adaptations in mdx mouse skeletal muscle. Hum Mol Genet 20:3478–3493
26. Ljubicic V, Burt M, Lunde JA et al (2014) Resveratrol induces expression of the slow, oxidative phenotype in mdx mouse muscle together with enhanced activity of the SIRT1-PGC-1α axis. Am J Physiol Cell Physiol 307: C66–C82
27. Miura P, Chakkalakal JV, Boudreault L et al (2009) Pharmacological activation of PPARβ/δ stimulates utrophin A expression in skeletal muscle fibers and restores sarcolemmal integrity in mature mdx mice. Hum Mol Genet 18:4640–4649
28. Amirouche A, Tadesse H, Lunde JA et al (2013) Activation of p38 signaling increases utrophin A expression in skeletal muscle via the RNA-binding protein KSRP and inhibition of AU-rich element-mediated mRNA decay: implications for novel DMD therapeutics. Hum Mol Genet 22:3093–3111
29. Gordon BS, Delgado Díaz DC, Kostek MC (2013) Resveratrol decreases inflammation and increases utrophin gene expression in the mdx mouse model of duchenne muscular dystrophy. Clin Nutr 32:104–111
30. Vianello S, Yu H, Voisin V et al (2013) Arginine butyrate: a therapeutic candidate for Duchenne muscular dystrophy. FASEB J 27: 2256–2269
31. Wishart DS, Feunang YD, Guo AC et al (2018) DrugBank 5.0: a major update to the DrugBank database for 2018. 46:D1074–D1082
32. Jahnke VE, Van Der Meulen JH, Johnston HK et al (2012) Metabolic remodeling agents show beneficial effects in the dystrophin-deficient mdx mouse model. 2:16
33. Ljubicic V, Jasmin BJ (2015) Metformin increases peroxisome proliferator-activated receptor γ Co-activator-1α and utrophin a expression in dystrophic skeletal muscle. Muscle Nerve 52:139–142
34. Charan J, Kantharia ND (2013) How to calculate sample size in animal studies? J Pharmacol Pharmacother 4:303–306
35. Love DR, Hill DF, Dickson G et al (1989) An autosomal transcript in skeletal muscle with homology to dystrophin. 339:55–58
36. McGreevy JW, Hakim CH, McIntosh MA et al (2015) Animal models of Duchenne muscular dystrophy: from basic mechanisms to gene therapy. Dis Model Mech 8:195–213

Part VII

Cell Models and Stem Cells

Chapter 27

Monitoring Plasma Membrane Injury-Triggered Endocytosis at Single-Cell and Single-Vesicle Resolution

Daniel C. Bittel and Jyoti K. Jaiswal

Abstract

Plasma membrane injury activates membrane trafficking and remodeling events that are required for the injured membrane to repair. With the rapidity of the membrane repair process, the repair response needs to be monitored at high temporal and spatial resolution. In this chapter, we describe the use of live cell optical imaging approaches to monitor injury-triggered bulk and individual vesicle endocytosis. Use of these approaches allows quantitatively assessment of the rate of retrieval of the injured plasma membrane by bulk endocytosis as well as by endocytosis of individual caveolae following plasma membrane injury.

Key words Muscle, Muscular dystrophy, Confocal, Membrane repair, Caveolae, Live cell imaging

1 Introduction

Mechanical strain associated with tissue function routinely stresses and even damages the cell's plasma membrane (PM) [1]. To ensure tissue health in the face of persistent plasma membrane injury, cells need to efficiently repair such injuries [2]. Studies from invertebrate egg cells and multiple mammalian cells have led to the current plasma membrane repair model that uses membrane trafficking events to enable membrane repair. These events include (1) calcium-stimulated exocytosis leading to addition of endomembrane to the injured PM and (2) shedding or endocytosis to remove the damaged parts of the PM [2–4].

Due to high mechanical strain and damage to skeletal muscle, PM repair is critical for the health of this tissue. This was identified by studies of muscles lacking the dysferlin protein, which causes limb girdle muscular dystrophy (LGMD2B) or "dysferlinopathy." Dysferlin deficit impairs myofiber PM repair and leads to muscle loss [5–7]. Similar to the synaptic vesicle fusion proteins such as synaptotagmins, dysferlin is also a transmembrane protein with calcium binding C2 domains. Thus, dysferlin is implicated in

Rika Maruyama and Toshifumi Yokota (eds.), *Muscular Dystrophy Therapeutics: Methods and Protocols*,
Methods in Molecular Biology, vol. 2587, https://doi.org/10.1007/978-1-0716-2772-3_27,

facilitating injury-triggered exocytosis to enable efficient PM repair [8–10]. Conversely, the role of endocytosis in membrane repair has most commonly been examined in the context of membrane injuries incurred via pore-forming toxins. These toxins induce plasma membrane injury and activate caveolar and non-caveolar endocytosis of the injured membrane, which is required for the removal of membrane pores required for the injured membrane to repair [11, 12]. Caveolar and non-caveolar endocytosis is also linked with skeletal myofiber membrane repair [12–15]. Accordingly, mutations in the Caveolin-3 gene induce muscular dystrophies called "caveolinopathies," characterized by exercise intolerance, myalgia, and rhabdomyolysis [16].

Therapies directed toward increasing stability and trafficking of PM improve muscle cell repair capacity and muscle tissue health in muscular dystrophies [9, 10, 17–19]. Thus, methods enabling the real-time monitoring of injured PM and endomembrane trafficking triggered by endo- and exocytosis are needed to better understand cellular response to injury, to determine how this is compromised in muscular dystrophies, and to evaluate therapies that can remedy these deficits. Here, we describe techniques for real-time imaging to monitor and quantify focal PM injury-triggered endocytosis in healthy muscle cells. These approaches can be applied to any adherent cultured cell and we have used it to study dystrophic patients and mouse muscle cells to assess if endocytic defects contribute to their poor repair, and test therapies that restore these deficits.

We previously described approaches for monitoring exocytosis of individual vesicles [9, 20]. Thus, here, we focus on approaches to allow monitoring bulk and individual endocytic events in single cells as they undergo repair from a focal PM injury. Epifluorescence microscopy is used together with a combination of fluorescent wheat germ agglutinin (WGA; a lectin that binds PM glycoproteins and glycolipids) and a quencher, bromophenol blue (BPB), to monitor kinetics of bulk PM endocytosis. This bulk endocytosis kinetics at cellular level can be paired with measurement of endocytosis of individual vesicles in the injured cells. We demonstrate individual vesicle endocytosis approach for caveolar endocytosis by real-time tracking of individual caveolae labeled using Caveolin-1 mRFP. Focal PM injury coupled with spinning disc confocal or total internal reflection fluorescence (TIRF) microscopy is used to track individual endocytic vesicles. While the procedures for whole cell bulk endocytosis and individual vesicle endocytosis are presented together, they can be utilized independently of each other. Accordingly, depending on the experimental question, appropriate microscopy approach (epifluorescence, spinning disc confocal, or TIRF microscopy) should be used. We describe the procedures here using muscle cells, but they can be utilized as such for other adherent cultured cells.

2 Materials

2.1 Cell Culture

1. Mouse or human myoblasts: Primary human or mouse muscle cells and immortalized mouse myoblast cell line (C2C12) are described previously [10]. Primary cells can be prepared as described elsewhere [21–23], while C2C12 cells are purchased from ATCC (#CRL-1772). The rest of the protocol below is applicable universally across other (non-muscle) cultured mammalian cells as well.
2. C2C12 myoblast growth media: DMEM (high-glucose – Gibco) supplemented with 10% fetal bovine serum (FBS – VWR) and 1% Pen/Strep (Gibco). For primary human myoblasts, cells are grown in Promocell growth media kit (Promocell #C-23160 – 15% FBS + 50 μg/mL Fetuin + 10 ng/mL rhEGF + 1 ng/mL rhBFGF + 10 μg/mL rhInsulin + 4 μg/mL Dexamethasone). Upon mixing, sterile filter media using a 0.20 μm vacuum-driven filter unit (VWR), and store all media at 4 °C.
3. 1× calcium- and magnesium-free PBS (Gibco).
4. Versene (Gibco).
5. Basic cell culture microscope (for monitoring and counting cell cultures).
6. 15 mL falcon tubes (VWR).
7. Centrifuge (to spin cells at 800 rpm for 4 min at room temperature to passage cells).
8. 10 cm culture dishes (Thermo Fisher) for cell propagation and passaging 6-well culture plates (Thermo Fisher).
9. #1.5 coverslips (for live-cell imaging – ThorLabs).
10. Coverslip holder or 6-well culture dishes for live-cell culture on coverslips (Thermo Fisher).
11. Hemocytometer (Sigma).
12. Curved forceps (Roboz).
13. Opti-MEM reduced serum media (Gibco).
14. Lipofectamine 2000 Transfection Reagent (Thermo Fisher).
15. Caveolin-1 mRFP plasmid (Addgene #14434).
16. Wheat-germ Agglutinin, Alexa Fluor-488 Conjugate (Thermo Fisher).
17. Bromophenol Blue (BPB) (Sigma).

2.2 Live-Cell Imaging

1. Epifluorescence microscope with spinning disc confocal and/or total internal reflection (TIRF) imaging capability and pulsed laser (e.g., *Ablate*!™ Intelligent Imaging Innovation) for targeted focal ablation.

2. SlideBook (Intelligent Imaging Innovations) or comparable software for automated control for image acquisition and laser ablation.
3. SlideBook (Intelligent Imaging Innovations), MetaMorph (Molecular Devices) or another suitable software package to acquire time-lapse videos and quantify images.
4. Stage top incubator or environmental chamber to maintain cells during imaging.
5. Sykes-Moore Chamber (Bellco Glass Inc.).
6. Cell Imaging Media (CIM) – HBSS (Sigma) adjusted to pH 7.4 (with 1 N HCL) and 1 mM calcium chloride (Sigma) added just prior to use.
7. Stock WGA solution: Dissolve contents of vial (5 mg) in 5 mL of 1× PBS, prepare 10 μl aliquots in microfuge tubes and store protected from light in −20 °C freezer.
8. Bromophenol Blue Quenching Media (20 mM stock: Dissolve 0.692 g BPB (Sigma) into 50 mL sterile water. Used for quenching at 5 mM working solution in CIM.
9. Immersion oil for objectives for 37 °C observations.
10. Vacuum aspiration hose/unit (in close proximity to the microscope).
11. Lens paper (Thermo Fisher).
12. Optical Lens Cleaner (Sparkle).

3 Method

3.1 Muscle Cell Culture

Culture cells in 10 cm dishes (or equivalent) to 75–80% confluence, ensuring that freshly thawed (or isolated) cells are passaged at least twice before being used for experiments.

1. Sterilize #1.5 glass coverslips using 70% EtOH at least an hour before plating the cells. Place the coverslip in individual 35 mm dishes of in a 6-well dish, ensuring the sterilized coverslip is fully dry before plating cells.
2. Wash the cells in the 10 cm culture dish twice with PBS (5 mL per wash), then add 3 mL of Versene per dish, allowing cells to detach for 2–10 min in the incubator.
3. While Versene-added cells are incubating, add 1 mL of appropriate growth media to each well of the coverslip-containing 6-well plate, and place in the incubator to CO_2 equilibration of the media pH prior to eventual cell addition.
4. When >60%+ of cells have detached, wash the dish by gently pipetting with 4 mL of growth media to detach all cells and transfer them to a 15 mL falcon tube.

5. Collect cells by centrifuging at 800 rpm for 4 min at room temperature, discard the supernatant, and re-suspend the pellet in 1 mL of growth media.
6. Count cells on hemocytometer, and add fresh growth media to bring the cells at 60,000 cells/mL.
7. Add 1 mL of cell suspension to each coverslip-containing dish/ well of a 6-well plate.
8. Distribute cells evenly by rocking the plate for 10 s, then place in the CO_2 incubator at 37 °C, 5% CO_2 for 24 h. *See* **Note 1**.

3.2 Bulk Membrane Endocytosis Studies

1. A day after plating the cells in Subheading 3.1, **step 8**, setup the epifluorescence microscope for time-lapse imaging in bright-field and epifluorescence modes.
2. Set up to capture images at 30 s intervals with laser injury occurring after the fourth capture and continue image capture post-injury to monitor bulk endocytosis over the next few minutes post injury (*see* Subheading 3.2, **steps 16** onward).
3. Gently aspirate the media out of the plate/well from 3.1.8 with cells growing on the coverslips and wash twice with pre-warmed CIM (1 mL/wash).
4. Transfer the coverslip (using curved forceps) to the Sykes-Moore chamber, place the O-ring and tighten the chamber to secure the coverslip.
5. Add 300 μL of the CIM (from Subheading 2.2, **item 6**), and use lens paper dipped in 70% alcohol to clean the bottom surface of the coverslip.
6. Mount the chamber on the microscope in the stage-top incubator (leaving the incubator top open) and use the 60× objective to focus on the cells to be imaged.
7. Using bright-field mode, identify up to four locations on the coverslip with cells to be imaged and save X, Y, Z coordinates of each location, to return to later when setting up epifluorescence imaging. *See* **Note 2**.
8. Return to the first saved location and set up epifluorescence imaging of Alexa Fluor 488 WGA fluorescence. *See* **Note 3**.
9. Mark the sites on the cell for focal laser injury—one site for each cell to be injured and keeping one cell uninjured to monitor photobleaching and correct for it post acquisition.
10. To set up focal injury, mark a 1–4 μm^2 region on the plasma membrane that will be irradiated with the pulsed (*Ablate!*™) laser for focal injury. Tune the laser power through the software to about 10% of the peak power and optimize the laser power to obtain membrane injury that is consistent, but non-lethal. *See* **Note 4**.

11. To label cells with Alexa Fluor 488 WGA mix 298.8 μL CIM from Subheading 2.2, **item 6** +0.2 μL WGA-Alexa 488 stock solution (*see* Subheading 2.2, **item** 7) to get 4 μg/mL WGA-Alexa 488 solution. *See* **Note 5**.
12. Through the open-stage top, remove CIM from the coverslip chamber using vacuum aspirator and replenish with 300 μL of WGA-Alexa 488 from Subheading 3.2, **step 11**. *See* **Note 6**.
13. Incubate for 2-min using timer, allowing for WGA to bind and label the cell surface before removing the unbound dye (Fig. 1a, step 1, Fig. 1b1). *See* **Note 7**.

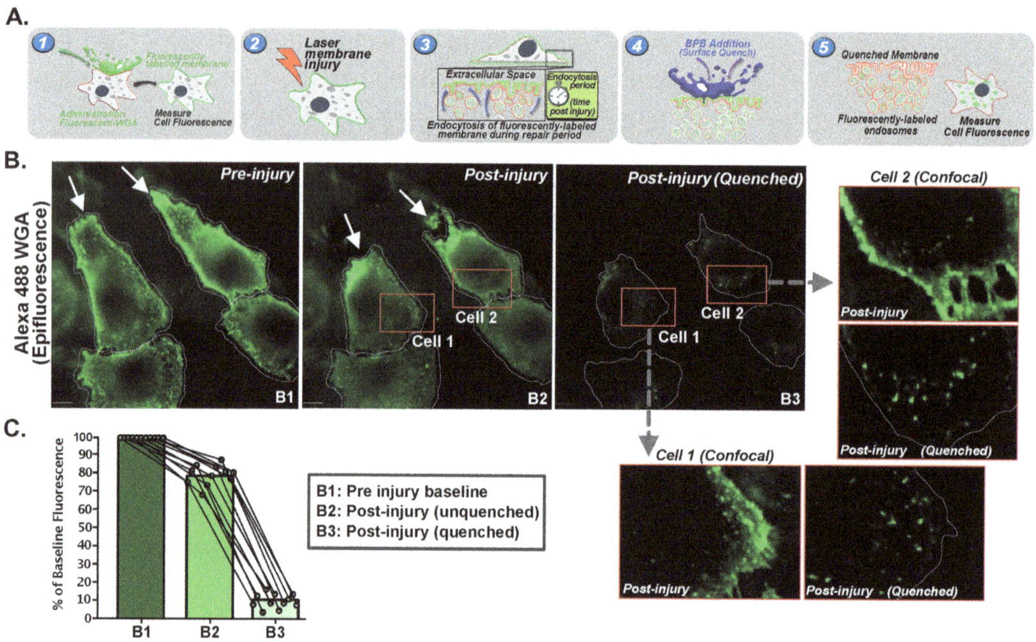

Fig. 1 Quantifying bulk membrane endocytosis. (**a**) Steps (*1–5*) required for membrane labeling with AF488 WGA (*1*), laser membrane injury (*2*), endocytosis of labeled membrane during repair (*3*), Quenching AF488 WGA with BPB (*4*), and measurement of cellular fluorescence indicating the endocytosed membrane (*5*). (**b**) Epifluorescence images of cells labeled with AF488 WGA. (**b**1) baseline WGA-membrane labeling (white arrows indicate site of laser injury). (**b**2) WGA labeling of the cells 5-min post-injury. (**b**3) WGA labeling of cells 5-min post-injury followed by BPB quenching. White dotted lines indicate cell boundaries and red boxes mark regions on two injured cells that are shown in the corresponding zoomed images acquired by spinning disc confocal microscopy. Notice that the WGA-labeled endosomes are distinctly visible upon BPB quenching of membrane fluorescence, after 5 min of post-injury repair period. (**c**) Plot showing cellular fluorescence of AF488 WGA imaged by epifluiorescence microscopy. Lines shown connect fluorescence intensity data points for individual cells quantified at pre-injury baseline and after 5 min of repair prior to and following BPB quenching. While there is a slight reduction in fluorescence intensity at 5-min post injury (middle column), BPB quenches most of the cellular fluorescence indicating only a small portion of the plasma membrane associated AF488 WGA was endocytosed

14. Return to the saved first (of the four) positions from 3.2.7 to confirm that the cell is in focus and the site for focal laser injury is placed on the desired position on the cell.
15. Repeat the above step for the other saved positions from 3.2.7, before returning to the first position to start imaging.
16. Start the time-lapse imaging as setup in Subheading 3.2, **step 2**. This will cause sequential imaging of each of the positions for 2 min prior to laser injury of all the marked cells and imaging of the injured cells will continue thereafter. *See* **Note 8**.
17. Independently, add 100 μL of the 20 mM stock BPB solution prepared in Subheading 2.2, **item 8** to the cells at either 3, 5, or 10 min following injury to image endocytosis that has occurred during repair at the respective time period post injury (Fig. 1a, step 4 and Fig. 1b2). *See* **Note 9**.
18. After 2 min of BPB addition, capture images for cells at each of the positions saved in Subheading 3.2, **step 6** (Fig. 1a, **step 5**, Fig. 1b3). *See* **Note 10**.
19. Proceed to Image Analysis Section to analyze bulk endocytosis (*see* Subheading 3.4). *See* **Note 11**.

3.3 Monitoring Caveolar Endosome Dynamics

1. Transfect cells after Subheading 3.1, **step 8**, with Caveolin-1 mRFP plasmid using Lipofectamine 2000 (or another preferred approach) as per manufacturer's instructions.
2. After allowing 24–48 h for transgene expression, setup time-lapse imaging in red channel on spinning disc confocal microscope as follows—time-lapse capture interval 500 ms, laser injury after tenth timepoint, and at least 600 timepoints in total (5 min).
3. Prepare coverslips for imaging by following Subheading 3.2, **steps 3–6**.
4. Use bright-field and epifluorescence mode to identify transfected cells to be imaged.
5. In confocal mode, focus at the cell surface/coverslip interface so the caveolae at the plasma membrane (not deep in the cell) are captured (Fig. 2). Ensure minimal laser exposure to prevent potential bleaching/phototoxicity to the cells. *See* **Note 12**.
6. Set up focal injury for the cells as described in Subheading 3.2, **step 10**.
7. Initiate time-lapse imaging as outlined in Subheading 3.3, **step 2**, the *Ablate!*™ laser will pulse to injure the cell at the tenth image (5 s after initiation), and caveolar dynamics will be recorded for 5–10 min post injury (or longer if needed). *See* **Note 13**.

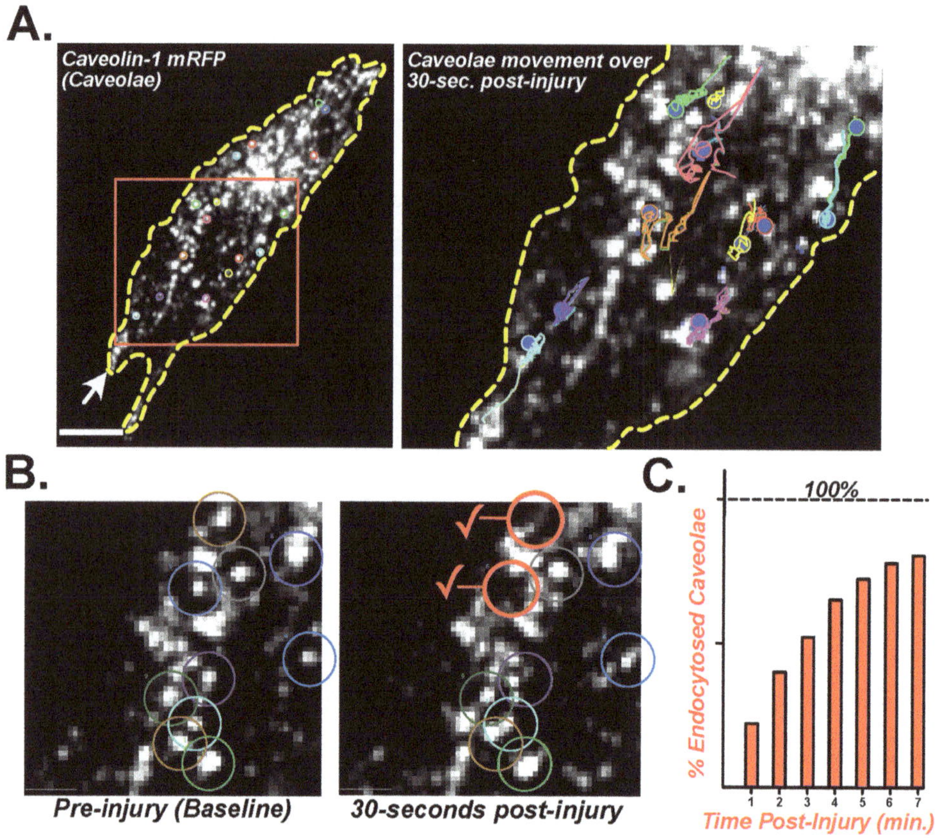

Fig. 2 Quantifying individual caveolar endocytosis. (**a**) Representative monochrome images of caveolin-1 mRFP labeled myoblast acquired by spinning disc confocal microscopy. Yellow dotted line shows cell outline, and white arrow marks the site of membrane injury. Red box marks the region of the cell zoomed in the image to the right. Note the mobile nature of caveolae in the injured cells as indicated by the tracks showing the mobility of the caveolar endosome over a period of 30-s post injury. The track for each endosome is displayed as individually colored line. (**b**) A portion of the cell showing Caveolin-1 labeled endosomes marked individually by 3 μm diameter-colored circles. If the endosome travels beyond this 3 μm region or disappears from the confocal plane for >10 s, they were noted as endocytosed and are marked in the image by red check mark. (**c**) Plot showing the fraction of the tracked Caveolin-1 endosomes that were endocytosed cumulatively over each 1-min interval over a specified time-period post-injury (7 min in this plot). For the cumulative plot, proportion of tracked caveolin-1 endosomes that are endocytosed is a sum of events that occurred in each successive minute and all the endocytic events that have occurred prior to that minute. This causes the binned timepoints to steadily increase over time reaching a plateau when no new endocytic events are observed in the subsequent time points

8. Move to another cell on the coverslip and repeat the above step. *See* **Note 14**.
9. Proceed to Image Analysis Section to analyze caveolar dynamics prior to and following focal laser injury of individual cells (*see* Subheading 3.5). *See* **Note 15**.

3.4 Quantification of Bulk Membrane Endocytosis

Open SlideBook software (or a comparable software of your choice) and open the imaging file generated in Subheading 3.2.

1. Open the time-lapse file from your first coverslip position imaged (From Subheading 3.2, **step 14**). *See* **Note 16**.
2. The first image will serve as baseline for injury-triggered change in WGA membrane labeling (Fig. 1a, step 1, and Fig. 1b).
3. With the analysis software's region drawing tool, trace the boundary of each cell carefully to include all membrane labeling.
4. Mark 3 background regions (a box of ~400 μm^2) in the imaged areas free of cells. *See* **Note 17**.
5. Use the analysis software to measure mean fluorescence intensity of each cell/region (both for the wide-field and confocal modality) across all of the time-lapse planes recorded.
6. Export all of the values to an excel file where each region imaged is marked and their quantifications are presented separately.
7. Repeat above **steps 1–5** in Subheading 3.4 for all the regions set up for imaging in Subheading 3.2, **step** 7. *See* **Note 18**.
8. Average the mean wide-field fluorescence of the three background regions from 3.4.4.
9. Subtract the average intensity value of background epifluorescence from the WGA intensity value of each cell, for each of the time-lapse image. *See* **Note 19**.
10. Open the post-BPB quenching images (from Subheading 3.1, **step 18** above—Fig. 1a, step 5, and Fig. 1b bottom images). Repeat **steps 3–9** in Subheading 3.4 above for this quenched image. *See* **Note 20**.
11. Using corrected fluorescence values measured prior to (3.4.9) and following BPB quenching (3.4.10) calculate the proportion of membrane internalized by dividing the baseline values (3.4.9) with quenched values (3.4.10). This value can be reported as a percentage by being multiplied with 100. (As depicted in Fig. 1c).

3.5 Quantification of Single-Vesicle (Caveolae) Endocytosis

Open SlideBook software (or a comparable software of your choice) and open the images generated for cells transfected with caveolin-1 in Subheading 3.3.

1. Open the confocal images in the time-lapse file from the cell imaged in Subheading 3.3, **step** 7 *and* threshold image to maximize visualization of all caveolae. Viewing the image in monochrome also aids in better visualization (Fig. 2).

2. Select 50–75 individual puncta within 30 μm of the injury site and mark them by drawing a region of 3-μm diameter around each punctum with the puncta being the center of this circle (Fig. 2b). *See* **Note 21**.
3. Select one punctum and advance through the time-lapse and record it as endocytosed punctum when it either travels beyond the 3 μm region boundary, and/or disappears out of the imaging plane for over 10 s (Fig. 2b marked with red checkmarks). *See* **Note 22**.
4. Upon meeting above-assigned features for movement, mark the specific caveolin puncta as "endocytosed" and record the timepoint at which it initiated mobility.
5. Vesicles that moved at any point of time in the first 30 s post-injury are cumulatively included to the endocytosed vesicle pool in the subsequent time points. *See* **Note 23**.
6. Once a vesicle is mobile, and noted, move on to the next caveolae. Repeat this for all caveolae, across the entirety of the time-lapse capture (in this case, 7 min as outlined in Subheading 3.3, **step 2**). *See* **Note 24**.
7. Repeat all the above steps for each of the other cell thus imaged to generate the averaged response of all the caveolae analyzed across all cells, representing the overall kinetics of injury-triggered endocytosis of individual caveolae.

4 Notes

1. To limit cells adhering to the well below the coverslip, gently press each coverslip down using a clean pipette tip to remove air bubbles and ensure that the coverslip is well settled at the dish bottom.
2. While not required, if the microscope setup allows taking confocal images, acquire these images for clearer visualization of WGA containing endosomes.
3. Use the shortest exposure with excitation light to minimize photodamage to the cell during the course of setup and imaging.
4. To minimize the impact of focal injury on compartments other than the plasma membrane, injure cells away from the nucleus (white arrows in Fig. 1).
5. Alexa Fluor-488 WGA concentration is kept low to label membrane with minimal background fluorescence, which avoids additional washes to remove unbound label.

6. Avoid touching the holder so as to ensure that the coordinates saved in the software (*see* Subheading 3.2, **step** 7) are not altered.
7. With a 2-min labeling, BPB treatment quenches all of the cellular WGA fluorescence, confirming minimal endocytosis in this timeframe.
8. If the microscope is setup to also take confocal images, appearance of WGA containing endosomes in the cells can be clearly seen (Fig. 1a, b zoomed images).
9. BPB will quench the fluorescence of only the labeled WGA that have not yet been endocytosed and are thus exposed on the cell surface, while WGA that has been endocytosed is protected from extracellular BPB. Thus, these appear as bright fluorescent endosomes within the cell (as observed in Fig. 1b).
10. If permitted by the microscope setup, in this and the previous step acquire confocal images to visualize WGA containing endosomes (Fig. 1, step 5).
11. Comparing the epifluorescence intensity of unquenched WGA fluorescence of each cell at baseline with the WGA fluorescence at the desired timepoint post injury provides the measure of the extent of endocytosis at that specific timepoint.
12. TIRF microscopy can also be utilized for this protocol as it allows imaging as little as within 100 nm of plasma membrane, which allows greater precision in monitoring of the axial movements of the caveolae. Elsewhere, we have detailed the approach to set up TIRF microscopy [24].
13. Cell injury is apparent by the slight retraction of the cell, if this does not occur, the time lapse should be restarted with higher Ablate! laser power.
14. Try to obtain injuries of 10–15 cells per coverslip, but limit total imaging time to ~1 h per coverslip, to avoid any stress on the cells in the coverslip caused by prolonged imaging.
15. While the above description is for caveolar endocytosis, an identical approach can be utilized to monitor endocytosis of clathrin-coated vesicles by making use of clathrin-mRFP (or another desired endocytic marker) in place of Caveolin-1 mRFP.
16. This file will contain fluorescent images acquired using widefield (and confocal, if that was setup in Subheading 3.2, **step** 7) illumination.
17. These intensity values will serve as the background for camera noise and nonspecific WGA staining on the coverslip surface.

18. With greater imaging volume of wide-field images, this data will capture all of the WGA fluorescence associated with a cell, while confocal image will provide clearer visualization of the internalized membrane.
19. This value will represent the mean baseline WGA fluorescence of the cell prior to any significant level of endocytosis.
20. These data will report the fluorescence associated with the cell after 3, 5, or 10 min of post-injury endocytosis. As the cell surface WGA-AF488 fluorescence will be quenched by BPB, the remaining fluorescence will represent internalized, WGA-labeled membrane.
21. If 50–75 caveolin puncta are not available within 30 μm of the injury site, mark as many clearly delineated caveolae that can be marked.
22. Occasionally, a caveolae may split into more than one. Classify these events as mobile (not endocytosed) caveolae.
23. The timing of endocytosis for each caveolae can be presented as binned individual data points to represent the endocytic kinetics (Fig. 2c).
24. During analysis, as a caveolae is analyzed, it can be marked by changing the color of the region marker so keep track of analyzed caveolae.

Use of approaches described above will provide quantitative measurement of the rates of endocytosis in resting cells and focal plasma membrane injury alters this. Such quantification will allow studying the contribution of specific endocytic machinery involving caveolin, clathrin, and accessory proteins as well as bulk endocytic processes in altering the repair ability of cells. We and others have demonstrated that plasma membrane repair is compromised in a number of muscle disease [5, 10, 25–28]. Therefore, we have described the protocol here using cultured muscle cells. Poor plasma membrane repair contributes to disease targeting other tissues as well [29], and enhanced plasma membrane repair in cancer cells contributes to their enhanced metastatic potential [30, 31]. The techniques to monitor endocytosis described above can be applied to these other cell types to study the involvement of endocytic processes in altered repair abilities of these cells. Such studies would improve our understanding of how the fundamental process of endocytosis enables regulating the vital process of maintaining the integrity of plasma membrane and how this is altered contributing to disease. This is required for efforts to develop therapeutic approaches for treating these diseases by designing targeted interventions that focus on regulating injury-triggered plasma membrane endocytosis.

Acknowledgments

This work was supported by NIH (NIAMS) grants, T32AR056993 and AR055686. Microscopy core used for this work is supported by NIH (NICHD) grant, U54HD090257.

References

1. Barbee KA (2005) Mechanical cell injury. Ann N Y Acad Sci 1066:67–84. https://doi.org/10.1196/annals.1363.006
2. Horn A, Jaiswal JK (2018) Cellular mechanisms and signals that coordinate plasma membrane repair. Cell Mol Life Sci 75(20): 3751–3770. https://doi.org/10.1007/s00018-018-2888-7
3. Cooper ST, McNeil PL (2015) Membrane repair: mechanisms and pathophysiology. Physiol Rev 95(4):1205–1240. https://doi.org/10.1152/physrev.00037.2014
4. Andrews NW, Almeida PE, Corrotte M (2014) Damage control: cellular mechanisms of plasma membrane repair. Trends Cell Biol 24(12):734–742. https://doi.org/10.1016/j.tcb.2014.07.008
5. Bansal D, Campbell KP (2004) Dysferlin and the plasma membrane repair in muscular dystrophy. Trends Cell Biol 14(4):206–213. https://doi.org/10.1016/j.tcb.2004.03.001
6. Han R, Campbell KP (2007) Dysferlin and muscle membrane repair. Curr Opin Cell Biol 19(4):409–416. https://doi.org/10.1016/j.ceb.2007.07.001
7. Doherty KR, McNally EM (2003) Repairing the tears: dysferlin in muscle membrane repair. Trends Mol Med 9(8):327–330. https://doi.org/10.1016/s1471-4914(03)00136-9
8. Codding SJ, Marty N, Abdullah N et al (2016) Dysferlin binds SNAREs (soluble N-ethylmaleimide-sensitive factor (NSF) attachment protein receptors) and stimulates membrane fusion in a calcium-sensitive manner. J Biol Chem 291(28):14575–14584. https://doi.org/10.1074/jbc.M116.727016
9. Bittel DC, Chandra G, Tirunagri LMS et al (2020) Annexin A2 mediates dysferlin accumulation and muscle cell membrane repair. Cells 9(9). https://doi.org/10.3390/cells9091919
10. Defour A, Van der Meulen JH, Bhat R et al (2014) Dysferlin regulates cell membrane repair by facilitating injury-triggered acid sphingomyelinase secretion. Cell Death Dis 5(6):e1306. https://doi.org/10.1038/cddis.2014.272
11. Husmann M, Beckmann E, Boller K et al (2009) Elimination of a bacterial pore-forming toxin by sequential endocytosis and exocytosis. FEBS Lett 583(2):337–344. https://doi.org/10.1016/j.febslet.2008.12.028
12. Corrotte M, Almeida PE, Tam C et al (2013) Caveolae internalization repairs wounded cells and muscle fibers. Elife 2:e00926. https://doi.org/10.7554/eLife.00926
13. McDade JR, Naylor MT, Michele DE (2021) Sarcolemma wounding activates dynamin-dependent endocytosis in striated muscle. FEBS J 288(1):160–174. https://doi.org/10.1111/febs.15556
14. Vassilopoulos S (2020) Unconventional roles for membrane traffic proteins in response to muscle membrane stress. Curr Opin Cell Biol 65:42–49. https://doi.org/10.1016/j.ceb.2020.02.007
15. Lo HP, Nixon SJ, Hall TE et al (2015) The caveolin-cavin system plays a conserved and critical role in mechanoprotection of skeletal muscle. J Cell Biol 210(5):833–849. https://doi.org/10.1083/jcb.201501046
16. Gazzerro E, Sotgia F, Bruno C et al (2010) Caveolinopathies: from the biology of caveolin-3 to human diseases. Eur J Hum Genet 18(2):137–145. https://doi.org/10.1038/ejhg.2009.103
17. Sreetama SC, Chandra G, Van der Meulen JH et al (2018) Membrane stabilization by modified steroid offers a potential therapy for muscular dystrophy due to dysferlin deficit. Mol Ther 26(9):2231–2242. https://doi.org/10.1016/j.ymthe.2018.07.021
18. Gushchina LV, Bhattacharya S, McElhanon KE et al (2017) Treatment with recombinant human MG53 protein increases membrane integrity in a mouse model of limb girdle muscular dystrophy 2B. Mol Ther 25(10): 2360–2371. https://doi.org/10.1016/j.ymthe.2017.06.025
19. Bittel DC, Sreetama SC, Chandra G et al (2021) Secreted acid sphingomyelinase as a potential gene therapy for limb girdle muscular dystrophy 2B. J Clin Invest 132:1

20. Defour A, Sreetama SC, Jaiswal JK (2014) Imaging cell membrane injury and subcellular processes involved in repair. J Vis Exp 85. https://doi.org/10.3791/51106
21. Spinazzola JM, Gussoni E (2017) Isolation of primary human skeletal muscle cells. Bio Protoc 7(21). https://doi.org/10.21769/BioProtoc.2591
22. Hindi L, McMillan JD, Afroze D et al (2017) Isolation, culturing, and differentiation of primary myoblasts from skeletal muscle of adult mice. Bio Protoc 7(9). https://doi.org/10.21769/BioProtoc.2248
23. Mamchaoui K, Trollet C, Bigot A et al (2011) Immortalized pathological human myoblasts: towards a universal tool for the study of neuromuscular disorders. Skelet Muscle 1:34. https://doi.org/10.1186/2044-5040-1-34
24. Johnson DS, Jaiswal JK, Simon S (2012) Total internal reflection fluorescence (TIRF) microscopy illuminator for improved imaging of cell surface events. Curr Protoc Cytom 12:Unit 12.29. https://doi.org/10.1002/0471142956.cy1229s61
25. Bittel AJ, Sreetama SC, Bittel DC et al (2020) Membrane repair deficit in facioscapulohumeral muscular dystrophy. Int J Mol Sci 21(15). https://doi.org/10.3390/ijms21155575
26. Chandra G, Defour A, Mamchoui K et al (2019) Dysregulated calcium homeostasis prevents plasma membrane repair in Anoctamin 5/TMEM16E-deficient patient muscle cells. Cell Death Discov 5:118. https://doi.org/10.1038/s41420-019-0197-z
27. Chandra G, Sreetama SC, Mazala DAG et al (2021) Endoplasmic reticulum maintains ion homeostasis required for plasma membrane repair. J Cell Biol 220(5). https://doi.org/10.1083/jcb.202006035
28. Jaiswal JK, Marlow G, Summerill G et al (2007) Patients with a non-dysferlin Miyoshi myopathy have a novel membrane repair defect. Traffic 8(1):77–88. https://doi.org/10.1111/j.1600-0854.2006.00505.x
29. Ammendolia DA, Bement WM, Brumell JH (2021) Plasma membrane integrity: implications for health and disease. BMC Biol 19(1): 71. https://doi.org/10.1186/s12915-021-00972-y
30. Sonder SL, Boye TL, Tolle R et al (2019) Annexin A7 is required for ESCRT III-mediated plasma membrane repair. Sci Rep 9(1):6726. https://doi.org/10.1038/s41598-019-43143-4
31. Jaiswal JK, Lauritzen SP, Scheffer L et al (2014) S100A11 is required for efficient plasma membrane repair and survival of invasive cancer cells. Nat Commun 5:3795. https://doi.org/10.1038/ncomms4795

Chapter 28

Evaluation of hiPSC-Derived Muscle Progenitor Cell Transplantation in a Mouse Duchenne Muscular Dystrophy Model

Minas Nalbandian, Mingming Zhao, and Hidetoshi Sakurai

Abstract

For cell therapy toward Duchenne muscle dystrophy (DMD), muscle progenitor cells derived from human-induced pluripotent stem cell (hiPSC-MuPCs) are recognized as a good candidate, and currently, cell transplantation of hiPSC-MuPCs is being tested with several DMD animal models. In this article, we describe an efficient method to dissociate, purify by cell sorting, transplant, and evaluate the transplantation efficacy of hiPSC-MuPCs.

Key words Human iPS cells, Muscle stem cell, Cell therapy, DMD, Skeletal muscle regeneration

1 Introduction

Duchenne muscular dystrophy (DMD) is a progressive degenerative disease that affects skeletal muscle tissue. DMD is caused by a mutation in the *DMD* gene, which prevents the DYSTROPHIN protein from being expressed in skeletal muscle. Lack of DYSTROPHIN expression in skeletal muscle leads to a loss in skeletal muscle tissue and respiratory failures at early ages.

One of the most promising therapies to treat DMD is the transplantation of healthy muscle stem cells [1]. Skeletal muscle stem cells possess the capacity to fuse with muscle fibers for skeletal muscle regeneration [2]. However, the transplantation of primary skeletal muscle stem cells from adult (satellite cells) is difficult for three major reasons: (1) the loss of quiescence and regenerative potential after in vitro expansion [3–5], (2) the possibility of adverse immune reactions by the patient [6], and (3) the need for skeletal muscle donors. Because they can be expanded and differentiated into several cell lineages including muscle stem cells, human-induced pluripotent stem cells (hiPSCs) present a

Rika Maruyama and Toshifumi Yokota (eds.), *Muscular Dystrophy Therapeutics: Methods and Protocols*,
Methods in Molecular Biology, vol. 2587, https://doi.org/10.1007/978-1-0716-2772-3_28,

promising alternative to satellite cells [7]. Moreover, hiPSCs can be genetically edited to correct the frame read of *DMD*, allowing autologous transplantation [8].

In recent years, many groups have succeeded in establishing transgene-free differentiation methods to generate hiPSC-derived muscle progenitor cells (hiPSC-MuPCs) [9–13]. We [9] and others [14, 15] have demonstrated that the developmental stage of hiPSC-MuPCs is closer to fetal skeletal muscle stem cells than adult muscle stem cells. Although hiPSC-MuPCs show excellent in vitro myogenic capacity, their in vivo regeneration capacity is moderate at best [9, 16]. Therefore, techniques for their transplantation need to be optimized, which includes evaluating their clinical potential using in vivo models. Here, we describe a method to evaluate the in vivo regeneration capacity of hiPSC-MuPCs along with purification methods using antibodies.

2 Materials

2.1 Dissociation and Cell Sorting

1. Collagenase G (Meiji, COLGS).
2. Collagenase H (Meiji, COLHS).
3. DMEM, high glucose, no glutamine (Thermo Fisher Scientific, 11960-044). To prepare DMEM with collagenase G and H, mix 10 mL DMEM with 50 μL collagenase G and 5 μL collagenase H.
4. TripLE Express Enzyme (TripLE, ThermoFisher, 12605036).
5. Hoechst 33342 (Hoechst, ThermoFisher, 62249).
6. HBSS (Life Technologies, 14065056). To prepare 1% BSA/HBSS, dissolve 10-g BSA (A9418, Sigma) in 1 L 1XHBSS.
7. To prepare 1% BSA/HBSS with Hoechst, mix 10 mL of 1% BSA/HBSS with 5 μL Hoechst. Keep the mixture on ice and cover from light.
8. FGFR4-PE antibody (Miltenyi, 130-110-036).
9. Cell strainer 40 μM (Falcon, 352340).
10. Y-27632 (Y), ROCK inhibitor (Nacalai Tesque, 08945-42). To prepare 10 mM Y-27632, resolve 25 mg Y-27632 in 7.8 mL sterilized water. Stock at 4 °C.
11. StemFit (AJINOMOTO, AK02N). To prepare 500 mL StemFit medium, mix solutions A, B, and C, and add 2.5 mL penicillin/streptomycin.
12. 5 mL polystyrene tubes and cell strainer (Falcon, 352235).

2.2 Cryo-Injury and Transplantation

1. Countess II FL Automated Cell Counter (Thermo Fisher Scientific, AMQAF1000).
2. DMEM (Nacalai, 08488-55).
3. 1 mL 27G syringe (Terumo, SS-10M2713).
4. Fine forceps.
5. Forceps with a circular tip (e.g., Fine Science Tools, 11022-14).
6. Fine scissors.
7. Microsurgical needle with a suture thread attached (Nescosuture, HR1604NA45-KF2).
8. NOG/mdx mice (Central Institute for Experimental Animals (CIEA), Japan). Obtaining these severe immunodeficient DMD model mice requires a material transfer agreement with CIEA.

2.3 TA Muscle Sampling and Histology

1. Fine scissors.
2. Fine forceps.
3. Cork (Sansyo, 94-0983).
4. Tragacanth gum powder (Fujifilm, 206-02242). To prepare the gum, mix using a lab spatula equals proportions of powder and water over parafilm until the mixture becomes a gum.
5. Isopentane (Fujifilm, 166-00615).
6. Cryostat (Leica, CM1950).
7. 4% paraformaldehyde (PFA) (Nacalai-Tesque, 09154-14): store at 2–8 °C.
8. PBS (-) (Nacalai-Tesque, 11482-15).
9. Blocking one (Nacalai-Tesque, 03953-95).
10. Can get signal solution B (Toyobo, NKB-601).
11. Triton X-100 (Nacalai-Tesque, 35501-15). To prepare PBS-T, mix 500 mL PBS (-) with 1 mL Triton X-100.
12. Human SPECTRIN mouse mono IgG2b (Leica, NCL-SPEC1).
13. Human NUCLEI mouse mono IgG1 (Millipore, MAB1281).
14. LAMININ α2 rat mono IgG (Alexis, ALX-804-190-C100).
15. Adhesive glass slides (Matsunami, APS-02).
16. Aqua-Poly/Mount solution (Polyscience, 18606-20).
17. Micro cover glass 24 × 24 (Matsunami, C024241).

3 Methods

3.1 Dissociation and Cell Sorting

Culture the hiPSCs as previously described [17] and differentiate them using transgenic free myogenic induction protocols [9–13]. The following dissociation protocol was optimized for cells cultured in 6-well dishes between 42–100 days and differentiated following the protocol described by Zhao et al. (2020) [9], but it has been done efficiently using other protocols as well [12, 13]. Because of the large heterogeneity of the cells in the hiPSC-MuPCs culture, cell purification must be performed before the transplantation. Currently, cell sorting with antibodies is the only published method for purifying hiPSC-MuPCs. In this protocol, we used FGFR4 as a surface marker for the cell sorting, since we have found it to be the most efficient marker for hiPSC-MuPC purification [18].

1. Wash the cultured cells with 1 mL PBS (-) per well and aspirate.
2. Add 1 mL DMEM containing collagenase H and G per well and incubate for 7 min at 37 °C.
3. Add 350 mL TrypLE per well and incubate for 7 min at 37 °C.
4. Dissociate the cells by gentle pipetting with 1 mL tips.
5. Add 2 mL of 1% BSA/HBSS per well and homogenize by gentle pipetting.
6. Pass the cell suspension through a 40-μm cell strainer into a 50 mL tube.
7. Centrifuge the cell suspension for 7 min at 1200 rpm and 4 °C. Aspirate the supernatant.
8. Resuspend the cell pallet with 1% BSA/HBSS (~1 mL per well) and count the cells (*see* **Note 1**).
9. Centrifuge for 7 min at 1200 rpm and 4 °C. Aspirate the supernatant.
10. Resuspend the cell pallet with an adequate volume of 1% BSA/HBSS and antibody (e.g., up to 1×10^7 cells in 100 μL with 11 μL FGFR4-PE antibody).
11. Incubate on ice and cover from light for 20 min.
12. Wash the cells with 5 mL of 1% BSA/HBSS.
13. Centrifuge for 7 min at 1200 rpm at 4 °C and aspirate the supernatant.
14. Wash the cells again with 5 mL of 1% BSA/HBSS.
15. Centrifuge for 7 min at 1200 rpm at 4 °C and aspirate the supernatant.
16. Resuspend with 1% BSA/HBSS containing Hoechst (~1 mL for 3×10^6 cells).

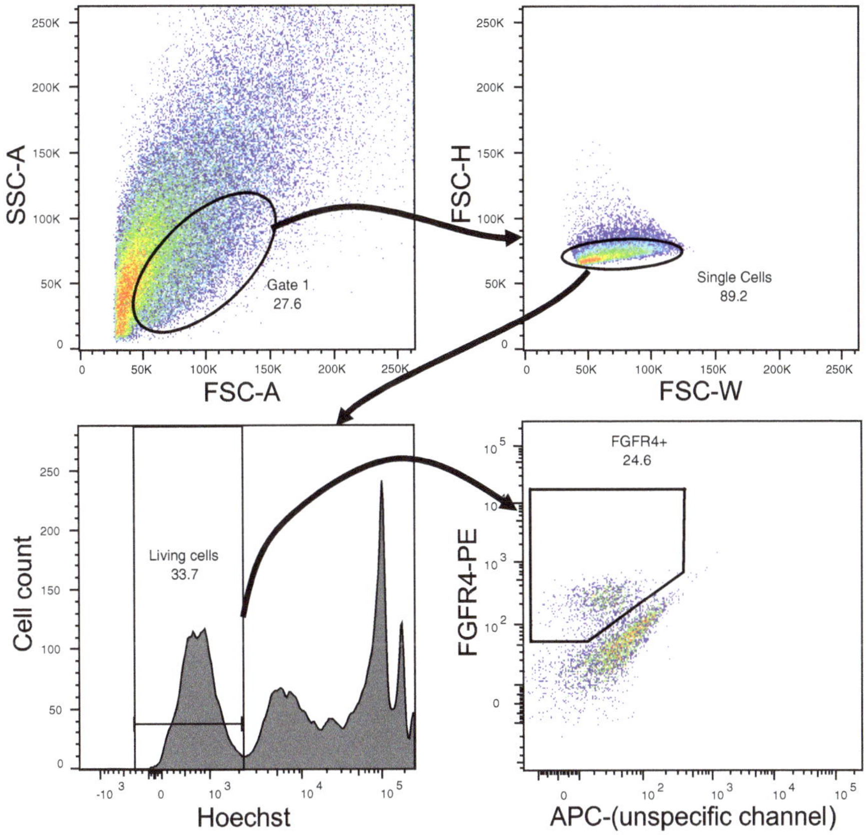

Fig. 1 Representative gaiting to purify iPSC-MuPCs. Plot of SSC (side scatter) versus FSC (forward scatter). The elliptical shape indicates the first gating (upper-left). The plot of FSC-H (forward scatter-height) versus FSC-W (forward scatter-width) was used to eliminate double cells. The elliptical shape indicates single-cell gating (upper-right). The Hoechst histogram was used to eliminate dead cells. The rectangle indicates live cells gating (lower-left). FGFR4-PE versus APC (unspecific channel) was plotted to sort FGFR4+ cells. The polygon indicates FGFR4+ cell gaiting (lower-right)

17. Pass the cells to a 5 mL polystyrene tube with cell strainer for cell sorting.
18. Perform cell sorting with the appropriate gating (Fig. 1) and collect the cells in collection tubes (15 mL tubes with 7 mL StemFit and 7 μL ROCK inhibitor added in each). During the cell sorting, set the sorter to keep the collection tubes and the sample tubes refrigerated at 4 °C to increase cell viability.

3.2 Transplantation

1. After sorting, centrifuge the collection tubes for 10 min at 1800 rpm and 4 °C and aspirate the supernatant.
2. Resuspend the cells in 1 mL DMEM and count them by trypan blue exclusion using a Countess automated cell counter.
3. Centrifuge for 5 min at 1200 rpm and 4 °C and aspirate the supernatant.

4. Resuspend the cells with DMEM at the adequate concentration for transplantation (1×10^5 cells in 30 μL).
5. Keep the cells on ice and bring them to your animal facilities.
6. Anesthetize the mice using isoflurane.
7. Spray ethanol on the mouse limbs.
8. Using the surgical scissors, make a small incision of about 5 mm in the skin of the mouse overlying the tibialis anterior (TA) muscle.
9. Using the small forceps, separate the skin and visualize the TA muscle.
10. Cool down the tip of a round tipped forceps for 30 s in a container with liquid nitrogen.
11. Press with the frozen round tip of the forceps over the exposed TA muscle for 12 s to induce cryoinjury (*see* **Note 2**).
12. Allow the muscle to thaw and repeat the cryoinjury two more times.
13. After the muscle thaws, inject the cells using a syringe (*see* **Note 3**).
14. Suture the skin using a microsurgical needle with the suture thread attached and keep the mouse warm until it has recovered from the anesthetic.

3.3 Euthanization and Histology

The engraftment of hiPSC-MuPCs can variate with the time of sampling. Usually, 2 weeks after the transplantation is enough to visualize engrafted fibers, but we recommend waiting for 4 weeks before euthanizing the mice to maximize the number of engrafted fibers.

1. Prepare gum with the cork to hold the TA muscle (*see* **Note 4**).
2. Euthanize the transplanted mice in a CO_2 chamber.
3. Bring the mouse carcass to the dissection table.
4. Position the mouse in a supine position on the dissection table.
5. Using scissors, cut the skin surrounding the ankle and peel off the skin from the limb to expose the TA muscle.
6. Using scissors and fine forceps, carefully remove the fascia covering the TA muscle.
7. Cut the distal tendon of the TA muscle, and while carefully holding it with fine forceps, start to peel the TA muscle until the proximal insertion.
8. Cut the TA muscle in the proximal insertion.
9. Vertically insert the TA muscle from the distal tendon into the gum and proceed to the cryopreservation.

10. Put isopentane in a glass beaker of around 3–4 cm depth. Pour liquid nitrogen into an insulated container and proceed to cool down the isopentane by setting the glass beaker into the liquid nitrogen (*see* **Notes 5** and **6**).
11. Freeze the TA muscle with the gum by submerging it into the cooled isopentane for 45 s.
12. Keep the frozen TA muscle at −80 °C until cryo-sectioning.
13. Before cryo-sectioning, pre-cool the cryostat and knife to −25 °C and place the samples in the cryostat for around 20 min for thermal equilibration.
14. Mount the sample (with the gum and cork attachment) to a metallic holder and perform the sectioning of the frozen TA muscle. Sections should be 10~20 μm in thickness and sequentially placed on glass slides until the tissue is processed.
15. Collect the sections on adhesive glass slides and allow the samples to dry at room temperature.
16. Store the glass slides in slide glass store boxes at −80 °C until immunostaining is performed.

3.4 Immunostaining and Analysis

1. Take the glass slides out of the −80 °C freezer and allow them to dry at room temperature for around 20 min.
2. Set the glass slides horizontally, fix them by applying 4% PFA over the TA muscle sections (approximately 400 μL), and incubate at room temperature for 20 min.
3. Remove the PFA by tipping the glass slides and then wash the glass slides (*see* next step).
4. To wash the glass slides, place them into a glass slide rack and submerge the rack in a glass staining dish containing PBS (-). Shake the dish at 4 °C for 5 min.
5. Discard the PBS (-) and wash with fresh PBS (-) two more times at 4 °C for 5 min.
6. Take out the glass slide rack from the glass staining dish and then the glass slides from the rack. Place the slides horizontally. Block the sections using Blocking one solution for 1 h (*see* **Note 7**).
7. Remove Blocking one by tipping the slides.
8. Incubate with first antibodies (e.g., h-NUCLEI 1:100, h-SPECTRIN 1:50, or LAMININ α2 1:50) diluted in Can get signal solution B overnight at 4 °C or for one hour at room temperature (*see* **Note 8**).
9. Remove the antibodies by tipping the slides and place the slides into a glass slide rack. Submerge the rack with the slides in a glass staining dish containing PBS-T and shake at room temperature for 10 min.

10. Discard the PBS-T and repeat the washing two times.
11. Take the slides from the rack and incubate the samples with secondary (1:500) antibodies and DAPI for 1 h at room temperature and covered from light.
12. Remove the secondary antibodies by tipping the slides and place the slides into a glass slides rack. Submerge the rack with the slides in a glass staining dish containing PBS-T and shake at room temperature for 5 min.
13. Remove the PBS-T and wash with PBS (-) two more times.
14. Carefully dry the slides with absorbent paper tissue without directly touching the TA muscle sections.
15. Put one drop of the mounting solution over the sections in the glass slides and place the Micro cover glass over the sections in the glass slide (*see* **Note 9**).
16. Acquire representative images of the samples and quantify the number of spectrin-positive myofibers using a microscope. Figure 2 shows representative results of NOG-mdx mice transplanted with 1×10^5 FGFR4+ cells (samples were obtained 28 days after the transplantation).

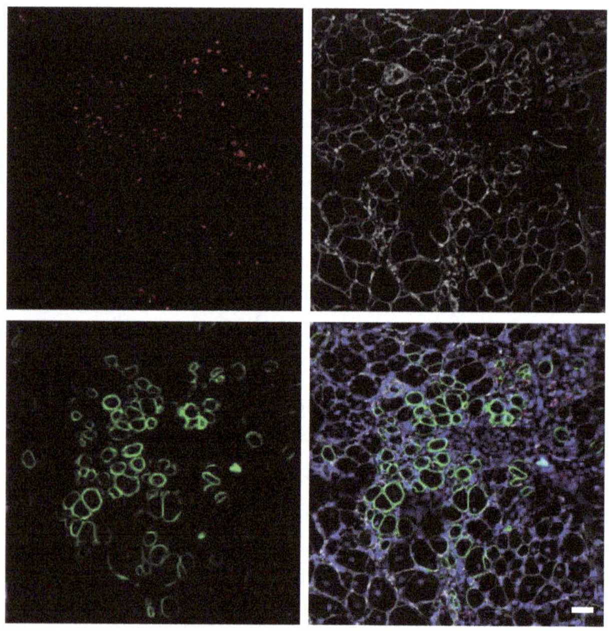

Fig. 2 Representative microscopy image of cross-sectional staining of TA muscle in a NOG mdx mouse transplanted with hiPSC-MuPCs. The TA muscle was first cryoinjured, and then 1×10^5 sorted FGFR4+ hiPSC-MuPCs were transplanted. Four weeks later, the mouse was euthanized, and the TA muscle was cryo-sectioned for immunohistochemical analysis. Scale bar 50 μm

4 Notes

1. Cells can be collected in a new 1.5 mL tube for easy handling.
2. Do not hold the cool forceps directly with your fingers. Wrap the forceps with enough paper to avoid self-injuring the skin.
3. Before injecting the cells, wait a few seconds (30 s) until the color of the cryoinjured zone returns to normal.
4. Cut the cork into cylinders ~0.5 cm high and label the sample on one of the faces of the cork. On the other face, cover with enough gum to insert the distal part of the TA muscle.
5. When isopentane starts to form ice crystals in the bottom of the beaker, remove the beaker from the liquid nitrogen until the ice has melted.
6. Work in a well-ventilated area when working with liquid nitrogen.
7. Be sure to cover the whole surface of the section with the blocking solution.
8. When incubating with the antibodies, place the slides horizontally with the sections facing up and cover the sections with a small square of parafilm to equally distribute the antibodies. Additionally, you may keep some wet tissues under the glass slides to avoid drying.
9. When placing the Micro cover glass over the sections, check that there are no air bubbles in the mounting solution. If bubbles are found, remove them by carefully aspirating with a pipette.

Acknowledgments

We would like to thank Peter Karagiannis for proofreading the manuscript and all lab members for their technical support.

References

1. Biressi S, Filareto A, Rando TA (2020) Stem cell therapy for muscular dystrophies. J Clin Invest 130(11):5652–5664. https://doi.org/10.1172/JCI142031
2. Relaix F, Zammit PS (2012) Satellite cells are essential for skeletal muscle regeneration: the cell on the edge returns centre stage. Development 139(16):2845–2856. https://doi.org/10.1242/dev.069088
3. Gilbert PM, Havenstrite KL, Magnusson KEG et al (2010) Substrate elasticity regulates skeletal muscle stem cell self-renewal in culture. Science 329(5995):1078–1081. https://doi.org/10.1126/science.1191035
4. Montarras D, Morgan J, Collins C et al (2005) Direct isolation of satellite cells for skeletal muscle regeneration. Science 309:2064–2067
5. Sacco A, Doyonnas R, Kraft P et al (2008) Self-renewal and expansion of single transplanted muscle stem cells. Nature 456(7221):502–506. https://doi.org/10.1038/nature07384
6. Cossu G, Previtali SC, Napolitano S et al (2015) Intra-arterial transplantation of HLA-matched donor mesoangioblasts in Duchenne muscular dystrophy. EMBO Mol Med 7(12):1513–1528. https://doi.org/10.15252/emmm.201505636

7. Takahashi K, Tanabe K, Ohnuki M et al (2007) Induction of pluripotent stem cells from adult human fibroblasts by defined factors. Cell 131(5):861–872. https://doi.org/10.1016/j.cell.2007.11.019
8. Li HL, Fujimoto N, Sasakawa N et al (2015) Precise correction of the dystrophin gene in duchenne muscular dystrophy patient induced pluripotent stem cells by TALEN and CRISPR-Cas9. Stem Cell Rep 4(1):143–154. https://doi.org/10.1016/j.stemcr.2014.10.013
9. Zhao M, Tazumi A, Takayama S et al (2020) Induced fetal human muscle stem cells with high therapeutic potential in a mouse muscular dystrophy model. Stem Cell Rep. https://doi.org/10.1016/j.stemcr.2020.06.004
10. Xi H, Fujiwara W, Gonzalez K et al (2017) In vivo human somitogenesis guides Somite development from hPSCs. Cell Rep 18(6):1573–1585. https://doi.org/10.1016/j.celrep.2017.01.040
11. Wu J, Matthias N, Lo J et al (2018) A myogenic double-reporter human pluripotent stem cell line allows prospective isolation of skeletal muscle progenitors. Cell Rep 25(7):1966–1981.e1964. https://doi.org/10.1016/j.celrep.2018.10.067
12. Shelton M, Metz J, Liu J et al (2014) Derivation and expansion of PAX7-positive muscle progenitors from human and mouse embryonic stem cells. Stem Cell Rep 3(3):516–529. https://doi.org/10.1016/j.stemcr.2014.07.001
13. Chal J, Oginuma M, Al Tanoury Z et al (2015) Differentiation of pluripotent stem cells to muscle fiber to model Duchenne muscular dystrophy. Nat Biotechnol 33(9):962–969. https://doi.org/10.1038/nbt.3297
14. Xi H, Langerman J, Sabri S et al (2020) A human skeletal muscle Atlas identifies the trajectories of stem and progenitor cells across development and from Human pluripotent stem cells. Cell Stem Cell 27:158–176.e110. https://doi.org/10.1016/j.stem.2020.04.017
15. Incitti T, Magli A, Darabi R et al (2019) Pluripotent stem cell-derived myogenic progenitors remodel their molecular signature upon in vivo engraftment. Proc Natl Acad Sci U S A 116(10):4346–4351. https://doi.org/10.1073/pnas.1808303116
16. Hicks MR, Hiserodt J, Paras K et al (2018) ERBB3 and NGFR mark a distinct skeletal muscle progenitor cell in human development and hPSCs. Nat Cell Biol 20(1):46–57. https://doi.org/10.1038/s41556-017-0010-2
17. Zhao M, Shoji E, Sakurai H (2018) In vitro evaluation of exon skipping in disease-specific iPSC-derived myocytes. Methods Mol Biol 1828:173–189. https://doi.org/10.1007/978-1-4939-8651-4_11
18. Nalbandian M, Zhao M, Sasaki-Honda M et al (2021) Characterization of hiPSC-derived muscle progenitors reveals distinctive markers for myogenic cell purification toward cell therapy. Stem Cell Rep. https://doi.org/10.1016/j.stemcr.2021.03.004

Chapter 29

Quantification of Muscle Satellite Stem Cell Divisions by High-Content Analysis

William Chen, Theodore J. Perkins, and Michael A. Rudnicki

Abstract

High-content screening is commonly performed on 2D cultured cells, which is high throughput but has low biological relevance. In contrast, single myofiber culture assay preserves the satellite cell niche between the basal lamina and sarcolemma and consequently has high biological relevance but is low throughput. We describe here a high-content screening method that utilizes single myofiber culture that addresses the caveats of both techniques. Our method utilizes the transgenic reporter allele *Myf5-Cre:R26R-eYFP* to differentiate stem and committed cells within a dividing couplet that can be quantified by high-content throughput immunodetection and bioinformatic analysis.

Key words High-content screening, Muscle satellite stem cell, Asymmetric self-renewal, Symmetric division, Myofiber culture, Niche

1 Introduction

The satellite stem cell population required for long-term regeneration has been identified as the subset of Pax7-expressing satellite cells that have never expressed the transcription factor *Myf5* [1–4]. We and others have demonstrated the necessity of satellite stem cell self-renewal in homeostatic maintenance and muscle repair [3, 5–12]. However, characterization of Pax7-expressing satellite cell biological processes has been challenging due to the rapid loss in stem cell identity once removed from the niche [1, 13–22]. Investigations of satellite stem cell biology and its regenerative potential must preserve the satellite cell niche between the basal lamina and sarcolemma to remain biologically relevant. Unfortunately, the gold standard single myofiber culture assay is a labor-intensive low-throughput method for investigating satellite stem cell biology due to its reliance on manual quantification [23–27].

Rika Maruyama and Toshifumi Yokota (eds.), *Muscular Dystrophy Therapeutics: Methods and Protocols*,
Methods in Molecular Biology, vol. 2587, https://doi.org/10.1007/978-1-0716-2772-3_29,

Recent technological advances in microscopy and detection algorithms have enabled the emergence of high-content and high-throughput screening techniques [28]. Although early uses of high-content screening began with drug screening and drug discovery, this powerful technique is increasingly used within basic science for the investigations into biological processes [29–31]. The advantage of high-content and high-throughput screening lies in its ability to identify changes in cell morphologies and biological processes in a relatively short amount of time by testing hundreds or thousands of conditions in parallel [29, 30, 32–34]. High-content, high-throughput screening is extremely versatile technique and can be adapted for use in nearly any biological system. Indeed, numerous biological assays have been developed for high-content screening using immortal cell lines, primary cells, and stem cells combined with fluorescent reporters and lineage tracing techniques to test small molecule compounds, peptides, RNAi, and CRIPSR [32–39].

To investigate muscle satellite stem cell biology in a high-throughput fashion, we have developed a high-content assay on a 96-well format utilizing the FDB myofibers from the *Myf5-Cre: R26R-eYFP* transgenic line to quantify satellite cell progenitor and stem cell divisions.

2 Materials

2.1 General Equipment

1. Standard humidified tissue culture incubator (37.5 °C, 5% CO_2 air).
2. Tissue-culture hood. Used for preparing culture media and dissection solutions.
3. Stereo dissecting wide-field microscope (Zeiss STEMI 2000-C) with a transmitted light base (Zeiss KL 1500 LCD).
4. Standard precision scale.
5. Bunsen burner.
6. Water bath (37.5 °C).
7. High-Content Instrument (Cellomics ArrayScan™).
8. MATLAB.

2.2 Surgical Tools

1. London Forceps: serrated 45° angled tip, 1 × 0.9 mm tip, stainless steel, 16-cm length (Fine Science Tools, cat. no.11080-02).
2. Dumont #7b Forceps: standard curved tip, 0.17 × 0.1 mm tip size, Inox, 11-cm length (Fine Science Tools, cat. no. 11270-20).
3. Extra Fine Bonn Scissors: non-serrated straight tip, 13 mm cutting edge, stainless steel, 8.5 cm length (Fine Science Tools, cat. no. 14084-08).

4. Vannas Spring Scissors: sharp tip, 5 mm cutting edge, 0.35 mm tip diameter, stainless steel, 9 cm length (Fine Science Tools, cat. no. 91500-09).
5. Providence-Hospital Hemostats: straight serrated tip, 1.5 mm tip width, 15 mm clamping length, stainless steel, 14 cm length (Fine Science Tools, cat. no. 13014-14).
6. Sterile disposable underpads, 17 × 24″ (COVIDIEN, cat. no. 7105).

2.3 Transgenic Mouse Lines

1. Breed *Myf5-Cre:R26R-eYFP* transgenic mice [3] according to institution regulations
2. Utilize young adult mice at 6–8 weeks of age (*see* **Note 1**).

2.4 Plastic and Glassware for FDB Myofiber Isolation and Culture

1. Standard 9″ glass Pasteur pipettes (VWR, cat. no. 14672-380).
2. Standard 5″ glass Pasteur pipettes (VWR, cat. no. 14672-200).
3. Create a wide-bore glass pipette that will be used to transfer and triturate the FDB muscles. Use a diamond pen to carve the end of a 9″ Pasteur pipette, creating an opening neared the flared base approximately 50 mm in diameter. Use a Bunsen burner flame to polish the glass edge that will prevent shearing of fibers and muscles.
4. A 5″ glass Pasteur pipette will be used to handle single myofibers and perform washes. Polish the glass edge directly on a Bunsen burner flame to approximately 15–20 mm in diameter.
5. Sterile polypropylene (PP) conical centrifuge tubes, 15-mL capacity (Corning, cat no. 352096).
6. Sterile polypropylene (PP) conical centrifuge tubes, 50-mL capacity (Corning, cat no. 352070).
7. Sterile 6-well flat-bottom plates (Corning, cat. no. 3516). Used for trituration and separation of FDB myofibers.
8. Sterile 96-well flat and clear bottom plates, tissue-culture treated (IBIDI, cat. no. 89626). Used for high-throughput applications in cell-based assays.
9. Sterile 10 mL Luer-Lok tip control syringes (BD, cat. no. 309659).
10. Sterile syringe filters, 45 μm, cellulose acetate (VWR, cat. no. 28145-481). Use for filtering culture media and solutions.

2.5 FDB Myofiber Isolation and Culture Reagents

1. DMEM (Dulbecco's modified Eagle's medium; high glucose, L-glutamine with 110 mg/mL sodium pyruvate. GIBCO, cat. no. 11995073) supplemented with 1% penicillin-streptomycin (HyClone, cat. no. SV30010).
2. Fetal bovine serum (FBS), premium quality, endotoxin <1, hemoglobin <25 (Wisent, cat. no. 080-150).
3. Collagenase Type I (Millipore Sigma, cat. no. C0130).

4. Chicken Embryo Extract (MP Biomedicals, cat. no. 92850145).
5. Fibroblast Growth Factor-Basic (Millipore Sigma, cat. no. GF003AFMG).
6. Myofiber digestion solution: Prepare 2 mL per mouse of 0.2% collagenase type I in DMEM, filter through a 45 μm sieve (*see* **Note 2**).
7. Quenching solution: Prepare fresh 5 mL per mouse of 10% FBS in DMEM, filtered through a 45 μm sieve. Aliquot 3–5 mL per 6-well plate (one well per mouse) and keep at 37 °C (*see* **Note 3**).
8. Wash solution: Prepare 10 mL DMEM per mouse and heat to 37 °C.
9. Myofiber growth medium: Prepare fresh DMEM supplemented with 20% FBS, 2% chicken embryo extract, 1% penicillin/streptomycin and 2.5 ng/mL bFGF, filtered through a 45 μm sieve.

2.6 Small-Molecule Compound Reagents

1. Create a 100× working stock plate of the small molecule library (*see* **Note 4**).
2. Create a 100× vehicle of solvent used to create the working stock (DMSO, PBS).
3. Create a 100× positive control for satellite cell asymmetric division: 10 μg/mL EGF (Miltenyi Biotec, cat. no. 130-093-825) in PBS (1×).
4. Create a 100× positive control for satellite cell asymmetric division: 5 μg/mL Wnt7a (R&D Systems, cat. no. 3008WN025/CF) in PBS (1×).

2.7 Immunofluorescence Reagents

1. 8% Paraformaldehyde (Cedarlane, cat no. 157-8).

   Prepare 2% PFA solution in ddH_2O.
2. Phosphate-buffered saline (PBS) solution 10× (Wisent, cat no. 311-010-FL).

   Prepare PBS (1×) solution in ddH2O.
3. Bovine serum albumin, BSA (Sigma, cat no. A7906).
4. Donor equine serum (Horse Serum, HS) (GE Life Sciences, cat no. SH30074.03)
5. Triton-X100 (Gibco, cat. no. T8787).
6. Immunofluorescence (IF) Blocking Solution: Prepare fresh blocking solution of 5% HS, 1% BSA, 0.5% triton in PBS (1×).
7. Prepare DAPI (4′,6-diamidino-2-phenylindole) nuclei stain solution: 0.4 μg DAPI per ml of PBS (1×).

3 Methods

3.1 Dissection of FDB Muscle

1. Euthanize mouse according to institution regulations. Prepare dissection site with a new surgical underpad (Fig. 1a).
2. Disinfect hindlimbs by spraying with 70% ethanol.
3. Use surgical scissors to make a circular incision of the skin along the ankle. Once this incision is complete, the skin of the foot will be separate from the leg, resembling a sock (Fig. 1b).
4. Create a straight incision along the skin overlaying the ventral part of the foot from the ankle to the base of the digits. This will alleviate pressure when degloving the foot (Fig. 1c).

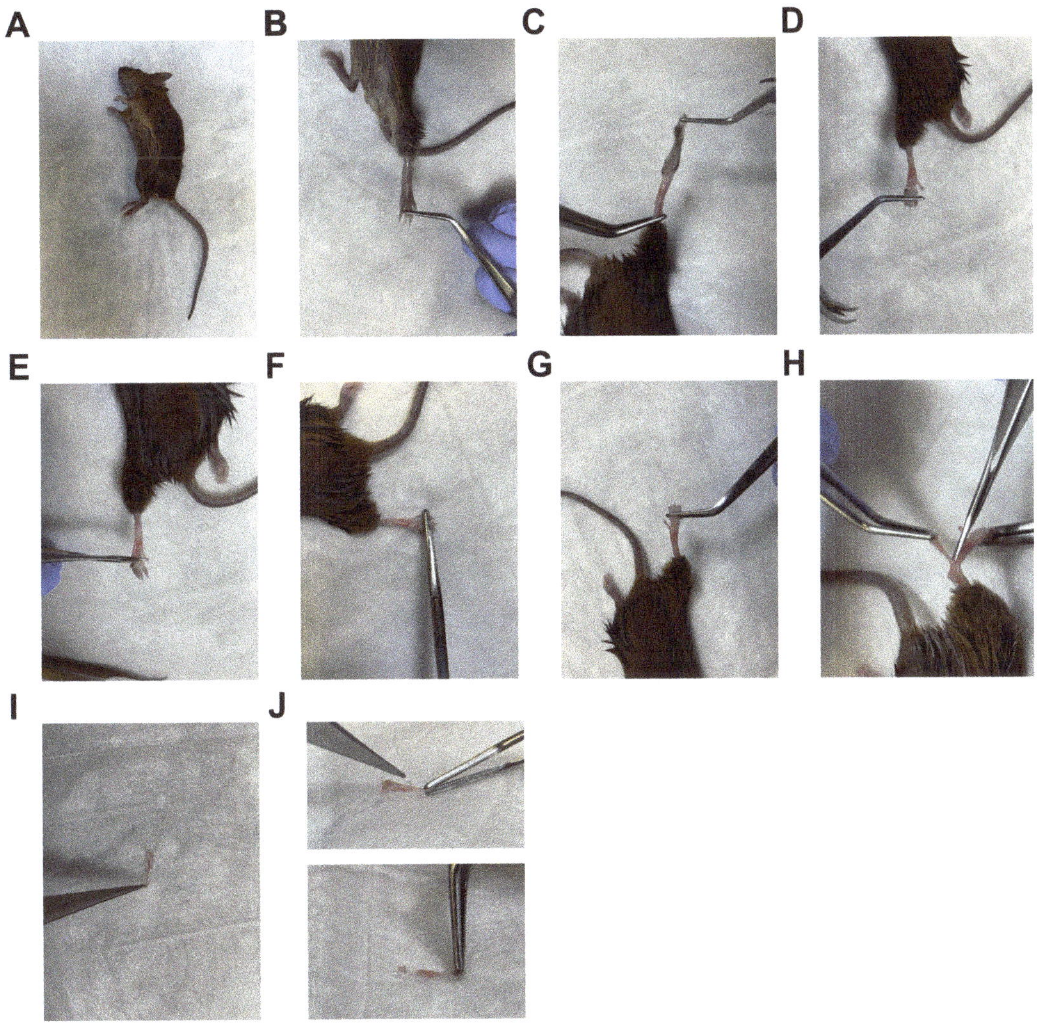

Fig. 1 Flexor digitorum brevis (FDB) dissection steps. Surgical dissection of the FDB muscle as described in Subheading 3.1 Dissection of FDB muscle

5. Use two angled forceps to grab the posterior skin flap of the ankle below the circular ankle incision and the ankle above the incision line (Fig. 1d). Pull the skin back gently to deglove the foot. Position the mouse on its stomach with the dorsal side of the feet facing up (Fig. 1e).
6. Use spring scissors to perform perpendicular cuts to sever the tendons along each of the five metatarsals connecting the FDB muscle, located at the base of the middle phalanges (Fig. 1f).
7. Use a hemostat clamp to clamp the digits of the foot and allow clamp to rest on the bench. Use Dumont forceps to grab the tendons and lift gently (Fig. 1g). Use a surgical scissor to perform small deliberate cuts along the connective tissue adhering the sides of the muscle to the foot, allowing the weight of the foot to gradually release the muscle from the foot between cuts. Continue making incisions to free the muscle until the tendon attached to the heel is reached (Fig. 1h).
8. Sever the tendons connecting the FDB to the heel (Fig. 1i).
9. The FDB muscle sits above the posterior plantar aponeurosis connective tissue and both the plantar aponeurosis and FDB inserts at the heel. Using forceps, brush the end of the tendon from the heel to separate the two tendons, grabbing each with a forceps. Gently pull apart the tendons to separate the FDB from the plantar aponeurosis connective tissue. The FDB is identified as the smaller tissue (Fig. 1j).
10. Deposit the isolated FDB muscle in a 15 mL falcon tube containing 2 mL myofiber digestion solution at room temperature.
11. Repeat steps 3–10 to isolate the second FDB muscle on the contralateral foot.

3.2 Isolation of FDB Myofibers

1. Transfer the falcon tube containing the 2 mL of myofiber digestion solution and isolated FDB muscles to a water bath heated to 37 °C.
2. Allow the FDB muscle to digest for 3 h, gently agitating the tube every 15 min (*see* **Note 5**) (Fig. 2).
3. When digestion is complete, uptake the FDBs in the wide-bore glass pipette, and transfer to a single well in a 6-well plate containing 2–3 mL of DMEM heated to 37 °C (*see* **Note 6**).
4. Use the wide-bore glass pipette to gently triturate the FDB muscle until fibers separate into single myofibers (*see* **Note 7**).
5. Use a polished 5″ glass pipette to remove tendons and debris.
6. Perform three washes with 2–3 mL of warm DMEM per wash, allowing fibers to fall to the bottom of the well before reuptake of medium with a polished 5″ glass pipette.
7. Resuspend myofibers in myofiber growth medium in preparation for dispensing into a 96-well plate (*see* **Note 8**).

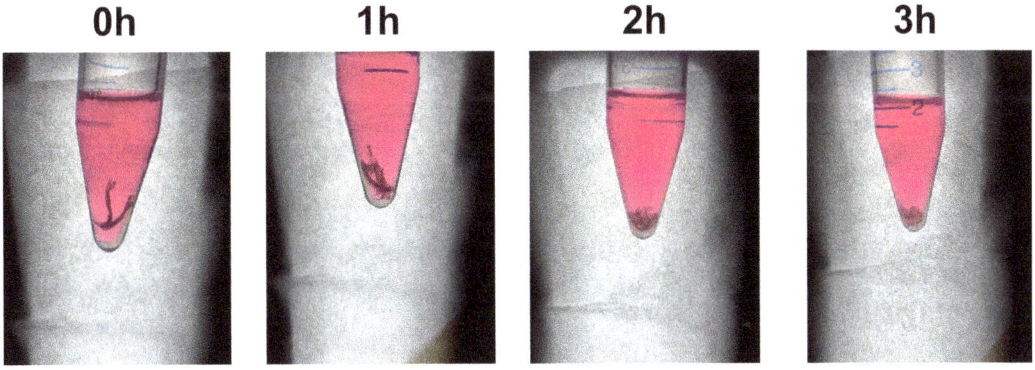

Fig. 2 FDB digestion time course. FDB muscles are digested in 0.2% collagenase type I in DMEM for 3 h. As the muscle bundle digests, single myofibers lift away from the main muscle

3.3 FDB Culture and Small-Molecule Plate Design

1. Prepare a 96-flat bottom well plate for culture by loading 10 μL of controls and small-molecule compounds at 10× concentration in myofiber growth medium (*see* **Note 9**).
2. Dispense 50 μL of FDB myofiber suspension into the 96-well plate using a repeat pipette. The total volume per well should be 60 μL at this step.
3. Load 40 μL of myofiber grown medium per well to top up each well to a final volume of 100 μL per well.
4. Culture myofibers at 37 °C for 42 h to allow for the first set of satellite cell divisions and proceed to fixation (*see* **Note 10**).

3.4 Fixation and Immunostaining of FDB Myofibers

1. Gently aspirate media with flame polished 5″ glass pipette and wash with warm PBS (1×). Allow myofibers to settle to the bottom of the well before aspirating PBS (*see* **Note 11**).
2. Fix myofibers with 100 μL of 2% PFA solution per well for 5 min at room temperature.
3. Perform three washes with PBS (1×).
4. Block myofibers in IF Blocking Solution for 2 h at room temperature or overnight at 4 °C.
5. Dilute primary antibodies in IF Blocking Solution (Table 1) and incubate myofibers overnight at 4 °C.
6. Perform three washes with PBS (1×).
7. Dilute secondary antibodies in IF Blocking Solution (Table 1) and incubate myofibers for 2 h at room temperature or overnight at 4 °C.
8. Perform nuclei stain with DAPI solution for 5 min at room temperature.
9. Perform three washes with PBS (1×).
10. Refill each well with 150–200 μL of PBS (1×) and proceed with high-content screen imaging.

Table 1
Antibodies used to immunodetect Pax7⁺ satellite cells and GFP⁺ committed cells in Myf5-Cre:R26R-eYFP FDB myofibers

Antibody	Host	Source	Identifier	Final concentration
Anti-Pax7 (IgG1)	Mouse	DHSB	https://doi.org/10.1016/S0925-4773(97)00097-X	0.5×
Anti-GFP	Chicken	Abcam	ab13970	10 μg/mL
DAPI	–	Sigma	D8417	0.4 μg/mL
Anti-Mouse IgG1 488	Goat	Invitrogen	A21121	2 μg/mL
Anti-Chicken 546	Goat	Invitrogen	A11040	2 μg/mL

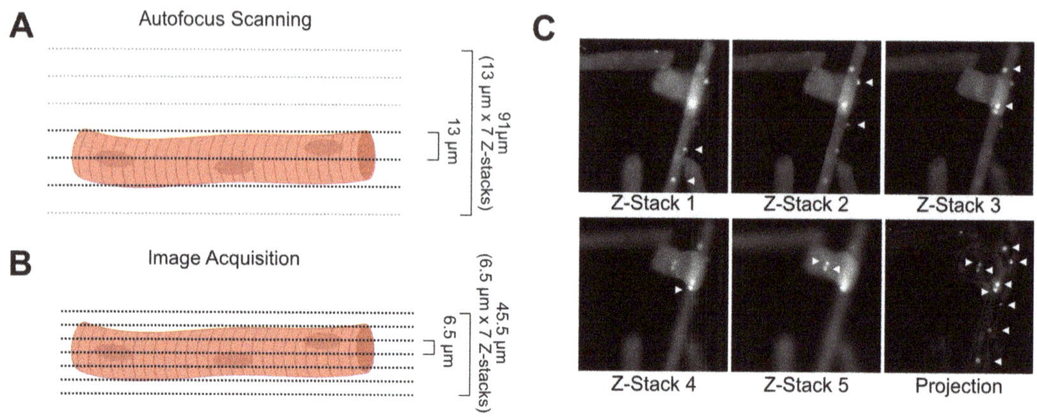

Fig. 3 Z-stack acquisition and projection image of satellite cells on myofibers. (**a**) Satellite cell positions are identified using an autofocus function to scan through the myofiber. (**b**) Z-stack images are taken at depth of field where satellite cells reside. (**c**) A projection image from the combined Z-stack images is used for downstream analysis

3.5 Programming Cellomics™ Arrayscan for Satellite Cell Detection

1. Create a Cellomics™ Arrayscan plate profile for the 96-well plates to be used for screening by following the Cellomics™ program on-screen instructions (*see* **Note 12**).
2. Within the Cellomics™ analysis program, select target activation within the bioapplications menu.
3. Set the autofocus function to utilize the DAPI channel, and scan for a depth of 91 μm by sampling seven z-stacks at 13 μm intervals prior to image acquisition as shown in Fig. 3 (*see* **Note 13**). Ensure that no images are acquired if zero DAPI-positive cells are detected.
4. Set image acquisition parameters to perform seven z-stacks spaced 6.5 μm apart for a total depth of field of 45 μm at 200× (objective 20×) magnification as shown in Fig. 3 (*see* **Note 14**).

Table 2
Cellomics™ Arrayscan detection script parameters

A. Image preprocessing				
Channel	**Object type**	**Method**	**Set value**	**Units**
1	Bright	Low pass filter	20	Pixels
2	Bright	3D surface fitting	15	Pixels
3	Bright	3D surface fitting	10	Pixels
B. Object identification				
Channel	**Algorithm**	**Set value**	**Max range**	**Units**
1	Smoothing	2	0–500	Pixel area
1	Thresholding	75	0–500	Pixels
1	Segmentation	2	1–246	Intensity
C. Object validation and selection				
Channel	**Parameter**	**Set range**	**Max range**	**Units**
1	Area	70–350	0–2000	Pixels
1	Shape (P2A)	0–1.28	0–2	Number
1	Shape (LWR)	0–1.76	0–3	Number
1	Variance intensity	0–400	0–600	Intensity
1	Average intensity	1800–6000	0–16,384	Intensity
2	Average intensity	0–15,000	0–16,384	Intensity
3	Average intensity	400–9000	0–16,384	Intensity

5. Set the z-stack image projection to utilize a maximum-minimum difference background subtraction algorithm as shown in Fig. 3 (*see* **Note 15**).
6. Input parameters for Image Preprocessing, Object Identification, and Object Validation for each channel: Pax 7 (Channel 1, CH1), YFP (Channel 2, CH2) and DAPI (Channel 3, CH3) using the parameters described in Table 2 and shown in Fig. 4 (*see* **Note 16**).
7. Export the following parameters in a CSV file: Well, Field, Cell, Top, Left, XCentroid, YCentroid, Object AvgIntenCh1, AvgIntenCh2, AvgIntenCh3.

3.6 Programming MATLAB-Based Script for Satellite Cell Couplet Detection

1. Move all Cellomics exports into the same folder as the MATLAB script and save the CSV file name as "new_excel.csv."
2. Run the MATLAB script labelled, *histogram_generation.m* to determine thresholds of detection for channels YFP (CH2) and DAPI (CH3) (*see* **Note 17**).

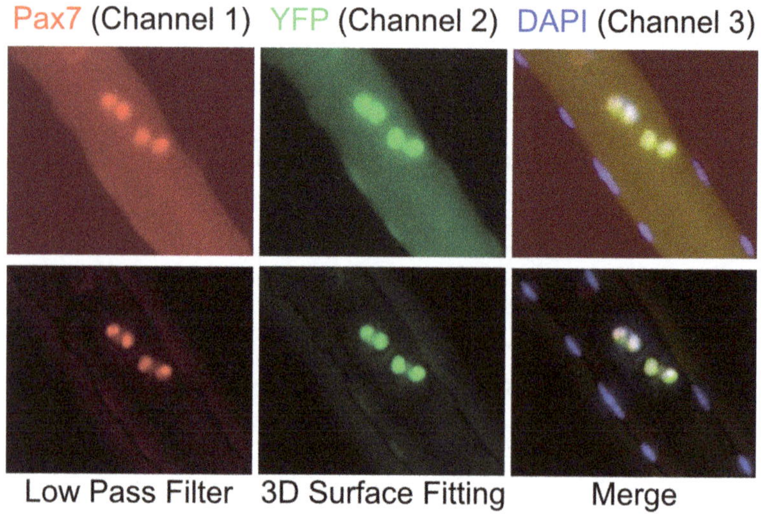

Fig. 4 Optimization of Arrayscan detection parameters. Optimizations before (top) and after (bottom) of Pax7 (Ch1), YFP (CH2) and DAPI (CH3)

3. Run the MATLAB script labeled, *Image_Divisions_Script.m* and enter threshold values for Channels YFP (CH2) and DAPI (CH3) to quantify dividing satellite cell couplets (*see* **Note 18** and Fig. 5).

3.7 High-Content Screen Analysis

1. Determine the mean, standard deviation, and standard error mean values of all experimental conditions.
2. Create an excel file to perform analysis on MATLAB output using the formulas described in Table 3 (*see* **Note 19**).

4 Notes

1. Approximately six to eight FDB muscles from three to four mice per 96-well plate and eight to ten FDB muscles from four to five mice per 384-well plate will yield sufficient myofibers for screening at a myofiber density of 50–100 myofibers per well.
2. For bulk preparation, we prepare collagenase solution in 50-mL conical tubes and aliquot into 15 mL falcon tubes following filtration. Aliquoted digestion solution may be kept frozen at −20 °C for up to one month. Avoid freeze-thaw cycles and ensure solution is warmed to room temperature before depositing isolated muscles for enzymatic digestion. We use 1 mL digestion media per FDB muscle, which equates to 2 mL per mouse, when isolating both left and right FDBs.
3. The FBS will quench the collagenase activity and prevent over digestion of myofibers. Additionally, this solution will prevent myofibers from adhering to plastic and glass surfaces. Thus,

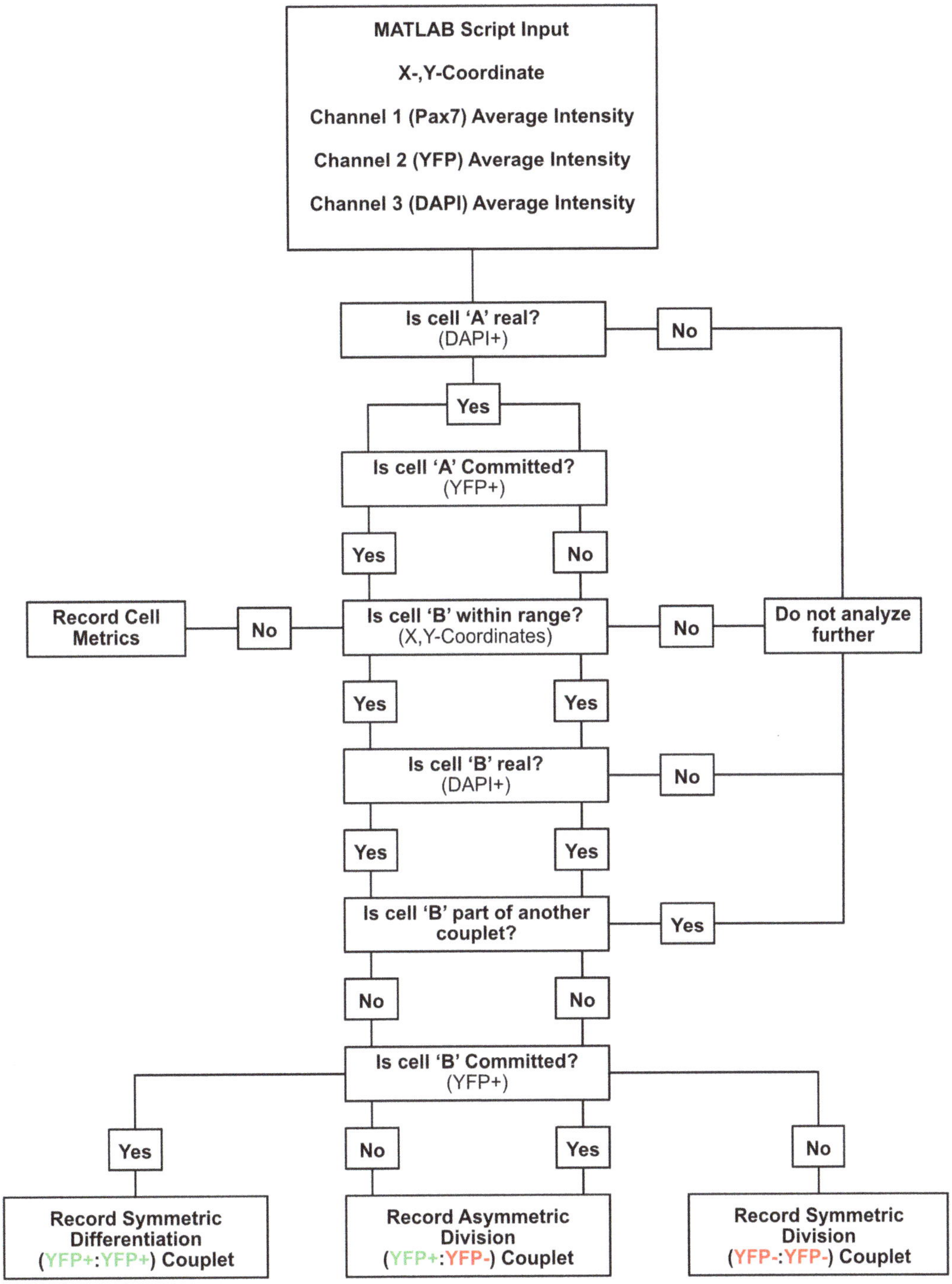

Fig. 5 Satellite cell detection logic. Schematic showing the logic tree used to identify satellite cells and classify the divisions as symmetric or asymmetric

Table 3
FDB HCS stem cell division analysis formula chart

MATLAB Output parameters	Description	
1. Well	Well number	
2. NCells	Total cells detected	
3. NType1=CH2+/CH3+	Number of cells that are YFP (CH2) and DAPI (CH3) positive	
4. NType2=CH2−/CH3+	Number of cells that are YFP (CH2) negative and DAPI (CH3) positive	
5. CH2+ symmetric pairs	Number of dividing couplets consisting of two YFP positive cells	
6. CH2+/CH2− asymmetric pairs	Number of dividing couplets containing a YFP positive and YFP negative cell	
7. CH2− symmetric pairs	Number of dividing couplets containing two YFP negative cells	
Metric	Description	Formula
8. Total cell divisions	Number of dividing couplets	= Sum (5,6,7)
9. Total stem cell divisions	Number of couplets containing at least 1x YFP- cell	= Sum (6,7)
10. Symmetric stem cell divisions	Symmetric stem cell divisions normalized over total stem cell divisions	= 5/8
11. Asymmetric stem cell divisions	Asymmetric stem cell divisions normalized over total stem cell divisions	= 6/8
12. Initial satellite cell population	The starting number of Pax7-expressing satellite cells at t=0	= 2 − 8
13. Total mitotic index	The rate of all satellite cell divisions	= 8/12
14. Stem cell mitotic index	The rate of stem cell (YFP-) divisions	= 9/12
15. Committed cell mitotic index	The rate of committed cell divisions	= 5/12
16. Ratio of asymmetric divisions	The number of asymmetric divisions over the number of symmetric divisions	= 11/10
17. Ratio of symmetric divisions	The number of symmetric divisions over the number of asymmetric divisions	= 10/11
18. Symmetric division fold change	The number of symmetric divisions compared to vehicle (veh)	= 10/10 (veh)
19. Asymmetric division	The number of asymmetric divisions compared to vehicle (veh)	= 11/11

additional quenching media or filtered FBS may be used to coat contact surfaces to prevent damage or loss of myofibers caused by adherence to these surfaces.

4. Master stock plates should be designed to accommodate for the edge effect by utilizing only the inner 60 wells. The diagonal wells should also be left blank for control conditions. Compounds may be screened at various concentrations: 1 μM, 100 nM, 10 nM, and 1 nM. Alternatively, the primary screen may be conducted at one concentration (typically 1 μM), with a secondary screen containing a shortlist of targets selected for dosage response curves.
5. Alternatively, myofibers may be placed in an agitator set at the lowest speed located inside an incubator set at 37 °C. The muscle has completed digestion when small myofibers begin to detach, and the muscle look bloated as shown in Fig. 2.
6. To ensure efficient quenching of the enzymatic collagenase and to prevent over digestion, complete the transfer of FDBs in one to two drops of the digestion solution.
7. To preserve myofiber integrity, ensure the DMEM medium is kept warm at 37 °C during trituration. This can be accomplished by periodically returning the plate to a tissue incubator, or replenishing the medium with fresh, warm DMEM. An appropriately digested FDB muscle should require only 5 min of trituration.
8. The resuspension volume may vary depending on number of fibers following isolation and number of wells to be screened. The target density of FDB myofibers should fall within 50–100 myofibers per well to allow for a monolayer culture of myofibers optimal for screening (*see* Table 4). It is important to allow for fibers to settle to the bottom of the well before aspirating. Aspiration of media should be performed under a dissection scope to ensure myofibers are not taken up. Wells may be loaded with a multichannel. The working volume per well for washes is 100–200 μL. Never allow fibers to dry, as this will impair the immunofluorescence signal quality.
9. The outer wells are prone to evaporative edge effects, which will affect the final concentration of small-molecule compounds. To avoid this issue, fill the outer 36 wells with water and use only the inner 60 wells for experimental conditions. Controls (positive, negative, vehicle) should be loaded in replicates diagonally across the plate or dispersed randomly. If this is not feasible due to the design layout of the master compound library, controls may be loaded in columns 2,3, and flipped orientation in columns 10,11. Dilute stock small-molecule compounds to 10× in myofiber growth medium.

Table 4
Flexor digitorum brevis properties

Characteristics	Flexor digitorum brevis (FDB)
Digestion time	3 h
N fibers per mouse	1000
N of $Pax7^{+}$ satellite cells per fiber	3–7
Myofiber length (mm) [27]	0.744 ± 0.01
Satellite cell density per mm	4.03–9.41
% $Pax7^{+}Myf5^{-}$ stem cells per total $Pax7^{+}$ satellite population	18 ± 2.3

10. Satellite cells in the *Myf5-Cre:R26R-eYFP* will undergo the first divisions between 36 and 42 h in culture and remain in close proximity.
11. It is important to allow for fibers to settle to the bottom of the well before aspirating. Aspiration of media should be performed under a dissection scope to ensure myofibers are not taken up. Wells may be loaded with a multichannel. The working volume per well for washes is 100–200 μL. Never allow fibers to dry, as this will impair the immunofluorescence signal quality.
12. A plate profile is created such that the Cellomics will position the camera at the bottom and center of each well, saving hours of run time and reduce reading errors.
13. Satellite cells require 3D imaging to visualize, as they are positioned at the periphery of FDB myofibers that are approximately 20–25 μm in diameter. This autofocus function allows the camera to quickly identify if any cells are present within the field of view.
14. A 45 μm field of depth accommodates any variability in plate level and thickness as well as overlapping (2–3) myofibers.
15. A maximum-minimum difference projection method utilizes differences in pixel brightness between adjacent sources and z-stacks to reduce background. Each channel is individually optimized based on immunofluorescent signal pattern and its purpose within the screening protocol.
16. The Cellomics™ analysis program first creates a mask around the areas of interest based on the parameters described in Table 2 for each individual channel: Pax7 nuclear (CH1), YFP cytoplasmic (CH2), and DAPI nuclear expression. If the cell is determined to be a real cell based on DAPI nuclear

expression (CH2), then the YFP cytoplasmic signal and X,Y coordinate is recorded. For the complete screen logic, refer to **Note 18** and Fig. 4.

17. The threshold of detection must be determined every time an immunofluorescent protocol is performed to maintain accurate and precise detection of satellite cells. Differences in immunofluorescent signal between runs may affect the detection of satellite cells and the calling of YFP status. The *histogram_generation.m* script will generate a histogram of each channel: Pax7 (CH1), YFP (CH2), DAPI (CH3). Each channel will contain a bimodal distribution of objects detected and not detected. The threshold of detection is the value between the two distributions.

18. The output will be exported to a file called *new_excel_Counts.* This file contains the parameters used in the script (couplet distance between cells and channel thresholds), a results summary that shows the quantification in each well of total cells, type of cells, and type of divisions detected, and a details section that shows the calling of couplet division based on X, Y coordinates. Our work through is a two-step process utilizing the Cellomics program to recognize satellite cells and a MATLAB-based script to quantify satellite cell dividing couplets. However, this process of satellite cell recognition and couplet detection may be combined into one and performed using other screening and detection software such as Columbus™ Image Data Storage and Analysis System, or Cell Profiler™ using the following logic described in Fig. 5: Identification and validation of satellite cells are first performed within the Cellomics™ script based on Pax7 (CH1) and DAPI (CH3) signal. The X,Y coordinates and YFP immunofluorescent signal is then exported for further analysis on MATLAB. The Matlab script determines whether a cell belongs to a dividing couplet based on its proximity to another cell. Cells within proximity to two other cells are excluded from couplet quantifications, because it is not possible to determine which cell it originated from. However, the committed/noncommitted status is recorded for additional screen metrics. Couplets are identified based on the combination of YFP expression within the two cells of each couplet.

19. To upscale the analysis, these calculations may alternatively be performed in R instead of excel. Although this screen was designed to quantify satellite stem cell divisions at 42 h, this system may be used to perform analysis on muscle satellite stem cells and myoblast kinetics at other various time points: 24-h activation, 48–72 proliferation, 96–120 h differentiation and myogenic markers such as MyoD and MyoG.

Funding

The studies from the Rudnicki lab were carried out with the support of grants from the Canadian Institutes of Health Research [FDN-148387], the US National Institutes for Health [R01AR044031], E-Rare-3 (Canadian Institutes of Health Research/Muscular Dystrophy Canada), the Foundation for Gene & Cell Therapy, and the Stem Cell Network.

References

1. Seale P, Sabourin LA, Girgis-Gabardo A et al (2000) Pax7 is required for the specification of myogenic satellite cells. Cell 102:777–786
2. Lepper C, Partridge TA, Fan C-M (2011) An absolute requirement for Pax7-positive satellite cells in acute injury-induced skeletal muscle regeneration. Development 138:3639–3646. https://doi.org/10.1242/dev.067595
3. Kuang S, Kuroda K, Grand F, Rudnicki MA (2007) Asymmetric self-renewal and commitment of satellite stem cells in muscle. Cell 129: 999–1010. https://doi.org/10.1016/j.cell.2007.03.044
4. Maltzahn J von, Jones AE, Parks RJ, Rudnicki MA (2013) Pax7 is critical for the normal function of satellite cells in adult skeletal muscle. Proc Natl Acad Sci 110:16474–16479. https://doi.org/10.1073/pnas.1307680110
5. Rocheteau P, Gayraud-Morel B, Siegl-Cachedenier I et al (2012) A subpopulation of adult skeletal muscle stem cells retains all template DNA strands after cell division. Cell 148:112–125. https://doi.org/10.1016/j.cell.2011.11.049
6. Sincennes M-C, Brun CE, Lin AYT et al (2021) Acetylation of PAX7 controls muscle stem cell self-renewal and differentiation potential in mice. Nat Commun 12:3253. https://doi.org/10.1038/s41467-021-23577-z
7. Shinin V, Gayraud-Morel B, Gomès D, Tajbakhsh S (2006) Asymmetric division and cosegregation of template DNA strands in adult muscle satellite cells. Nat Cell Biol 8: 677–687. https://doi.org/10.1038/ncb1425
8. Troy A, Cadwallader AB, Fedorov Y et al (2012) Coordination of satellite cell activation and self-renewal by Par-Complex-Dependent asymmetric activation of p38α/β MAPK. Cell Stem Cell 11:541–553. https://doi.org/10.1016/j.stem.2012.05.025
9. Dumont NA, Wang YX, von Maltzahn J et al (2015) Dystrophin expression in muscle stem cells regulates their polarity and asymmetric division. Nat Med 21:1455–1463. https://doi.org/10.1038/nm.3990
10. Wang Y, Feige P, Brun CE et al (2019) EGFR-Aurka signaling rescues polarity and regeneration defects in Dystrophin-deficient muscle stem cells by increasing asymmetric divisions. Cell Stem Cell. https://doi.org/10.1016/j.stem.2019.01.002
11. Bernet JD, Doles JD, Hall JK et al (2014) p38 MAPK signaling underlies a cell-autonomous loss of stem cell self-renewal in skeletal muscle of aged mice. Nat Med 20:nm.3465. https://doi.org/10.1038/nm.3465
12. Chang NC, Sincennes M-CC, Chevalier FP et al (2018) The Dystrophin Glycoprotein complex regulates the epigenetic activation of muscle Stem Cell Commitment. Cell Stem Cell 22:755–768.e6. https://doi.org/10.1016/j.stem.2018.03.022
13. Rayagiri SS, Ranaldi D, Raven A et al (2018) Basal lamina remodeling at the skeletal muscle stem cell niche mediates stem cell self-renewal. Nat Commun 9:1075. https://doi.org/10.1038/s41467-018-03425-3
14. Kuang S, Gillespie MA, Rudnicki MA (2008) Niche regulation of muscle satellite cell self-renewal and differentiation. Cell Stem Cell 2: 22–31. https://doi.org/10.1016/j.stem.2007.12.012
15. Gilbert PM, Corbel S, Doyonnas R et al (2012) A single cell bioengineering approach to elucidate mechanisms of adult stem cell self-renewal. Integr Biol 4:360–367. https://doi.org/10.1039/C2IB00148A
16. Choo H-J, Canner JP, Vest KE et al (2017) A tale of two niches: differential functions for VCAM-1 in satellite cells under basal and injured conditions. Am J Physiol-cell Ph 313: C392–C404. https://doi.org/10.1152/ajpcell.00119.2017
17. Quarta M, Brett JO, DiMarco R et al (2016) An artificial niche preserves the quiescence of

muscle stem cells and enhances their therapeutic efficacy. Nat Biotechnol 34:752–759. https://doi.org/10.1038/nbt.3576

18. Sampath SC, Sampath SC, Ho AT et al (2018) Induction of muscle stem cell quiescence by the secreted niche factor Oncostatin M. Nat Commun 9:1531. https://doi.org/10.1038/s41467-018-03876-8
19. Lukjanenko L, Jung MJ, Hegde N, et al (2016) Loss of fibronectin from the aged stem cell niche affects the regenerative capacity of skeletal muscle in mice. Nat Med 22:897–905. https://doi.org/https://doi.org/10.1038/nm.4126
20. Verma M, Asakura Y, Murakonda BSRS et al (2018) Muscle satellite cell cross-talk with a vascular Niche maintains Quiescence via VEGF and Notch signaling. Cell Stem Cell 23:530–543.e9. https://doi.org/10.1016/j.stem.2018.09.007
21. Baghdadi MB, Castel D, Machado L, et al (2018) Reciprocal signalling by Notch–Collagen V–CALCR retains muscle stem cells in their niche. Nature 557:714–718. https://doi.org/10.1038/s41586-018-0144-9
22. Eliazer S, Muncie JM, Christensen J et al (2019) Wnt4 from the Niche controls the Mechano-properties and Quiescent State of muscle stem cells. Cell Stem Cell. https://doi.org/10.1016/j.stem.2019.08.007
23. Pasut A, Jones AE, Rudnicki MA (2013) Isolation and culture of individual myofibers and their satellite cells from adult skeletal muscle. JoVE e50074. https://doi.org/10.3791/50074
24. Shefer G, Yablonka-Reuveni Z (2005) Isolation and culture of skeletal muscle myofibers as a means to analyze satellite cells. Method Mol Biol (Clifton, NJ) 290:281–304. https://doi.org/10.1385/1-59259-838-2:281
25. Stuelsatz P, Keire P, Yablonka-Reuveni Z (2017) Isolation, culture, and immunostaining of skeletal muscle myofibers from wildtype and Nestin-GFP Mice as a means to analyze satellite cell. Method Mol Biol (Clifton, NJ) 1556: 51–102. https://doi.org/10.1007/978-1-4939-6771-1_4
26. Danoviz M, Yablonka-Reuveni Z (2012) Skeletal muscle satellite cells: background and methods for isolation and analysis in a primary culture system. 798:21–52. https://doi.org/10.1007/978-1-61779-343-1_2
27. Tarpey MD, Amorese AJ, Balestrieri NP et al (2018) Characterization and utilization of the flexor digitorum brevis for assessing skeletal muscle function. Skelet Muscle 8:14. https://doi.org/10.1186/s13395-018-0160-3
28. Carpenter AE (2007) Image-based chemical screening. Nat Chem Biol 3:461–465. https://doi.org/10.1038/nchembio.2007.15
29. Szymański P, Markowicz M, Mikiciuk-Olasik E (2011) Adaptation of high-throughput screening in drug discovery—toxicological screening tests. Int J Mol Sci 13:427–452. https://doi.org/10.3390/ijms13010427
30. Thomsen W, Frazer J, Unett D (2005) Functional assays for screening GPCR targets. Curr Opin Biotechnol 16:655–665. https://doi.org/10.1016/j.copbio.2005.10.008
31. Zanella F, Lorens JB, Link W (2010) High content screening: seeing is believing. Trends Biotechnol 28:237–245. https://doi.org/10.1016/j.tibtech.2010.02.005
32. Kiger A, Baum B, Jones S et al (2003) A functional genomic analysis of cell morphology using RNA interference. J Biol 2:27. https://doi.org/10.1186/1475-4924-2-27
33. Liu T, Sims D, Baum B (2009) Parallel RNAi screens across different cell lines identify generic and cell type-specific regulators of actin organization and cell morphology. Genome Biol 10:R26. https://doi.org/10.1186/gb-2009-10-3-r26
34. Boutros M, Ahringer J (2008) The art and design of genetic screens: RNA interference. Nat Rev Genet 9:554–566. https://doi.org/10.1038/nrg2364
35. Chen S, Do JT, Zhang Q et al (2006) Self-renewal of embryonic stem cells by a small molecule. Proc Natl Acad Sci 103: 17266–17271. https://doi.org/10.1073/pnas.0608156103
36. Nazlamova L, Thomas NS, Cheung M-K et al (2020) A CRISPR and high-content imaging assay compliant with ACMG/AMP guidelines for clinical variant interpretation in ciliopathies. Hum Genet:1–15. https://doi.org/10.1007/s00439-020-02228-1
37. Barbuti P, Antony P, Santos B et al (2020) Using high-content screening to generate single-cell gene-corrected patient-derived iPS clones reveals excess Alpha-Synuclein with familial Parkinson's disease point mutation A30P. Cell 9:2065. https://doi.org/10.3390/cells9092065
38. Valtonen S, Vuorinen E, Kariniemi T et al (2020) Nanomolar Protein–Protein interaction monitoring with a Label-Free Protein-Probe technique. Anal Chem. https://doi.org/10.1021/acs.analchem.0c02823
39. Li Y, Liu M, Xie S (2020) Harnessing phage display for the discovery of peptide-based drugs and monoclonal antibodies. Curr Med Chem 27. https://doi.org/10.2174/0929867327666201111144353

Part VIII

Antibody

Chapter 30

Systemic Delivery of a Monoclonal Antibody to Immunologically Block Myostatin in the A17 Mouse Model of OPMD

Alberto Malerba, Pradeep Harish, and Linda Popplewell

Abstract

Oculopharyngeal muscular dystrophy (OPMD) is a late-onset rare muscle disease affecting approximately 1 in 80,000 individuals worldwide. However, it can affect as much as 1:600 individuals in some populations due to a strong founder effect. The muscle pathology is characterized by progressive eyelid drooping (ptosis), swallowing difficulties (dysphagia), and limb weakness at later stages of disease progression. The genetic defect is associated with significant fibrotic deposition and atrophy in affected muscles. No treatments are available to cure the disease. Only surgical techniques to correct ptosis and swallowing are currently possible, though they carry a risk of recurrence. Myostatin is a negative regulator of muscle growth, and several strategies to downregulate its expression have been developed with the aim of improving muscle mass and strength in muscular pathologies. We recently showed that weekly systemic treatment of the A17 murine model of OPMD with a monoclonal antibody for myostatin improves body and muscle mass, increases muscle strength, and reduces muscle fibrosis. Here, we describe the methodology for repeated intraperitoneal delivery of myostatin antibody in the murine model. Furthermore, we detail the most relevant analyses to assess histopathological and functional improvements of this treatment in this mouse model.

Key words Monoclonal antibodies, Oculopharyngeal muscular dystrophy, A17 mouse model, Intraperitoneal injection, Muscle atrophy, Fibrosis

1 Introduction

Oculopharyngeal muscular dystrophy (OPMD) is a rare autosomal dominant, degenerative muscle disease affecting about 1/700 Bukhara Jews, 1/1000 French Canadian peoples, and 1/80,000 people in Europe. The disease usually presents in the fifth decade of life [1]. OPMD is due to mutations in the poly(A)-binding protein nuclear-1 (*PABPN1*) gene that codes for a protein that has pivotal roles in many biological pathways [2]. The expansion at the

Rika Maruyama and Toshifumi Yokota (eds.), *Muscular Dystrophy Therapeutics: Methods and Protocols*, Methods in Molecular Biology, vol. 2587, https://doi.org/10.1007/978-1-0716-2772-3_30,

N-terminus to up to 18 alanines makes the protein prone to misfolding and aggregation leading to formation of intranuclear inclusions that trap a multitude of other small molecules (e.g., RNAs, proteins) [3, 4]. The primary pathological effects manifest as progressive eyelid drooping (ptosis) and swallowing difficulties (dysphagia) [5]. There are currently no pharmacological treatments available for OPMD. Patients have access only to medical intervention where surgical techniques are used to correct the ptosis and improve swallowing in moderate to severely affected individuals. No treatments are available to improve the defects in proximal muscles that severely affect patient's ambulation and quality of life.

Myostatin acts as a negative regulator of muscle mass, and naturally occurring myostatin null mutants (such as the Piedmontese cattle, Texel sheep, and the bully whippet dogs) display a remarkable hypermuscular phenotype [6]. Inhibition of myostatin has been reported to increase muscle mass and myofiber size by cytoplasmic growth [7]. A large number of strategies (e.g. myostatin propeptide, gene therapy, gene editing, ligand traps, exon skipping, and monoclonal antibodies) have been investigated to attenuate myostatin activity in both preclinical and clinical settings [8–11], especially in cases of cachexia, sarcopenia, muscular atrophy, and muscular dystrophies such as Duchenne muscular dystrophy (DMD) [12–19].

The most used mouse model of OPMD, the A17 mouse, was created by inserting the cDNA of bovine expanded PABPN1 with 17 alanines under the control of a muscle-specific human skeletal actin (HSA) promoter [20, 21]. This mouse model shows strong overexpression of expanded PABPN1 protein, substantial amount of aggregates in all skeletal muscles (i.e., 30–40% of total nuclei), muscle atrophy, and reduced muscle strength, and it has been extensively used to screen drugs that reduce the amount of aggregates [20, 22–25] and to test new genetic approaches [26, 27].

We recently showed that administration of an antibody specific for myostatin (RK35, Pfizer) induces substantial benefit in the A17 mouse model of OPMD in terms of body and muscle mass enhancement, muscle strength increase, and fibrosis reduction as compared to the saline-treated control.

In this chapter, we describe in detail the methodology for the systemic weekly intraperitoneal administration of a monoclonal antibody for myostatin. We list methods to assess the functional improvements in animals as the measurement of forelimbs strength and the use of in situ muscle physiology. Protocols to assess the histopathological improvements by histology and immunofluorescence to analyze muscle size, muscle fibrosis, and myofiber typing are also described in detail.

2 Materials

2.1 RK35 Antibody

1. RK35 antibody (Pfizer) is provided resuspended in PBS: 137 mM NaCl, 2.7 mM KCl, 8 mM Na_2HPO_4, and 2 mM KH_2PO_4, and it is kept at −80 °C until use (*see* **Note 1**).

2.2 Animals

1. All animal procedures are performed in accordance with the UK Animals (Scientific Procedures) Act, 1986.
2. A17 and FvB mice are bred in a standard 12 h light/dark cycle with free access to food and water. A17 mice are available from The Jackson laboratory (FVB-Tg(ACTA1-PABPN1*A17) 1Drub/DrubJ, #016193, https://www.jax.org).

2.3 Genotyping of A17.1 Mice

1. Ear punch pincer, hole 2 mm OD Stainless Steel or similar instruments to collect ear notches.
2. 70% ethanol.
3. 1.5 mL tubes.
4. DNA extraction buffer: 500 mM KCl, 100 mM Tris–HCl pH 8.3, 0.1 mg/ml gelatin, 0.45% v/v Tween 20, 0.45% v/v Nonidet P40, 5 mg/ml Proteinase K; this buffer is prepared as 10X stock as follows: mix 3.73 g KCl and 1.21 g Trisma-Base with 60 mL of Milli-Q water and stir. While stirring, add 450 μL of Tween 20 and 450 μL of Nonidet P40. Add 10 mL of a solution of 1 mg/ml gelatin in Milli-Q water (previously autoclaved and cooled) to the rest of the buffer. Mix well and adjust the pH of the buffer to pH 8.3 with 1 M HCl. Bring to volume of 100 mL with Milli-Q water. The buffer is sterilized by autoclaving for prolonged storage. Afterward, proteinase K 5 mg/ml is added, and solution is mixed well and stored at −20 °C. Dilute the 10X solution with water before using.
5. Dry block incubator.
6. Box with ice.
7. GoTaq Polymerase (Promega M7805).
8. Gel electrophoresis prepared using agarose 2%, 1× TAE buffer. The TAE buffer is 40 mM Trizma base, 20 mM glacial acetic acid, 1 mM Na_2EDTA, pH 7.6.
9. Gel imager.
10. PCR primers (bovine).
    (a) PABPN1 (forward): GAACCAACAGACCAGGCATC
    (b) PABPN1 (reverse): GTGATGGTGATGATGACCGG
    (c) 18S (forward): TTGACGGAAG GGCACCACCAG
    (d) 18S (reverse): GCACCACCACCCACGGAATCG

2.4 Intraperitoneal Injection

1. RK35 monoclonal antibody (as described in Subheading 2.1).
2. 0.9% saline.
3. BD Microfine Insulin 1 mL syringes with 29 × 1/2″ G needles attached.
4. 70% v/v ethanol in water.
5. Clean cages with bedding, food, and water.

2.5 Postmortem Tissue Harvesting and Processing

1. CO_2 chamber for euthanasia.
2. Liquid nitrogen (N_2).
3. Isopentane, isopentane container.
4. Ice to store blood samples.
5. Microcentrifuge.
6. 70% v/v ethanol in water.
7. BD 1 mL syringes.
8. BD 26 × 25 mm G (brown) needles.
9. Dissection tools (i.e., scalpels, scissors, tweezers, forceps).
10. Raymond lamb cork disc for cryostat use (20 mm × 3 mm), cork holder.
11. Optimal cutting temperature (OCT) embedding medium.
12. 1.5-mL tubes.
13. Scales to weigh the mice and the samples harvested.

2.6 Immunohistochemistry Staining and Histological Analyses

1. OTF 5000 cryostat (Bright).
2. Superfrost® Plus microscopy slides, coverslips (VWR).
3. 4% Paraformaldehyde (PFA) in water: add 4 g of PFA to 50 mL of H_2O. Add 1 mL of 1 M NaOH and stir gently on a heating block at ~60 °C until the is dissolved. Add 10 mL of 10X PBS and allow the mixture to cool to room temperature. Adjust the pH to 7.4 with 1 M HCl (~1 mL), then adjust the final volume to 100 mL with H_2O. Filter the solution through a 0.45 μm membrane filter to remove any particulate matter. Make the PFA solution fresh prior to use, or store in aliquots at −20 °C for several months. Avoid repeated freeze/thawing.
4. 100% acetone.
5. Dako (hydrophobic) pen.
6. PBS10X: 1.37 M NaCl, 27 mM KCl, 80 mM Na_2HPO_4, and 20 mM KH_2PO_4 in water.
7. PBST: 0.05% Tween-20 in PBS.
8. Blocking reagents for immunostainings: use 10% goat serum in PBST for Collagen VI staining and 1% goat serum and 0.1% triton ×100 in PBS10X for PABPN1 staining.

9. KCl solution: 1 M KCl, 30 mM HEPES, 65 mM PIPES, 10 mM EDTA, 2 mM MgCl2. Adjust pH to 6.9.
10. Primary antibodies: rabbit monoclonal anti-PABPN1 (Abcam, ab75855); rat monoclonal anti-Laminin (Sigma-Aldrich L0663); rabbit polyclonal anti-collagen VI (Abcam ab182744).
11. Secondary antibodies (Life Technologies): goat anti-rabbit or anti-rat conjugated with Alexa488 or Alexa568.
12. DAPI (4′,6-diamidino-2-phenylindole, Sigma Aldrich).
13. PPD (p-Phenylenediamine) solution (Sigma Aldrich): 0.1% PPD dissolved in water, stored at −20 °C and protected from light by using aluminum foil.
14. Mowiol solution: stir 4.8 mL glycerol and 2.4 g mowiol in 6 mL water for 2 h, at room temperature; add 12 mL of 0.2 M Tris buffer pH 8.5 and 0.02% sodium azide. Incubate the solution in 50–60 °C water bath for 10 min, stirring frequently. Centrifuge at 5000× *g* for 15 min and finally collect and aliquot the supernatant and store at −20 °C until use. Do not disturb the pellet.
15. Mounting solution: mix 1 part of PPD solution with nine parts of Mowiol solution, maintain at room temperature, protect from light and use within a day.
16. Picrosirius red 0.3% W/V (prepared in saturated solution of picric acid). The solution should be filtered before the use.
17. DPX mounting medium for histology (Sigma-Aldrich).
18. 16. Fluorescence microscope (e.g., Axio Observer D1 microscope or Nikon Eclipse Ni-E upright).
19. Image acquisition and analysis software.

2.7 ELISA for Myostatin Expression

1. Quantikine ELISA immunoassay (cat number DGDF80, R&D system).
2. Sample Activation Kit 1 (cat number DY010, R&D system).

3 Methods

3.1 Genotyping of Mice (from Ear Notches)

1. Each ear notch should be collected in separate 1.5 mL tube and immediately frozen at −20 °C. Carefully open tube.
2. Add 50 μL of 1× DNA extraction buffer, and spin samples at 3000 g for 2 min and make sure that the ear notch is in the solution (use end of a pipette tip if necessary).
3. Heat the samples at 55 °C for 3–5 h until the tissue is dissolved.
4. Heat at 100 °C for 10 min to inactivate the Proteinase K.

5. Cool down the samples to room temperature on ice.
6. Prepare the master mixes for PCRs of PABPN1 and 18S.
7. Program of PCR for both PABPN1 and 18S:

   95 °C, 2′00;

   40 cycles x

   - 95 °C, 0′30
   - 60 °C, 0′30
   - 72 °C, 0′30

   72 °C, 10′00;

   04 °C hold.

   The PCR bands are ~300 bp for PABPN1 and ~100 bp for 18S.

3.2 Intraperitoneal Injection

1. Weigh mouse and calculate the dose of antibody needed in a total volume of 200 μL/mouse (maximally 10× μL of body weight in grams).
2. Dilute RK35 myostatin blocking antibody in sterile 0.9% saline at a desired concentration (tested at 10 mg/kg) immediately prior to injection and keep in ice during the procedure.
3. Load the antibody solution into syringe and ensure there are no bubbles by tapping the syringe and purging trapped air.
4. Restrain the mouse by cuffing along the scruff, invert and tilt the mouse 45 degree downward superiorly. Inject the mouse intraperitoneally 1 cm above the genitals, 3–4 mm to the right of the "linea alba" (to avoid the liver) and carefully staying below the sternum. Intraperitoneal injection of RK35 antibody showed to successfully increase the muscle size when administered weekly at 10 mg/kg for 10 weeks (Fig. 1).
5. Wipe place of injection with 70% ethanol and return mouse to the cage.
6. The mice are regularly monitored for health.

3.3 Functional Tests

Functional improvement in mice can be assessed up to 2 weeks before the experiment terminates. TREAT-NMD has published online protocols (http://www.treat-nmd.eu/resources/research-resources/dmd-sops/) for the three functional tests we use in A17 mice:

1. Behavioral and locomotor measurements using open-field animal activity monitoring system (SOP ID No.: DMD_M.2.1.002).
2. Use of grip strength meter to assess the limb strength of *mdx* mice (SOP ID No.: DMD_M.2.2.001).

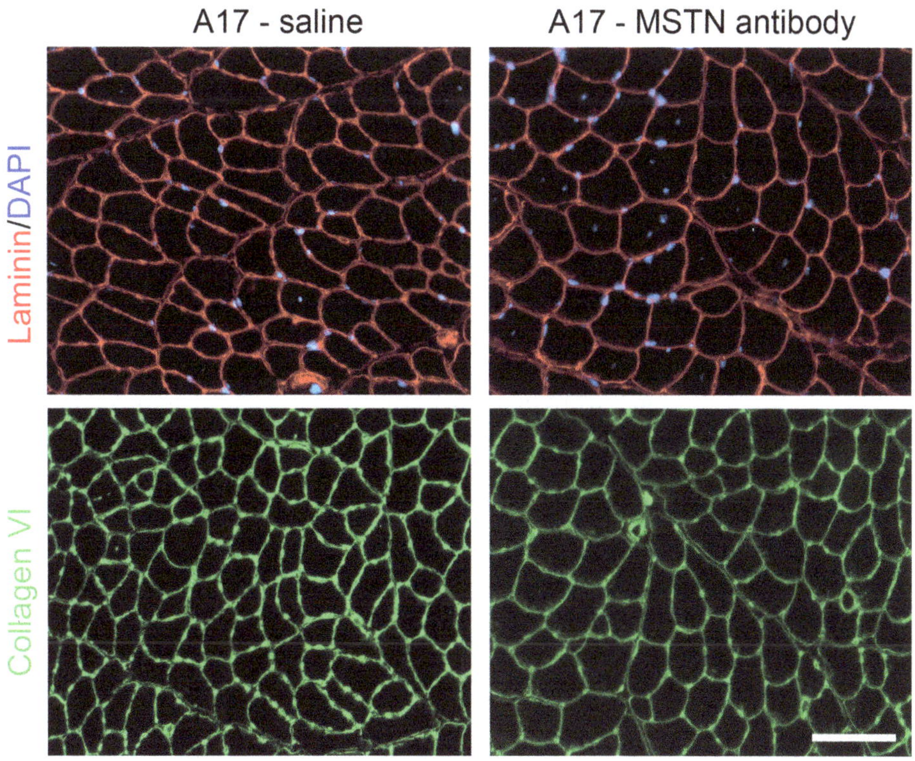

Fig. 1 Treatment with RK35 antibody increases myofiber size and reduce fibrosis in OPMD mice: Mice were administered with a weekly regimen of either saline or the RK35 antibody (10 mg/kg IP) for 10 weeks from 12 weeks of age. Muscle samples were sectioned and immunostained for laminin/DAPI (upper panel) and collagen VI (bottom panel). The mice treated with the antibody displayed an increased myofiber diameter and a reduced collagen deposition. Scale bar, 100 μm

3. Measuring isometric force of isolated mouse skeletal muscles in situ (TREAT-NMD, SOP ID No.: DMD_M.2.2.005).

The SOP DMD_M.2.1.002 is performed in the last 2 weeks before the end of the experiment. SOP DMD_M.2.2.001 is used after the behavioral and locomotor measurements as it could affect this analysis. The SOP DMD_M.2.2.005 is performed under terminal anesthesia before harvesting the muscles and blood samples. These protocols are optimized for the study of DMD murine models, and therefore, some changes were introduced to the SOP DMD_M.2.2.001 and DMD_M.2.1.002 to reflect a difference in mouse strain responsiveness. Changes are reported in **Note 2**.

3.4 Postmortem Tissue Processing

1. If the mouse is analyzed by SOP DMD_M.2.2.005, proceed with neck dislocation to confirm death. Alternatively weigh mouse before sacrificing by CO_2 exposure and confirm the death by neck dislocation.
2. Collect total blood by cardiac puncture into 1 mL syringe (*see* **Note 3**). This has to be done before neck dislocation: if the

mouse is under terminal anesthesia after the SOP DMD_M.2.2.005, or immediately after CO_2 exposure in all other cases.

3. Transfer the blood into an ice-cold 1.5 mL tube, keep on ice overnight prior to serum extraction. Then centrifuge at 2000 × *g* for 10 min at 4 °C. Transfer the supernatant to fresh tubes and repeat centrifugation. Finally collect the supernatant, aliquot as needed (50 μL is sufficient for most purposes) and store at −80 °C.
4. From each mouse, we collect the diaphragm, extensor digitorum longus, gastrocnemius, soleus, and tibialis anterior. Tissues from one side of the body are frozen immediately in liquid N_2 for RNA and protein extraction, while tissues from the other side are embedded in OCT and subsequently frozen in liquid N_2-cooled isopentane for cryo-sectioning (*see* **Note 4**).
5. All samples are kept at −80 °C until use.

3.5 Immunostaining for PABPN1, Laminin, Collagen VI

Muscles are sectioned and co-immunostained for PABPN1 and laminin to assess the percentage of nuclei-containing insoluble aggregates or immunostained for collagen VI to assess the amount of muscle fibrosis.

1. Frozen tissue is sectioned on an OTF 5000 cryostat (or similar) at 10-μm thickness through the muscle length. Cutting temperature is between −20 and −18 °C.
2. Collect transverse sections onto Superfrost® Plus microscopy slides, keep at −80 °C for long-term storage (*see* **Note 5**) or proceed to immunostaining (as below).
3. Air-dry samples for 20 min if sections were previously frozen.
4. Fix in ice-cold acetone (for collagen VI staining) or 4% PFA (for PABPN1/laminin staining) for 10 min.
5. Air-dry for 10 min (for collagen VI staining).
6. Limit the sections to be stained with a Dako (hydrophobic) pen and wait 5 min to dry.
7. Wash quickly with PBS 10X (for PABPN1/laminin staining) or 1× PBST (for Collagen VI staining).
8. Add a KCl solution for 1 h at room temperature (RT) (only for PABPN1/laminin staining).
9. Wash quickly with PBS 10X (only for PABPN1/laminin staining).
10. Add the suitable blocking reagent (depending on the staining) for 1 h, at room temperature.
11. Incubate with rabbit polyclonal anti-collagen VI, (1:400) 1 h RT or with a mix of PABPN1 antibody Abcam ab75855 (1:400) and rat monoclonal anti-laminin (1:800) overnight, at 4 °C diluted in blocking solution.

12. Wash 3 × 5 min in 1× PBST.
13. Incubate with goat anti-rabbit Alexa488 (1:500) for collagen VI staining and a mix of goat anti-rat Alexa568 (1:500) and goat anti-rabbit Alexa488 (1:500) for PABPN1/laminin staining, diluted in 10% goat serum in PBST for 1 h, at room temperature. Protect from light afterward.
14. Wash 3 × 5 min in PBST.
15. Incubate with 1 μg/ml DAPI diluted in 1× PBS for 5 min, at RT.
16. Rinse with PBST.
17. Add mounting solution (50–100 μL/slide) and cover with coverslips.
18. Store at 4 °C.

3.6 Picrosirius Red Staining

Muscles are sectioned and stained with picrosirius red to assess the amount of muscle fibrosis. Picrosirius red solution stains in yellow the fiber cytoplasm and in red the tissue containing high amount of collagen proteins I and III.

1. Air dry slides for 30 min.
2. Fix 10 min with 4% PFA.
3. Wash twice 5 min in tap water then once 5 min in distilled water.
4. Wash in 100% ethanol for 5 min.
5. Air dry sections (*see* **Note 6**).
6. Transfer the slides in picrosirius red solution 0.3% for 1 h.
7. Wash quickly in distilled water.
8. Fix staining in 2 wash of 0.5% v/v acetic acid in tap water for 5 min each.
9. Dehydrate in 3 wash of 100% ethanol for 5 min each.
10. Clarify for 10 min in xylene.
11. Mount the slides in DPX mounting medium for histology.

3.7 Analyses of Histological and Immunofluorescence Samples

Immunostainings are visualized under a fluorescence microscope with a camera, and images are analyzed with an imaging software like NIH ImageJ and the MuscleJ macro (*see* **Note 7**). Ideally, a mosaic image of the whole muscle section is acquired by overlapping and stitching individual images of that section together. Alternatively, five to six images 200× magnification are taken randomly in the muscle section (*see* **Note 8**).

For the analysis, the ImageJ software (downloadable from https://imagej.nih.gov/ij/) is used.

1. To quantify Intranuclear inclusions (INI): use the ImageJ software with the threshold function to highlight the INI and quantify the number. The same approach is used with DAPI positive nuclei. The ratio of such values provides the INI/nuclei percentage (*see* **Note 9**).
2. To quantify the cross-sectional area, images for laminin staining are processed and analyzed using CellProfiler (downloadable from https://cellprofiler.org/) as described in the TREAT-NMD SOP ID No.: DMD_M.1.2.001.
3. To quantify muscle fibrosis, the area covered by Collagen VI or Sirius red positive staining should be analyzed: use ImageJ and select the area covered by the section. Then using the threshold function, highlight the regions of the section positive for the Collagen VI or Sirius red stainings and evaluate as the percentage on the total area.

3.8 ELISA for Myostatin Expression

1. Blood sera from treated mice are collected by cardiac puncture under terminal anesthesia as described in Subheading 3.4, **step 5** and kept at −80 °C until used.
2. Blood sera are thawed in ice and subjected to sample inactivation before proceeding with the ELISA sandwich. This step is critical to accurately measure myostatin levels. Sample activation is performed using manufacturer's protocol.
3. The ELISA kit is used under manufacturer's recommended protocol.

4 Notes

1. RK35 monoclonal antibody for myostatin (Pfizer) has been used to describe these experiments; however, other monoclonal antibodies for myostatin are available from other companies, and the conditions of storing should be verified for any different antibody.
2. A17 and relative background strain FvB mice are extremely more active and learn repetitive actions very quickly compared to the dystrophic *mdx* mice. Therefore, A17/FvB mice learn very quickly the grip strength assessment (SOP DMD_M.2.2.001), and after the first day of measurement, they leave the mesh very early, or they do not grab it at all. Therefore, a very short acclimatation is performed the week before the assessment with only two tests per mouse for a single day, and no acclimatation is done on the day of the test. The experiment is performed for three consecutive days, but only the first day of test provides reliable data. Similarly, in the SOP

DMD_M.2.2.002, a shorter acclimatation time is done before the 1 h long measurement, and mice are left in the cage for only 10 min before capturing the activity data.

3. The volume of total blood collected depends on the time necessary to get to this stage after the death. We usually can collect ~300–500 μL/mouse.
4. Cover samples on cork discs with silver foil before immersing in liquid N_2 for additional tissue protection.
5. Transverse sections can be kept at room temperature during sectioning for few hours but must be stored at −80 °C afterward until the staining.
6. Slides have to be completely dried otherwise cytoplasm of myofibers will be colored in orange.
7. Select the biggest section on the slide, or the nearly biggest one, which shows the best sectioning and staining in both treated and control samples. This is important to make the analyses comparable between samples.
8. Analysis should be performed in a blinded manner and by a single operator to reduce variability among samples and avoid biased results.
9. Alternatively (but definitely more time-consuming), INI and DAPI+ve nuclei can be counted manually using the cell counter plugin of ImageJ.

References

1. Bouchard JP, Brais B, Brunet D et al (1997) Recent studies on oculopharyngeal muscular dystrophy in Quebec. Neuromuscul Disord 7 (Suppl 1):S22–S29
2. Banerjee A, Apponi LH, Pavlath GK et al (2013) PABPN1: molecular function and muscle disease. FEBS J 280(17):4230–4250
3. Gidaro T, Negroni E, Perie S et al (2013) Atrophy, fibrosis, and increased PAX7-positive cells in pharyngeal muscles of oculopharyngeal muscular dystrophy patients. J Neuropathol Exp Neurol 72(3):234–243
4. Tome FM, Fardeau M (1980) Nuclear inclusions in oculopharyngeal dystrophy. Acta Neuropathol 49(1):85–87
5. Trollet C, Gidaro T, Klein P et al (1993 [updated 2014 Feb 20]) Oculopharyngeal muscular dystrophy. GeneReviews® [Internet]
6. Lee SJ (2004) Regulation of muscle mass by myostatin. Annu Rev Cell Dev Biol 20:61–86
7. Sartori R, Schirwis E, Blaauw B et al (2013) BMP signaling controls muscle mass. Nat Genet 45(11):1309–1318
8. Bogdanovich S, Perkins KJ, Krag TO et al (2005) Myostatin propeptide-mediated amelioration of dystrophic pathophysiology. FASEB J 19(6):543–549
9. Campbell C, McMillan HJ, Mah JK et al (2017) Myostatin inhibitor ACE-031 treatment of ambulatory boys with Duchenne muscular dystrophy: results of a randomized, placebo-controlled clinical trial. Muscle Nerve 55(4):458–464
10. Bhattacharya I, Pawlak S, Marraffino S et al (2018) Safety, tolerability, pharmacokinetics, and pharmacodynamics of domagrozumab (PF-06252616), an antimyostatin monoclonal antibody in healthy subjects. Clin Pharmacol Drug Dev 7(5):484–497
11. Wagner KR, Fleckenstein JL, Amato AA et al (2008) A phase I/IItrial of MYO-029 in adult subjects with muscular dystrophy. Ann Neurol 63(5):561–571
12. Bogdanovich S, Krag TO, Barton ER et al (2002) Functional improvement of dystrophic muscle by myostatin blockade. Nature 420(6914):418–421

13. Zhang L, Rajan V, Lin E et al (2011) Pharmacological inhibition of myostatin suppresses systemic inflammation and muscle atrophy in mice with chronic kidney disease. FASEB J 25(5):1653–1663
14. Siriett V, Salerno MS, Berry C et al (2007) Antagonism of myostatin enhances muscle regeneration during sarcopenia. Mol Ther 15(8):1463–1470
15. Kang JK, Malerba A, Popplewell L et al (2011) Antisense-induced myostatin exon skipping leads to muscle hypertrophy in mice following octa-guanidine morpholino oligomer treatment. Mol Ther 19(1):159–164
16. Lu-Nguyen N, Dickson G, Malerba A (2018) Systemic intravenous administration of antisense therapeutics for combinatorial dystrophin and myostatin exon splice modulation. Methods Mol Biol 1828:343–354
17. Lu-Nguyen N, Ferry A, Schnell FJ et al (2019) Functional muscle recovery following dystrophin and myostatin exon splice modulation in aged mdx mice. Hum Mol Genet 28(18): 3091–3100
18. Lu-Nguyen N, Malerba A, Popplewell L et al (2017) Systemic antisense therapeutics for dystrophin and myostatin exon splice modulation improve muscle pathology of adult mdx mice. Mol Ther Nucleic Acids 6:15–28
19. Malerba A, Kang JK, McClorey G et al (2012) Dual myostatin and dystrophin exon skipping by morpholino nucleic acid oligomers conjugated to a cell-penetrating peptide is a promising therapeutic strategy for the treatment of Duchenne muscular dystrophy. Mol Ther Nucleic Acids 1:e62
20. Davies JE, Wang L, Garcia-Oroz L et al (2005) Doxycycline attenuates and delays toxicity of the oculopharyngeal muscular dystrophy mutation in transgenic mice. Nat Med 11(6): 672–677
21. Trollet C, Anvar SY, Venema A et al (2010) Molecular and phenotypic characterization of a mouse model of oculopharyngeal muscular dystrophy reveals severe muscular atrophy restricted to fast glycolytic fibres. Hum Mol Genet 19(11):2191–2207
22. Davies JE, Rose C, Sarkar S et al (2010) Cystamine suppresses polyalanine toxicity in a mouse model of oculopharyngeal muscular dystrophy. Sci Transl Med 2(34):34ra40
23. Davies JE, Rubinsztein DC (2011) Overexpression of BCL2 rescues muscle weakness in a mouse model of oculopharyngeal muscular dystrophy. Hum Mol Genet 20(6):1154–1163
24. Davies JE, Sarkar S, Rubinsztein DC (2006) Trehalose reduces aggregate formation and delays pathology in a transgenic mouse model of oculopharyngeal muscular dystrophy. Hum Mol Genet 15(1):23–31
25. Malerba A, Roth F, Harish P et al (2019) Pharmacological modulation of the ER stress response ameliorates oculopharyngeal muscular dystrophy. Hum Mol Genet 28(10): 1694–1708
26. Malerba A, Klein P, Bachtarzi H et al (2017) PABPN1 gene therapy for oculopharyngeal muscular dystrophy. Nat Commun 8:14848
27. Malerba A, Klein P, Lu-Nguyen N et al (2019) Established PABPN1 intranuclear inclusions in OPMD muscle can be efficiently reversed by AAV-mediated knockdown and replacement of mutant expanded PABPN1. Hum Mol Genet 28(19):3301–3308

Correction to: Cardiac and Skeletal Muscle Pathology in the D2/mdx Mouse Model and Caveats Associated with the Quantification of Utrophin

Tahnee L. Kennedy and Hannah F. Dugdale

Correction to:
Chapter 4 in: Rika Maruyama and Toshifumi Yokota (eds.), *Muscular Dystrophy Therapeutics: Methods and Protocols*,
Methods in Molecular Biology, vol. 2587,
https://doi.org/10.1007/978-1-0716-2772-3_4

The corresponding author of the chapter informed that he realised the reference to Kennedy (2018) paper was not included in Figure 1 legend, which needs to be added for potential copyright reasons. The legend of Figure 1 should read as '**Fig. 1** Considerations for utilising the *D2/mdx* mouse model for assessing new therapeutics. Depiction of the experimental design requirements, samples processing and analysis to correctly interpret the data generated using the *D2/mdx* mouse model. Image adapted from Kennedy et al. [1].

Also a potential acknowledgement needs to be added.

Acknowledgement: As Kennedy et al. [1] referenced in Fig. 1, the work flow described there was used in a study performed at the University of Oxford, under Prof. Kay Davies. Original experiments were funded by the MRC.

The original version of this chapter has been revised.

The updated original version of this chapter can be found at
https://doi.org/10.1007/978-1-0716-2772-3_4

Rika Maruyama and Toshifumi Yokota (eds.), *Muscular Dystrophy Therapeutics: Methods and Protocols*,
Methods in Molecular Biology, vol. 2587, https://doi.org/10.1007/978-1-0716-2772-3_31,

INDEX

A

B

C

Rika Maruyama and Toshifumi Yokota (eds.), *Muscular Dystrophy Therapeutics: Methods and Protocols*, Methods in Molecular Biology, vol. 2587, https://doi.org/10.1007/978-1-0716-2772-3,

E

F

G

H

N

O

P

Milton Keynes UK
Ingram Content Group UK Ltd.
UKHW051029011223
433412UK00004B/25

9 781071 627747